# Near Field Optics

# NATO ASI Series

**Advanced Science Institutes Series**

*A Series presenting the results of activities sponsored by the NATO Science Committee, which aims at the dissemination of advanced scientific and technological knowledge, with a view to strengthening links between scientific communities.*

The Series is published by an international board of publishers in conjunction with the NATO Scientific Affairs Division

| | |
|---|---|
| **A Life Sciences**<br>**B Physics** | Plenum Publishing Corporation<br>London and New York |
| **C Mathematical and Physical Sciences**<br>**D Behavioural and Social Sciences**<br>**E Applied Sciences** | Kluwer Academic Publishers<br>Dordrecht, Boston and London |
| **F Computer and Systems Sciences**<br>**G Ecological Sciences**<br>**H Cell Biology**<br>**I Global Environmental Change** | Springer-Verlag<br>Berlin, Heidelberg, New York, London, Paris and Tokyo |

**NATO-PCO-DATA BASE**

The electronic index to the NATO ASI Series provides full bibliographical references (with keywords and/or abstracts) to more than 30000 contributions from international scientists published in all sections of the NATO ASI Series.
Access to the NATO-PCO-DATA BASE is possible in two ways:

– via online FILE 128 (NATO-PCO-DATA BASE) hosted by ESRIN, Via Galileo Galilei, I-00044 Frascati, Italy.

– via CD-ROM "NATO-PCO-DATA BASE" with user-friendly retrieval software in English, French and German (© WTV GmbH and DATAWARE Technologies Inc. 1989).

The CD-ROM can be ordered through any member of the Board of Publishers or through NATO-PCO, Overijse, Belgium.

**Series E: Applied Sciences - Vol. 242**

# Near Field Optics

edited by

Dieter W. Pohl
IBM Research Division,
Zurich Research Laboratory,
Rüschlikon, Switzerland

and

Daniel Courjon
Laboratoire d'Optique P.M. Duffieux,
Université de Franche-Comté,
Besançon, France

Springer Science+Business Media, B.V.

Proceedings of the NATO Advanced Research Workshop on
Near Field Optics
Arc-et-Senans, France
October 26-28, 1992

A C.I.P. Catalogue record for this book is available from the Library of Congress.

DOI 10.1007/978-94-011-1978-8

---

*Printed on acid-free paper*

---

This book contains the proceedings of a NATO Advanced Research Workshop held within the programme of activities of the NATO Special Programme on Nanoscale Science as part of the activities of the NATO Science Committee.

Other books previously published as a result of the activities of the Special Programme are:

NASTASI, M., PARKING, D.M. and GLEITER, H. (eds.), *Mechanical Properties and Deformation Behavior of Materials Having Ultra-Fine Microstructures.* (ASIE 233) 1993 ISBN 0-7923-2195-2

VU THIEN BINH, GARCIA, N. and DRANSFELD, K. (eds), *Nanosources and manipulation of Atoms under High Fields and Temperatures: Applications.* (E235) 1993 ISBN 0-7923-2266-5

LEBURTON, J.-P., PASCUAL J. and SOTOMAYOR TORRES, C. (eds.), *Phonons in Semiconductor Nanostructures.* (E236) 1993 ISBN 0-7923-2277-0

AVOURIS P. (ed.), *Atomic and Nanometer-Scale Modification of Materials: Fundamentals and Applications.* (E239) 1993 ISBN 0-7923-2334-3

BLÖCHL, P. E., JOACHIM, C. and FISHER, A. J. (eds.), *Computations for the Nano-Scale.* (E240) 1993 ISBN 0-7923-2360-2

POHL, D. W. and COURJON, D. (eds.), *Near Field Optics.* (E242) 1993 ISBN 0-7923-2394-7

SALEMINK, H. W. M. and PASHLEY, M. D. (eds.), *Semiconductor Interfaces at the Sub-Nanomater Scale.* (E243) 1993 ISBN 0-7923-2397-1

# Table of Contents

## Optical Forces



## Theory

## Plasmons in Near-Field Optics



## Instrumental Techniques



## Near-Field Infrared Microscopy



## STM and Light

## Related Techniques

## Preface

This volume contains papers presented at the NATO Advanced Research Workshop (ARW) on Near Field Optics, which was the first conference devoted exclusively to this new area of optics. The ARW was held at Arc et Senans near Besançon, France, from October 26 to 28, 1992. This location, a renovated industrial monument dating back to the time of Louis XVI, ideally accommodated the approx. 70 participants. Remote from urban areas, it is a place that fosters scientific reflection and meditation.

Forty communications were presented in six sessions (including two opening lectures) and 16 posters. Among those, 47 contributions were submitted for publication in the NATO Proceedings. They were accepted after a strict but constructive peer refereeing process.

The topics presented and discussed ranged from experimental near-field microscopy to theoretical modeling and also included the combination of SNOM/AFM, the near-field optical aspects of gratings, light forces and theory as well as hybrid and related techniques.

To bring some order into the variety of topics presented in this volume, the papers are grouped mainly according to the experimental and theoretical techniques employed. The layout roughly follows the order of presentation at the workshop. The poster contributions were combined with those of the thematically corresponding oral presentations. A certain degree of arbitrariness was unavoidable in this process, for which the editors apologize.

This international meeting was organized under the auspices and with the financial support of NATO, the Centre National de la Recherche Scientifique (France), the French Program ULTIMATECH, the Institut des Microtechniques, the University of Besançon, and the Ministère de la Recherche et de l'Espace (France). The SPIE and the Société Française d'Optique also acted as sponsors. Both organizers and participants gratefully acknowledge the generous support.

We thank all the authors for their contributions, the referees for their critical reviewing work, and our local colleagues for their help before, during and after the workshop. Particular thanks go to the secretaries of the workshop, both of the University of Besançon (France) as well as of the IBM Research Division, Zurich Research Laboratory (Switzerland).

The editors thank in particular the NATO Science Committee, whose logistic and financial support played a stimulating role and ensured the excellent level of the meeting. They also wish to thank H. Rohrer, who opened the conference and introduced the subject in the general framework of nanoscale science and technology.

Dieter Pohl — Daniel Courjon

1. N. Van Hulst
2. T.J. Silva
3. K. Wickramasinghe
4. P. Moyer
5. A. Korpel
6. M. Vaez-Iravani
7. F. Creuzet
8. U.C. Fischer
9. M. Spajer
10. G. Granet
11. M. Specht
12. H. Heinzelmann
13. R. Berndt
14. Ch. Girard
15. D. Courjon
16. M. Nieto-Vesperinas
17. G. Devant
18. G. Lampel
19. M. Hrynevych
20. G. Kolb
21. O. Marti
22. F. De Fornel
23. D. Van Labeke
24. A.C. Boccara
25. S.I. Bozhevolnyi
26. M. Ohtsu
27. T. Hartmann
28. C. Baur
29. H.M. Hertz
30. P. Tomanek
31. D.W. Pohl
32. R. Kopelman
33. M. Völcker
34. A. Roberts
35. A. Dereux
36. H. Bielefeldt
37. R. Williamson
38. F. Baida
39. J.P. Vigneron
40. F. Depasse
41. M.H.P. Moers
42. D. Maystre
43. S. Leblanc-Held
44. X. Bouju
45. R. Scholten
46. K. Lieberman
47. H.-U. Danzebrink
48. A. De Marco
49. F. Keilmann
50. C. Jahncke
51. M. Paesler
52. A. Trabucco de Marco
53. S. Alvarado

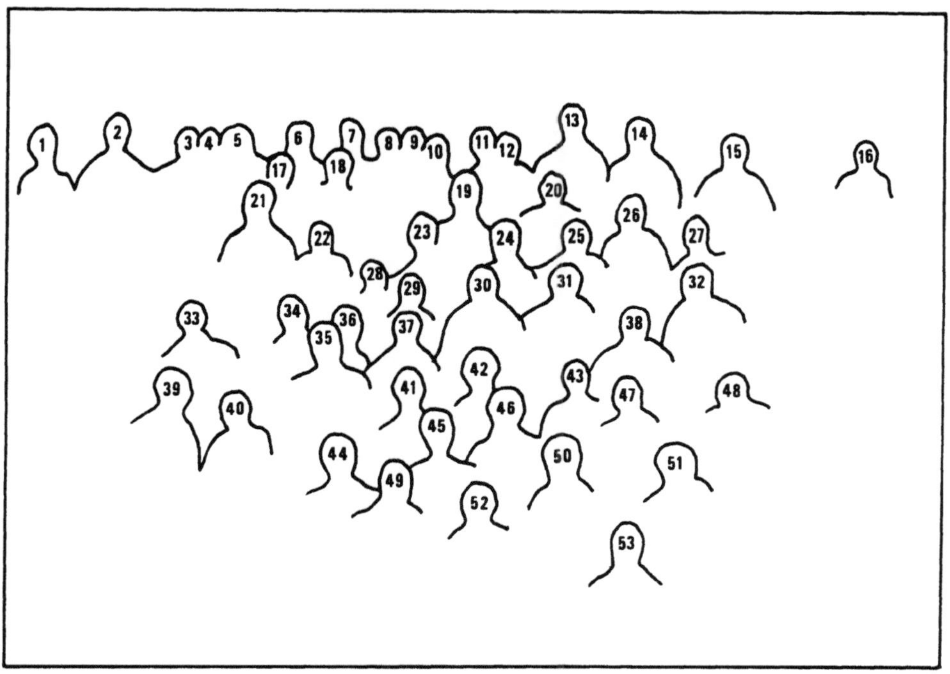

## Glossary of Some Names and Acronyms Used in NFO Microscopy

**SNOM**
**NSOM** Scanning near-field and near-field scanning optical microscope, respectively: General names for NFO microscopes or microscopies; in a narrower sense, the aperture NFO microscope. The two acronyms were invented independently at the two places where NFO microscopy was originally developed, viz. Rüschlikon and Cornell. The abbreviation SNOM emphasizes the analogy to other scanning probe microscopies (SPM) where the letter P (or X) may stand for any kind of interaction, for instance, tunneling (STM). NSOM, on the other hand, stresses the near-field effect as compared to the more technical aspect of scanning.

**Optical stethoscope, NFOS** Early names for SNOM, no longer in use

**STOM**
**PSTM** Scanning tunneling optical and photon scanning tunneling microscope or microscopy. These acronyms refer to NFO microscopy by tapping an evanescent wave produced by total reflection with a glass tip. The two acronyms relate to the two places of origin, Besançon (STOM) and Oak Ridge/Dijon (PSTM).

**EFOM** Evanescent field microscope, same STOM/PSTM, introduced by the Twente group.

**SPNM** Scanning plasmon near-field microscopy (see paper by J.D. Pedarnig *et al.*, p. 271).

Compiled by D.W. Pohl

## Some Remarks on the History of Near-Field Optics

D. W. Pohl
*IBM Research Division,*
*Zurich Research Laboratory,*
*8803 Rüschlikon, Switzerland*

ABSTRACT. The historical background of present-day near-field (NF) optical activities is briefly reviewed. First demonstrations of subwavelength imaging capability were made in the early 1980's but systematic development did not begin until the past few years.

### 1. Prehistory

The peculiarities of near fields (NF) have been recognized for a long time. Hertz was one of the first to notice that the field of a radiating dipole, at very small distance, decays with $1/r^3$ rather than with $1/r$. The field next to the dipole is hence much larger than what a naive extrapolation of the detectable far-field (FF) value would suggest. Sommerfeld in 1909 already calculated the influence of an NF perturbation on the radiation characteristics of a dipole antenna near ground.

In classical optics, NF effects traditionally played a minor role, being needed for the explanation of only a few phenomena, such as the scattering from small particles (Mie 1908), the transmission of narrow apertures (Bethe 1944, Bouwkamp 1950), and the quenching of fluorescence near metallic surfaces (see, for instance, Förster 1946, Kuhn 1970, and Drexhage 1974).

And then, way back in 1928, there was a British scientist, E.H. Synge with a truly prophetic vision: He suggested the use of narrow apertures for optical microscopy beyond the limit of diffraction (Synge 1928)...The proposed apparatus comes extremely close to today's near-field optical (NFO) microscopes — except, maybe, for the screen around the aperture which he considered to be flat rather than of the highly convex shape used in modern scanning probe microscopes. Apparently, Synge was soon forgotten, although similar but more casual suggestions have been made several times since then, without his name ever being mentioned.

The other direct forerunners of NF optical microscopy were Ash and Nichols (1972) who demonstrated microwave subwavelength imaging capability, and achieved a resolution in one dimension of $\lambda/60$. Again, this first attempt was not further pursued although the extension to optical wavelengths was one of the goals of Ash's group (E.A. Ash, private communication).

A third direct root of NFO microscopy may be traced back to the submicroscopic pattern replication with visible light by Zingsheim (1977) and Fischer and Zingsheim (1981). They exploited the fluorescence quenching mentioned above to transfer patterns from a metallic film to a dye layer in a kind of a contact imaging process.



*D.W. Pohl and D. Courjon (eds.), Near Field Optics, 1–5.*

## 2. A Personal Reminiscence

The developments described above came to my attention only a couple of years after NFO had entered my life. This happened in the fall of 1981: Gerd Binnig had just given a seminar talk on his first results with the scanning tunneling microscope. Afterwards, a hot discussion evolved at the coffee machine, and it went something like this:

- *"What else could one do with such a tip?"* somebody began.
- *"One might place a small light source at the apex,"* I threw in, drawing on my background in laser physics.
- *"Haha, funny idea, to better see the electrons tunnel..."*
- *"No, you Spassvogel,"* I replied impulsively, *"to do optical microscopy with nanometer resolution, using, say, a transparent tip with metallic coating and a pinhole at the end."*
- *"Hm, sounds great, but how will you ever make such a hole and get it at the right position?"*
- *"... And just imagine how little light you would get through!"*
- *"... And even if you did succeed, you will never beat the diffraction limit ... forget it!"*

I had almost forgotten it indeed when one day, towards the end of '81, Winfried Denk knocked at my office door, looking for a job as a student in our laboratory. Less than a year later, a first scan trace was obtained with a rudimentary NFO microscope, an aperture-type instrument operated in transmission. Its step-like structure with extremely steep slopes (Fig. 1) yielded first evidence of the experimental feasibility of near-field optical microscopy and the resolution to be expected. The date on the record happens to be 22. 10. 1982, almost exactly ten years before this first workshop on near field optical microscopy. After another year of small scale experimentation, our first publication was submitted (Pohl et al. 1984). In order to illustrate the underlying principle and also for the fun of it, we invented the slogan "optical stethoscopy."

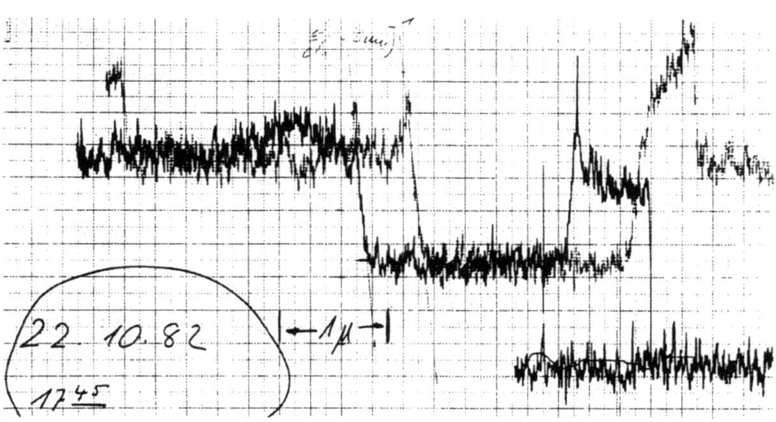


**Figure 1.** The first rudimentary SNOM scan trace (Denk, Pohl and Lanz, unpublished) across a half-transparent strip of metal on a glass substrate. Lighter and darker records correspond to forward and backward motions of the piezo-scanner. Note the steep slopes. The experimental setup was similar to that described in the publication by Pohl et al. (1984).

## 3. Early NFO Microscopy

The next step was taken together with Urs Dürig who produced the first NFO images with an aperture scanning NFO microscope (Pohl et al. 1985, Dürig et al. 1986). The resolution — ca. 20 nm laterally — and reproducibility obtained is by no means inferior to any results available today.

Having now become a serious instrument, we felt the need for a more serious name than "stethoscope," preferentially an acronym. The first choice was NFO-S, the "S" standing for scanner or stethoscope; later, we changed it to SNOM (scanning near-field optical microscopy) to stress its membership in the family of new scanning probe microscopies (SXM, with "X" standing for tunneling, forces, capacitance, near-field optics, and so on).

While we believed to be alone with our stethoscope, other activities in NFO were independently evolving at a number of places: Ulli Fischer (1985) had begun to study the scattering from narrow holes in a dark field scheme, a development which later on culminated in the first reflection SNOM (Fischer et al. 1988) and experiments with individual particle plasmons (Fischer and Pohl 1989). G.A. Massey (1984) published theoretical considerations on infrared microscopy with a scanned slit which he verified with experimental scan traces shortly afterwards (Massey et al. 1985).

The most systematic and strongest independent activities in these years were those of Lewis, Isaacson, Harootunian, and Betzig at Cornell. They introduced the microlithographic hole (Lewis et al. 1984) and later the micropipette to NFO microscopy (Harootunian et al. 1986; Betzig et al. 1987), the latter being a particularly fruitful concept that opened a new way not only for the production of narrow apertures but also, at a later stage, for the development of fluorescent NFO probes (Lieberman et al. 1990).

The careful studies of the Cornell group with the collection mode geometry (inverted path of light) greatly contributed to the acceptance of NFO microscopy (for which they introduced the acronym 'NSOM', the letters having the same meaning as in SNOM) in the scientific community. Their objects were very well characterized, and they varied the operating conditions in a systematic way. The beautiful results which were published by Betzig and Trautman (1992) recently and were presented in greater detail at this workshop may be seen as the present end-point and highlight in the development of aperture NFO microscopy.

## 4. More Recent Work

NFO remained an almost singular branch of microscopy during most of the '80s, the few active groups being countable on one hand. It may be that the groups interested in SXM were too fascinated by the atomic resolution that can be achieved by STM and atomic force microscopy, but not with SNOM (Actually – who knows?...). Towards the end of the previous decade, however, new life came into the scene, stimulated to some extent by the insight that NFO microscopy can give more chemical information than any other SXM technique. Three novel developments added to the variety of NFO techniques presently pursued: photon tunneling through an evanescent wave region, plasmon microscopy, and photon mapping during scanning (electron-) STM.

The first technique, was developed independently by two groups: Courjon et al. (1989) who named it scanning tunneling optical microscopy (STOM) and Reddik et al. (1989) in collaboration with de Fornel et al. (1989) who called it photon scanning tunneling microscopy (PSTM). A third name, evanescent field optical microscopy (EFOM) was introduced by van Hulst et al. (1991). In STOM/PSTM/EFOM, a glass fiber stylus is scanned across the object surface illuminated from the other side such that total reflection occurs. This scheme is easy to implement and provides a well-defined distance dependence, in close analogy to STM.

The second recent focus are NFO microscopies based on the creation or perturbation of propagating surface plasmons. These excitations are extremely sensitive to thin absorbing layers and topographic irregularities, see, for instance Räther (1988). They are therefore excellently suited for high resolution microscopy as discussed in recent publications (Möller et al. 1991, Specht et al. 1992, see also Fischer and Pohl 1989) and by a number of authors in the present proceedings volume.

The third technique, invented by Coombs et al. (1988), provides optical images of almost atomic resolution by recording the light emitted during STM from metal and semiconductor surfaces; it was extended to semiconductor superlattices by Abraham et al. (1990).

Last but not least, a word about SNOM theory: In order to understand the properties of NFO devices, the electric fields in the NF zone have to be calculated. The low symmetry of the SNOM geometry imposes complicated boundary conditions in general, resulting in the need for numerical solutions of Maxwell's equations. In addition, it is essential in most cases to include the transition into the FF since the NF is usually produced by irradiation from an external light source, and the effect of the NFO interaction is detected in the FF.

In the past few years, fortunately, a wave of interest in these problems has evolved, including groups from Australia via Japan and Europe to the US, see, for instance, Labani et al. (1990). Besides the determination of NF distributions for SNOM geometries, questions emerged regarding field enhancement by plasmon excitation, field-induced forces and the coupling between NFs and FFs. These activities begin to produce results for realistic experimental conditions on resolution, contrast, and optimum operating conditions — predictions that are sorely needed by experimentalists and will be instrumental in the further development of NFO microscopy and other NFO techniques.

In the last few paragraphs, I began to depart from history and enter the present state of NFO — time to finish this article, to apologize for possible errors and incompleteness, to thank A. Dereux and H. Heinzelmann who both presently work at Rüschlikon for their assistance and for helpful discussions, and to make way for the new results presented at this first workshop on Near Field Optics at Arc et Senans, Oct. 26 - 28, 1992.

## References

Abraham, D.L., Veider, A., Schönenberger, C., Meier, H.P., Arent, D.J. and Alvarado, S.F., Appl. Phys. Lett. **56**, 1564-1566 (1990)

Ash, E.A. and Nichols, G., "Super-Resolution Aperture Scanning Microscope," Nature **237**, 510-516 (1972)

Bethe, H.A., Phys. Rev. **66**, 163-182 (1944)

Betzig, E., Isaacson, M., and Lewis, A., Appl. Phys. Lett. **51**, 2088-2090 (1987)

Betzig E., and Trautman, J.K., Science 257, 189 (1992)

Bouwkamp, C.J., "On Bethe's Theory of Diffraction by Small Holes," Philips Research Reports 5, 321-332 (1950); "On the Diffraction of Electro-Magnetic Waves by Small Circular Disks and Holes," *ibid.*, 401-422 (1950)

Coombs, J.H., Gimzewski, J.K., Reihl, B., Sass, J.K., and Schlittler, R.R., J. Microscopy **152**, Pt. 2, 325-336; Z. Phys. **B72**, 497-501 (1988)

Courjon, D., Sarayeddine, K., and Spajer, M., Opt. Commun. **71**, 23-28 (1989)

de Fornel, F., Goudonnet, J.P., Salomon, L., and Lesniewska, E., "An Evanescent Field Optical Microscope," in *Proc. SPIE*, Vol. 1139, (SPIE, Washington, 1989) pp. 77-84

See, for instance, Drexhage, K.H., "Interaction of Light with Monomolecular Dye Layers," in *Progress in Optics*, ed. by E. Wolf, Vol. XII (North-Holland Publishing Co., Amsterdam, 1974) pp. 163-232

Dürig, U., Pohl, D.W. and Rohner, F., J. Appl. Phys. **59**, 3318-3327 (1986)

Fischer, U.Ch., and Zingsheim, H.P., J. Vac. Sci. Technol. **19(4)**, 881-885 (1981)

Fischer, U.Ch., J. Vac. Sci. Technol. B **3**, 386-390 (1985)

Fischer, U.Ch., Dürig, U.T., and Pohl, D.W., Appl. Phys. Lett. **52**, 249-251 (1988)

Fischer, U.Ch., and Pohl, D.W., Phys. Rev. Lett. **62**, 458-461 (1989)

Förster, T.H., Die Naturw. **33**, 166-175 (1946)

Harootunian, A., Betzig, E., Isaacson, M., and Lewis, A., Appl. Phys. Lett. **49**, 674-676 (1986)

Kuhn, H., J. Chem. Phys. **53**, 101 (1970).

Labani, B., Girard, C., Courjon, D., and van Labeke, D., "Optical Interaction between a Dielectric Tip and a Nanometric Lattice: Implications for Near-Field Optical Microscopy," J. Opt. Soc. Am. 7, 936-943 (1990)

Lewis, A., Isaacson, M., Harootunian, A., and Murray, A., Ultramicroscopy **13**, 227-231 (1984)

Lieberman, K., Harush, S., Lewis, A., and Kopelman, R., Science **247**, 59-61 (1990)

Massey, G.A., Appl. Optics **23**, 658-660 (1984)

Massey, G.A., Jeffrey, A., Davis, S.M., and Omon, E., Appl. Optics **23**, 658-660 (1985)

Mie, G., Ann. d. Physik IV **25**, 377-445 (1908)

Möller, R., Albrecht, U., Boneberg, J., Koslowski, B., Leiderer, P., and Dransfeld, K., "Detection of Surface Plasmons by Scanning Tunneling Microscopy" J. Vac. Sci. Technol. **B9**, 506-509 (1991)

Pohl, D.W., Denk, W., and Lanz, M., Appl. Phys. Lett. **44**, 651-653 (1984)

Pohl, D.W., Denk, W., and Dürig, U., "Optical Stethoscopy: Imaging with $\lambda/20$," in *Micron and Submicron Integrated Circuit Metrology*, ed. by K.M. Monahan, Proc. SPIE **565**, (SPIE, Washington, 1985) pp. 56-61

Räther, H., "Surface Plasmons on Smooth and Rough Surfaces and on Gratings," in *Springer Tracts in Mod. Phys.*, Vol. 111 (Springer-Verlag, Berlin, 1988)

Reddik, R.C., Warmack, R.J., and Ferrell, T.L., Phys. Rev. B **39**, 767-770 (1989)

Sommerfeld, A., Ann. d. Physik IV **28**, 665-737 (1909)

Specht, M., Pedarnig, J.D., Heckl, W.M., and Hänsch, T.W., "Scanning Plasmon Near-Field Microscope," Phys. Rev. Lett. **68**, 476-479 (1992)

Synge, E.H., "A Suggested Method for Extending Microscopic Resolution into the Ultra-Microscopic Region," Phil. Mag. **6**, 356-362 (1928)

van Hulst, N.F., de Boer, N.P., and Bölge, B., "An Evanescent Field Optical Microscope," J. Micr. 163, 117-130 (1991)

Zingsheim, H.P., *The Physical Deposition of Information on a Molecular Scale,"* Ber. d. Bunsenges. 11, 1185-1188 (1976)

# PRINCIPLES AND APPLICATIONS OF NEAR-FIELD SCANNING OPTICAL MICROSCOPY (NSOM)

E. BETZIG
*AT&T Bell Laboratories*
*600 Mountain Ave.*
*Murray Hill, NJ 07974*

ABSTRACT. The technology of near-field scanning optical microscopy has advanced to the point where it can be applied to a broad range of problems on a routine basis. Several recent examples are shown to support this assertion.

## 1. NSOM Technology

There are several key elements to a near-field optical imaging system, but two of the most important are the subwavelength sized probe of visible light and the mechanism whereby this probe is kept at a fixed distance from the specimen. Some time ago, we reported [1] a new probe formed from an aluminum (Al) coated, adiabatically tapered, single mode optical fiber. The key to the success of this probe lay in its ability to efficiently transfer light to the aperture region via dielectric guiding. The efficiency of this process relative to earlier near-field probes is shown in Fig. 1. In the pipette configuration [2], a large number of modes are excited, all ofwhich (except for the fundamental $HE_{11}$ diectric mode) couple to radiation modes as they reach their respective cutoff diameters. They therefore do not contribute at all to the desired signal through the aperture, but instead are retroreflected, absorbed in the coating, or

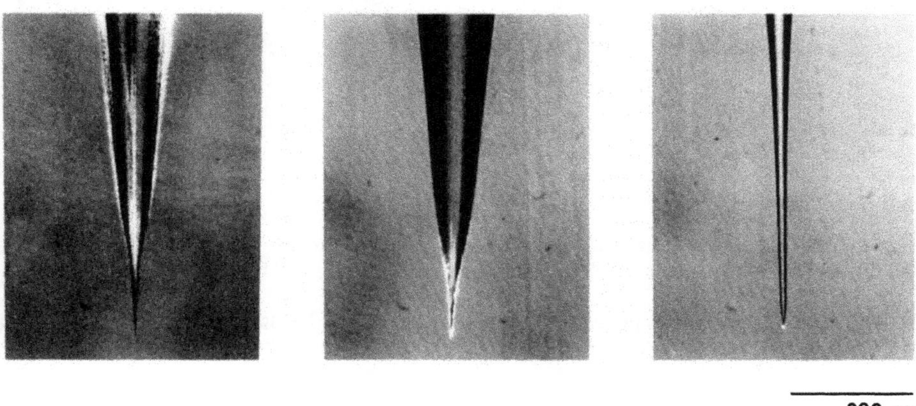


Figure 1. Optical micrographs of three *uncoated* near-field probes illustrate their relative efficiencies in delivering incident light to the tip: a) standard pipette; b) fluid filled pipette; c)adiabatically tapered single mode fiber. The bright speckle indicates where power is lost to radiation modes in each case.



*D.W. Pohl and D. Courjon (eds.), Near Field Optics, 7–15.*

transmitted as unwanted background radiation via pinholes. Significantly better guiding with a larger degree of coupling to the fundamental mode can be achieved with a high refractive index fluid in the pipette bore to approximate the structure of a tapered multimode fiber (Fig. 1b). Indeed, we have observed an ~30x increase in signal for a typical Al coated pipette with an ~ 100 nm diameter aperture after the addition of such fluid [3]. However, the *most* efficient guiding is obtained by initially coupling the bulk of the input energy to the fundamental mode, such as is done in a single mode optical fiber. If the ensuing taper is made sufficiently gradual so that this mode does not also lose energy to radiation modes (i.e., if the taper is "adiabatic"), then the light will be transmitted to the tip region with high efficiency (Fig. 1c), and the requisite subwavelength confinement can be achieved with an Al coating. Furthermore, the high overlap integral between the dielectric bound mode and the metallic confined mode will insure good coupling within the transition region. The result is a structure with several orders of magnitude more brightness than in any other near-field probe of demonstrably similar dimensions, and resolution (down to ~12 nm) limited not by the achievable tip diameter (see, for example, Fig. 2), but only by the optical penetration depth within the coating. The probe also combines versatile contrast [4] with an unmatched level of reliability, as befitting a technique with aspirations of broad acceptance.

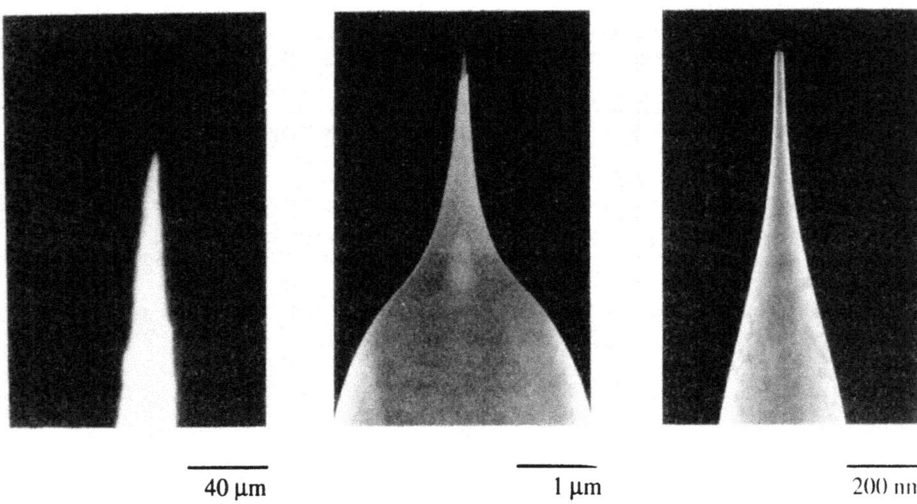


Figure 2. This probe, optimized for use in shear force microscopy, nevertheless illustrates the small tip diameters achievable by the fiber pulling process.

More recently, we demonstrated that this probe could itself be used as a vertically oriented resonant cantilever to measure small shear forces between the probe end face and the sample [5], thereby resulting in a versatile and easily implemented distance regulation mechanism. Just as in attractive mode atomic force microscopy (AFM), a number of methods can be used to sense the oscillatory motion of the tip, including position sensitive detection using the light emitted by the aperture (Fig. 3a), position sensitive detection using light from a secondary source focused onto and scattered by the probe (Fig. 3b), and differential interferometry [6]. As pointed out by Vaez-Iravani [7], the aperture method results in considerable coupling between the optical and topographic signals on samples with significant variations in transmissivity, but in some cases, including the fluorescence and polarization applications to follow, the transmitted signal is very uniform and no such difficulties arise. On the other hand, this method is *preferable* in cases where another light source cannot be used (either due to photobleaching, undesirable carrier generation, or crosstalk with a luminescence signal), or in situations where the motion at the apex can differ considerably from that a few microns further up the shank (such as can happen with exotic tip shapes (e.g., Fig. 2) or during operation at higher order resonance frequencies). In practice, we use the aperture detection scheme when necessary, and the external focusing and scattering arrangement otherwise. Both methods are sufficiently sensitive to image even soft samples non-invasively with very high resolution, as demonstrated in Fig. 4. However, the scattering method provides a properly decoupled topographic signal, and yet does not entail the complexity of the differential interferometric approach. It is also readily adapted to reflection mode NSOM by placing the detector on the same side of the sample as the probe, as has been demonstrated by ourselves and others [8].

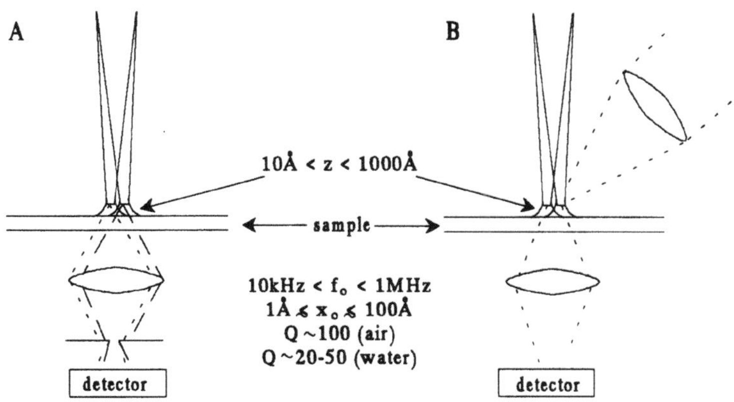


Figure 3. Two methods for the detection of the lateral dither motion essential to shear force feedback: a) aperture detection; b) scattering detection.

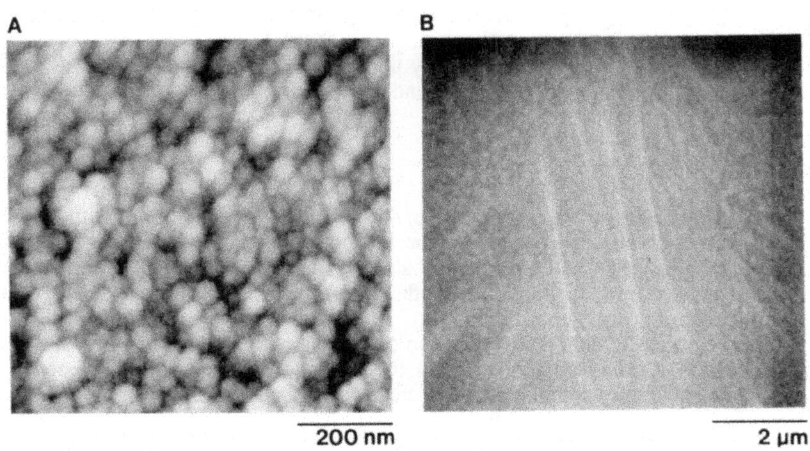


Figure 4. Topographic images of soft samples obtained via shear force microscopy: a) polystyrene beads (mean diameter ~19 nm) spin coated onto glass (from [5]); b) a Swiss mouse 3T3 fibroblast cell cultured onto glass. The dither detection was through the aperture method in a) and the scattering method in b).

While it is indeed possible to decouple optical effects from the topographic signal, it is frequently not possible to do the reverse. The problem is that the point of closest approach on the probe will shift when scanning over a rough surface, which can result in significant variations in the coupling of light from the aperture as its vertical position shifts within the near field. This effect is illustrated in the simultaneously acquired shear force and near-field images in Fig. 5. The NSOM image clearly has two contributions - a low resolution component representing the '"true" resolution and, offset from this, a high resolution pattern, identical to the shear force image, resulting from the topographic coupling. It is therefore clear that, when using an independent distance regulation mechanism, any claims of high resolution in NSOM are suspect unless the features of interest can be shown to be completely uncorrelated to the topographic data. This criterion is certainly met in the applications below.

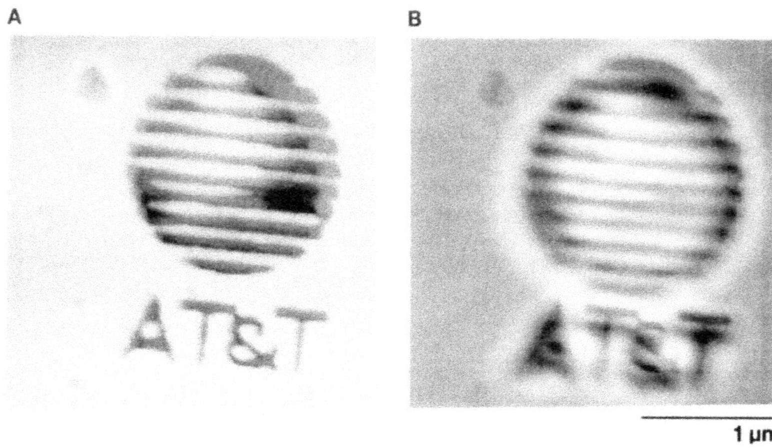


Figure 5. Simultaneous a) shear force and b) NSOM images of a test pattern demonstrate the spurious introduction of high spatial frequencies in the optical image arising from the vertical motions induced by the feedback mechanism. The shear force resolution is high and the intrinsic NSOM resolution is low for probably the same reason - a sharp projection from the coating which can reach into crevices yet prevents a closer approach by the aperture.

## 2. Applications

With the development of the tapered fiber probe and its integration with shear force feedback, NSOM is at last a reliable tool for the optical characterization of surfaces on a nanometric scale. However, its versatility is also enhanced by our demonstration that most optical contrast mechanisms, many refined with years of effort, can be extended to the near-field. Several examples are shown in Fig. 6. Details of this work can be found elsewhere [4,9], but here it suffices to note that whereas in certain cases the contrast can be understood in terms of a straightforward extrapolation from the far field, in other instances probe-sample interactions can significantly complicate this picture. Such interactions must always be kept in mind when interpreting any near-field image.

In the past year, we have endeavored to demonstrate the power of near-field optical contrast through a number of applications. Simple absorption contrast, for example, has been employed to image histological specimens, such as tissue sections from the hippocampus region of the monkey brain [9,10]. So far we have labeled these samples with toluidine blue, a common optical stain, in order to generate contrast. By choosing a false color table to mimic the colors

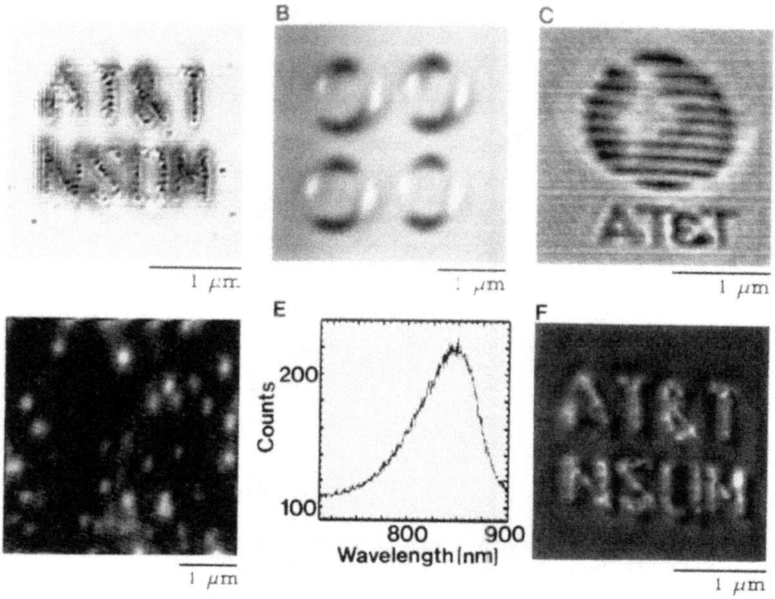


Figure 6. Optical contrast in the near field: a) absorption; b) polarization; c) refractive index; d)fluorescence (two color version can be found in [9]); e) spectroscopy; and f) reflectivity.

as observed by conventional optical microscopy, the resulting images can be easily interpreted, and a number of features readily identified, as shown in the example in Fig. 7. Although the resolution of this image is well beyond the limit of diffraction, it remains far less than that obtainable by transmission electron microscopy (TEM). Compared to TEM, however, an NSOM system is inexpensive and easy to operate. Furthermore, when integrated with conventional optics, it can be used to quickly pick out and image multiple regions of interest from a much larger field of view. Consequently, we are now investigating possible applications in both research and clinical pathology where these advantages may make NSOM the method of choice.

One of the perhaps surprising features of the tapered fiber probe is that the input polarization can be manipulated to generate any desired output polarization from the aperture, almost always with a high degree of purity [11]. We have used this capability to demonstrate near-field magneto-optic (MO) imaging and ultra-high density data storage [12], as illustrated in Fig. 8. The sample in this case was a platinum/cobalt multilayer film of the variety developed for commercial MO recording. Indeed, the large bits over the majority of the image were written by far-field optics, and are representative of the size used currently in the MO storage industry. The detailed shape and edge structure of these bits (both of which are important in the determination of the read-back noise) are clearly resolved with NSOM, making it a potentially useful tool for the characterization of MO media and the conventional read/write process. In addition, however, we have found that by using a higher power pulsed beam, we can heat the medium and also record individual bits on a nanometric scale. The pattern at the top of Fig. 8 was written in this manner, and consists of ~100nm diameter bits on ~200nm centers. Elsewhere, bits as small as ~60 nm have been formed reproducibly, and packing densities up to ~45 Gbits/sq. in. have been attained. This is ~100x the density of current optical storage devices, and ~300x that of magnetic hard drives. Of course, numerous

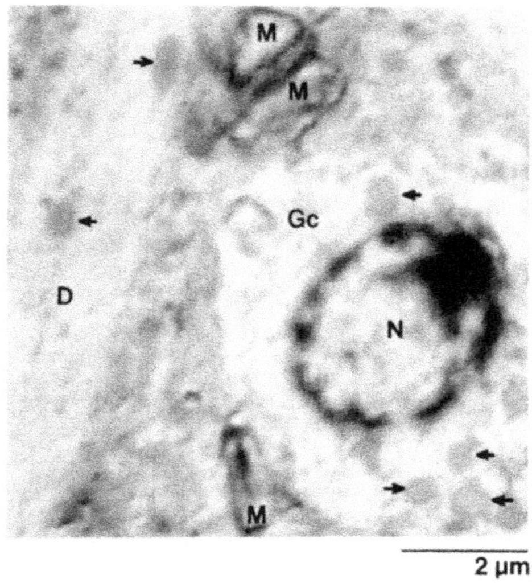


Figure 7. Near-field absorption image of a stained tissue section from the hippocampus region of the monkey brain (D, pyramidal cell dendrite; Gc, glial cell; M, myelinated axon; N, nucleus, arrows point at mitochondria) (from [9]).

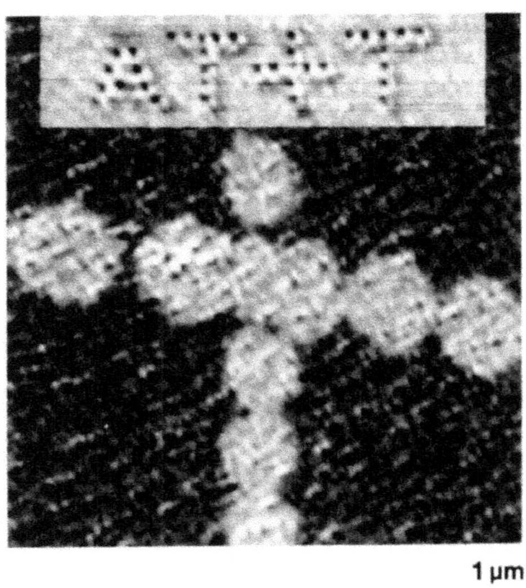


Figure 8. A bit pattern (top) recorded with near-field optics in a magneto-optic thin film is compared with a set of much larger bits written by conventional means (adapted from [12]).

other parameters in addition to areal density are important from a commercial standpoint, including read/write speed, tracking, and fly height, and all present vexing engineering challenges. Nevertheless, we continue to investigate the commercial potential of near-field MO storage, as it presents certain unique advantages, among which is the ability to borrow heavily from existing and future magnetic and optical storage technologies.

Among all of the optical contrast mechanisms we have been able to adapt to the near field, fluorescence may prove to be the most important, largely because of its implications for biology. In particular, an enormous amount of work has gone into the development of a wide range of fluorescent labeling techniques, such as fluorescent analogs, fluorescently tagged antibodies, and fluorescent *in situ* hybridization of DNA. These methods exhibit exceptional specificity, are largely non-invasive, can be used to detect single macromolecular species and, in some cases, can be used to monitor intracellular conditions such as pH, $Na^+$ concentration, or even electrostatic potential. A hint of this power is presented in the images of a fixed and air dried mouse fibroblast cell in Fig. 9 [13,14]. In the course of locomotion, these cells extrude flat and thin protrusions known as lamellipodia, and the scan region in Fig. 9 is over

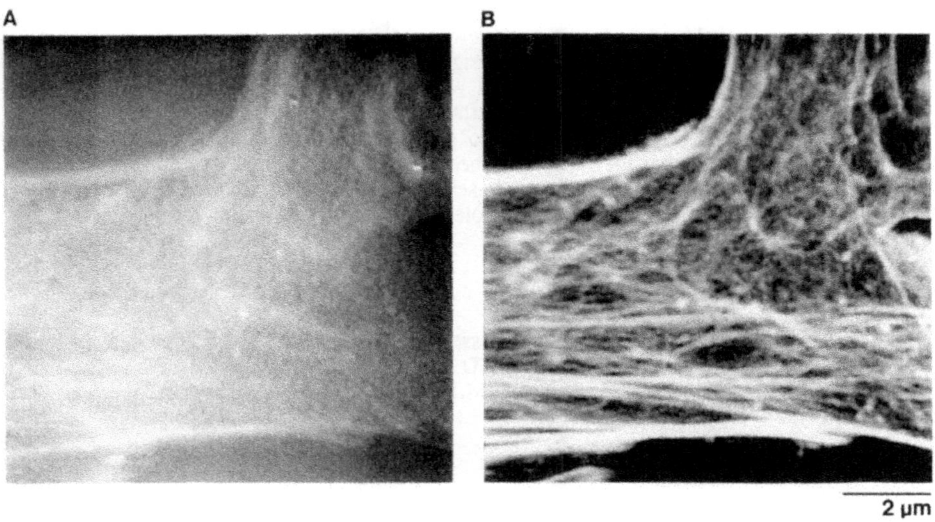


Figure 9. a) shear force topographic and b) near-field fluorescence images of a lamellipodium from a fixed mouse fibroblast cell. The cytoskeletal actin network immediately underneath the cell membrane has been labeled with rhodamine-phalloidin (from [13]).

one such structure. Central to this locomotive process is the organization of cytoskeletal actin, and it is this cytoskeletal network which is visualized through fluorescence NSOM in Fig. 9b. Comparison with the shear force topographic image in Fig. 9a illustrates an important point: although the resolution of the shear force image is perhaps superior, the contrast generated by fluorescence labeling provides far more detail of biological interest. Indeed, despite considerable efforts to apply STM and AFM to biology, the scientific return has thus far been meager, in large measure because the contrast mechanisms are neither as interesting nor as powerful as those available to NSOM. Furthermore, the specificity of fluorescence labeling as applied to NSOM results in images that are more easily interpretable than those of STM and

AFM, where there is always the difficulty of conclusively identifying the source of the observed topography. We are now applying fluorescence NSOM to a study of wound healing mechanisms in a continuous layer of these cells, and we continue to work towards the goal of dynamic near-field fluorescence microscopy of living cells.

## 3. The Future

The past history of NSOM has largely centered on the development of near-field technology. Although such work will continue (as evidenced by the bulk of the experimental contributions in this volume), I believe we are entering a new era in which much of the most exciting work will involve applying the existing techniques to a broad range of scientific and technological problems[9]. This is not to suggest that further technical innovation will not occur. Indeed, I anticipate that the variety of near-field probes will increase largely because of the demands of specific applications, just as has proved the case with AFM (e.g., magnetic force microscopy, electrostatic force microscopy, etc.). However, our primary goal should always be to see that NSOM becomes an indispensable, broadly accepted tool in a number of different disciplines, and we should avoid the trap of technology for technology's sake. Consequently, when we are developing new near-field methods, we must always ask ourselves what are the limitations of the existing techniques that our methods are designed to overcome, what are the *real, practical* obstacles in implementing these methods, and what new limitations do we introduce in our quest to address the original problem. Furthermore, we should always work to see that new methods are quickly and effectively applied to problems of practical interest, for it is only through such applications that they will truly become accepted. Indeed, if those of us in the near-field community universally adopt such self-critical standards, I predict that near-field optics will quickly grow to have a substantial role in science and technology.

## 4. Acknowledgements

This work would not be possible without the essential contributions of many people, including J.K. Trautman, R.J. Chichester, P.L. Finn, T.D. Harris, and J.S. Weiner (AT&T Bell Labs), F. Naftolin and C. Leranth (Yale Univ.), and M.H. Kryder, C.-H. Chang, F. Lanni, and D.L. Taylor (Carnegie Mellon Univ.).

## 5. References

1. Betzig, E., Trautman, J.K., Harris, T.D., Weiner, J.S., and Kostelak, R.L. (1991) 'Breaking the diffraction barrier: optical microscopy on a nanometric scale', Science 251, 1468-1470.
2. Betzig, E., Lewis, A., Harootunian, A., Isaacson, M., and Kratschmer, E. (1986) 'Near-field scanning optical microscopy (NSOM): development and biophysical applications', Biophys. J. 49, 269-279.
3. Betzig, E., Weiner, J.S., and Kostelak, R.L., unpublished results. Copies available upon request.
4. Trautman, J.K., Betzig, E., Weiner, J.S., DiGiovanni, D.J., Harris, T.D., Hellman, F., and Gyorgy, E.M. (1992) 'Image contrast in near-field optics', J. Appl. Phys. 71, 4659-4663.
5. Betzig, E., Finn, P.L., and Weiner, J.S. (1992) 'Combined shear force and near-field scanning optical microscopy', Appl. Phys. Lett. 60, 2484-2486.
6. Toledo-Crow, R., Yang, P.C., Chen, Y., and Vaez-Iravani, M. (1992) 'Near-field differential scanning optical microscope with atomic force regulation', Appl. Phys. Lett. 60, 2957-2959.
7. Vaez-Iravani, M. and Toledo-Crow, R. (1993) 'Amplitude, phase contrast, and polarization imaging in near-field scanning optical microscopy', elsewhere in this volume.

8. Moyer, P.J. and Paesler, M. (1993) 'Reflection shear force / near-field scanning optical microscopy', elsewhere in this volume.
9. Betzig, E. and Trautman, J.K. (1992) 'Near-field optics: microscopy, spectroscopy, and surface modification beyond the diffraction limit', Science 257, 189-195.
10. Work in collaboration with F. Naftolin and C. Leranth, Dept. of Obstetrics and Gynecology, Yale University.
11. Betzig, E., Trautman, J.K., Weiner, J.S., Harris, T.D., and Wolfe, R. (1992) 'Polarization contrast in near-field scanning optical microscopy', Appl. Opt. 31, 4563-4568.
12. Betzig, E., Trautman, J.K., Wolfe, R., Gyorgy, E.M., Finn, P.L., Kryder, M.H., and Chang, C.-H. (1992) 'Near-field magneto-optics and high density data storage', Appl. Phys. Lett. 61, 142-144.
13. Betzig, E., Chichester, R.J., Lanni, F., LaRocca, G., and Taylor, D.L., manuscript in preparation.
14. Work in collaboration with F. Lanni, G. LaRocca, and D.L. Taylor, Center for Light Microscope Imaging and Biotechnology, Carnegie Mellon University.

# NEAR FIELD OPTICAL AND EXCITON IMAGING, SPECTROSCOPY AND CHEMICAL SENSORS

RAOUL KOPELMAN, WEIHONG TAN, ZHONG-YOU SHI and DUANE BIRNBAUM
*Department of Chemistry*
*The University of Michigan*
*Ann Arbor, MI 48109-1055, USA*

ABSTRACT. Subwavelength optical probes have been fabricated with micropipettes and nanofabricated optical fiber tips. A new near-field nanotechnology, photo-nanofabrication, has been developed, leading to a thousandfold miniaturization of immobilized Fiber Optical Chemical Sensors and to a billionfold decrease in necessary sample volume. Submicrometer pH sensors have been prepared by attaching a copolymer covalently to an activated nanofabricated fiber tip surface via near-field photo-polymerization. The sensors have demonstrated their excellent detection limit (zeptomoles) and fast response time (milliseconds). Applications of these subwavelength optical probes also include scanning tips, exciton probes and nanospectroscopy. Biological single cell analysis, supertip energy transfer, interfacial Kasha effect and near-field scanning spectroscopy illustrate our progress.

## 1. Introduction

Recently there has been an explosion of interest in scanning tip microscopy. The development of specialized subwavelength optical probes is critical for Near-field Scanning Optical Microscopy(NSOM), Molecular Exciton Microscopy(MEM), Near-field Scanning Spectroscopy(NSS) and Fiber Optical Chemical Sensors (FOCS). We have fabricated different subwavelength optical probes, both passive and active, with or without specific chemical sensitivity, by using micropipettes and nanofabricated optical fiber tips (Lieberman *et al.*, 1990; Kopelman *et al.*, 1992; Tan et al., 1992a). One of the most successful applications of those probes is the development of submicrometer FOCS. FOCS has played an increasingly important role in chemical and medical analysis (Wolfbeis, 1991; Wise et al, 1991). They have several advantages compared to electro-chemical sensors, *e.g.*, sensitivity, selectivity, and non-invasiveness. However, so far their physical size has been limited by that of commercial optical fibers. Conceptually it has often been assumed that the wavelength of light is the lowest limit for their ultimate size. Theoretically, single molecule sensor should be possible. The smaller the sensor, the better the detection limit, the sensitivity and response time. Also, smaller sizes are preferable for intracellular research—they enable non-invasive and spatially-resolved measurements.

Existing optical fiber sensors have sizes of hundreds of microns and response times of seconds or minutes, while ultramicroelectrodes have subsecond response times. In this paper we describe a



*D.W. Pohl and D. Courjon (eds.), Near Field Optics*, 17–24.

new and controllable nanotechnique, photo-nanofabrication, based on nanofabricated optical fiber tips and near-field photopolymerization, to make nanometer optical fiber sensors with extremely short response times. The same technique can be used to prepare smaller light and exciton probes for NSOM, NSS and MEM. We applied the subwavelength optical and exciton probes in biological single cell analysis, supertip energy transfer, interfacial Kasha effect and near-field scanning spectroscopy.

## 2. Experimental

Optical fiber tips have been used in many areas(Vogelmann et al, 1988; Betzig et al, 1991; Kopelman et al, 1992) and can be fabricated either by heating and stretching or by chemical etching. The apparatus used for fiber tip nanofabrication in this work has been described previously (Tan *et al.* 1992b). The fiber tip is produced by drawing an optical fiber in a micropipette puller while heating it with a $CO_2$ laser. Commercial optical fibers can be tapered to as small as 200Å. The sides of the pulled fiber tips are coated with aluminum, leaving the end face as a transmissive aperture. Submicrometer optical fiber pH sensors are prepared by incorporating fluoresceinamine into an acrylamide-methylenebis (acrylamide) copolymer that is attached covalently to a silanized fiber tip surface by near-field photopolymerization. The pH sensitive copolymer is excited by a 488 nm Ar+ laser beam.

Nanometer crystal light and exciton sources were prepared with organic crystals or doped polymers grown inside the very tip of a micropipette from a solution. We have used a great variety of crystals and doped polymers, such as anthracene, DPA (9, 10-diphenyl-anthracene) , perylene, BASF dyes and their doped PMMA or polystyrene polymers. These active probes were excited by different laser lines from UV to visible.

## 3. Results

### 3.1 PHOTO-NANOFABRICATION

Photo-nanofabrication is a new and novel nanofabrication technique based on near-field optics. It is a controllable nanotechnology which can produce nanometer sized optical and exciton probes with or without a specific chemical sensitivity. Using the near-field optics principle, photo-nanofabrication controls the size of the luminescent materiel grown at the top of a light transmitter or conductor, such as a micropipette or an optical fiber tip, by photochemical or photo-biological reactions. These reactions are initiated and driven by an appropriate wavelength of light. The luminescent materiel will be synthesized only in the presence of light and "bonded" only to the area where light is emitted. We notice that the key to photo-nanofabrication is near-field photopolymerization, in which the size of the active luminescent probe is defined by the light emitting aperture. We thus believe that the photo-chemistry only occurs in the near-field region, where the photon flux is the highest (Tan *et al.*, 1992a; Tan *et al.*, 1992b).

The size of the luminescence materiel grown on the aperture of the light conductor will be equal to or smaller than that of the aperture. This has been shown in our ultramicro optical fiber sensor preparation (Tan et al., 1992b). By using multi-step photochemical synthesis, we can miniaturize optical and exciton probes to sizes smaller than the aperture sizes of the original light conductors. Also, by using a multi-dye system for photochemical synthesis, we can make multi-functional probes with extremely small size. The ultimate goal of our photo-nanofabrication technique is to produce optical, exciton and sensor probes with extremely small sizes (for example, molecular size) for NSOM, NSS, MEM and FOCS by controllable molecular engineering. These probes should play an important role in single cell analysis, single molecule detection and other key biotechnologies, such as DNA sequencing.

## 3.2 SUBMICROMETER FOCS

One of the most successful applications of photo-nanofabrication to date is the preparation of submicrometer FOCS. For the fabrication of the submicrometer sensor, the reaction temperature, polymerization time, laser coupling efficiency had to be optimized in order to grow submicrometer sized polymers on the fiber tip surfaces. We produced many different size sensors, down to 0.1 μm. Fig. 1 shows the schematic drawing of a submicrometer FOCS. FOCS could be made as small as 200Å.

The submicrometer pH sensors were tested using "Nuclepore" porous polycarbonate membranes with sizes of 2-10 μm. Before testing, the membranes were first immersed inside a pH buffer solution, then taken out and put on a microscope viewing stage where the sensor was aligned with a specific hole. By driving the Z translational stage, the sensor could be inserted into one of the holes and detect the pH of the buffer solution at that hole. Fig. 2 gives the results. The fluorescence intensity of our submicrometer sensor is one of the highest among all reported optical fiber sensors, even though the sensors used here are the smallest ever reported. One major reason is the very high excitation efficiency in the near-field of the submicrometer FOCS.

## 3.3. SENSOR RESPONSE TIMES AND SENSITIVITY

The miniaturization of the sensor results in very fast response times. The size of the sensor is between 0.1 to 1 μm, and no mechanical confinement is used. Thus the analytes have immediate access to the dye on the sensor tip. This gives our sensors the shortest response times, well below 100 ms (Tan et al, 1992a).

The submicrometer pH sensor has a very good detection limit. We notice that the smallest volume that gave definite pH measurements had less than 3000 hydrogen ions (at pH=8). With the above performance values and considering the high fluorescence intensity recorded in our work ($10^5$ cps), we believe that there should be no problem to downsize the sensor to tens of nanometers.

## 3.4. BIOLOGICAL APPLICATIONS

The feasibility of the miniaturized optical fiber sensor and the new internal calibration technique (Tan *et al.* 1992a) has been tested with several biological samples. The first biological application of submicrometer FOCS has been demonstrated for 10 day and 12 day old rat embryos (Nau *et al.*, 1987), placed on the stage of an inverted fluorescence microscope, as shown in Fig. 3. One or two laser (Ar+) lines are used for excitation. The emission spectrum is taken with an OMA. The analysis is based on ratios of fluorescence intensities at different wavelengths of the same spectrum, or on ratios of fluorescence intensities at different wavelengths of two different spectra obtained by two different excitations( ratios of ratios), providing for internal calibration (Tan *et al.*, 1992a). By using internal calibration, the intra-embryo pH values were 7.55 for 10 day rat embryos

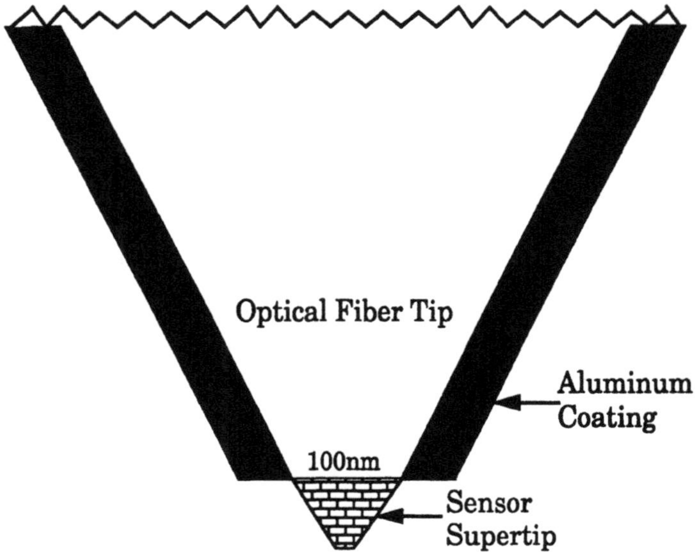


**Figure 1. Schematic Drawing of Submicron Optical Fiber Chemical and Biological Sensor**

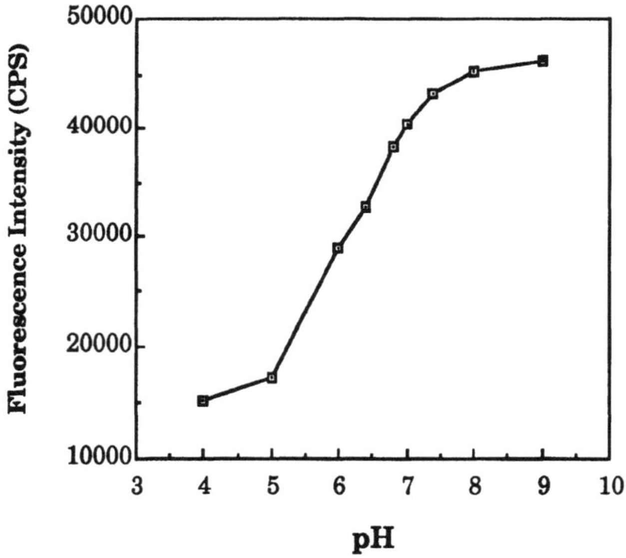


**Figure 2. pH test of submicron sensor in micron-sized polycarbonate membrane holes**

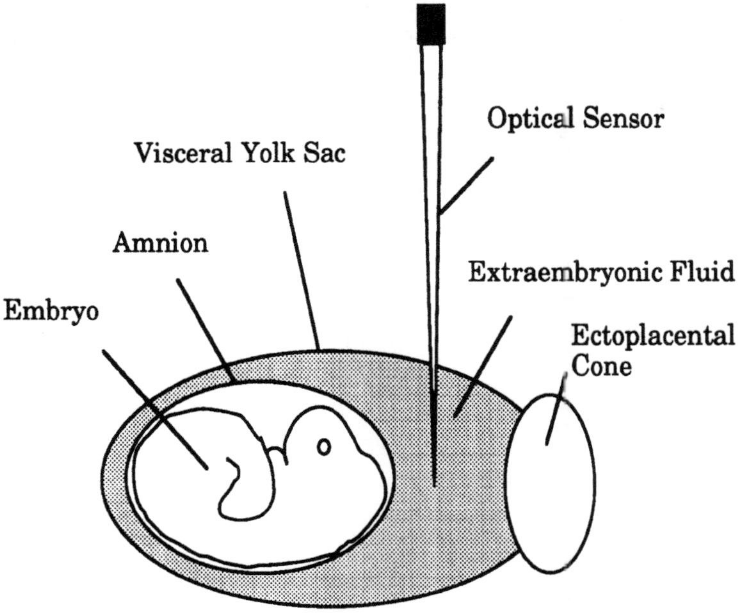


**Figure 3. Schematic representation of the rat conceptus with its associated tissues, showing placement of the ultramicrofiberoptic sensor inside the extraembryonic fluid compartment.**

and 7.27 for 12 day rat embryos, respectively. These values are in good agreement with the reported results for "homogenized" embryo samples, where more than 1000 embryos had to be crushed. In contrast, only one single embryo was needed for the pH measurements via miniaturized FOCS. Also, in a dynamic response experiment a drug (diamide) was introduced to the external environment of a rat embryo. The pH change inside the rat embryo was monitored and a rapid pH drop was observed when the diamide solution was introduced.

3.5. EXCITON TIPS AND SUPERTIPS

Using both micropipettes and FOCS tips, we prepared a series of optical supertips. Fig. 4 shows the relation between the tip, the supertip and the active center. For the highest resolution, as required for NSOM and MEM, this active center consists of a single molecule, or molecular cluster, that does the imaging. This molecule is the energy donor site for a probe to sample Förster-transfer, or the spin-orbit interaction site for a probe to sample Kasha effect. As much as the supertip is part of the tip, the active center is part of the supertip. To demonstrate the feasibility of the above ideas we used an anthracene or DPA crystal tip and a monolayer Langmuir-Blodgett (LB) film containing a dye molecule at low concentration. From a distance of approximately 0.5

microns, the anthracene tip is just a blue light source. This emitted blue light is absorbed by the dye containing film, resulting in a yellow-orange dye fluorescence. This absorption is fairly inefficient and is barely dependent on the Z distance (crystal tip to LB film), as the number of dye molecules in the beam is practically independent of Z (Z < 0.5 micron). However, once the anthracene tip practically "touches" the LB film (coming within the Förster radius of about 5 nm), the mechanism of energy (exciton) transfer dominates the LB film excitation process. These results lend credence to our theoretical modeling of nanometer or subnanometer optical probes.

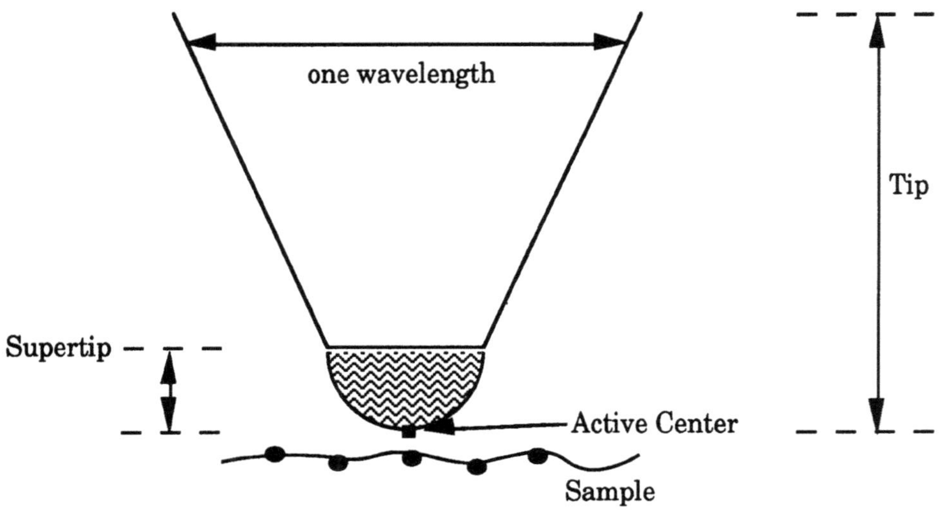


**Figure 4. Schematic drawing of tip, supertip and active center**

## 3.6. EXTERNAL HEAVY ATOM (KASHA) EFFECT AT INTERFACE

This effect (Kasha, 1953) is based on an *interatomic* spin-orbit coupling, where one atom (the heavy one) has a high degree of spin-orbit coupling ("relativistic effect") and the other atom (the light one) has very little of it on its own. The effect is "internal" when both atoms belong to the same molecule and is "external" when they are not. The "effect" is empirically observed spectroscopically, as a change in intensity or lifetime of an absorption or emission. For instance, a molecule such as anthracene has a fluorescence quantum efficiency of near unity because the first excited singlet state cannot transfer its energy to the lower lying first excited triplet state due to spin selection rules. However, upon "induced" spin-orbit coupling such singlet-triplet transfer becomes possible both internally (with the aid of vibrational quanta that absorb the extra energy) or radiatively. Empirically one observes a "quenching" of the fluorescence (both intensity and lifetime are reduced). These effects have been observed before only when both molecules are neighbors (or collide) inside the same phase, *e.g.*, in liquid solution (Pope et al., 1982). The interaction is extremely short range, *e.g.*, 5 Å or less. Our plan is to employ this effect at the *interface* of two distinct phases, *i.e.*, the scanning supertip and the DNA sample (where certain base atoms are chemically substituted with heavy atoms, *e.g.*, Hg or I.

Our experiment involved a solid film of perylene in contact with a liquid solution (water) containing heavy atoms (NaI). We have observed significant fluorescence quenching effects for two situations: 1) The perylene film is at the bottom of an optical fiber tip that is scanned towards contact with the water solution (and back out again). 2) The perylene is a thin solid film floating on the water solution. As seen in Fig. 5, the quenching is significant. We emphasize that blank experiments were run with aqueous NaCl solutions and showed no quenching. Furthermore, as soon as the optical fiber tip (with the perylene "supertip") is pulled out of the water solution, the fluorescence is back to its old value. This eliminates the possibility that the perylene was dissolved, or chipped off, or reacted chemically with the iodide (or formed some permanent van der Waals complex with it). This is the first observation of an interfacial Kasha effect. The interfacial effect occurs in the near-field zone and illustrates the principle of Scanning exciton microscopy. The goal is to achieve single molecular sensitivity.

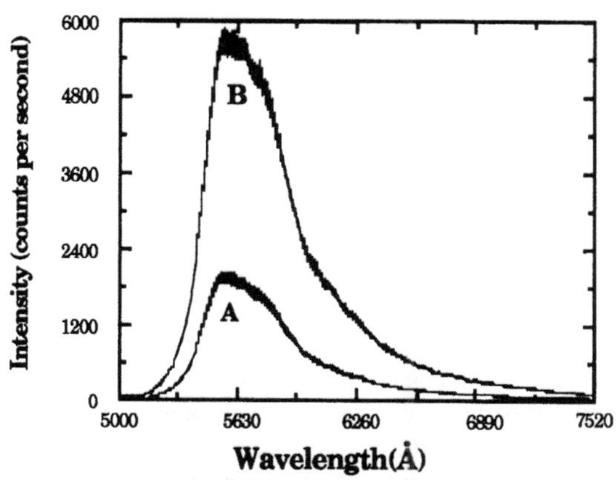


**Figure 5 Emission spectra of perylene interface with NaI solution(A) and blank(B)**

3.7. SPATIALLY AND SPECTRACALLY RESOLVED IMAGING

Basically, this approach adds spectral resolution to the near-field optical microscopy. Sample runs were performed with mixed molecular crystals and with molecularly doped polymers. The spatially resolved spectra give new information (Fig. 6). For instance, the molecularly doped polymer sample, which appears to be homogeneous (little scattered light visually and under the microscope) turns out to be quite heterogeneous or structured on the submicrometer level. From the results it is obvious that there are submicrometer crystallites in the polymer matrix (while in bulk spectra all features are "washed-out"). One can easily distinguish the perylene crystal from the perylene (single molecule) doped PMMA film. The different spectra are taken at increments of 40nm horizontal scanning.

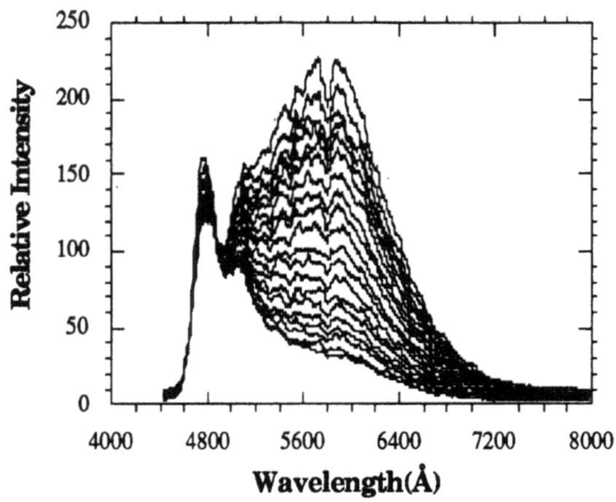


**Figure 6. Nano-spectra of perylene-doped PMMA (Increments of 40nm)**

## 4. Acknowledgments

We thank C. Harris, S. Smith, B. Thorsrud and R. Zenobi for their help and useful discussions. We acknowledge the support from the U. S. Department of Energy grant DE-FG02-90ER61085 and NSF grant DMR-9111622.

## 5. References

Betzig, E., Trautman, J. K., Harris, T. D., Weiner, J.S., and Kostelak, R. L. (1991). Breaking the diffraction barrier: optical microscopy on a nanometric scale, Science, 251, 1468.

Kasha, M. (1952) "Collisional perturbation of spin-orbital coupling and mechanism of fluorescence quenching. A visual demonstration of the perturbation", J. Chem. Phys. 20, 71.

Kopelman, R. Smith, S. Tan, W., Zenobi, R., Lieberman, K and Lewis, A. (1992) "Spectral Analysis of Surfaces at Subwavelength Resolution", SPIE, 1637, 33.

Lieberman, K, Harush, S., Lewis, A. and Kopelman, R. (1990) "A Light Source Smaller than the Optical Wavelength", Science 247, 59-61.

Nau, H. and Scott, W. J. (1987) "Teratogenicity of Valproic Acid and Related Substances in the Mouse: Drug Accumulation and pHi in the Embryo During Organogenesis and Structure-Activity Considerations", Arch. Toxicol., Suppl. 11, 128.

Pope, M., and Swenberg, E. (1982) Electronic Processes in Organic Crystals, Oxford Univ. Press, New York.

Tan, W., Shi, Z-Y., Smith, S., Birnbaum, D. and Kopelman, R. (1992a) "Submicrometer Intracellular Chemical Optical Fiber Sensors" Science 258, 778.

Tan, W., Shi, Z-Y. and Kopelman, R. (1992b) "Development of Submicrometer Chemical Fiber Optic Sensors", Anal. Chem., 64(22).

Vogelmann, T. C., Knapp, A. K., McClean, T. M. and Smith, W. K. (1988). "Measurements of light within thin plant tissues with fiber optic microprobes", physiologia Plantarum, 72, 623.

Wise, D. and Wingard, L. (1991). Biosensors with Fiberoptics, Humana Press, Clifton, New Jersey.

Wolfbeis, O.S. (1991) Fiber Optical Chemical Sensors and Biosensors, Vol. 1, CRC, Boca Raton, FL, p. 413.

# AMPLITUDE, PHASE CONTRAST, AND POLARIZATION IMAGING IN NEAR-FIELD SCANNING OPTICAL MICROSCOPY

**M. Vaez-Iravani and R. Toledo-Crow**
Center for Imaging Science,
Rochester Institute of Technology,
P.O. Box 9887,
Rochester, N.Y. 14623, USA.

**ABSTRACT.** The design and theory of operation of a variety of new techniques in NSOM are discussed. The basic framework of all these modalities is the force-regulated NSOM. A number of results are presented to demonstrate the power of this simultaneous microscope. Amongst the new techniques introduced are linearized polarizing microscopy, as well as amplitude, and phase, interference contrast imaging modalities. In each case results are presented to demonstrate the potential of the techniques. Finally, a new approach to Kerr/Faraday polarization imaging is introduced, which is based on the behavior of orthogonal states of light. This has direct application in magneto-optic imaging.

## 1. Introduction

A major problem hindering the development of near field scanning optical microscopy (NSOM), and photon scanning tunneling microscopy (PSTM) [1-9] has been the lack of a universally applicable feedback signal to regulate the tip/sample spacing. In addition, the NSOM has been plagued by a rather low level sample light. The development of pulled fiber tips [7] has alleviated this problem to some extent. However, the NSOM can still benefit from enhanced signal detection techniques. Furthermore, such effects as phase contrast imaging have remained largely unexplored. In this paper, we preset the design, and explain the operation, of a number different techniques in NSOM which overcome the above problems.

## 2. Tip/Sample Distance Regulation: Combined NSOM/Force Microscopy

The usual approach to regulating the tip/sample separation, viz the use of the detected light level in a feedback system, is not universally applicable. This is due to the fact that i) both sample topography and transmissivity would influence this signal, and ii) the lack of a signal in imaging opaque parts of the sample would result in a crash.

Recently, force regulation of the tip/sample distance has been introduced [7,8,9], in which the sample topography is deduced from the simultaneously obtained force image of the sample. Unlike tunneling



*D.W. Pohl and D. Courjon (eds.), Near Field Optics, 25–34.*

regulated systems [3], this approach is equally applicable in imaging dielectrics. Fig. 1 shows our implementation of this approach. At the heart of the technique there is a micro-pipette, or a pulled optical fiber, covered everywhere with gold or aluminum, except at its tip. This system operates in two simultaneous modalities of force microscope, and NSOM. To obtain the NSOM image, the output of a He-Ne laser is transmitted through the sample and is directed toward a photomultiplier tube (PMT), where it is detected (collection mode). Alternatively, one can illuminate the sample using light through the aperture, and the transmitted light through the sample can be directed to a PMT with an objective lens (illumination mode).

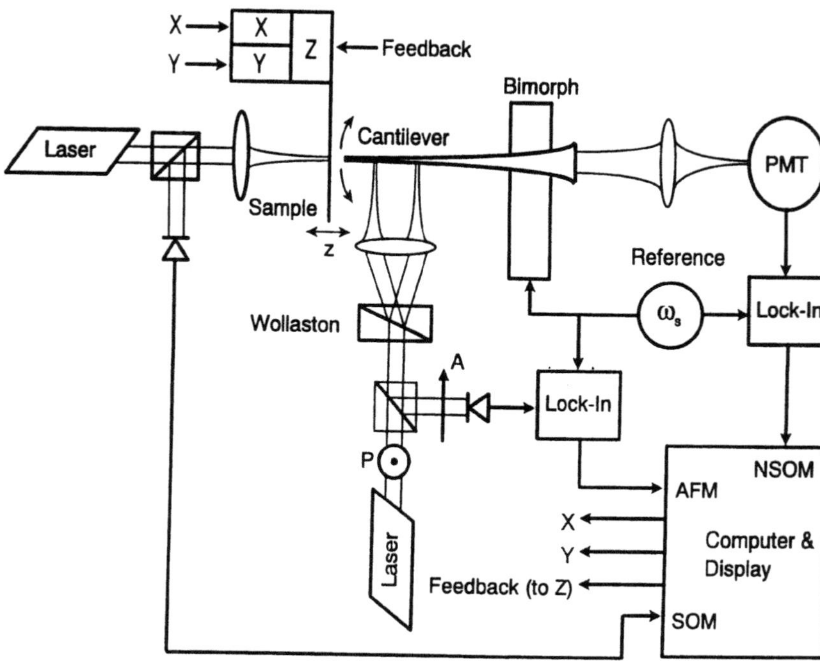


Fig. 1. Combined force microscope/NSOM

The force microscopy capability of the system is based on the detection of the sideways vibration of the cantilever using a differential optical interferometer [7]. A direct consequence of using this format is that the sample size is limited by the working distance of the objective. In some case, therefore, it may be advantageous to tilt the sample, by a few degrees, with respect to the tip. Typically, in our experimental arrangements, we have utilized 85°-90° orientation between the tip and the sample. The detection laser used is a linearly polarized He-Ne (6328 nm; 0.5 mW output). The separation between the two foci in our system is about 100 $\mu$m, and the first focal spot is usually about 30-50 $\mu$m from the very tip. Depending on the type of objective that is used, we have obtained focal spot diameters of the order of 3-5 $\mu$m. The resonance frequency for the tip is typically in the range 75-82 kHz, and the tip oscillation amplitude is set to 2-5 nm. Thus far, all our experiments have been performed by monitoring the amplitude of the tip oscillation in a feedback loop, with the reference set to an amplitude corresponding to 75% of the free oscillation signal.

Our force detection mechanism is totally external to the sample and is, therefore, not affected by the sample transmissivity, resulting in independent force and NSOM modalities. This is in contrast to the method of sensing the position of the tip by imaging it through the sample [8]. For i) in that method, one requires (at least) a partially transmissive sample, thus excluding samples with totally opaque parts, and ii) the AC signal used to sense the tip position is indistinguishable from that due to the differential optical transmissivity of the sample. Thus, the position sensing signal in that method is not pure and is, in fact, coupled with the sample optical characteristics. Nevertheless, that system is perfectly acceptable in imaging samples which don not show significant variations in their transmissivity across the field. Finally, we note that our system has been used extensively to image a variety of samples [10].

## 3. Pure Linear Polarizing Microscopy in the Near Field

It has been shown that in the near field, the basic approach adopted in conventional polarizing microscopy, viz. viewing the sample between a pair of crossed polarizes, can successfully be employed [11]. This technique results in an output current which can be written as [12]:

$$I_{out} \propto [t_1^2 + t_2^2 - 2t_1 t_2 \cos(\delta\phi)] \tag{1}$$

here $t_1$ and $t_2$ are the amplitudes of the 45° resolved components of the emergent light, and $\delta\phi$ is the phase difference between these components due to sample birefringence. We note that i) for a small $\delta\phi$, $\cos(\delta\phi)=1-(\delta\phi)^2/2$, resulting in a quadratic dependence of the signal on the birefringence, and severely limiting the sensitivity; ii) for $t_1=t_2=t$, the output signal is proportional to $t^2$, i.e. the resulting signal, and therefore the image, are not pure, and the sample transmittance is superimposed on the output image. In near field systems employing small apertures, a further problem can arise due to the partial depolarization of light. We have found, experimentally, that this depolarization could be as high as 50% in some cases. This appears to be highly dependent on the tip shape, particularly the symmetry of the metallic coating of the sidewall, or the presence of protrusions in the tip area. The depolarized light contributes a signal, which is again simply a representation of the sample transmittance, not its polarization effects.

We have resolved these issues by adapting our basic force regulated NSOM, so as to include a pair of polarizers, and a sinusoidally driven electro-optic modulator (a Pockels cell), together with some additional signal detection equipment [12,13]. The light from a linearly polarized He-Ne laser is passed through the Pockels cell, whose induced axes are aligned at 45° to the incoming polarization of the light. The light transmitted through the sample is detected by a PMT, in front of which a second polarizer is inserted with its pass axis crossed with respect to that of the first polarizer.

The additional sinusoidal phase modulation imparted to the sample light by the Pockels cell results in the generation of a spectrum of harmonics of the Pockels frequency, all of which are proportional to $t_1t_2$. The ratio, $V_r$, of the first and second harmonic signals is $V_r = -\tan(\delta\phi)\{J_1(\theta)/J_2(\theta)\}$, which represents a pure polarization signal, regardless of the transmissivity [12]. The terms $J_i$ are the ith order Bessel functions of the first kind. At the same time, for small $\delta\phi$, $\tan(\delta\phi)$ is a linear function of $\delta\phi$, which implies a much greater sensitivity compared with the result of viewing the sample between crossed polarizers without the Pockels cell. In addition, the use of AC signals ensures an automatic discrimination against any depolarized light out of the aperture.

Here, we present a series of images taken from a chrome checker-board pattern on a quartz substrate. In these experiments, the light source for the optical branch was a 5 mW, linearly polarized, He-Ne laser. The Pockels cell drive frequency was 60 kHz. Fig. 2 (a) shows the first harmonic image of the checker-board, and 2(b) is the corresponding, simultaneously obtained, second harmonic image. The ratio image (i.e. $\tan(\delta\phi)$), shown in Fig. 2(c), indicates the polarization structure of the sample. As expected the contrast is primarily evident in the edge areas

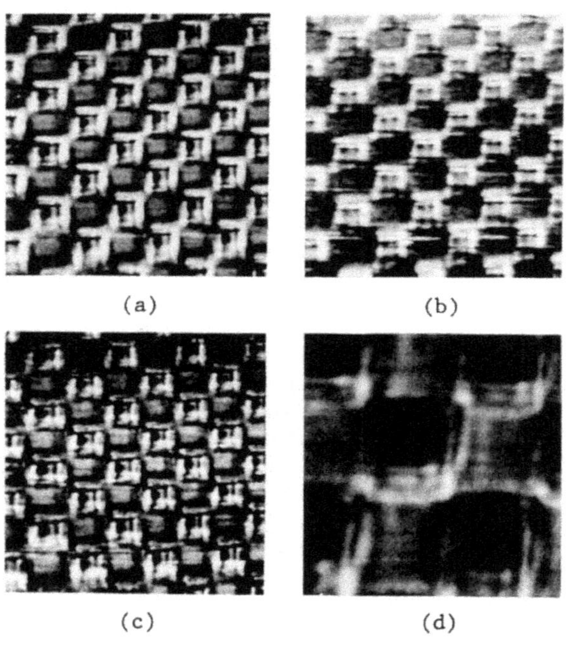

(a) (b) (c) (d)

Fig. 2. 12X12 $\mu$m first (a) and second (b) harmonic, and ratio (c) images of a chrome checker-board on quartz; (d) is a 4X4 $\mu$m ratio image.

and the corners. The details of these polarization variations are better shown in Fig. 2(d), which displays the ratio image of a small re-scanned portion of the same sample. To obtain these images, we used a "hybrid" tip, in which a cleaved polarization preserving fiber (without jacket) was inserted into a micro-pipette tip until it was stopped by the inner sloped surface of the latter. The outside of the micropipette was covered with gold, as usual. This design was adopted primarily to avoid the necessity to change the fiber in case of a crash. The images were obtained in the illumination mode.

## 4. Interference Contrast Microscopy in the Near Field

Thus far, the issue of phase contrast microscopy in NSOM has not received sufficient attention. Furthermore, although the introduction of pulled fiber tips has substantially enhanced the available sample light level, any effect that would increase the system sensitivity, would be greatly beneficial. We have recently introduced a new technique [14], based on the pseudo-heterodyne principle, which i) dramatically increases the signal strength and its detectability, and ii) provides a phase contrast modality.

The system is shown in Fig. 3, whose NSOM part is configured as a Mach-Zehnder interferometer, with the sample in one branch. The tip used here is a pulled polarization preserving fiber. The light from a linearly polarized He-Ne laser is split by a beam splitter. The transmitted portion goes through a Faraday isolator, and is launched into the fiber, whose length is piezo-electrically stretched at the frequency of $f_m$. The reflected part is directed onto a second beam splitter, where it is made to interfere with the probe light transmitted through the sample. The output of the detector is given by:

$$I_{out} \propto \{A^2+B^2+2AB\cos[\theta \sin\omega_m t+\phi+\gamma]\} \quad (2)$$

where A represents the sample light, and B the reference light. Typically, one has $B \gg A$. Here, $\omega_0$ is the optical frequency, $\theta$ is the amount of phase modulation by the piezo-stretcher, $\phi$ is the sample phase, and $\gamma$ is the environmentally dependent phase fluctuation. In the absence of interference (i.e. with the conventional set-up), one detects $A^2$ only. In the present system, we detect an AC signal at $f_m$, due to the interference term $AB\cos(\theta \sin\omega_m t+\phi+\gamma)$. Therefore, we have achieved a signal enhancement by approximately B/A, which can be very large. This approach also increases the shot noise associated with the large reference beam power. However, in the shot noise limited case, the heterodyne (or pseudo-heterodyne) detection can achieve a greatly superior S/N over the dark current limited, regular, PMT detection [14]. It is worth noting that the roots of this work are in the bulk-optics differential heterodyne interference microscope [15]. However, the basic framework of the present system, and its mode of operation, are very different from that employed previously.

The interference signal is subject to instability due to the environmentally varying phase term, $\gamma$. To eliminate the instability, we resort to an active stabilization method, in which the signal at $2f_m$ is detected, and is fed back into the drive voltage of the fiber piezo-modulator, to attempt to eliminate the $2f_m$ signal [16]. The image is obtained by taking the signal at $f_m$ at each sample point. In general, therefore, one has two bandwidths to deal with: the image bandwidth,

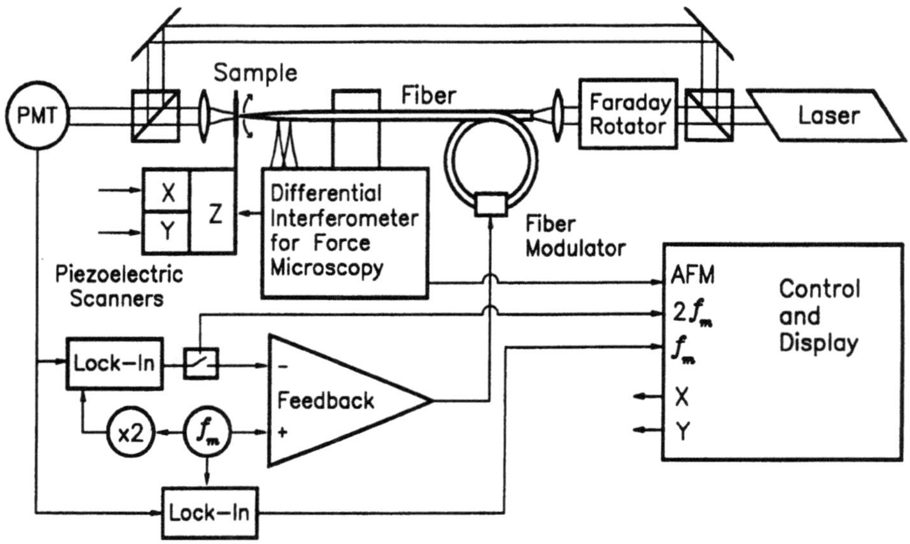


Fig. 3. Near field, pseudo-heterodyne, interference optical microscope

determined by the pixel time, and the fiber feedback bandwidth, determined by us, as desired. Depending on the relationship between these two bandwidths, three distinct regimes of operation are identified, each giving rise to a separate imaging modalities [14]:

i) *Fast fiber feedback loop*, i.e loop bandwidth > image bandwidth. In this case we expand the interference term in Eq. (2) to obtain:

$$2AB\cos[\theta\sin\omega_m t+\phi+\gamma]=2AB\{\cos(\theta\sin\omega_m t)\cos(\phi+\gamma)-\sin(\theta\sin\omega_m t)\sin(\phi+\gamma)\} \quad (3)$$

Since only the first cosine term includes the second harmonic signal ($2f_m$), the elimination of this signal through the use of the feedback loop amounts to setting $\cos(\phi+\gamma)=0$, which at the same time maximizes the term $\sin(\phi+\gamma)$. The amplitude of the detected signal at $f_m$ is thus proportional to $ABJ_1(\theta)$, which is an enhanced amplitude signal, and is devoid of any sample phase information.

ii) *Slow feedback loop*, i.e. loop bandwidth << image bandwidth. In this case the slow loop can only eliminate the low frequency components of the term $\cos(\phi+\gamma)$, dominated by the value of $\gamma$, which tends to set $\gamma=\pi/2$. The signal at $f_m$ is proportional to $ABJ_1(\theta)\cos(\phi)$, i.e. in this mode the signal represents the sample phase information as well as its amplitude. We note, however, that the signal at $2f_m$ is not actually eliminated in this mode, and is proportional to $ABJ_2(\theta)\sin(\phi)$, and can be used for imaging.

iii) *Open feedback loop*. This is the fastest mode of operation, in which we apply no correction signal, and simultaneously detect the signals at $f_m$ and $2f_m$, using two lock-in amplifiers. By squaring these signals, adding them, and taking the square root of the result, we obtain:

$$V \propto AB([J_1(\theta)\sin(\phi+\gamma)]^2+[J_2(\theta)\cos(\phi+\gamma)]^2)^{1/2} \tag{4}$$

If through the adjustment of the drive voltage to the fiber piezo modulator, we arrange for $J_1(\theta)=J_2(\theta)$, we obtain $V \propto (AB)$, i.e. once again we have a pure amplitude signal, regardless of any instabilities or phase information, together with all the advantages of pseudo-heterodyne detection.

This system has been applied to imaging a variety of samples, in different modalities, and employing both the transmission, and reflection formats. Here, we show a number of transmission micrographs obtained from a chrome lithographic mask on quartz [14]. All the images presented here were obtained using a pulled fiber tip in the illumination mode. Fig. 4(a) shows the conventional near field transmission image of a number of tracks. One can identify the track edges; however, the signal is clearly very weak, and the image is dominated by the random fluctuations of the PMT. To obtain this image, we arranged for the sample light to be of the order 10 pW, and the bandwidth was set at 1 kHz, resulting in a poor S/N. Fig. 4(b) shows the interference image of the same sample, using the same amount of sample light, and bandwidth, as in 4(a), but using a reference beam power of 50 $\mu$W. The signal improvement is obvious. This image was taken in a *slow loop* mode, using a correction loop bandwidth of 3 Hz (image bandwidth was 1 kHz). In addition to imaging the tracks, the image shows a number of other effects, which are clearly phase related phenomena.

Fig. 5(a) shows the $f_m$, and 5(b) the $2f_m$ image of a different part of the sample, taken in the open loop mode. These images are obviously dominated by the fluctuations of the environmentally varying phase term, $\gamma$. Fig. 5(c) shows the result of the operation of Eq. (4). We note that all the spurious fluctuations in 5(a) and 5(b) have disappeared, and the intricate details of the sample, including small dust particles are clearly visible.

## 5. Analyzer-less, photodiode-detected, low light polarizing microscopy

In this section we propose and demonstrate a new technique to perform low light level polarizing microscopy (including Kerr/Faraday microscopy), which is based on the fact that orthogonally polarized beams do not interfere. Our method consists of a pseudo-heterodyne interference set-up, in which the polarization of the reference beam and that of the sample beam (for a given point) are orthogonal. Should the sample show any Faraday (or Kerr) rotation, the sample beam will have a component along the reference beam, giving rise to a signal at the modulation frequency of the fiber stretcher. This signal has two important features: i) it is the

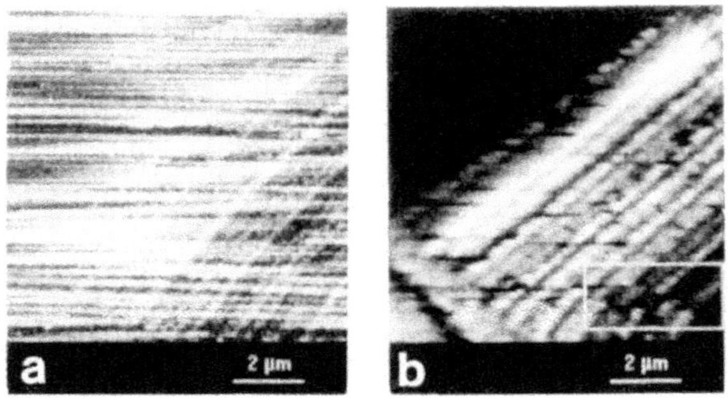


Fig. 4. Image of chrome tracks: regular transmission (a); stabilized interference contrast image (b)

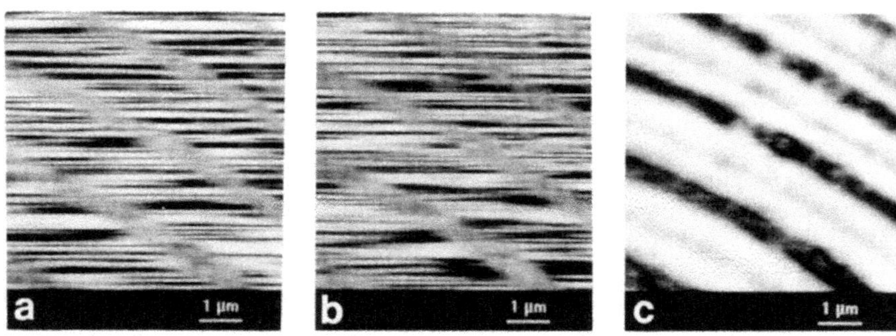


Fig. 5. Interference (open loop) images of chrome tracks: first harmonic (a) and second harmonic (b); (c) is result of operation in Eq.(4)

result of the pseudo-heterodyne interference with a strong reference beam, thus eliminating any problems associated with low light level detection; ii) it automatically has a *linear* dependence on the rotation angle, rather than a square law dependence, as is dictated by Malus's law in a conventional set-up. We note that there is no need for an analyzer. As a result of these features, the signal so detected represents a great improvement in S/N, enabling a simple photodiode to be used, instated of a PMT. Any problems associated with environmentally varying phase terms are resolved using the methods discussed above.

To demonstrate this concept, we performed an experiment in which a sample light beam of 1 nW was interfered with a reference beam of 100$\mu$W. The initial polarization of the two beams was orthogonal. The signal at the fiber modulation frequency was monitored as a function of the angle of rotation of the polarization of one light beam with respect to the other.

No analyzer was used in this experiment, and the detector employed was a simple photodiode. Fig. 6 shows the result of this experiment for a detection bandwidth of 1 kHz. We note that i) the output is linear with respect to the rotation angle and ii) the sensitivity for these conditions is about 0.6 degree. This is about 1 order of magnitude below the theoretical sensitivity for a linearized shot-noise limited detection scheme. The excess noise was due to imperfect polarizers, and some excess Johnson noise. Nevertheless, this level of sensitivity for a photodiode is quite remarkable. The implication of such a technique in magneto-optic detection systems is obvious [17].

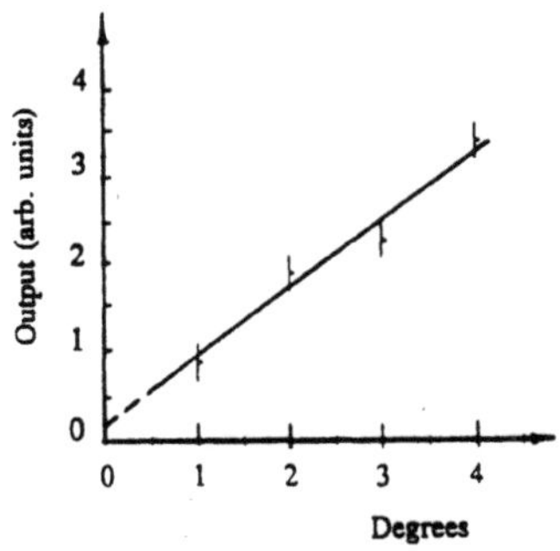


Fig. 6. Output of a photodiode in an interference polarizing experiment

## 6. Conclusions

In this paper, we have discussed a number of concepts in the NSOM. We have established the viability of the force regulated system as a universally applicable tip/sample regulation mechanism. We have also described techniques to perform linear polarizing microscopy independent of sample transmissivity/reflectivity. Pseudo-heterodyne techniques have been demonstrated which achieve ultra-high sensitivity detection of sample light, leading to both phase contrast, and pure amplitude imaging. We have further introduced a new technique to perform low light level, linear Kerr or Faraday polarizing microscopy using a simple photodiode. In each case, we have provided experimental results to support our claims, and to demonstrate the potential of the techniques. We believe that the work presented in this paper resolves some important problems in near field microscopy, and that the stage is clearly set for a rapid development of the field in many applications.

## Acknowledgement

This work was supported by a grant from the Whitaker Foundation.

## References

1. Pohl, D.W., Denk, W., Lanz, M. (1984), "Optical stethoscopy: image recording with resolution λ/20", Appl. Phys. Lett. 44, 651-653
2. Harootunian, A., Betzig, E., Isaacson, M., and Lewis, A. (1986), "Super-resolution fluorescence near-field scanning optical microscopy" Appl. Phys. Lett. 49, 674-676

3. Fischer, U. Ch., During, U.T., and Pohl, D.W. (1988),"Near-field optical scanning microscopy in reflection",Appl.Phys.Lett. **52**, 249-251
4. Reddick, R.C., Warmack, R.J., and Ferrell, T.L. (1989), "New form of scanning optical microscopy", Phys. Rev. B **39**, 767-770
5. Courjon, D., Vigoureux, J.M, Spajer, M. and Sarayeddine, K. (1990), "External and internal reflection near field microscopy: experiment and results", Appl. Opt. **29**, 3734-3740
6. van Hulst, N.F., Sergerink, F.B., and Bolger, B. (1991), "An evanescent field optical microscope", in H.K. Wickramasinghe (ed.), Scanned Probe Microscopy, AIP, New York, p. 79
7. Yang, P.C., Chen, Y., and Vaez-Iravani, M. (1992), "Attractive Mode Force Microscopy with Optical Detection in an Orthogonal Tip/Sample Configuration", Jour. Appl. Phys. **71**, 2499-2503
8. Betzig, E., Finn, P.L., and Weiner, J.S. (1992), "Combined shear force and near-field scanning optical microscopy", Appl. Phys. Lett. **60**, 2484-2486
9. Toledo-Crow, R., Yang, P.C., Chen, Y., and Vaez-Iravani, M. (1992), "Near-field differential scanning optical microscope with atomic force regulation", Appl. Phys. Lett. **60**, 2957-2959
10. Vaez-Iravani, M., Toledo-Crow, R., and Chen, Y. (1992), "Correlative imaging in scanning near field optical microscopy", Jour. Vac. Sci. Tech. (in press)
11. Betzig, E., Trautman, J.K., Weiner, J.S., Harris, T.D., and R. Wolfe (1992), "Polarization contrast in near-field scanning optical microscopy", Appl. Opt. **31**, 4563-4568
12. Vaez-Iravani, M. and Toledo-Crow, R. (1992), "Pure Linear Polarization Imaging in Near Field Scanning Optical microscopy", Appl. Phys. Lett. (submitted)
13. See, C.W. and Vaez-Iravani, M. (1986), "Linear Imaging in Scanning polarization and Interference Contrast Optical Microscopy", Electron. Lett. **22**, 1079-1081
14. Vaez-Iravani, M. and Toledo-Crow, R. (1992), "Phase Contrast and amplitude pseudo-heterodyne interference near field scanning optical microscopy", Appl. Phys. Lett. (in press)
15. See, C.W., Vaez-Iravani, M., and Wickramasinghe, H.K. (1985), "Scanning Differential Phase Contrast Optical Microscope: Application to Surface Studies", Appl. Opt. **24**, 2373-2379
16. Nonnenmacher, M., Vaez-Iravani, M., and Wickramasinghe, H.K. (1992), "Attractive Mode Force Microscopy Using Feedback Controlled Fiber Interferometer", Rev. Sci. Instrum. **63**, 5373-5776
17. Betzig, E., Trautman, J.K., Wolfe, R., Gyorgy, E.M., Finn, P.L., Kryder, M.H., and Chang, C.H. (1992), "Near Field Magneto-Optics and High Density Data Storage", Appl. Phys. Lett. **61**, 142-144

# A SCANNING NEAR-FIELD OPTICAL MICROSCOPE (SNOM) FOR BIOLOGICAL APPLICATIONS

T. HARTMANN, R. GATZ, W. WIEGRÄBE, A. KRAMER, A. HILLEBRAND, K.LIEBERMAN[1], W. BAUMEISTER, and R. GUCKENBERGER
*Max-Planck-Institute for Biochemistry*
*D-8033 Martinsried*
*Germany*

ABSTRACT. We have built a scanning near-field optical microscope (SNOM) for the investigation of planar model membranes.The lipid-protein films are either transferred onto a solid substrate ("Langmuir-Blodgett" films) or are examined at the water-air interface of a Langmuir trough.The SNOM is built onto the stage of an inverted fluorescence microscope. It allows parallel scanning tunneling microscope (STM) imaging: From comparison of SNOM and STM images of test structures the tips are characterized. For the distance control on liquids, we use a capacitive signal; we have succeeded in maintaining a 100 nm distance between a tip and the surface of a liquid without getting into contact. We present the design of the instrument in some detail, together with first parallel SNOM and STM images.

## 1. Introduction

We aim at investigating planar lipid and protein layers, either transferred onto solid substrates ("Langmuir-Blodgett" films) or, preferably, *in situ* on the water-air interface of a Langmuir trough. Our particular interest in such systems arises from transmission electron microscopy of proteins, more specifically from electron crystallography where two-dimensional crystalline arrays are a prerequisite. In this context, crystals smaller than the resolution limit of the conventional light microscope are already useful. Recently significant progress has been made in using monolayers of functionalized lipids to bind and concentrate proteins at an interface, thus promoting 2-D crystallization (for a review see [1]). It is desirable to have a means of monitoring the crystallization process and to identify and locate crystalline patches. The flatness of such samples makes them ideal for scanning near-field optical microscopy. (For a review on this type of microscopy see [2].)

A strong contrast-mechanism is provided by the fluorescence of dye-labelled lipid and protein molecules. Accordingly, our scanning near-field optical microscope (SNOM) is built onto the stage of an inverted fluorescence microscope. One can look at this setup as a conventional light microscope, equipped with an "optical overdrive".

Our instrument allows parallel SNOM and scanning tunneling microscope (STM) imaging. This is particularly convenient for working on metallic test samples, where the SNOM images can be compared with the higher resolution STM images.

In the following we introduce the instrument, describe the preparation of tips and metallic test samples, present first parallel SNOM and STM images, and we discuss signals which can be used for distance control on liquids.

[1] Department of Applied Physics, Hebrew University, Jerusalem 91904



*D.W. Pohl and D. Courjon (eds.), Near Field Optics,* 35–44.

## 2. Experimental Setup

### 2.1. DESIGN CRITERIA

The following design criteria apply when the properties of our objects are taken into account. The resulting design of our SNOM is shown schematically in fig. 1.

2.1.1. *SNOM or PSTM ?* Due to the flatness of our samples they are basically equally suitable for scanning near-field optical microscopy and for photon scanning tunneling microscopy (PSTM, [3, 4]). However, we want to exploit fluorescence as a contrast mechanism. A limitation in using fluorescence is the limited number of photocycles that a dye-molecule can undergo before suffering photobleaching. With SNOM one excites fluorescence only very locally and collects a high percentage of the emitted photons. PSTM, however, would require a comparatively large area to be illuminated, from which only a small fraction of fluorescence excitations is used for image formation. Therefore, the SNOM has clearly advantages for our application.

2.1.2. *Tip-sample geometry.* Since our samples are transparent, we can afford to choose a transmission geometry for the SNOM. Since we want to investigate surfaces of liquids, the scanner is placed above the sample, pointing towards the surface and an objective lens is located below in order to collect the light. For constant optical conditions one would prefer to leave the tip at a fixed position and move the sample. Nonetheless, in our present setup we are moving the tip, since the sample will certainly be too vibration-sensitive.

2.1.3. *Light source.* An $Ar^+$-Laser has been chosen as a light source, since it offers several wavelengths in the blue and the green allowing the use of different dyes and easy change of excitation wavelengths. A water-cooled laser has been chosen, with the advantages over an air-cooled laser of less optical noise, less vibrations, and easier handling of the cooling medium. On the other hand the water-cooled instruments are usually more overpowered, and consequently more of the optical power has to be attenuated by grey filters. For non-fluorescence experiments, a red HeNe laser is employed.

2.1.4. *The tip.* The ideal near-field light source is bright and small. Our near-field light source is a sub-wavelength metallic aperture which is illuminated from behind. The light passing through this aperture is strongly attenuated [5], which makes the transmitted intensity small. An upper limit for the incoming light is given by the absorption of the light by the aperture and the risk of its thermal destruction. We have chosen tapered, metal-coated optical fibres as tip structures [6]. With monomode optical fibres the undesirable heating of the aperture is minimized, since the light is already confined to a core with only a few μm in diameter.

2.1.5. *Combination with a light microscope.* There are two reasons for combining the SNOM with a conventional light microscope: The combination allows to survey the specimen at low magnifications and to locate the tip position. Moreover, the comparison of the conventional light micrographs with the SNOM images allows to judge their fidelity. The objective lens serves for direct imaging and alternatively as the collection optics in the near-field mode. For detecting the SNOM signal the microscope is equipped with components as used in microphotometry.

2.1.6. *Parallel STM control and imaging* (cp.[7]). For distance control, the SNOM requires another near-field signal. STM control is accurate and easy to achieve. However, for conducting the tunneling current one has to apply a thin conductive coating to the front part of the fibre tip (at the expense of some light loss).

SNOM tips can be characterized by comparison of STM and SNOM images. The absolute

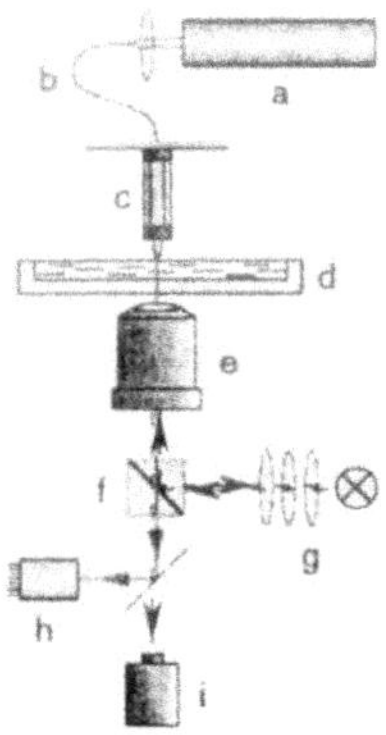

Figure 1. Schematic drawing of the optical parts of the SNOM setup. The light path for the SNOM is marked by grey arrows, and for epi-fluorescence microscopy by black arrows. (a) $Ar^+$-laser, (b) monomode optical fibre, (c) piezo scanner, (d) transparent sample; here: Langmuir trough for scanning across a water-air interface, (e) objective lens, (f) dichroic mirror and fluorescence filters, (g) illuminator for epi-fluorescence microscopy, (h) photomultiplier, (i) CCD-camera.

Figure 2. View of the modified microscope stage. Size: The base plate measures 235 mm x 220 mm. The sample stage has been removed for the sake of clarity. It normally rests upon three 1/2 inch piezo tubes incorporated in the base plate. For details see text (2.2.1.)

distance determined by means of STM on conducting samples allows one to explore the distance dependence of various optical and electronic signals potentially useful for non-conducting or liquid samples (see 5.3.).

## 2.2. DESCRIPTION OF THE INSTRUMENT

2.2.1. *The microscope stage.* The central part of the instrument consists of the modified stage of an inverted fluorescence microscope, carrying the mechanical parts of the SNOM. It is shown in fig. 2: The scanner tube with the fibre tip points downwards towards the objective lens. The piece of the optical fibre carrying the tip is fed through the scanner tube. This fibre is either directly connected to the fibre coupler of the laser, or it is coupled to another fibre on the microscope stage. In this fibre-fibre coupler the two fibre ends are guided on two parallel highly polished bars towards each other and clamped. The coupling is improved by means of index matching oil.

The maximum scan range is 12 µm x 12 µm x 3 µm. The scanner is mounted on a triangular plate which also carries the preamplifier for the tunneling current. The plate itself rests on top of three piezo tubes which are of the same type as the (1 inch) scanner piezo. This helps to reduce thermal drift along the vertical axis. The three piezoes serve two purposes: They allow remote control of the lateral scanner position via a slip-stick motion [8]; a high voltage of up to 500 V applied to their inner electrodes provides a vertical range of 7 µm for fine positioning, useful for tunneling approach, for drift compensation, and for long-range signal-distance curves. The "carrier" tubes which are enclosed by plastic pieces for mechanical protection, are mounted on two platforms of a DC motor-driven lift for rough vertical positioning and approach.

For lateral movement of the sample it was desirable to have another backlash-free remote control facility. Since the slip-stick motion is not suitable for heavy loads, we use a modified version of it: Three piezoes carry the sample in the desired direction, place it down onto three rest points, and move back without the sample, ready for another transport cycle. The base plate has to be thick enough for incorporation of these (1/2-inch) piezoes and for minimizing vibrations. Like most of the mechanical parts, it is made from aluminium. The base plate is 25 mm thick; a circular cutout in the bottom part allows to turn the objective revolver without changing focus, even when carrying objective lenses with working distances down to 0.1 mm. The SNOM stage is covered by a plexiglass box which minimizes air-flow and allows control of the humidity.

2.2.2. *Optical components and light detection.* For fluorescence applications we use a 1 W $Ar^+$-Laser (multiline) with a prism wavelength selector ($\lambda$ = 514 nm, 502 nm, 497 nm, 488 nm, 477 nm, 458 nm). For first experiments we have used a 10 mW red HeNe Laser ($\lambda$ = 632 nm; polarized light). Coupling of the laser light into the monomode fibre is done with a standard fibre coupler (Photon Control Microblock).

The inverted fluorescence microscope is a NIKON Diaphot-TMD, with a twin-port adapter and with microphotometry equipment: an adjustable selected area diaphragm (useful for exclusion of stray light in SNOM), an optional rotating analyzer, a beamsplitter for CCD-camera and photomultiplier and optionally additional filters.

The detector is a side-on type photomultiplier, selected for 25 cps dark rate at room temperature (Hamamatsu R4632), in a single photon counting system (preamplifier, window discriminator) from SMT (type WD 100). Counting is done on the MacAdios card of the computer (see 2.2.3.) At high light levels the signal from the photomultiplier preamplifier is used as an analog signal, low-passed and connected to one of the AD-converter inputs.

2.2.3. *Analog and digital control.* The electronics of the microscope is essentially based on a home-built STM control unit. The analog electronics provides a number of channels for the parallel processing of signals, and it has some special features which are of particular interest for SNOM work (cp. 4.2. and 5.3.): It provides a simple lock-in amplifier for modulation techniques and facilities for compensation of cross-talk. It allows to work with "high" specimen voltages of up to ±10 V and tunneling currents down to 5 pA. The feedback for *z*-regulation can be switched from tunneling current to other signals.

Scan generation and parallel acquisition of several images is controlled by means of a program written by us which runs on a Macintosh Quadra 900 computer. Both forward and backscan images are recorded (hysteresis!). Signal-distance curves can also be recorded. The interfacing between computer and analog electronics is established via a MacAdiosII card with DA converter daughterboards (12-bit AD conversion, 16-bit and 12-bit DA conversion).

## 3. Materials and Methods

### 3.1. TIPS

3.1.1. *Tip fabrication.* We produce our SNOM tips by etching monomode optical fibres in hydrofluoric acid (HF). For the first experiments we used fibres FS-SN-3221 from 3M for the red HeNe laser.

Before etching, the plastic coat of the fibre end is softened in methylene chloride for about 15 sec and stripped off. The fibre is cleaved so that it extends only few mm beyond the plastic coat. The fibre is moved a few times through droplets of acetone and methanol for cleaning. Several fibres are mounted at a time and their ends are dipped into 48% HF. Typical etching times are between 45 and 70 min. The surface of the HF is covered with ca. 0.5 mm silicon oil. Thus etching

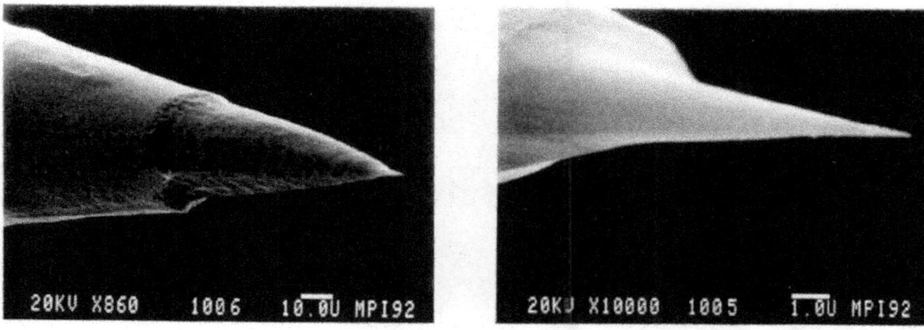


Figure 3. Optical fibre, tapered by etching in 48% HF and imaged with a SEM, at two different magnifications. The fibre axis was tilted 20° away from the beam direction and accordingly appears with reduced length in the images.

of the fibre from the vapour phase is avoided, and the point at which the taper starts is well-defined. It is very convenient for fibre handling to choose this point just about 1 mm beyond the end of the plastic coat. After etching the tips are dipped into distilled water and cleaned with organic solvents.

After etching the fibres are mounted in brass cylinders, glued and contacted with electrically conductive paint for use as STM tips. For metal coating by vacuum evaporation we used chromium (nominal thickness ≈ 5 nm, from a 45° angle; for good adhesion of the coat to the glass), aluminium (nominal thickness ≈ 250 nm, from a 90° angle; for strong absorption of light) and platinum/carbon (nominal thickness ≈ 3 nm, from a 60° angle; to have a conducting surface at the foremost part of the fibre to be used as a STM tip). Aluminium can be replaced by chromium for sake of convenience.

3.1.2. *Tip characterization.* A first idea of the quality of the etched fibre tips is obtained by inspecting them side-on in the light microscope. Occasionally we inspect them head-on, with laser light coupled into the fibre in order to see whether the tip is at the location of the fibre core, and how symmetric the appearance is. The quality of the coated tips can be judged roughly from the amount of emitted light, and more stringently from the comparison of parallel SNOM and STM images. For characterizing tip shapes and radii more directly we examine them in a SEM (fig. 3).

### 3.2. TEST SAMPLES

A test sample for parallel STM-SNOM imaging has to be conductive and at the same time it should have strong absorption contrast. The sample should be flat, to minimize the influence of corrugation on the optical signal. Our test specimens are glass cover-slips coated with 4 nm and in some areas with 2 nm of Pt/C. Such films provide enough conductance for STM, and 50% vs. 70% transmission of light for SNOM. To render sharp edges, a thin mask is needed which can easily be removed and which exhibits a variety of structures on a proper size scale. For this purpose we use salt deposits.

Clean cover slips are rendered hydrophilic in a plasma cleaner. They are dipped into concentrated sodium chloride solution, pulled out again and allowed to dry - either quickly in the vacuum of the plasma cleaner, or slowly at ambient pressure. The dry sample is coated with 2 nm of Pt/C. The sample is rinsed with distilled water. Finally, another 2 nm of Pt/C are deposited. The samples show a variety of round, square, zigzag, and "fractal" structures from sub-µm sizes up to several microns. A light micrograph of the sample is shown in fig. 4.

Figure 4. Test specimen, viewed through a conventional light microscope at low magnification. The pattern was prepared by vacuum evaporation using a "NaCl mask". Bright areas correspond to 2 nm of platinum/carbon, dark areas to 4 nm (for details see 3.2.)

## 4. Results

### 4.1. PARALLEL IMAGING IN STM AND SNOM MODE (Cp. fig. 5)

For STM feedback, relatively high specimen voltages, between 6 V and 10 V, and small tunneling currents between 5 pA and 50 pA were chosen. Under these conditions ("field-emission") the distance between fibre and sample surface is large (typically 10 nm to 60 nm) [9]. This makes the tunneling more stable, and vibrations cause less problems. The distance dependence of the tunneling current is less extreme than at usual tunneling voltages of around 1 V.

For SNOM, the analog photomultiplier signal has been used so far for image recording because of the high light levels. In the beginning the light level was adjusted by defocusing the fibre coupling. Since this makes the optical signal extremely vibration-sensitive, we have optimized the fibre coupling and attenuated the laser beam with grey or polarization filters instead.

Simultaneous tunneling and near-field optical images of a 12 µm square region of the salt crystal test sample are shown in figure 5. The optical image shows a perfect correlation to the topographic (tunneling) image: Holes in the film show up bright, while several small salt crystals which remained on the slide during the preparation procedure (higher and brighter in the STM image) appear dark in the SNOM image. The streak in the centre of the images could probably be caused by accidental touching of the tip with the sample surface.

### 4.2. DISTANCE-DEPENDENCE OF SIGNALS

Since we cannot use tunneling distance control on insulating samples and on liquids, we have explored other signals which may be used for maintaining a tip-sample distance of 100 nm or less.

We have recorded the distance dependence of the optical signal by moving the tip from tunneling contact to several µm distance. We have seen ten maxima with a distance of $\lambda/2$ between them, indicating interference. The difference between minima and maxima is of the order of 5 % of the total signal and decreases only slowly with distance.

We have also investigated capacitive signals: A voltage of 10 V was applied between tip and sample, and the *z*-position of the tip was modulated in a sinusoidal fashion, at a frequency f in the 10 kHz range and with an amplitude of ± 50 nm or larger. The tip picks up A.C. current, since tip

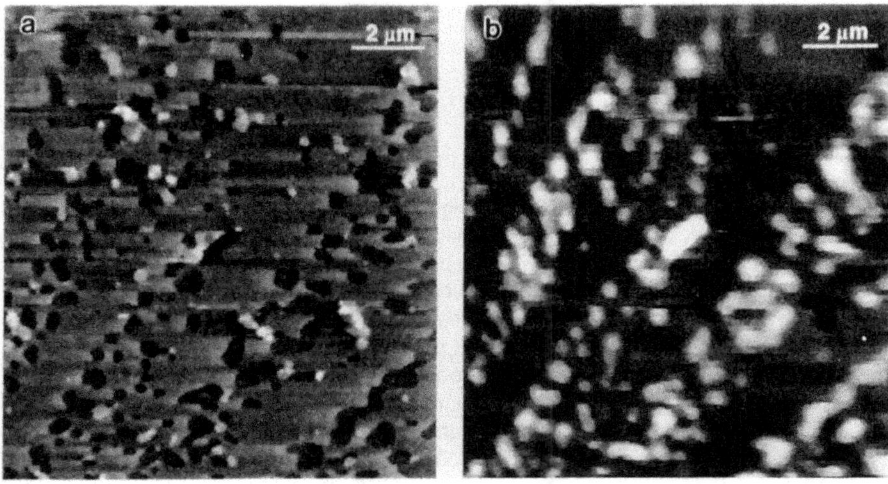


Figure 5. Parallel STM-SNOM images: (a) Height signal from STM; STM parameters: +9 V bias voltage, 50 pA constant tunneling current; dark patches represent depressions of 2 nm. (b) Light signal from SNOM; the light level in the dark areas corresponds to 70 % of level in the bright areas. The light source was a HeNe laser with $\lambda$ = 632 nm; numerical aperture of the objective lens was 0.55.

and sample represent a capacitor with a modulated electrode distance d. This signal is proportional to the applied voltage, to the electrode area, and to the effect of modulation on the term (1/d). In practice, the signal due to the very end of the tip is small compared to the "background" from other parts of the tip and from the tip holder. Therefore we filter out the second harmonic (2f) by a resonant circuit. The second harmonic signal is caused by the non-linear distance dependence of the capacitance; it varies strongly only when the tip comes close to contact. This is demonstrated in fig. 6. For distance control the rectified 2f-signal is fed into the

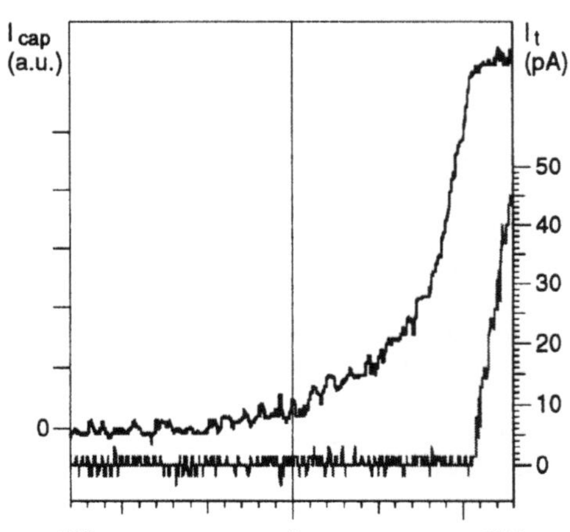


Figure 6. Distance dependence of the capacitive second harmonic signal (Icap, dark line) and the D.C. (tunneling) current signal (It, grey line). Signals were taken one after the other, under identical conditions: A metal-coated optical fibre tip close to a Pt/C film; ± 75 nm distance modulation at 10 kHz, 10 V specimen voltage. "0 nm" marks the point where Icap was regulated to a preset value. For taking the curves distance regulation was switched off. Absolute distance information is obtained from the onset of the tunneling current: The distance at this point is about 100 nm.

feedback loop for approach and *z*-regulation. Thus it was possible to approach the surface and maintain distance control on a droplet of glycerol mixed with NaCl solution.

## 5. Discussion

### 5.1. RESOLUTION

Mechanically our instrument is capable of reaching a resolution of better than 2 nm laterally and better than 1 nm vertically, as checked by STM imaging of evaporated platinum/carbon films. This means that the relatively large dimensions of the microscope do not limit our SNOM resolution, although the SNOM head is more sensitive to acoustic vibrations than our smaller STM heads.

It is a doubtful procedure to derive the actual resolution of a SNOM from edges or from periodic structures. The many different sized structures on our test sample, however, allow one to determine directly the size and distance of two structures at which they can still be resolved. In the SNOM images with our current tips object features with a centre to centre distance of ca. 350 nm are clearly resolved (cp. fig. 5.) On that scale the correspondence between SNOM and STM images is perfect. Of course, we aim at higher resolution, which means that better tips have to be manufactured.

### 5.2. TIP PREPARATION

Basically, there are two methods for preparing tapered optical fibres for use in SNOM: Pulling the fibres while heating them with a carbon-dioxide laser, or chemical etching in HF-containing solutions. Pulling appears to be the preferable method: It is supposed to give more reproducible, well-defined shapes and smoother surfaces, and it is a quick and simple procedure once a set of puller parameters is established.

Etching in HF depends on different etchabilities of the doped fibre core and the cladding (see e.g. [10]), wherefore each type of fibre behaves differently. Yet, etching has also some advantages: It can produce short tapers, which is more favourable for the throughput of light. The taper can start shortly behind the plastic cladding, which is favourable for mechanical stability and handling. Moreover, the taper can be shaped by changing the immersion depth during the etching process. For resonance techniques one can think of etching a thin cylindrical part of fibre allowing for mechanical vibrations in the desired frequency range, and a rapid taper allowing for high light intensities.

### 5.3. DISTANCE CONTROL

The SNOM image in fig. 5b was taken using tunneling distance control. For insulators and liquids we have to resort to other signals: optical, force, or capacitive.

5.3.1. *Optical.* The distance dependence of the light emitted from the fibre can in principle be used with any type of sample. If there is a strong increase or decrease of the transmitted light when coming close to the surface, one may work with constant light feedback. If the light signal shows interference-like maxima and minima with distance, one can regulate to a maximum or to a minimum via modulation techniques. However, it is difficult to identify the last extremum, and this distance (which for interference is always $\lambda/4$ or smaller) may still be too large for near-field purposes. So far, we could see interference-like modulations in the optical signal, but no drastic changes when coming into tunneling contact. We might still see such near-field effects with finer tips (see e.g. [11] for such curves.)

5.3.2. *Force.* A variety of force signals has been developed in scanning force microscopy (SFM). For a review see e.g. [12]. For our purpose a non-contact/resonance mode with small forces would be preferred (see [13] for this type of force detection). One type of resonance mode, known as "shear force", has been developed in combination with SNOM [14, 15]. The application on liquid surfaces appears critical. On all solid supports we should face less problems.

5.3.3. *Capacitive.* A capacitive signal can be used in the case of conductors (i.e. also electrolytes), even when they are covered with thin non-conducting layers. An insulating spacer layer may even be necessary for fluorescent molecules on a metal in order to avoid fluorescence quenching. The samples posing the most serious problems are surfaces of liquids. We have accomplished distance control on the surface of a liquid by modulating the *z*-position of the tip and using the second harmonic signal of the resulting A.C. current. We are well aware that we exert forces on the surface when we apply our modulation technique, and the successful glycerol experiments (see 4.2.) may have been facilitated by the small area and the high viscosity of the liquid droplet.

Particularly with liquid surfaces it may be necessary to split the task of distance control: We need a point of reference close to the surface ("approach signal"); as soon as the reference distance is established, any distance-dependent signal can be used for regulation ("feedback signal"). Candidates for feedback signals are the first harmonic signal obtained by modulation of either the specimen voltage or of the tip-sample distance. The second harmonic signal can be used for both feedback and approach. For the task of the approach on liquids - crude as it may appear - one could touch the surface, take a record of the piezo voltages (i.e. the height) at which contact occurs, withdraw the tip and adjust the (capacitive or optical) feedback signal accordingly.

With "large" liquid surfaces additional problems may arise from vibrations (surface waves) and drift (due to evaporation of the liquid). Our strategy for damping waves on a water surface is to work with thin layers of water (0.1 mm to 0.2 mm). The presence of a lipid layer will have an additional damping effect and also reduces evaporation.

## 6. Summary

We have built a scanning near-field optical microscope (SNOM) for the investigation of lipid and protein layers, either on an air-liquid interface (Langmuir trough) or transferred onto a solid substrate (Langmuir-Blodgett films). The instrument allows evaluation of current signals for STM imaging and for distance control. The apparatus is mechanically stable as seen from reproducible STM imaging of nm-sized structures (platinum/carbon films). First parallel SNOM and STM images have been taken. For improving the resolution of the SNOM images the tips (tapered optical fibres) have to be improved. In our initial experiments we have shown that we can control the distance between the tip and the surface of a liquid in the 100 nm regime by exploiting a capacitive signal.

## Acknowledgements

We are grateful to M. Spajer (Besançon) for helpful advice on fibre etching, to J. Murphy (Martinsried) for taking the SEM images, and to M. Radmacher (TU Munich) for providing us with helpful software tools. This work was supported by a grant from the Deutsche Forschungsgemeinschaft (SFB 266-B4).

## References

1. Jap, B. K., Zulauf, M., Scheybani, T., Hefti, A., Baumeister, W., Aebi, U., and Engel, A. (1992) '2D crystallization: from art to science', Ultramicroscopy 46, 45-84.
2. Pohl, D. W. (1991) 'Scanning near-field optical microscopy (SNOM)', in Sheppard, C. J. R. and Mulvey, T. (eds.), Advances in Optical and Electron Microscopy, Vol. 12, Academic, London, pp. 243-312.
3. Reddick, R. C., Warmack, R. J., Chilcott, D. W., Sharp, S. L., and Ferrell, T. L. (1990) 'Photon scanning tunneling microscopy', Rev. Sci. Instrum. 61, 3669-3677.
4. Courjon, D., Sarayeddine, K., and Spajer, M. (1989) 'Scanning tunneling optical microscopy', Opt. Commun. 71, 23-28.
5. Lieberman, K., Harush, S., Lewis, A., and Kopelman, R. (1990) 'A light source smaller than the optical wavelength', Science 247, 59-61.
6. Betzig, E., Trautmann, J. K., Harris, T. D., Weiner, J. S., and Kostelak, R. L. (1991) 'Breaking the diffraction barrier: Optical microscopy on a nanometric scale', Science 251, 1468-1470.
7. Lieberman, K. and Lewis, A. 'Simultaneous scanning tunneling and optical near-field imaging with a micropipette', Appl. Phys. Lett., to be published.
8. Besocke, K. (1987) 'An easily operable scanning tunneling microscope', Surf. Sci. 181, 145-153.
9. Guckenberger, R., Hacker, B., Hartmann, T., Scheybani, T., Wang, Z., Wiegräbe, W., and Baumeister, W. (1991) 'Imaging of uncoated purple membrane by scanning tunneling microscopy', J. Vac. Sci. Technol. B 9, 1227-1230.
10. Pangaribuan, T., Yamada, K., Jiang, S., Ohsawa, H., and Ohtsu, M. (1992) 'Reproducible fabrication technique of nanometric tip diameter fiber probe for photon scanning tunneling microscope', Jpn. J. Appl. Phys. 31, L 1302-L1304.
11. Pohl, D. W., Fischer, U. Ch., and Dürig, U. T. (1988) 'Scanning near-field optical microscopy (SNOM)', J. Microscopy 152, 853-861.
12. Meyer, E. and Heinzelmann, H. (1992) 'Scanning Force Microscopy (SFM)', in Wiesendanger, R. and Güntherodt, H.-J. (eds.), Scanning Tunneling Microscopy II, Springer, Berlin, Heidelberg, pp. 99-149.
13. Albrecht, T. R., Grütter, P., Horne, D., and Rugar, D. (1991) 'Frequency modulation detection using high-$Q$ cantilevers for enhanced force microscope sensitivity', J. Appl. Phys. 69, 668-673.
14. Betzig, E., Finn, P. L., and Weiner, J. S. (1992) 'Combined shear force and near-field scanning optical microscopy', Appl. Phys. Lett. 60, 2484-2486.
15. Toledo-Crow, R., Yang, P. C., Chen, Y., and Vaez-Iravani, M. (1992) 'Near-field differential scanning optical microscope with atomic force regulation', Appl. Phys. Lett. 60, 2957-2959.

# REFLECTION SHEAR FORCE/NEAR-FIELD SCANNING OPTICAL MICROSCOPY (NSOM)

P.J. MOYER and M.A. PAESLER
*North Carolina State University*
*Department of Physics*
*Box 8202*
*Raleigh, NC 27695-8202*

ABSTRACT. The spatial resolution afforded by near-field scanning optical microscopy (NSOM) is primarily a function of tip size, tip morphology, and tip-sample separation. Combining scanning force microscopy with NSOM allows one to maintain a small tip-sample separation distance and, consequently, improve NSOM resolution. This provides, simultaneously, a topographic perspective of the sample as well as an NSOM image. We present in this paper an instrument that provides shear force and reflection NSOM images. In addition, with this microscope, we can also obtain Raman scattering information with comparable resolution. Data is provided on nonlinear optical crystals and gold-on-silicon samples.

## 1. Introduction

The introduction of the scanning tunneling microscope (STM)[1] led to the subsequent development of the atomic force microscope (AFM)[2] and near-field scanning optical microscope (NSOM)[3]. Scanning force microscopy provides one with a perspective view of the sample surface. Certain materials can be imaged with atomic resolution. The operation of these instruments is quite simple. The probe is raster scanned across the surface and at each point the force is measured, typically by the deflection of a cantilever, onto which the probe is mounted. If the mechanical properties of the cantilever are sensitive enough, small electrostatic and van der Waals forces 'bend' the cantilever. In most cases, the compliance of the cantilever is such that it has a very low vertical spring constant. In NSOM, one uses a tapered optical fiber that has been appropriately coated with Al so that a small optical aperture (approximately 50 nm) has been formed at the tip. Light focused into the back end of the fiber is then guided to the tip and creates a subwavelength-size light source. If this source is placed close enough to the sample and scanned along the surface, the transmitted or reflected signal from the sample surface provides an optical image with resolution that is not limited by the effects of diffraction, which limits classical optical microscopy. Here, we discuss a combination force/reflection NSOM instrument that provides information regarding both topography and optical information.

Several important realizations have been made concerning the imaging theory of NSOM techniques. First of all, the near-field of the subwavelength source is very much collimated to the diameter of the aperture for a distance along the aperture normal approximately equal to the aperture radius.[4] Secondly, the mode structure of the fiber can be calculated from known waveguide theory and appropriate perturbation assumptions.[5,6] With this information, it is clear that to using an optical fiber for the probe is advantageous.[5,6] Also, one wants to keep this probe as close to the surface as possible and at a constant distance from the surface. This will provide a more homogeneous image of the optical properties of the sample. NSOM, in the geometry used for these experiments, does not have an inherent feedback mechanism to serve this purpose. The force measurement allows this distance to be held constant so that the NSOM resolution can be improved.



*D.W. Pohl and D. Courjon (eds.), Near Field Optics*, 45–50.

## 2. Experimental Results and Discussion

Betzig et al. have developed a combination shear force/NSOM instrument for the purpose of maintaining constant tip-sample separation.[7] This instrument operates in transmission mode. In this paper, we present a combined reflection shear force/NSOM instrument that operates in the same manner. The benefit of this type of microscope is that a wider variety of samples can be studied. We have investigated transparent samples as well as metals and semiconductors. A schematic diagram of the instrument is shown in figure 1. The tip is dithered laterally at a high resonant frequency of the probe. The reflected image of the tip off of the sample is projected onto a pinhole. The dithered tip results in the image oscillating in the plane of the pinhole. As the probe is lowered to the sample, the force interaction between the tip and sample lower the amplitude of the tip at this resonant frequency. A frequency response curve of a typical force/NSOM probe is shown in figure 2. We typically operate at resonances between 100 and 200 kHz. This can be varied with tip fabrication. The tips are made by using a $CO_2$ laser puller designed for pulling micropipettes.[8] Tip shank can be tailored primarily with incident laser power and pull strength. Notice at the high resonance frequency, the Q, or quality factor, is measured to be larger than 300. The calculated spring constant is approximately 75 N/m.

Figure 3 indicates how the AC amplitude of the tip decreases with decreasing tip-sample distance. In this curve, the AC amplitude of the photomultiplier output has actually been inverted to resemble the shear force of the interaction. We are currently studying the nature of this interaction. This is important to understand since the optical image is a convolution of the force interaction and the nature of the near-fields of the probe. One must consider whether the interaction force is conservative, dissipative, or both. If it is conservative,

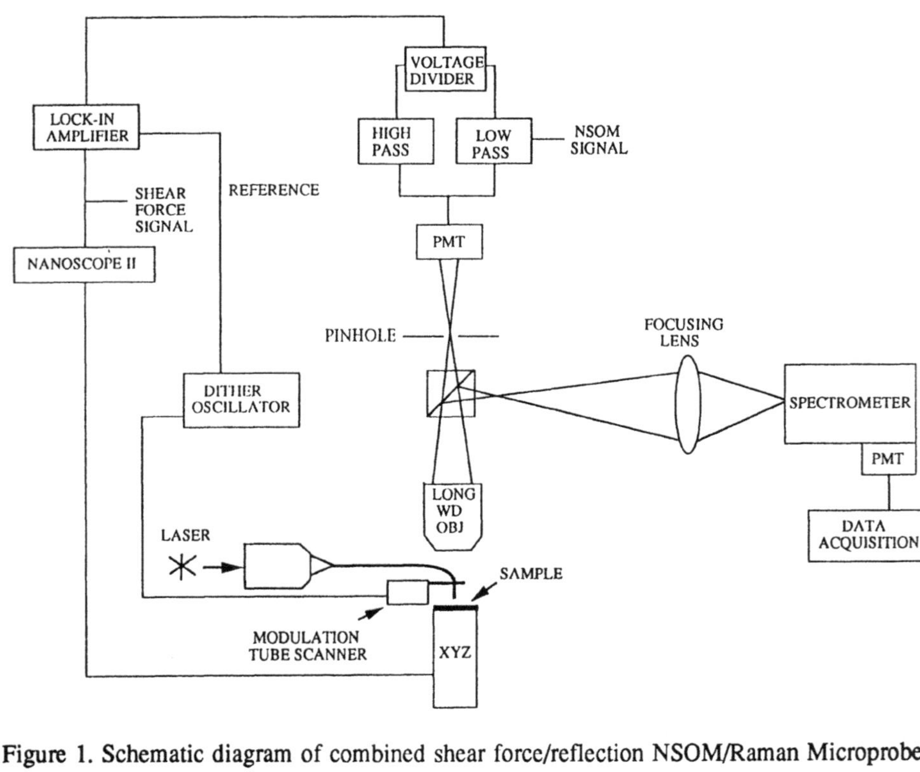


Figure 1. Schematic diagram of combined shear force/reflection NSOM/Raman Microprobe.

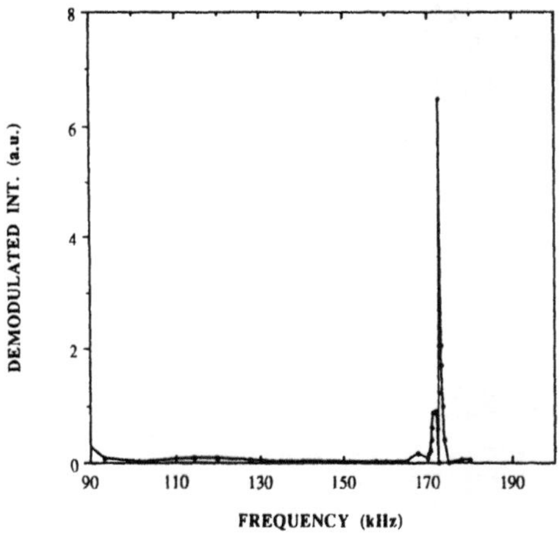


Figure 2. Frequency response curve for shear force/NSOM probe. Typical operation is performed at between 100 and 200 kHz.

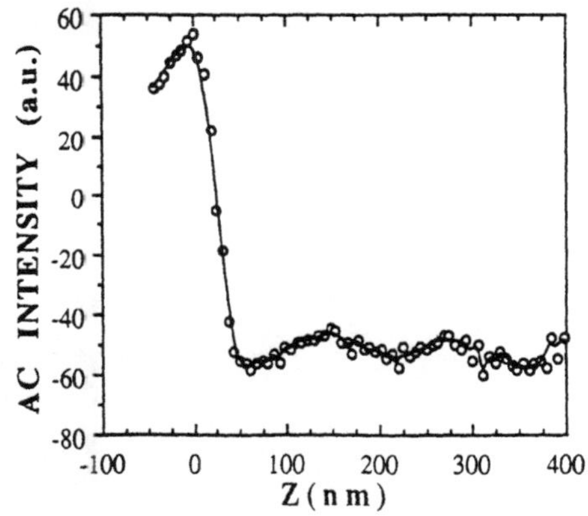


Figure 3. AC intensity (inverted) vs. tip-sample separation distance.

the decrease in amplitude is due to the change in the resonant frequency upon experiencing the interaction.[9] However, if this were the case, one should be able to place the tip about 20 nm from the surface, vary the driving frequency and return to the original tip amplitude by sweeping through the new probe resonant frequency. This is not observed. In fact, the amplitude decreases and cannot be increased by sweeping the driving frequency. This indicates that the interaction may be primarily dissipative. This can be verified by obtaining the same type of resonance curve as shown in figure 2. If the interaction is primarily dissipative, the Q factor should decrease. This type of quantitative investigation is currently underway. One can also get an idea for the type of interaction by imaging materials that are relatively soft, such as biological specimens. Comparison with AFM images of the same sample will shed light on the forces since those of AFM are better known. Further evidence of dissipative force detection is an observed 'jump-to-contact' effect. That is, as the tip is lowered slowly with the piezo, the amplitude decreases abruptly (but remains finite and not zero) and a larger applied piezo voltage is necessary to return the tip to free vibration. That is, the behavior is hysteretic. This indicates a capillary meniscus effect on the tip which would lead to viscous damping of the tip oscillation. The capillary action is due to contaminants on the tip and sample. In brief, the interaction is dissipative, the tip could lose energy to 1)drag through water vapor or contaminants on the tip and sample, 2) plastic deformation to the tip and/or sample, or 3) elastic deformation of tip and/or sample.

A shear force image of a calibration grating is shown in figure 4. This sample consists of Au islands on a Si substrate. As a comparison to a more proven force imaging technique, figure 5 shows an AFM image of the same sample. The shear force image measures up quite well to that of the AFM. With a grating of 1 micron, as this one, the resolution limit of shear force is not tested. Obviously, the amplitude of the lateral oscillation (~ 5 nm), along with the width of the probe, places a lower bound on the resolution. Steep sidewall features, such as those encountered in integrated circuit fabrication, would also be a good test of this type of shear force operation.

We have previously shown that it is possible to obtain spectroscopic information of the sample with the fiber in the near-field with resolution better than that afforded by conventional optical techniques.[10] In a slightly different optical geometry, we provide here results of Raman spectra obtained with the shear force/NSOM microscope. The sample is a nonlinear optical material (courtesy of Dupont). This particular sample is KTP. This

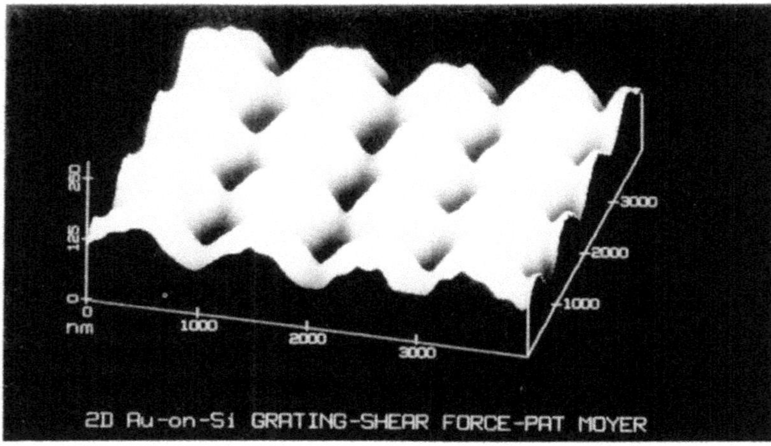


Figure 4. Shear force perspective image of a 2-D Au-on-Si calibration grating.

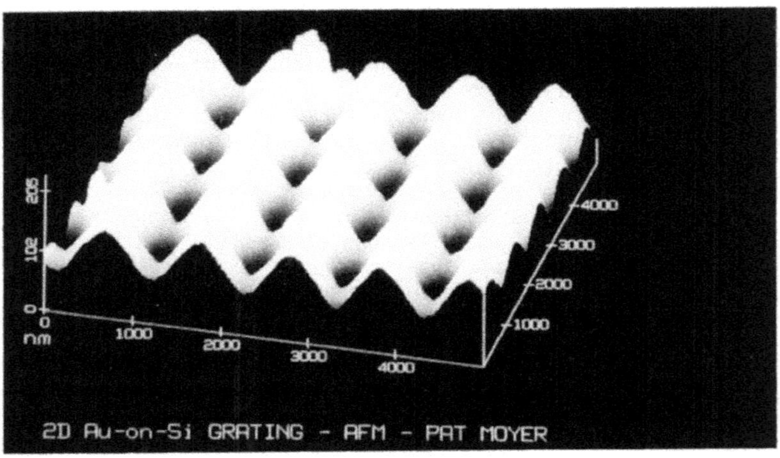


Figure 5. AFM perspective image of a 2-D Au-on-Si calibration grating.

material has a high second-order susceptibility and demonstrates surface second harmonic generation. Since nonlinear optical properties are very dependent on crystal structure and symmetries, the stress induced in these materials during formation and diffusion is critical. A good noninvasive way to characterize material stress is Raman scattering. Thus, the capability to obtain stress information with one hundred nanometer resolution is valuable. Figure 6 shows a shear force image of KTP. The sample is extremely smooth. Figure 7 shows the Raman scattering study of this sample. The top curve is a conventional micro-Raman spectrum of the sample. The unit cell is quite large as the number of Raman peaks indicates. The middle curve is a Raman spectrum of the fiber itself. The broad peak at about 500 $cm^{-1}$ is due to the amorphous $SiO_2$ phase of the fiber material. The lower Raman spectrum is the shear force/NSOM spectrum. It consists essentially of the KTP spectrum superimposed on that of the fiber. This spectrum was obtained by illuminating the sample through an Al

Figure 6. Shear force perspective image of nonlinear optical crystal KTP.

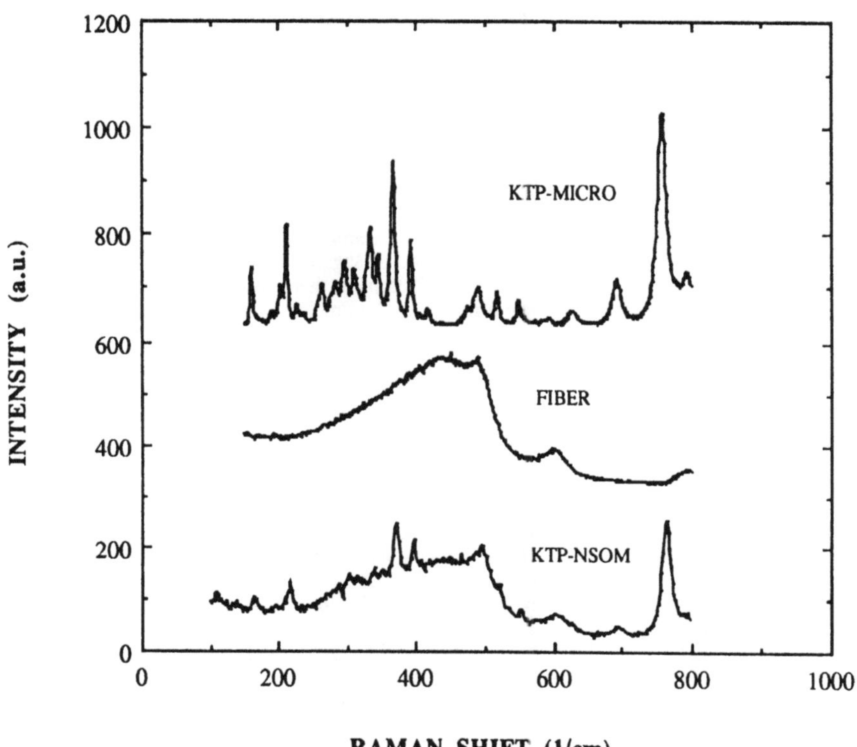


Figure 7. Raman spectra of nonlinear optical crystal KTP (courtesy of DuPont) using the shear force/reflection NSOM/Raman instrument.

coated fiber with an aperture size of approximately 150 nm. Current investigations study the possibility of improving the Raman scattering efficiency by coating with Ag instead of Al. Ag is typically used for surface enhanced Raman scattering (SERS) where 514.5 nm excitation is utilized. Thus, if the probe, with the Ag located at the its tip, is brought into close proximity to the sample, the localized plasmon enhancement may increase the Raman intensity of the sample, as is the case in conventional SERS experiments.

## 3. Conclusions

Since the development of the STM, many techniques have been developed that operate quite similarly. That is, they employ a small probe (typically < 20 nm in radius) and measure surface properties including mechanical, electrical, optical, ion conductance, thermal, and capacitance. these techniques are used for a very wide range of aplications. Combining some of these capabilities into a single instrument can provide complementary information as well as improving the individual operation. the time is certainly right for force and NSOM microscopy to unite due to the obvious potential of each. We have demonstrated the operation of a scanning force/reflection NSOM/Raman microprobe with specific examples. Due to the infancy of this force technique, careful consideration must be given concerning the actual force detection. We have outlined, what we deem to be, some important considerations and insights into the shear force interaction. This is crucial to interpreting the force images, the deconvolution of the optical information, and the limits of the force imaging capabilities. With a better understanding of these mechanisms, this technique can also be valuable in determining, mor quantitatively, the contrast mechanisms of NSOM. Finally, this type of shear force imaging is highly directional, due to the very low lateral spring constant and high vertical spring constant. AFM probes, typically, do not have this property. Thus, in AFM, vertical and lateral forces are coupled in terms of the cantilever deflection and subsequent force detection. Lateral force detection, with the method used here, as introduced by Betzig et al. [7], could provide useful information not readily available with the more conventional AFM due to the isolation of this force component.

## 4. Acknowledgements

The authors wish to thank Phil Russel of North Carolina State University and Eric Betzig of AT&T Bell Laboratories for many helpful discussions and suggestions. Also, thanks to John Thornton of N.C. State University for his work on AFM contributions to this paper. This work was supported by the Army Research Office and the corporate affiliates of the Precision Engineering Center at N.C. State University, particularly Digital Instruments.

## 5. References

1. Binnig, G., Rohrer, H. and Gerber, Ch. (1982) 'Surface studies by scanning tunneling microscopy', Phys. Rev. Lett. **49**, 57.
2. Binnig, G., Quate, C.F. and Gerber, Ch. (1986) 'Atomic force microscopy', Phys. Rev. Lett. **56**, 930.
3. For a comprehensive review, see Pohl, D.W. (1991) 'Scanning near-field optical microscopy' in Advances in Optical and Electron Microscopy, Academic Press, New York, pp. 243-312.
4. Betzig, E., Harootunian, A., Lewis, A. and Isaacson, M. (1986) 'Near-field diffraction by a slit: Implications for superresolution microscopy', Applied Optics **25**, 1890.
5. Buckland, E.L., Moyer, P.J. and Paesler, M.A. (1993) 'Resolution in collection-mode scanning optical microscopy', J. Appl. Phys. **73**, 1018.
6. Betzig, E., Trautmann, J.K., Harris, T.D., Weiner, J.S. and Kostelak, R.L. (1991) 'Breaking the diffraction barrier: Optical microscopy on a nanometric scale', Science **251**, 1468.
7. Betzig, E., Finn, P.L. and Weiner, J.S. (1992) 'Combined shear force and near-field scanning optical microscope', J. Appl. Phys. **60**, 2484.
8. Sutter Instruments Company, Novato, CA.
9. Sarid, D. (1991) Scanning Force Microscopy, Oxford University Press, New York.
10. Paesler, M.A., *et al.* (1990) 'Analytical photon scanning tunneling microscopy', Phys. Rev. B **42**, 6750.

# Optical Tunneling through an Adjustable Liquid Metal Gap

D.W. POHL,[1] D. COURJON,[2] C. BAINIER,[2] A. DEREUX[1]*
and H. HEINZELMANN[1]**

[1] *IBM Research Division*
*Zurich Research Laboratory*
*CH-8803 Rüschlikon*

[2] *Laboratoire d'Optique P.M. Duffieux*
*Université de Franche-Comté*
*UFR des Sciences et des Techniques*
*Rte. de Gray*
*F-25030 Besançon Cedex*

ABSTRACT. First results were obtained with a *scanning near-field optical microscope (SNOM)* whose gap is filled with a droplet of a liquid metal alloy. The transmission of light through a metal layer may be considered a photon tunneling process. As to be expected on theoretical grounds, we find that the transmission vs. gap width varies on a scale of 5 to 10 mm, which is much more rapid than in any other NFO scheme. A correspondingly high lateral resolution may be anticipated. Successful operation of the device requires good wetting between the liquid metal, probe and sample.

## 1. Introduction

Aperture scanning near-field optical microscopy (SNOM = NSOM) offers very high lateral resolution [1-3] but has the disadvantage of having a weak gap width dependence. In operation, the gap width therefore has to be controlled with an auxiliary mechanism, e.g. tunneling [1,2] or detection of interaction forces [3]. The opposite situation exists in scanning tunneling optical microscopy (STOM [4] = PSTM [5,6]): The evanescent wave involved provides a strong, exponential dependence of the signal upon distance but the receiving area and hence lateral resolution is less well defined.

A scheme that combines the advantages of both techniques is described in the present communication. The basic new idea is to fill the gap between a transparent probe tip and the sample surface with a heavily absorbing fluid, in particular a liquid metal (LM). From the point of view of aperture-SNOM the LM film becomes semitransparent at the apex when the gap is sufficiently small thus forming an aperture of variable width. In terms of STOM/PSTM, the LM acts as an optical tunnel barrier with an extremely small penetration depth when compared to the usual STOM/PSTM scheme. Tunneling occurs at any angle of incidence, and very high lateral resolution may be anticipated.

*On leave from Faculté des Sciences, Dépt. de Physique, Rue de Bruxelles 61, B-5000 Namur
**On leave from Physik-Institut, Universität Basel, Klingelbergstr. 82, CH-4056 Basel



D.W. Pohl and D. Courjon (eds.), Near Field Optics, 51–58.

First experiments, performed with standard STOM/PSTM setups and a Ga-In alloy as LM support these ideas but also indicate unexpected difficulties resulting, presumably, from wetting problems.

## 2. Theory

The capability to *tunnel* across some kind of *potential barrier* is common to any form of wave-like propagation. Inside the barrier, the amplitude of the probability function evanesces, typically in an exponential way. Mathematically, this corresponds to an imaginary wave vector component in the direction of propagation(⊥), more precisely

$$Re\, k_{\perp}^2 < 0 \tag{1}$$

Condition (1) is a consequence of the wave equation which both electrons and photons obey. In its stationary, scalar, and Fourier-transformed representation, the wave equation is simply written

$$(\mathbf{k}_{\perp}^2 + \mathbf{k}_{\parallel}^2 - q_{1,2}^2)\hat{A} = 0 \ . \tag{2}$$

Here $k_{\perp,\parallel}$ denote the wave vector components in the direction of propagation (assumed perpendicular to the interfaces of the tunnel barrier) and in the plane normal to it. $\hat{A}$ denotes the amplitude of the probability function in the case of electrons and of the relevant electric field component in the case of photons, and

$$q_{1,2}^2 = \begin{cases} (2m/\hbar^2)\,[E - V_{1,2}] & \textit{electrons} \\ (\omega^2/c_0^2)\ \varepsilon_{1,2} & \textit{photons}\ . \end{cases} \tag{3}$$

In this expression, all the symbols have their usual meaning; the subscripts 1 and 2 refer to the media allowing propagation and the tunnel barrier in between, respectively. It is obvious from (2) that the tunnel condition (1) requires

$$Re\,(\mathbf{k}_{\parallel}^2 - q^2) > 0 \ . \tag{4}$$

This means that either $\mathbf{k}_{\parallel}$ has to be very large or $|q|$ very small, in particular < 0. The first option is exploited in STOM/PSTM for photons and plays a role in electron STM (e-STM) near contact on graphite and similar layer compounds. Under standard e-STM operating conditions, however, the second condition is satisfied. The arrangement to be discussed here allows for PSTM under these conditions for the first time.

For electrons, the simplest tunnel barrier is formed by an insulator gap between two metal pieces (*MIM* junction) with $V_1 - E > 0$ (metal), and $V_2 - E < 0$ (gap). For photons, an analogous barrier can be formed between two insulators with $\varepsilon_1 > 0$ by a thin metallic layer with $\varepsilon_2 < 0$. Such an arrangement may be called an *IMI* junction, in analogy and as counterpart to the above-mentioned MIM. The tunnel barrier becomes adjustable like the vacuum gap in e-STM by using a liquid metal (LM) film as barrier.

An interesting difference between e-STM and PSTM is caused by the complex value of the dielectric constant which has no analogy in the electron energy and potential expression of Eq. (3). It corresponds to absorption of energy and results in a modulation of the exponential decay curve. Another difference concerns the angle of incidence which can be chosen at convenience for light/IMI but not for electrons/MIM because of their short free path length in a solid medium. The nearly unperturbed propagation of photons in the IMI structure thus adds a new parameter to tunnel studies; it allows, in particular, the decay length to be minimized by oblique incidence.

## 3. Material Requirements

Several metals, in particular alloys, are known to be liquid at or near room temperature, e.g. the In/Ga eutectic alloy with a melting point of 15.7 °C [7], In/Hg alloys with up to 70% indium [8], pure gallium (which melts at 29 °C) or mercury.

In order to maintain a stable fluid film between tip and sample even at very small gap widths, it is necessary that the metal wets tip and/or sample surface to some extent. Our observations favor Ga-In and In-Hg in this respect since clean dry glass and similar materials such as freshly cleaved mica are readily wetted, as are metallic films.

The wetting requirement imposes restrictions on the manifold sample surfaces that can be investigated with this technique; however, there is a large variety of surface preparation techniques to achieve wettability. The variety of liquid alloys is also very large if one includes ternary and more complicated compositions.

The approximate optical dielectric constants (at ≈2 eV) and resulting penetration depths $z_0$ (see below) are compiled in Table I.

**Table 1:** Optical properties of Ga, In, Hg

| MATERIAL | REF. | DIEL. CONSTANT | $z_0$ [nm] |
|---|---|---|---|
| Ga | 9 | $-10+i15$ | 15 |
| In | 10 | $-20+i8$ | 7 |
| | 11 | $-100+i30$ | 7 |
| Hg | 12 | $-30+i30$ | 10 |
| Ga-In | 9,10 | $-11+i20$ | 10 |
| | 9,11 | $-30+i30$ | 10 |

Unfortunately, the literature data for In diverge strongly [10,11]; in calculations, the smaller $\varepsilon$ of Ref. 10 was chosen to remain on the conservative side. The optical constant of the Ga-In alloy was estimated by linear interpolation with regard to the atomic ratio of the two constituents.

## 4. Calculated Transmission of IMI Junctions

The transmission of an IMI junction can be estimated on the basis of the well-known theory of wave propagation in stratified conducting media [13]. Figure 1(a) shows the transmissivity as a function of thickness for the three elements in Table 1 and normal incidence. The finite imaginary part of $\varepsilon$ is included in these calculations. The transmission falls off more abruptly at oblique incidence, cf. Fig. 1(b), calculated for $\varepsilon = -11+i20$ (Ga-In). The decay is not strictly exponential, in particular at very small gap widths.

The decay length obtained from an exponential fit over a large distance range is hence not necessarily suited for practical applications where the transmission values of, say, 10 to 100%

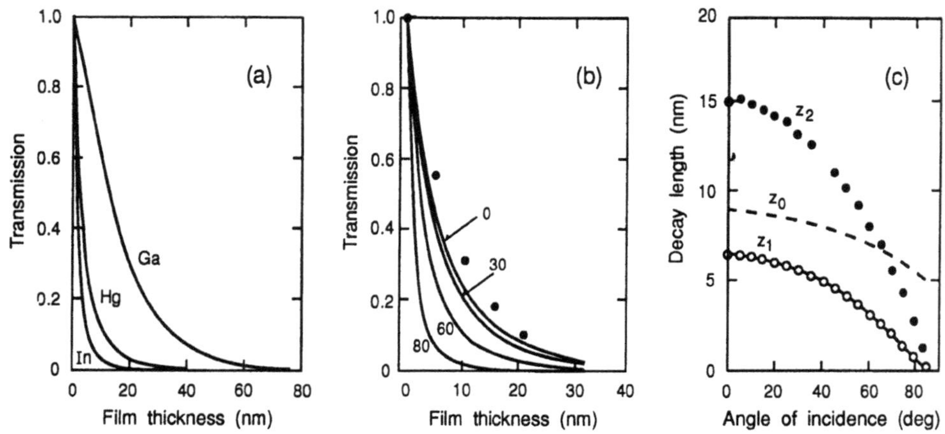


Fig. 1. Calculated transmission of metal films: (a) Ga, In, Hg at normal incidence, stratified layers (data from Refs. 9, 10, 12). (b) Ga-In (data from Refs. 9, 10) various angles of incidence; curves: stratified layers, dots: coupling of 90° prism to substrate as a function of gap width. (c) Decay lengths in Ga-In according to different criteria: solid line: fit to an exponential curve, dashed line: $1/e$, dotted line: $1/e^2$ intensity distances.

are of major interest. The deviation from exponential behavior is obvious from Fig. 1(c) which shows the *inverse exponential intensity decay constant* $z_0 = 1/2k_\perp$ (solid line) together with the *intensity* $1/e$ *gap width* $z_1$ (dashed line), and $z_2$, the *intensity* $(1/e^2)$ *gap width* (dotted line). For exponential decay, one would have $z_0 = z_1 = z_2/2$. It is seen that there are sizable differences between the three quantities. At large angle of incidence, the $1/e$ lengths of In and Hg can be as small as 2 nm!

For such mesoscopically small values, the validity of the concept of a local dielectric function is questionable [14] and the actual transmissivity may differ considerably from the calculated values. Differences between theory and experiment might also result from the pointed shape of the tip. More exact numerical calculations on the basis of the vector Green's function approach [15] indicate, however, that the deviations from the stratified media results

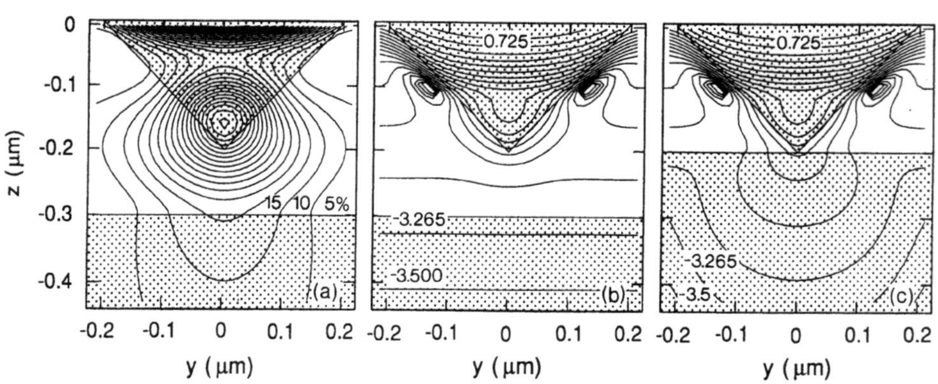


Fig. 2. Coupling of 488 nm light from a 90° glass prism to a glass flat through (a) a vacuum and (b,c) a Ga-In LM layer. Contours represent lines of constant electrical energy density (lin. scale in (a), log. scale in (b,c)). Field perpendicular to the plane of the image ('TE').

are not very dramatic. The data points in Fig. 1(c) show the degree of coupling of a 90° prism to a nearby transparent substrate obtained with the latter method. The geometry and the energy distribution in and around the prism are depicted in Fig. 2.

Note that, at contact, the arrangement is equivalent to a two-dimensional aperture of vanishing width and 45° wedge angles. The very narrow transmissive zone supports our expectation of high lateral resolution capability of IMI-SNOM. Actually, from the analogy to (electron) STM, the resolution of IMI-SNOM can be assumed to equal the decay length.

## 5. Experimental Setup and Operation

Experiments were performed both at U. Besançon (BES) and at IBM Rüschlikon (RUE). The BES setup was the same as in previous investigations, cf. Fig. 5 of Ref. 4. The RUE apparatus differs from that of BES in minor aspects only (Fig. 3): It employs an argon laser as light source ($\lambda \approx 500$ nm) instead of a red diode laser; it also allows light to be sent to the surface of the sample at three fixed angles of incidence, viz. 0° (normal incidence), 45°, and 75°. This is achieved by means of the 90° glass prism and the auxiliary mirror shown in the figure. The test sample is either a thin glass or mica flat in optical contact with the prism or the prism surface itself. The whole setup is installed in a dry box to avoid capillary condensation effects. All glass fiber tips were prepared at BES.

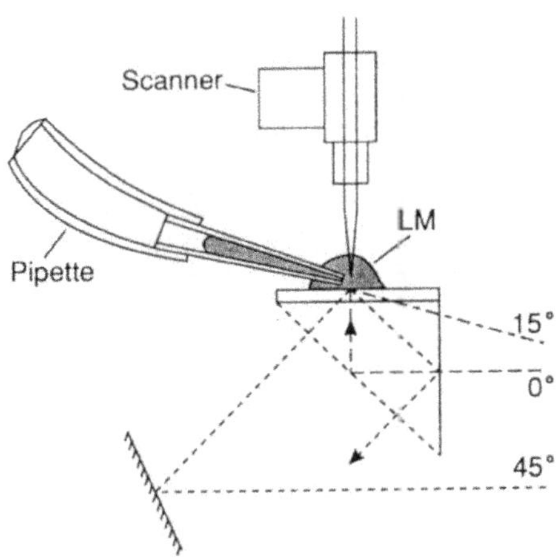


Fig. 3. Central part of the RUE experimental setup.

Without application of the liquid metal film, the instruments can be operated in the established STOM mode [4]. Figure 4(a) shows, as an example, the flux-distance relation observed with the RUE instrument for 45° and 75° before filling the gap with liquid metal. The experimental decay lengths are 100 nm and 40 nm, respectively, in good agreement with theory. The reproducibility is evident from the almost coinciding multiple traces which correspond to subsequent approach and retreat cycles.

The liquid Ga/In eutectic alloy was loaded in a pipette which, when pressurized with a hand balloon, ejects a drop of the alloy right onto the spot where the fiber tip approaches the sample surface. In order to install a stable fluid metal layer between tip and sample, it is important to avoid oxidation/corrosion while the drop starts to fill the volume of the gap.

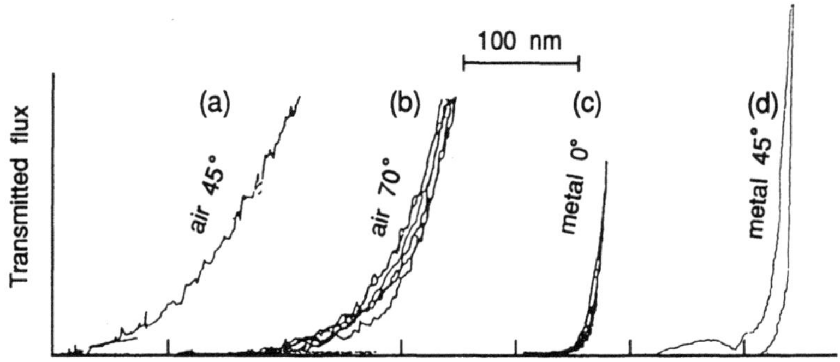


Fig. 4. Experimental transmission: (a,b) air gap, 488 nm, 45° and 75°, (c,d) Ga-In gap, 488 nm, 0° and 670 nm, 45°. The vertical scale is arbitrary and different for each of the four records.

Wetting is promoted by careful cleaning and drying, reduction of air humidity, and by applying thin adhesive metal coatings such as a hue of chromium and/or gold.

When these conditions are fulfilled and approach is made, an extremely steep relation between transmitted flux and gap width is reproducibly found. Figures 4(b,c) show the distance relation for normal incidence and $\lambda$ = 488 nm obtained at RUE. The $(1/e)$ decay length is 7.5 nm, in good agreement with the value estimated on the basis of Ref. 11 (and not 10!). The corresponding results of BES are shown in Fig. 4(d). They are obtained for 45° and $\lambda$ = 670 nm. The intensity-$1/e$ distance is 9.5 nm, again in good agreement with expectation. The slight jump at maximum intensity might be an indication of humidity effects since the BES setup is operated at ambient conditions.

It was also possible to stabilize the light flux with a feedback loop acting on the $z$-axis of the piezo scanner. Three examples of the instrumental performance are shown in Fig. 5. In Fig. 5(a), the stability of the light flux for four different reference settings is recorded; in Fig. 5(b), a voltage ramp is applied as reference which varies exponentially in time. In Fig. 5(c),

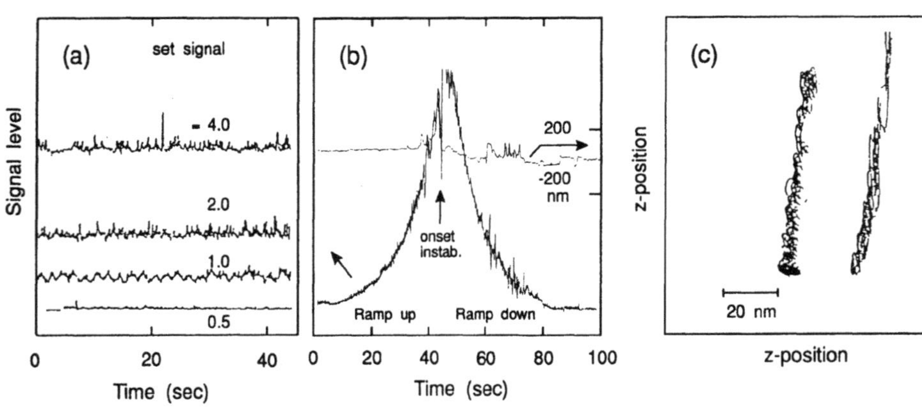


Fig. 5. Stabilized operation: (a) light level vs. time for different reference settings; (b) light level and position signal while the reference is ramped up and down with (excessively) large amplitude; (c) light level vs. position signal for a small reference ramp (second record was offset relative to the first one).

the light level following the reference as in the previous examples, is recorded as a function of $z$-position. In Fig. 5(b), the reference level was raised to a very high value such that contact occurred. This is marked by an itinerant instability, in particular on the $z$-position trace. In Fig. 5(c), the first ramp resulted in a steeper rise of the light/distance curve than the second one. Such variations were found frequently and are tentatively attributed to insufficient adhesion between metal film and tip. It is expected that the use of vacuum deposition techniques will eliminate this problem.

## 6. 'Capillary Evaporation'

In a number of experimental runs, the recorded transmission was unstable, increasing at fixed position in an uncontrollable way. Upon retraction, the transmission usually remained at a fairly high level over distances ranging from 100 nm up to several microns. The origin of this behavior is not yet clear; it might be caused by a spontaneous breakup of the LM film due to insufficient wetting, a process that might be called *capillary evaporation* constituting the complement to the well-known *capillary condensation* that occurs, for instance, between tip and sample in a humid atmosphere.

The energy balance with regard to the interfacial energies (LM/vacuum, LM/glass, glass/vacuum) indeed indicates that the system tip/LM/sample can gain energy by forming of a void if the interfacial forces are sufficiently repulsive. To illustrate this notion, consider a cylindrical piston approaching a flat surface through a layer of mercury. If the mercury is driven out of the volume between piston and substrate as indicated in Fig. 6(a), the interfacial energy changes by

$$\Delta E = \pi r z(p_0 r + 2\sigma_{HgV}) - 2\pi r^2(\sigma_{HgGl} - \sigma_{GlV}) \ . \qquad (5)$$

Here the σ's denote the interface tensions between Hg, glass, and vacuum, respectively; $r$ and $z$ are the piston radius and the gap width. Since Hg forms nearly spherical droplets on glass, one may assume $\sigma_{HgGl} - \sigma_{GlV} \approx \sigma_{HgV}$ whose literature value is 0.465 N/m. Due to the large surface tension of Hg the work against the outside pressure $p_0$ can be ignored for all microscopic considerations. The energy balance then mainly depends on the ratio $\Re_i$ of interface areas created and annihilated by a breakup of the gap. The critical distance for

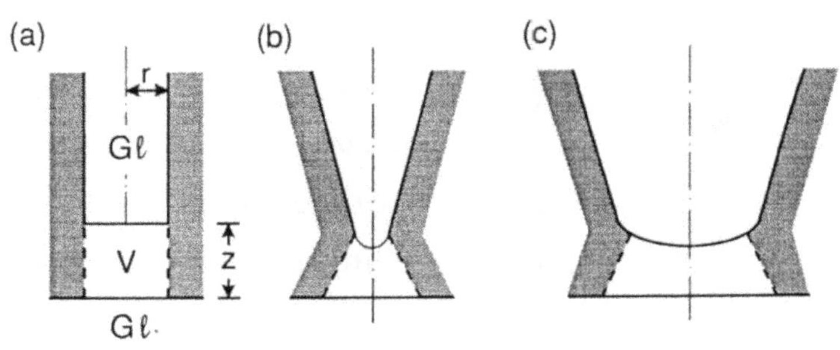


Fig. 6. Capillary evaporation: (a) elementary configuration: piston next to substrate (b) pointed tip, more stable than (c) a blunt one.

capillary evaporation, defined by $\Delta E = 0$, is simply $d_{crit} \approx r$. The formation of a void is equivalent to a first-order phase transition; hence considerable hysteresis may be expected.

The surface energy of the alloy is not known but is certainly smaller than that of Hg. Our observations indicate that a delicate balance might exist in this case: Arrangements with very pointed tips and large $\Re_i$ [Fig. 6(b)] would be stable while blunt ones with small $\Re_i$ would favor the breakup [Fig. 6(c)].

## 7. Outlook

On the basis of the accumulated experience with STM and with "conventional" STOM, the observed distance relation allows the prediction of a lateral resolution of at least 5 to 10 nm with good chances to undercut this value by a factor of two to three which would result in a resolving power of 2 to 5 nm. This would open the door to optical studies of truly mesoscopic and even molecular structures. Similarly, when exploiting the capability of (laser) light to modify a material [3], signatures of that size might be created. IMI-SNOM hence bears potential for optical storage with sub-10-nm bit size.

### References


[1] Pohl, D.W., Denk, W., and Dürig, U. (1985) *Optical stethoscopy: imaging with λ/20*, in "Micron and Submicron Integrated Circuit Metrology" (ed. K. M. Monahan) Proc. SPIE 565, 56-61, see also D.W. Pohl: "Scanning Near-Field Optical Microscopy (SNOM)," *Advances in Optical and Electron Microscopy*, ed. by C.J.R. Sheppard and T. Mulvey, Vol. 12, (Academic Press, London, UK)

[2] Betzig, E., Isaacson, M., Lewis, A. (1987) *Collection mode nearfield scanning optical microscopy*, Appl. Phys. Lett. 51, 2088-2090

[3] Betzig, E., Trautman, J.K., Wolfe, R., Gyorgy, E.M., Finn, P.L., Kryder, M.H. and Chang, C.-H. (1992) *Near-field magneto-optics and high density data storage*, Appl. Phys. Lett. (to be publ. July 13, 1992)

[4] Courjon, D., Sarayeddine, K., and Spajer, M. (1989) *Scanning tunneling optical microscopy*, Opt. Commun. 71, 23-28

[5] Reddik, R.C., Warmack, R.J., and Ferrell, T.L. (1989) *New form of scanning optical microscopy*, Phys. Rev. B 39, 767-770

[6] de Fornel, F., Goudonnet, J.P., Salomon, L., and Lesniewska, E. (1989) *An evanescent field optical microscope*, Proc. SPIE 1139, 77-84

[7] Hansen, M. and Anderko, K. (1958) *Constitution of Binary Alloys*, Vol. 2, (McGraw Hill, New York)

[8] Elliot, R.P. (1965) *Constitution of Binary Alloys, First Supplement*, 531-532 (McGraw Hill, New York)

[9] Kofman, R., Cheyssac, P. and Richard, J. (1977) *Optical properties of Ga monocrystal in the 0.3 to 5 eV Range*, Phys. Rev. B 16, 5216-23

[10] Mathewson, A.G. and Myers, H.P. (1971) *Absolute values of the optical constants of some pure metals*, Phys. Scripta 4, 291-292

[11] Koyama, R.Y., Smith, N.V. and Spicer, W.E. (1973) *Optical properties of indium*, Phys. Rev. B 8, 2426-32

[12] Inagaki, T., Arakawa, E.T. and Williams, M.W. (1981) *Optical properties of liquid mercury*, Phys. Rev. B 23, 5246-62

[13] See, for instance, Born, M. and Wolf, E. (1959) *Principles of Optics*, Chapts. 13.4, Eq. (24) (Pergamon Press, London)

[14] See, for instance, Feibelman, P.J. (1982) *Surface electromagnetic fields*, Progress in Surf. Sci. 12, 287-408

[15] Dereux, A. and Pohl, D.W. (1992) *The 90° prism as a model SNOM probe: Near-field, photon tunneling, and far-field properties*, these proceedings

# RECENT EXPERIMENTAL RESULTS WITH THE PSTM:
# - OBSERVATION OF A STEP ON A QUARTZ SURFACE.
# - SPATIAL SPECTROSCOPY OF MICROWAVEGUIDES.

F. DE FORNEL, E. BOURILLOT, P. ADAM, L. SALOMON,
G. CHABRIER, J.P. GOUDONNET
Laboratoire Physique du Solide, associé au CNRS,
Equipe Optique Submicronique,Université de Bourgogne,
Faculté Sciences Mirande, BP 138, 21004 Dijon Cedex, France.

ABSTRACT. The Photon Scanning Tunneling Microscope (PSTM) is based on the frustration of the total internal reflected beam by the end of an optical fiber. Till today it has been used to obtain topographic information generally for smooth samples. In this paper we report two different kinds of experimental results. First, when the sample is in the form of a step, our measurements demonstrate how the images, obtained in the constant intensity mode, depend on the orientation of the incident beam of light with respect to the step. Next, we show that the first derivative of the collected intensity with respect to the probe-sample distance at each point of the sample yields to a new kind of image named hereafter "spatial spectroscopic image". With this spectroscopy the distribution of the refractive index variation of the sample can be found out. Images of microwaveguides in topographic mode as well as in spectroscopic mode are presented and discussed.

## I. INTRODUCTION

The Photon Scanning Tunneling Microscope (PSTM) as well as the Near Field Optical Microscope (NFOM) provide images with a resolution below the Rayleigh limit[1-6].But these images can differ strongly from the real structure, especially when the sample surface is not very smooth [6-8]. In fact, the image is a representation of the isointensity lines collected by the fiber probe. The images also show how the incoming light is diffracted by the structure. A step on a surface results in a local discontinuity of the slope of the surface that can create strong local diffraction effect if the sample is illuminated either in total internal reflection or in external reflection or in transmission.



D.W. Pohl and D. Courjon (eds.), Near Field Optics, 59–70.

Such local diffraction in external reflection or in transmission has been observed and calculated for the classical microscope [9-11]. In these cases diffraction causes some oscillations in the vicinity of the image of the discontinuity.

For a better understanding of the PSTM image formation, we have studied a sample with a single step. The images were taken in the constant intensity mode. These were obtained for s and p polarizations corresponding to different values of the angle of incidence of the incoming light.

For homogeneous samples the images are correlated to the geometry of the sample, but for inhomogeneous samples the images contain a mixture of information about the topography and the distribution of the inhomogeneities [12,13]. We have then shown that by using spatial spectroscopy, the PSTM is able to give images of the distribution of the refractive index variations.

## II. ROLE OF THE ORIENTATION OF THE INCIDENT BEAM OF LIGHT

The sample studied is a single step, 50 nm height made of silica ( refractive index 1.458 at $\lambda$=0.633 $\mu$m). Suprasil II of Heraus was chosen as the material to avoid the scattering inside the bulk of the sample. Fig. 1 displays the three different ways to illuminate the sample.

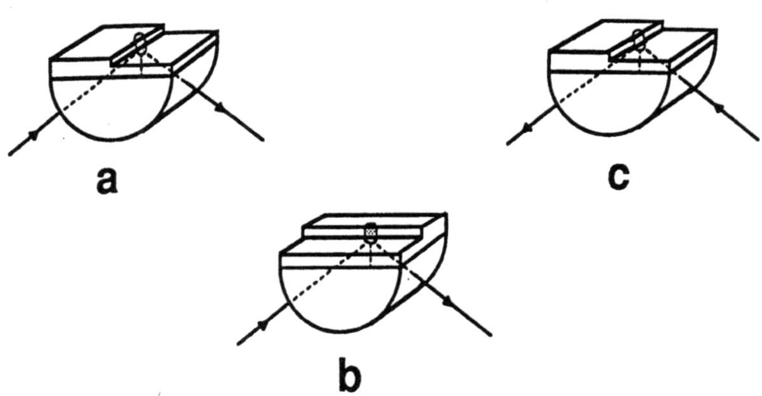


Fig. 1 Different ways to illuminate the sample

All images, shown is this section, were taken in the constant intensity mode. The optical source is a He-Ne polarized laser and the angle of incidence was fixed at 50°. The tip was made by chemically etching a standard multimode fiber. The experimental set up is described in details elsewhere [12,14].

Fig. 2 shows the effects of the orientation of the incident beam of light with respect to the step.

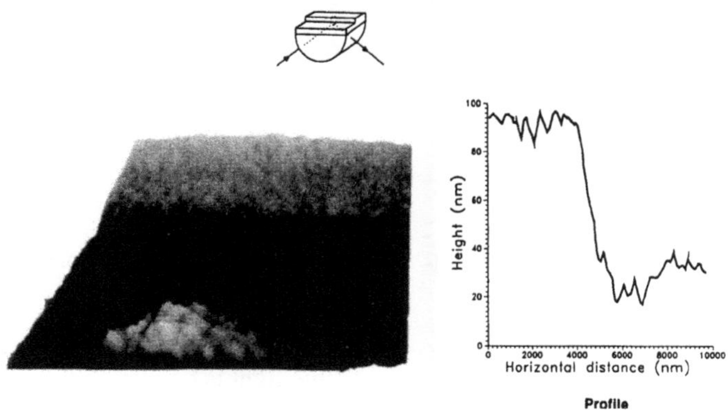


Fig. 2a Images of the step. The scanning range is 20μmx20μm, angle of incidence θ=50°, in p polarization .

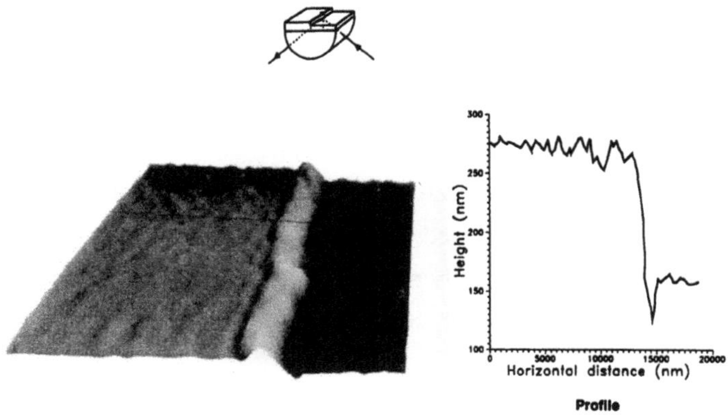


Fig. 2b Images of the step. The scanning range is 20μmx20μm, angle of incidence θ=50°, in p polarization .

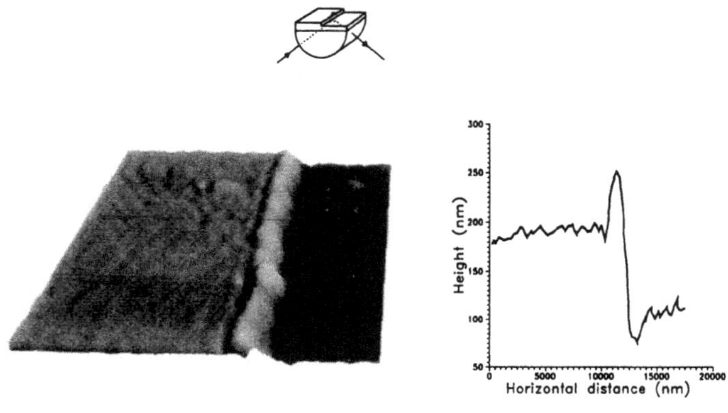


Fig. 2c Images of the step. The scanning range is 20μmx20μm, angle of incidence θ=50°, in s polarization.

If the plane of incidence of the incoming light is parallel to the edge of the step, no diffraction phenomenon is noticeable and the measured step height is in agreement with its actual value (Fig. 2a)

Images look greatly different when the plane of incidence is perpendicular to the edge as illustrated in Fig. 2b and 2c, which also clearly show how the images look different when the direction of propagation of the light is reversed. If the direction of light is as in Fig.1b the image has a dip at the base and a bump at the top. If the direction of the light is reversed, the image has only a dip at the base of the step. The width of the bump or of the dip is of the order of few microns. Their heights are of the order of the step height when s polarized light is used.

The distance between the tip and the surface can also play an important role on the image formation. Fig.3. is an illustration of the effect of the tip-surface distance on the images. If the probe is moved far away from the surface, images with giant corrugation are obtained and finally the step disappears. The corrugation observed are only due to the dffraction of the light by the step.

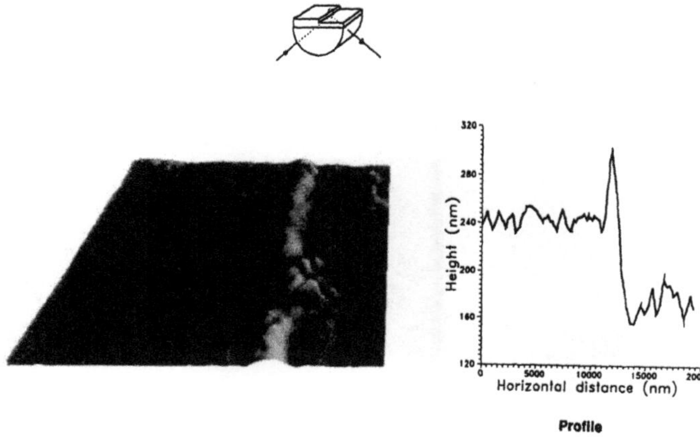


Figure 3a Effect of the tip-surface distance Z on the images, for p polarized light, $\theta=50°$, $\lambda=0.633\mu m$, $Z=Zo$, the scan range is $20\mu m \times 20\mu m$.

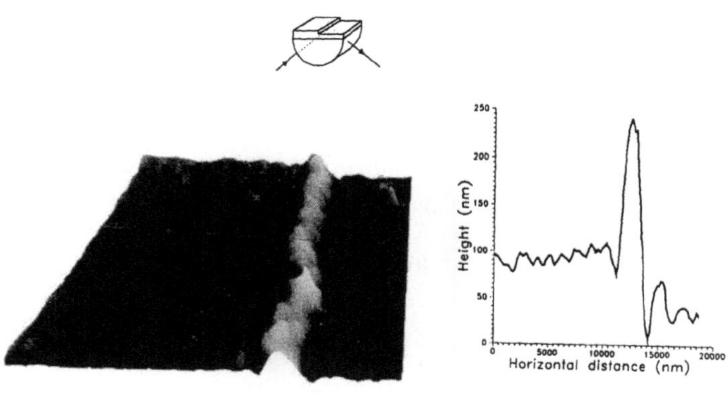


Figure 3b Effect of the tip-surface distance Z on the images, for p polarized light, $\theta=50°$, $\lambda=0.633\mu m$, $Z=Zo+90nm$, the scan range is $20\mu m \times 20\mu m$.

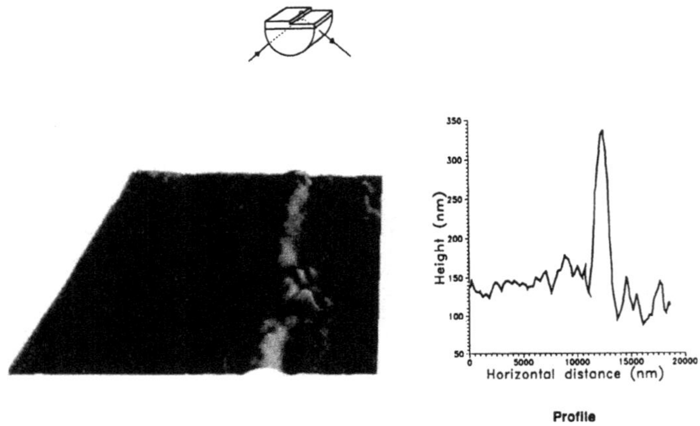


Figure 3c Effect of the tip-surface distance Z on the images, for p polarized light, $\theta=50°$, $\lambda=0.633\mu m$, $Z=Zo+130$ nm, the scan range is $20\mu m \times 20\mu m$.

To compare the different images, all measurements were made with the same optical fiber tip.The tip surface separation was throughout kept greater than 10 nm so that the tip did not touch the sample surface.

## III. PSTM SPATIAL SPECTROSCOPY

For inhomogeneous samples, the images are a mixture of information about the topography and the local inhomogeneities. Considering total internal reflection of light at a surface, one can write down the following expression for the intensity I(z) of the evanescent field versus the distance to the surface z:

$$I(z) = I_o \exp(-2\, z / d_p)$$

where dp is expressed by

$$d_p = \frac{\lambda_0}{2\pi} \frac{1}{\sqrt{n_1^2 \sin^2\theta - n_2^2}}$$

with $\lambda o$ the wavelength of the light in vacuum and $n_1$ and $n_2$ the indices of the prism and the outside medium. $\theta$ is the angle of incidence of the incoming light. The first derivative of this intensity with z is related to dp as :

$$dI(z) / dz = -(2 / d_p) I(z).$$

This slope is related to the local refractive index value of the sample.

The previous expressions are no more suitable when an optical fiber collects a part of the intensity of the evanescent field by frustration. By using the simple multilayer model it has been shown that the intensity collected by the fiber is not a monotonic function of the refractive index of the sample surface [12]. Calculations, taking in account the shape of the fiber tip are in progress [15].

To get the local slope dI(z)/dz of the intensity, we modulate z at a frequency above the cutoff frequency of the feedback (and below the resonance frequency of the piezo-transducer). Thus it is possible to simultaneously obtain the image of the surface at constant intensity mode and the image of dI(z)/dz at each point of the sample. To obtain the slope we have to fix the amplitude of the modulation much smaller than the depth of penetration dp. In our case, this amplitude was generally taken as 20nm. The experimental set up is shown in Fig. 4

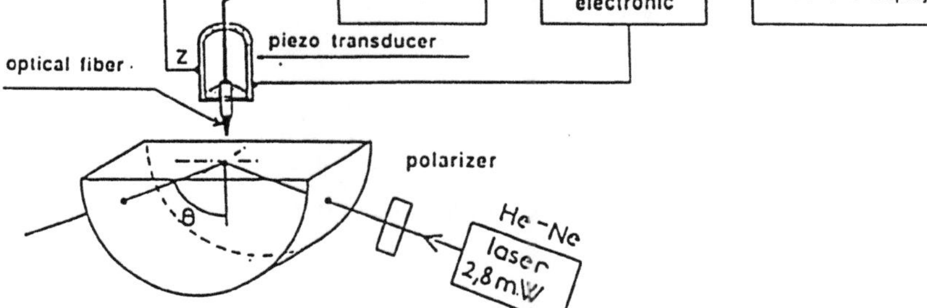


Fig. 4 Experimental set-up used to get "spatial spectroscopic" images

The image shown in Fig. 5 (b), is the "the spatial spectroscopic image" of a grating whose period and amplitude are respectively 2mm and 50nm. The sample was illuminated in p polarization with the plane of incidence parallel to the grating axis.

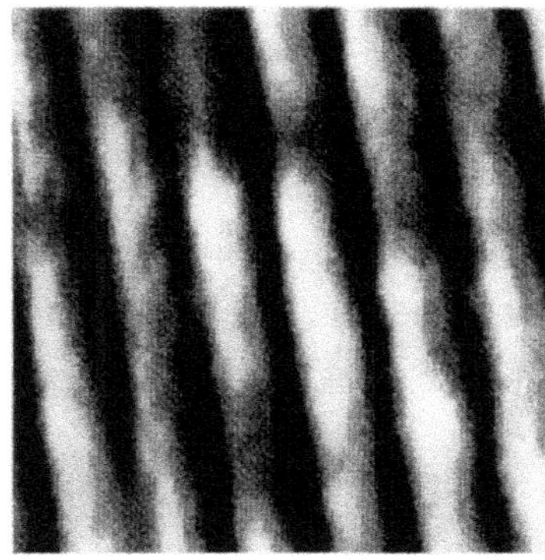

Fig.5a Topographic image of a grating. The scan range is 3μmx3μm

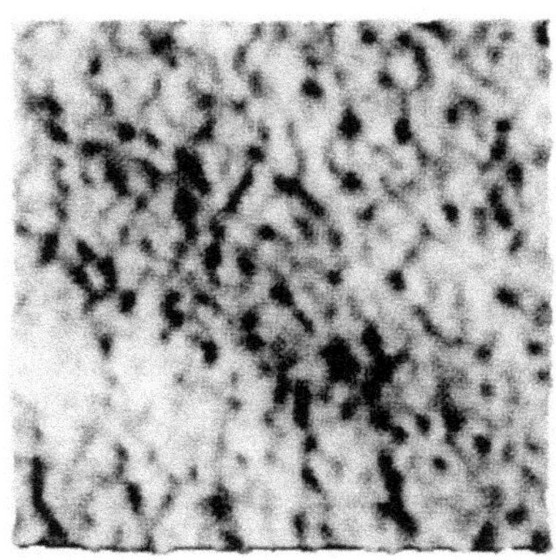

Fig.5b Spatial spectroscopic image of a grating. The scan range is 3μmx3μm

We can notice that the spatial spectroscopic image does not give information about the grating,i.e., the geometry of the sample in our experimental conditions does not contribute to the spectroscopic image.

For an inhomogeneous sample, the spatial spectroscopic image shows corrugations partly related to the local variation of the optical refractive index, (Fig 5). The sample is a microwaveguide made by diffusion of Ag salt inside a glass slide. The full description of this sample and the images with the PSTM have been previously reported [12]. Fig. 6 is an image taken of the microwaveguide in the spatial spectroscopic mode .

Fig. 6 Spatial spectroscopic image of a microwaveguide. Scan range is 40 μm x 40 μm, θ=60°.

By comparison with the PSTM image of the sample, it appears that the geometry of the surface does not generate corrugation on the spatial spectrocopic image.

For a better analysis of these results the figures 7 and 8 give a profile of images taken perpendicular and parallel to the microwaveguide illuminated in S polarisation .

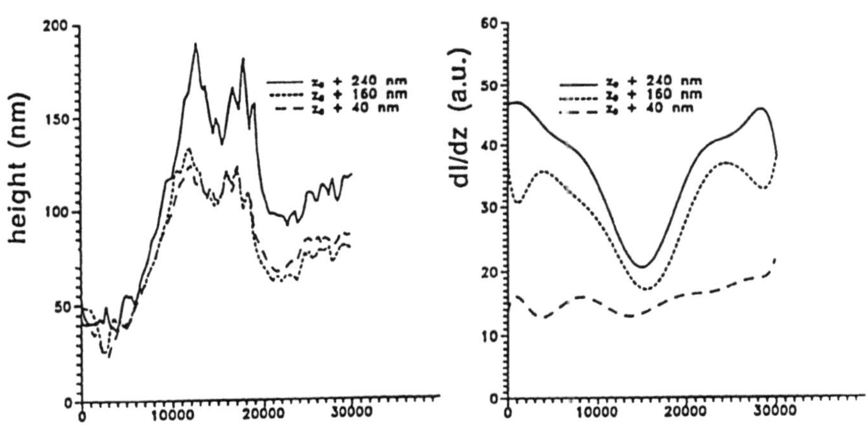


Fig. 7 Comparison of the profile of the (a) topographic mode and (b) spatial spectroscopic mode of the microwaveguide illuminated parallel to the guide axis

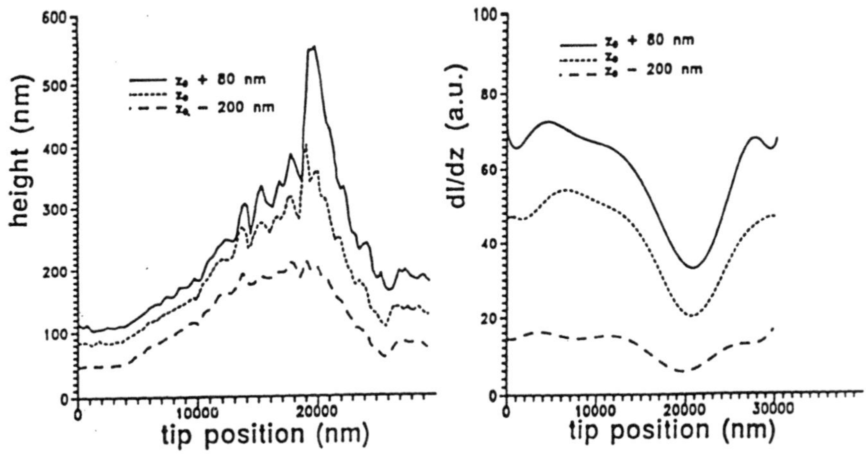


Fig. 8 Comparison of the profile of the (a) topographic mode and (b) spatial spectroscopic mode of the microwaveguide illuminated orthogonally to the guide axis

The profile of topographic image looks very complex. The corrugations in the image of the sample is simultaneously created by the geometry of the sample, the local index variation [12,13] and due to some interferences inside and on the surface of the sample [14,15]. The corresponding profile of the spatial spectroscopic image shows a smoother aspect. By comparison with the index profile of the microwaveguide, we observe that the width of the guide obtained from the spatial spectroscopic image is in agreement with its actual width. The same figure also shows the effect of the tip-sample distance on the topographic image as well on the spatial spectroscopic image. We observe that the variation of the tip-surface distance affects the amplitude but not the width of the profile of the spatial pectroscopic image.

## IV. CONCLUSION

This paper mainly presents some interesting experimental results. In the first part, for a better understanding of the formation of the images, we have studied images of a single step object illuminated in several configurations. Our results show the effect of the tip-surface separation and of the orientation of the edge of the step with respect to the plane of incidence of the incoming light. It appears that oriention of the plane of incidence of the light on such a sample plays a very important role in the formation of images. These results are to be taken into consideration while analyzing the images of sample with local step discontinuities. The complete calculation of the diffration by a step illuminated in total internal reflection has not yet been carried out. However our results confirm the results reported by Van Labeke et al.[17].

In the second part of this paper we have shown that by using local spectroscopy, the PSTM is able to furnish information on the refractive index distribution of microwaveguides. Correlation of the spectroscopic image to the actual refractive index distribution is under investigation and will be reported elsewhere.

## ACKNOWLEDGMENTS

The authors gratefully acknowledge Mme H.Launois and M. Ph. Guerin, from L2M Laboratory, for providing the sample with a step on the surface and Ms A. Kervorkian and D. Persegol, from the GeeO Laboratory, for providing the microwaveguide.

The authors thank I. Hossein for fuitful discussions and Y. Lacroute for help in making the experimental setup. This work, performed within the CNRS ULTIMATECH program, has been also supported by the MRT (Ministère de la Recherche et de la Technologie), the Burgundy Region.

## REFERENCES

[1] U. Durig, D. W. Pohl and F. Rohrer, J. Appl. Phys., 59, 3318 (1986).

[2] E. Betzig, A. Harootunian, M. Isaacson and E. Kratshmer, Biophys. J., 49, 269 (1986).

[3] R. C. Reddick, R. J. Warmack, T. L. Ferrell, Phys. Rev. B, 39, 767, (1989).

[4] F. de Fornel, J. P. Goudonnet, L. Salomon and E. Lesniewska, Proc. SPIE 1139, 77 (1989).

[5] D. Courjon, K. Sarayeddine and M. Spajer, Opt. Comm., 71, 23 (1989).

[6] N. F. Van Hulst, N. P. de Boer and B. Bölger, J. of Microscopy, 163, 117, (1991).

[7] F. de Fornel, L. Salomon, P. Adam, E. Bourillot, J. P. Goudonnet, M.Nevière, Ultramicroscopy, 42, 422 (1992).

[8] J.P. Goudonnet, L. Salomon, F. de Fornel, E. Bourillot, P. Adam, M. Nevière, P. Guérin, SPIE, Vol. 1639, Scanning Probe Microscopies, 12, 23, (1992).

[9] E.Ash: "Scanned image microscopy", Academic Press, London (1980)

[10] J.W. Goodmann: "Introduction à l'optique de Fourier et à l'holographie", Masson, Paris (1972)

[11] F. Depasse, D. Courjon: Appl. Optics, 30,11(1991)

[12] E. Bourillot, F. de Fornel, L. Salomon, P. M. Adam, J. P. Goudonnet, J. of Optics (Paris), 23, 57 (1992).

[13] D.P.Tsai,H.E.Jackson,R.C.Reddick,S.H.Sharp,R.J.Warmack : Appl. Phys. Lett., 56, 16 (1990)

[14] L.Salomon: Thesis, Dijon (1991)

[15] M. Neviere :" Workshop on Near Field Optics", Arc et Senans, France, (1992)

[16] N.F. Van Hulst, F.B. Segerink, B. Bolger: Opt. Comm.87, 214 (1992)

[17] D. Van Labeke, D. Barchiesi:" Workshop on Near Field Optics", Arc et Senans, France (1992)

# COMBINATION OF A FIBER AND A SILICON NITRIDE TIP AS A BIFUNCTIONAL DETECTOR; FIRST RESULTS AND PERSPECTIVES

F. BAIDA, D. COURJON and G. TRIBILLON
*Laboratoire d'Optique P.M. Duffieux, URA 214 CNRS*
*Université de Franche-Comté, UFR Sciences, route de Gray*
*25030 Besançon cedex*
*France*

ABSTRACT. In this communication we propose the association of a monomode fiber and a silicon nitride tip working both as a force detector and a near field detector. The optical signal detected by the tip is transmitted by the fiber towards a photomultiplier. Simultaneously or sequentially, the topography of the object is measured by means of a second beam reflected on the cantilever. For discriminating the two signals, each beam is modulated at a given frequency. A possible alternative is the use of two different wavelengths. Our first results show glass surface images generated by optical near field and force detection. The efficiency of silicon nitride tips could be strongly improved by piercing the very tip by ion etching. The first tentatives are very promising since holes of about 20 nanometers have been obtained.

## 1. Introduction

On the contrary to far field microscopy, no particular focussing plane can be determined in near field microscopy. This particularity is certainly one of the main problem encountered in near field optics. Without defined analysis plane, it is not possible to imagine a strategy for solving the inverse problem, i.e., the retrieving of object features from image data. It is thus necessary to go further in the understanding of the interaction between light and matter to define first some standards easily reproduced from one laboratory to another one. Two solutions have been developed for the last years. The first one consists of moving the tip in a given plane very close to the object surface. Its first advantage is the simplicity of the technique. Its main drawback is the impossibility of keeping the tip close enough to the object to ensure a near field detection (see figure 1a). The second problem is the risk of contact with the high parts of the object. The second way widely used in STM and AFM is based on controlling the distance between tip and sample by using a feedback control of the detected intensity. This technique known as *detection in constant intensity mode* should lead to higher resolutions since the tip moves closer to the object on average. Unfortunately, the feedback can increase the noise level and the correlation between the detected field and the real topography is not obvious, except for very low relief objects [1]. Some groups have recently proposed an alternative to the second solution: instead of trying to maintain the object distance by means of the intensity itself, the solution consists of combining a force microscope with the near field microscope [2,3]. By this way, the tip can be maintained at a small distance of the object. It is evidently not free from disadvantages, the first one being the mixing of parameters. This mixing introduces the risk of detecting a combination of optical phenomena and of force effects. Nevertheless the first results are sufficiently promising to incite the development of similar



*D.W. Pohl and D. Courjon (eds.), Near Field Optics, 71–78.*

hybrid techniques. In this communication, we propose a variant of these techniques. In our device, the tip is directly in contact with the object and consequently the field detected will be the near field just on the surface.

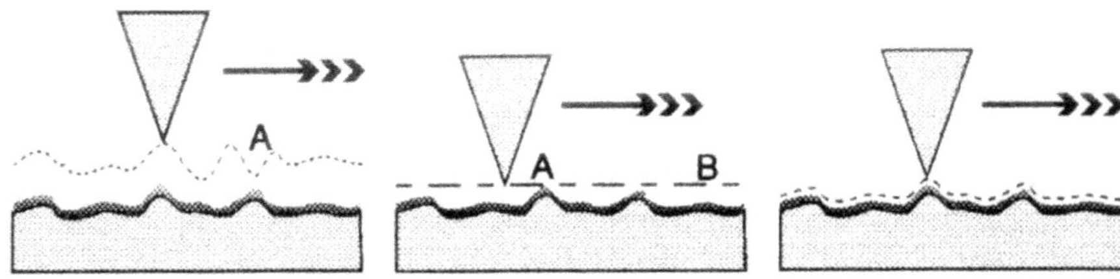


(a) (b) (c)

Figure 1. Three possible modes of xyz scanning. In (a), the tip is rather far from the surface. The tip follows the intensity isolines by feedback control. The correlation between topography and optical field variations is weak. For example, in A, we note a kind of contrast reversal. In (b), the tip moves in a given plane. Only in very localized region of the scanning, (region A), the tip will be in the object near field. The overall detection will be achieved in far field regime (region B). In (c), the tip position is controlled by means of an external signal, for instance a force detection (van der Waals, shear or repulsive force).

## 2. Principle

This microscope uses a combination of a fiber and a silicon nitride tip similar to the one used in AFM. The near field microscope is of STOM/PSTM type (scanning tunneling optical microscope). An approaching technique has been published recently [4]. It uses optical deflection for measuring the forces. In our case (figure 2), the system is very compact. The fiber plays two roles simultaneously: first, it transmits the beam diffracted by the extremity of the tip towards a remote sensor and second, it lights the microcantilever to generate an interference phenomenon usable for determining the distance between tip and sample. The fundamental difference with other techniques is that the tip is in contact with the object. Although such a configuration is often considered as a drawback because the contact with the tip risks to scratch the object surface, it is the only way for determining the field distribution on the object. This technique could be interesting for instance in the case of index variations introduced by dopants in semiconductors. At such exceedingly short distances of the object, the field variations should be mainly generated by index (or more precisely polarizability) rather than object topography. When the tip touches the surface, it necessarily follows the object profile which corresponds in first approximation, to an iso-intensity line (this hypothesis neglects the wall effects).

## 3. Problems to solve

Since the same fiber ensures two different functions simultaneously, the first condition to fulfil is the non mixing of the two signals. Several problems have to be solved. First, the light beam coming from the object, must not interfere between the cantilever and the fiber extremity. Second, the detection stage must be able to detect separately the two beams.

To estimate the validity of our hypotheses and statements, a simulation has been carried out (fig.3 and 4). This modeling needs assuming that the reflected intensity over the cantilever (AFM distance control) is very large in comparison with the STOM signal. The ratio has been estimated to about 20 (actually, it is often very much smaller).

We will first consider that the tip is so small that the beam coming from the fiber is simply reflected over the lever. For computing the detected intensity from both the AFM control and the STOM, we suppose that before moving, the object is far enough from the tip (a few wavelengths). The distance tip-object then decreases till the object touches the tip (step 1). Then the object going on, it drags the cantilever with it (step 2). During step 2, the distance between the fiber and the combination object-cantilever will obviously decrease.

For computing the AFM signal, we have developed a macroscopic approach based on the transmission of a light beam through dielectric multilayers [5]. The absorption of $Si_3N_4$ has been taken into account ($2.5\ 10^4$/m) [6].

Concerning the STOM signal detection, the problem is more complex since we could take into account the diffraction of the evanescent wave by the extremity of the tip. For the sake of simplicity, we will only consider the normal rays diffracted by the tip. This approximation is justified by noting that only paraxial rays can be transmitted by the fiber because of their monomode properties.

Starting from this hypothesis, the mathematical development is the same as for the AFM control study.

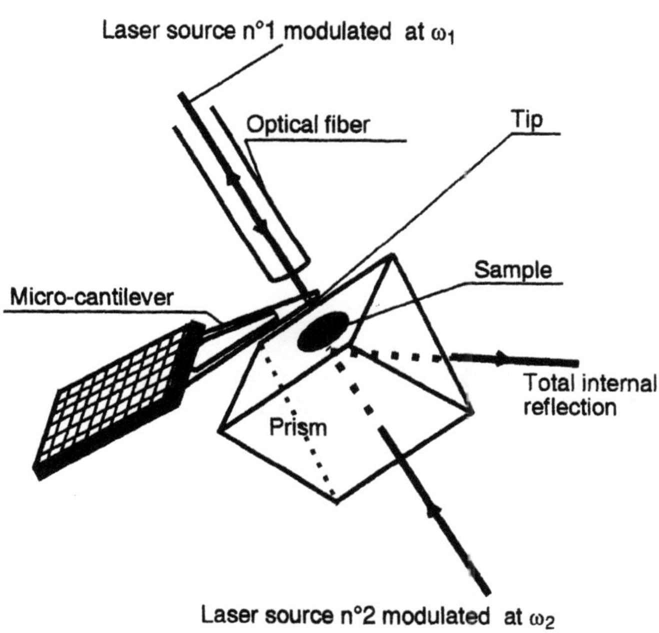


Figure 2. Schematic diagram of the hybrid microscope.

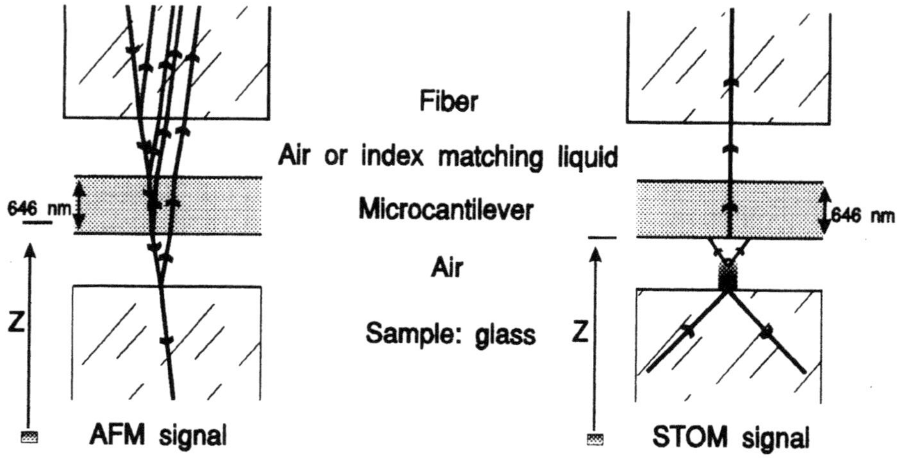


Figure 3. Scheme of light beam propagation between the different interfaces, for the two types of beams (AFM and STOM).

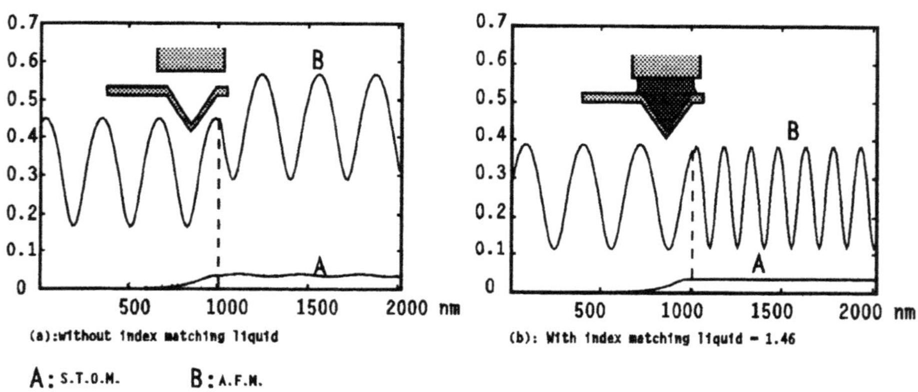


Figure 4. Theoretical approach curves assuming air (on the left) or liquid (on the right) between cantilever and fiber. The vertical dashed lines correspond to the contact point between object and tip.

In fig.4, A corresponds to the STOM response whereas, B is the AFM control signal. The fiber index is 1.46, the silicon nitride index is 2.1 [7]. It is clear that the difference of intensity between A and B can be a problem for detecting the two beams with the same photomultiplier simultaneously. Various solutions can be imagined for separating the beams. The first one consists of modulating the beams at two different frequencies (by means of an acousto-optic cell, or directly in the case of a laser diode source). Another way consists of using two different wavelengths that can be split by means of a suitable dichroic mirror or filter. The last solution can prevent the saturation of the photomultiplier by the AFM signal. The third way, which is certainly the most simple, consists of reducing the AFM signal to get the same magnitude of the STOM signal. However, this solution will affect the S/N ratio of the AFM signal.

The theoretical results show an excellent independence of the two beams in the working region (region on the right of the vertical dashed line). The residual ripple observed on the STOM signal is due to the interference of the beam between the cleaved face of the fiber and the upper part of the cantilever. For cancelling almost completely this parasitic oscillation, we propose to fill the gap between fiber and cantilever with an index matching liquid of same value as the fiber core. In this case, the STOM ripple could disappear completely.

## 4. Experiments

The STOM/PSTM technique is now well known. The setup is described in figure 5. The precise setting of the fiber position over the cantilever is achieved by means of a xy piezoelectric actuator. The xyz scanning of the object is done by means of a piezoelectric tube as usually. Two laser sources are used: a laser diode is used as STOM illumination system, whereas a HeNe laser generates the AFM control beam. In these preliminary experiments, the AFM and the STOM images are recorded sequentially.

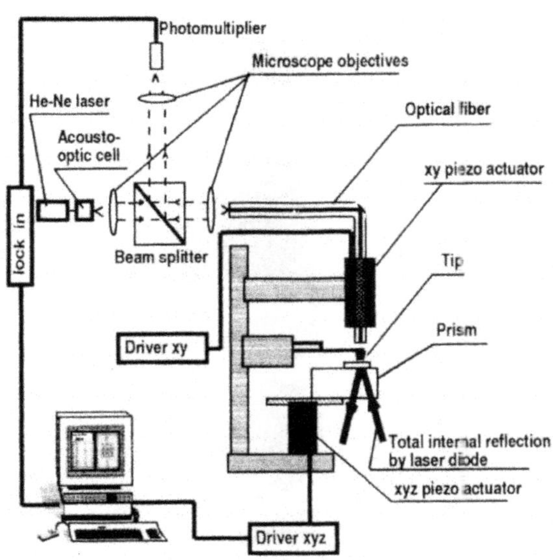


Figure 5. Schematic setup of the AFMxSTOM microscope.

The cantilevers come from Park Instrument. They are not metallized to be sufficiently transparent to the near field. The AFM images have been recorded both in constant force mode and in constant z altitude mode. In the first case, the distance between the cantilever and the fiber extremity is kept constant by means of a feedback control of the AFM signal. The available signal is then the variations of the z-voltage applied on the piezoelectric tube. In the second case, the z-voltage does not vary. The signal is the variations of the modulation of the interference fringes between tip and fiber.

The test object is simply the face of the prism used to generate the evanescent field. It has not been dehydrated. The first experiments have consisted in approaching the tip to the surface of the glass plate and recording the optical and AFM signals sequentially. In the AFM experiment, we used a metallized cantilever in order to increase the signal on noise ratio. Figure 6 shows the two detected signals.

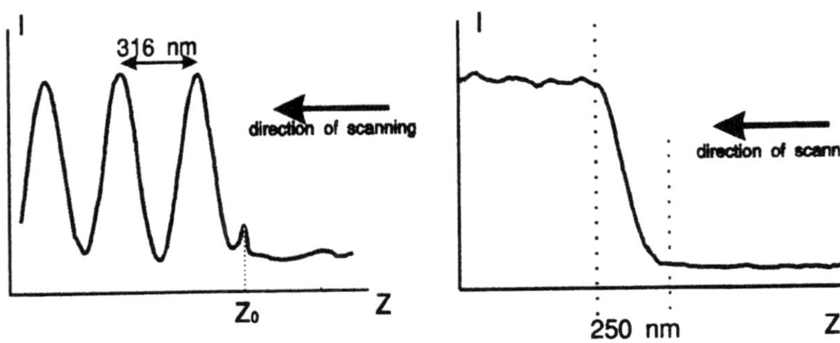


Figure 6. Experimental approach curves. On the left, AFM signal, on the right, STOM signal.

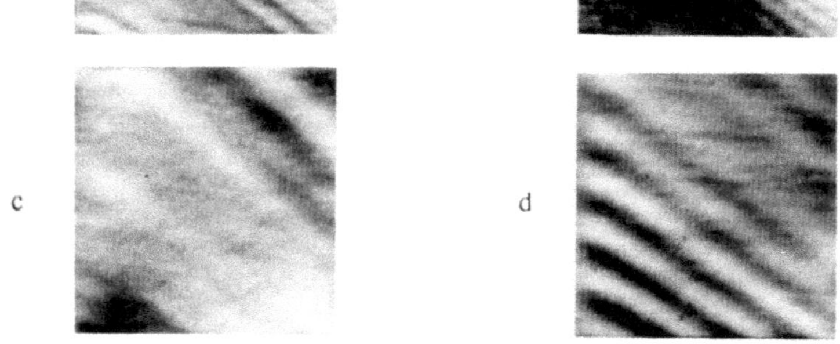


Figure 7. Top view of AFM (a and c) and STOM (b and d) images of an optically flat glass plate without feedback control. The scanned area is 2.7x2.7 $\mu m^2$ for images a and b and 1.35x1.35 $\mu m^2$ for images c and d.

The second step in our investigation was the recording of two-dimensional images of the glass surface. Figure 7 shows such AFM and optical mappings. We note a good agreement between the two types of images. The experiment has been reproduced without any problems several times.

## 5. Next developments

The use of microcantilevers has got a few advantages compared to etched fibers. First, it limits the risk of destruction. It can be then used many times before changing. The second interest is the easy combination with an AFM as shown in this paper. Finally, it is a manufactured product always available. Unfortunately, its geometrical properties do not fit well the near field detection. Because of the large aperture angle of the tip (70 degrees), a large section of the tip will convert the evanescent field into a propagating one, leading to a low image resolution. A solution could consist of metallizing the outerpart of the tip and then piercing by ion etching the tip extremity. The first attempts in this way have been made. Unfortunately the generated pinholes are too small to conduct the light beam (see figure 8) [9].

Finally, although this study has been carried out with a STOM version, any other scanning near field microscope could be used in combination with an AFM.

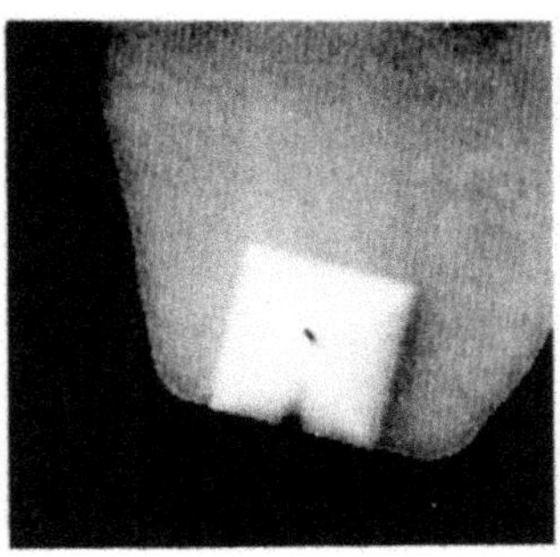

Figure 8. Photomicrograph of a silicon nitride tip after making a 20 nm hole by ionic etching at the very tip (black spot).

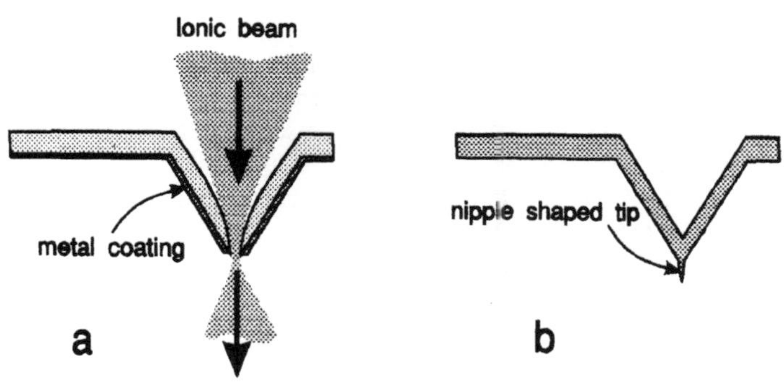


Figure 9. Possible solution for enhancing the lateral resolution. (a), after metallization of the lower tip side, making of a conical shaped hole in the extremity of a silicon nitride tip. (b), use of a supertip and a non metallized cantilever.

By using a converging ionic beam the hole can be shaped as indicated in figure 9a. With this cone-shaped etching, the light beam could go through the tip without losses during the propagation. Finally another promising solution could be the use of nipple slapedtips as shown in figure 9b [8].

## 6. Conclusion

The choice of an interferometric detection using a monomode fiber is certainly not the most popular way for measuring the displacement of the cantilever. This technique allows to use very small diameter piezoelectric tubes and consequently allows to realize very compact systems. A second interesting characteristic of the fiber is its ability to angularly filter the incident beam. The presented results are the first ones we have obtained with this hybrid AFMxSTOM method. It is clear that the optical field varies sufficiently on the object surface to lead to an optical mapping. If these results are confirmed by future attempts, this technique could partially solve the dilemma of the non separability of the field variations due either to index or to topography variations. First, because the simultaneous knowledge of topography and field distribution allows to discriminate the influence of the relief in the optical image. Second, at so short distances, the relief will not affect strongly the field. Consequently, the images will be essentially correlated to the index mapping of the object.

This work has been supported by the 'programme interdisciplinaire' ULTIMATECH/CNRS, France.

## References

1 Courjon, D., Bainier, C. and Spajer, M. (1992) 'Imaging of submicron index variations by scanning optical tunneling', J. Vac. Sci. Technol. B, 10, 1-3.
2 Betzig, E., Finn, P.L. and Wiener, J.S. (1992) 'Combined shear force and near-field scanning optical microscopy', Appl. Phys. Lett., 60, 2484-2486.
3 Toledo-Crow, R., Yang, P.C., Chen, Y. and Vaez-Iravani, M. (1992) 'Near-field differential scanning microscope with atomic force regulation', Appl. Phys.Lett., 60, 2957-2959.
4 Van Hulst, N., Moers, M., Noordman, O., Faulkner, T., Segerink, F., Van der Werf, K., de Grooth, B. and Bölger, B. (1992) 'Operation of a scanning near field optical microscope in reflection in combination with a scanning force microscope', SPIE Proceedings 'Scanning probe microscopies', 1639, 36-43.
5 Vigoureux, J.M. (1991) 'Polynomial formulation of reflection and transmission by stratified planar structures', J.O.S.A. A, 8, 1697-1701.
6 Nazar, F.M. (1979) 'Absorption in silicon nitride films', Short notes in JPN. J.Appl. Phys.,18, 1181-1182.
7 Allegrini, M., Ascoli, C., Baschieri, P., Dinelli, F., Frediani, C., Lio, A. and Mariani, T. (1992) 'Laser thermal effects on atomic force microscope cantilevers', Ultramicroscopy, to be published.
8 Lee, K. L., Abraham, D. W., Secord, F. and Landstien, L., (1991) 'Submicron Si trenth profiling with an electron-beam fabricated atomic force microscope tip', J. Vac. Sci. Technol. B, 9, 3562-3568.
9 Roberts, A. (1992) 'Theoretical considerations in optical near-field microscopy: probe-sample interaction and probe positioning effects', in this issue.

# COMBINED PHOTON SCANNING TUNNELING MICROSCOPE AND ATOMIC FORCE MICROSCOPE USING SILICON NITRIDE PROBES

M.H.P. MOERS, R.G. TACK, O.F.J. NOORDMAN,
F.B. SEGERINK, N.F. v. HULST & B. BÖLGER.
*Department of Applied Physics, University of Twente*
*P.O.Box 217, 7500AE Enschede*
*The Netherlands*

ABSTRACT. A Photon Scanning Tunneling Microscope is presented in which $Si_3N_4$ cantilevers are used as optical probes. This method allows close contact scanning without tip destruction. Simultaneous measurement of the optical signal and the deflection of the cantilever permits a comparison between the optical image and the topography. The optical signal as a function of the distance clearly shows the exponential behaviour of the evanescent field. Several structures have been examined, giving a lateral resolution in the optical image of about 40 nm.

## Introduction.

Both Scanning Near field Optical Microscopy (SNOM) and Photon Scanning Tunneling Microscopy (PSTM) are refered to as near field optical techniques. They are, however, based on different principles. In SNOM an aperture is scanned over a surface, acting as a sub wavelength sized lightsource and/or detector [1,2,3]. PSTM, on the other hand, takes advantage of the sub-wavelength information contained in the evanescent field, which is present at an interface with a total internal reflection [4,5,6]. Both techniques are capable of high resolution optical imaging.

Although the principles are different, the techniques suffer from identical problems. The lateral resolution is mainly determined by the size of the probe, and by the probe-sample distance. Both parameters are difficult to minimize. Furthermore, the tips usually are fragile and will be destroyed by a single touch at the surface.

Several feedback mechanisms are used to keep the probe within close distance of the surface. Apart from the 'classical' feedback on tunneling current (SNOM) [7] or optical intensity (PSTM) [4,5,6], feedback on shear force has been introduced [8,9]. This method is closely related to the mode of operation of a dynamic atomic force microscope [10].

In our laboratory a slightly different approach has been followed. $Si_3N_4$ cantilevers are used as optical near field probes [11]. These cantilevers, with integrated pyramidal tips, are commercially available for Atomic Force Microscopy (AFM). Several features make them suitable for near field microscopy. First, they are optically transparent down to 300 nm, with a refractive index of 2.0. Second, they are microfabricated with a tip radius of 30-50 nm. And finally, the low force



*D.W. Pohl and D. Courjon (eds.), Near Field Optics, 79–86.*

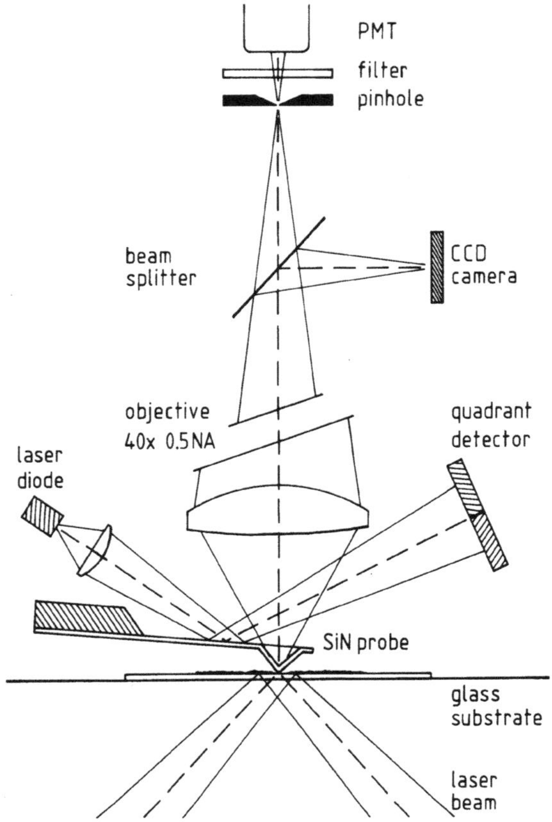


Figure 1. The set-up showing the combination of a PSTM, an AFM and a conventional microscope.

constant of the cantilever allows high speed close-contact scanning of the probe, without destruction of the tip.

The PSTM set-up described in this paper makes use of $Si_3N_4$ cantilevers to frustrate the evanescent field. An Atomic Force Microscope is incorporated, allowing a simultaneous measurement of the topography. Besides these two scanning probe microscopes the apparatus includes a third, conventional, microscope to preselect the scan area. Results of first measurements on test structures, optical thin films and chromosomes will be shown.

## The PSTM set-up.

The heart of our PSTM set-up is the $Si_3N_4$ probe (fig.1) [12]. A pyramidal tip (4×4 μm base, 3 μm high) is mounted at the end of a long (200 μm), thin (0.6 μm), V-shaped cantilever (force constant 0.06 N/m). This tip converts the evanescent waves, present above the sample surface, into radiative waves. The evanescent field is generated by the total internal reflection of a 5 mW HeNe laser beam, weakly focused to 100 μm, on the sample-air interface. The sample is placed on the substrate with index-matching fluid and is raster-scanned by a three axes piezo positioner (Microcontrol). The light, generated by frustrated total internal reflection (FTR), is collected by an extra long working distance objective (40×, NA=0.5, 10mm). A 75 μm pinhole in the image plane discriminates between light coming from the apex and light scattered by the sample. The light from the tip is detected by a photomultiplier tube (PMT).

Part of the light is split towards a CCD camera. This allows simultaneous viewing of tip and sample, which is essential for preselecting the scan area [13].

An Atomic Force Microscope is incorporated in the set-up by means of a beam deflection system [14]. A laser diode is focused on the upper side of the cantilever, and the reflected beam is detected by a quadrant detector. Small deflections of the cantilever induce a shift of the laser spot on the detector. Two modes of operation are possible for the AFM. While scanning in open loop, the cantilever follows the topography of the sample, and the deflection is monitored. Alternatively, in the closed loop mode the deflection is preset to a certain value, corresponding to a constant force, with a feedback loop controlling the position of the sample relative to the probe.

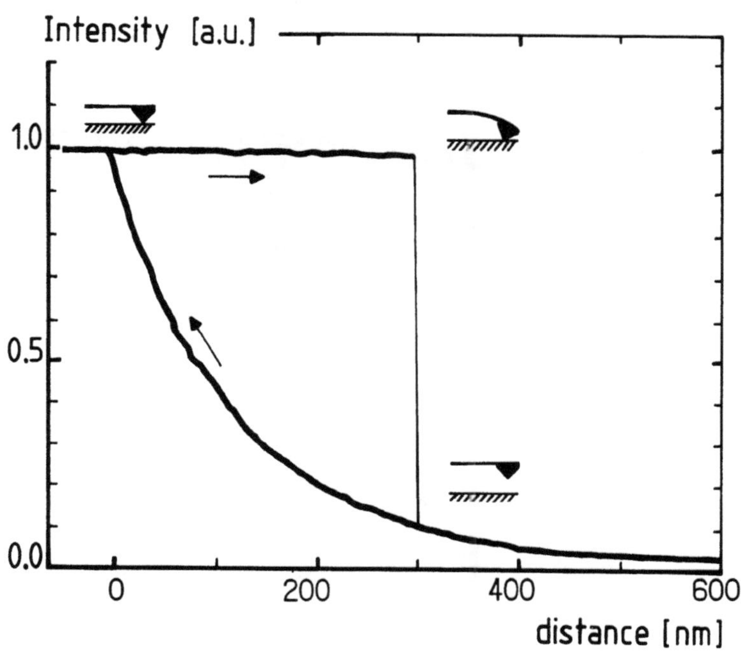


Figure 2. The FTR intensity as a function of the probe-sample distance. On approaching, the exponential behaviour is clearly observed. On retracting, the tip sticks to the surface due to adhesive forces of the water film.

## Results and discussion

*Optical signal versus distance curve.* The exponential behaviour of the evanescent field [6] is clearly observed when the probe approaches a glass substrate, as shown in figure 2. An intensity level of approximately 0.1 nW is reached when the tip touches the surface. All PSTM images shown in these proceedings have been obtained in this contact regime, with feedback on force. On retracting the tip, contact is maintained due to adhesive forces of the water film on the surface [15]. The cantilever typically bends over 300 nm (strongly depending on the relative humidity), corresponding to 18 nN adhesion force, until contact is released. This causes a drastic decrease in intensity.

*Water film.* The effect of the water film is clearly demonstrated on days with a high relative humidity (>70%). On these occasions a droplet of liquid, up to 1 μm in size, is observed with the conventional microscope at the contact area. The size of this droplet is strongly depending on the relative humidity. Although it does not effect the AFM, the working of the PSTM strongly depends on the humidity. Probably the area of FTR is determined by the size of the water droplet.

*Grating.* Figure 3 displays a 1 μm period symmetric $Si_3N_4$ grating, with 200 nm deep tracks. The structure is clearly resolved in the force image, although the low parts seem to be narrower than the high parts, due to the tip convolution problem. The grating is also visible in the PSTM image, albeit with a worse resolution.

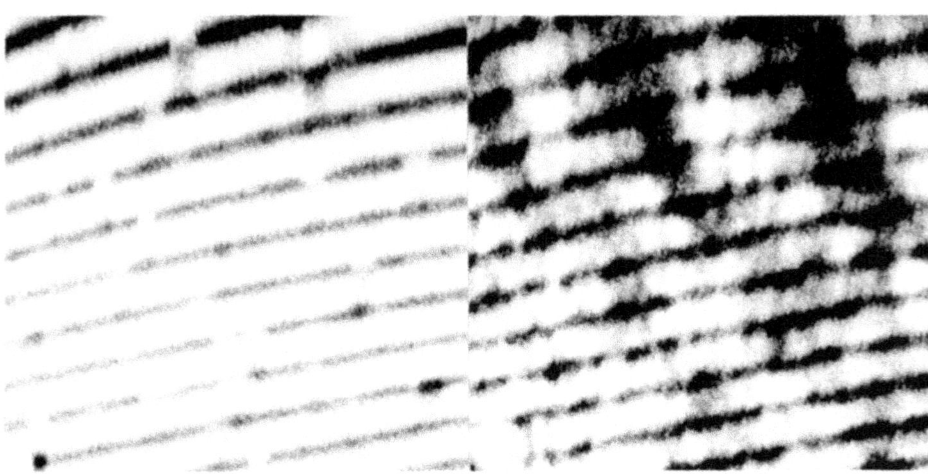

Figure 3. A 11 × 11 μm scan of a 1 μm period grating.
a) The AFM image, obtained in closed loop mode.
b) The corresponding PSTM image.

*Chromosome.* Preliminary results of biological objects with the PSTM-AFM combination are shown in figure 4. The inclusion of the conventional microscope is essential, especially for finding localized objects like cells and chromosomes. Figure 4a displays the topography of a metaphase chinese hamster lung chromosome. Although biological objects are relatively soft samples, this does not seem to affect the quality of the image. No material is removed over several scans, with forces in the order of 10 nN. The corresponding PSTM image shows two fundamental problems of PSTM on structures with strong surface topography. The laser beam is incident from the left. On this side of the chromosome an interference pattern is observed. This is an interference between the incident surface wave and the reflected wave from the chromosome [16]. One should notice that this interference pattern is restricted to the surface, so it can not be seen in the far field. To the right, on the 'shadow' side, radiative waves cause a pattern which is extending for about 4 μm beyond the structure, due to the disturbance of the surface wave. Possibly the conditions for total internal reflection are not matched at the downward slope of the chromosome.

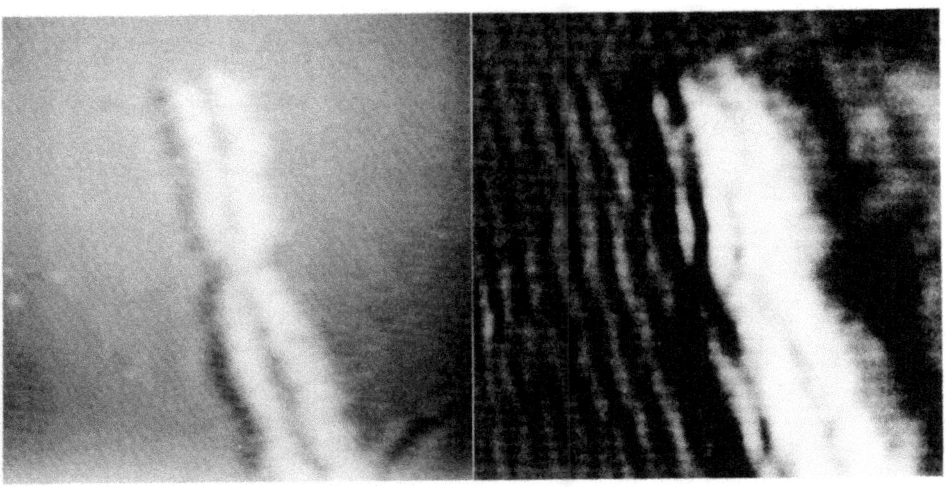

Figure 4. A 15 × 15 μm scan of a chromosome.
a) The AFM image clearly showing the topography.
b) The corresponding PSTM image showing interference and scattering.

*Supertips.* In order to reduce the contact area between the probe and the surface, the use of 'supertips' can be considered. In a Scanning Electron Microscope (SEM) a thin carbon needle is grown on a $Si_3N_4$ tip by local electron-beam induced deposition of residuents [17]. This supertip can be 20-100 nm wide and up to several microns long. The advantage of these tips is threefold. In the first place the coupling between surface and probe is better localized, as can be seen by the size of the light spot. In the second place the apex angle is much less than for normal tips, which improves the resolution when imaging steep structures. And finally, the adhesion forces are reduced.

It seems remarkable that FTR is also possible with supertips, because it is not clear whether the carbon structure is optically transparent. The obtained intensity is, however, of the same order of magnitude as with normal $Si_3N_4$ tips. Of course, any structure placed in the evanescent field will scatter, whatever the material it is made of. Optical transparent material is only necessary to guide the scattered light to the detector (tapered fibers) or to enable imaging with the objective ($Si_3N_4$ cantilevers).

*Indium Tin Oxide.* Figure 5 shows a 1x1 μm image of an Indium Tin Oxide film (Baltracon), obtained with a supertip. The granular structure of crystallites, 30-100 nm in size, is clearly resolved in the force image. In the optical image similar structures are observed, although the grains appear dark. This can be due to the refractive index and limited transmittivity of the crystallites. In the area between the grains the tip couples to the substrate resulting in a rather sharp light rim around the grains. Crystallites, 50 nm in size, are clearly resolved.

On the other hand, the topography could force the tip out of the evanescent field, affecting the efficiency of coupling the light into the probe. If this is true, one should realize, however, that this is an effect which is inherent to PSTM (in fact, to all scanning probe microscopes). It is not a limitation of the use of $Si_3N_4$ probes. The lateral resolution of the PSTM image seems to be about 40 nm, better than obtained with normal $Si_3N_4$ cantilevers.

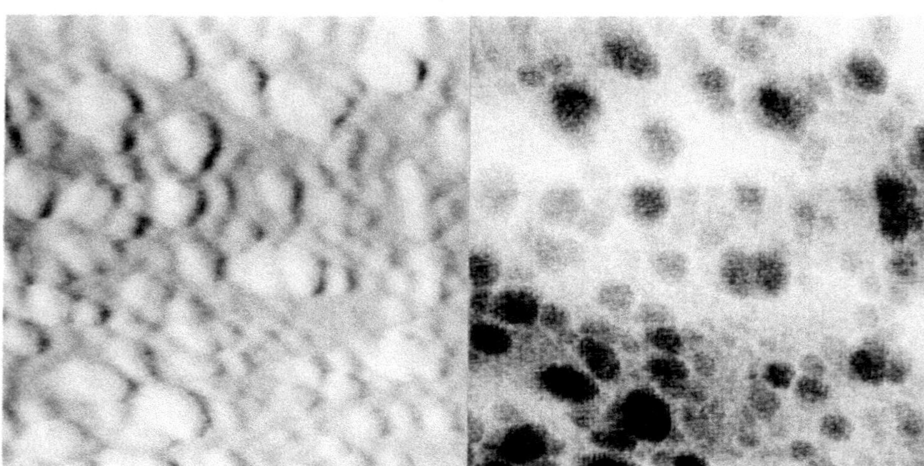

Figure 5. A 1 × 1 μm scan of Indium Tin Oxide.
a) The AFM image showing the granular structure of ITO.
b) The corresponding PSTM image.

## Conclusions

The operation of a Photon Scanning Tunneling Microscope with a $Si_3N_4$ cantilever as optical probe has been demonstrated, as well as the problems inherent to the technique. Routine close contact scanning without tip destruction is possible. Simultaneous operation of the Atomic Force Microscope allows comparison between the optical image and topography.

The optical signal versus distance curve indicates that the frustrated evanescent field is detected. Problems arise first then, when a sample is placed on the substrate. Interference and scattering are major problems of Photon Scanning Tunneling Microscopy, as well as the coupling between topography and optical signal. These problems have not completely been solved yet.

Indium Tin Oxide, a granular structure with dimensions well below the optical wavelength, gives good results. With supertips, a resolution in the optical image of about 40 nm is obtained. The effect of adhesion due to the water film seems to be reduced by the use of supertips.

## Acknowledgements

This work is mainly supported by the Dutch Foundation for Fundamental Research (FOM). The authors wish to thank K. v.d. Werf for his help with the construction of the AFM and B. Otter for the fabrication of the supertips.

## References

1. Pohl, D.W., Denk, W. and Lanz, M. (1984) 'Optical stethoscopy: Image recording with resolution $\lambda/20$', Appl.Phys.Lett. 44(7), 651-653.
2. Betzig, E., Isaacson, M. and Lewis, A. (1987) 'Collection mode near-field scanning optical microscopy', Appl.Phys.Lett. 51(25), 2088-2090.
3. Betzig, E, Trautman, J.K., Harris, T.D. and Weiner, J.S. (1991) 'Breaking the diffraction barrier: Optical microscopy on a nanometric scale', Science 251, 1468-1470.
4. Reddick, R.C., Warmack, R.J. and Ferrell, T.L. (1989) 'New form of optical microscopy', Phys.Rev. B39(1), 767-770.
5. Courjon, D., Sarayeddine, K. and Spajer, M. (1989) 'Scanning tunneling optical microscopy', Opt.Comm. 71, 23-28.
6. v.Hulst, N.F., de Boer, N.P. and Bölger, B. (1991), 'An evanescent-field optical microscope', J.Micr. 163(2), 117-130.
7. Dürig, U., Pohl, D. and Rohner, F. (1986), 'Near-field optical scanning microscopy with tunnel-distance regulation', IBM J.Res.Develop. 30(5), 478-483.
8. Betzig, E., Finn, P.L. and Weiner, J.S. (1992), 'Combined shear force and near-field scanning optical microscopy', Appl.Phys.Lett. 60, 2484-2486.
9. Toledo-Crow, R., Yang, P.C., Chen, Y. and Vaez-Iravani, M. (1992), 'Near field differential scanning optical microscope with atomic force regulation', Appl.Phys.Lett. 60, 2957-2959.
10. Den Boef, A.J. (1991), 'The influence of lateral force in Scanning Force Microscopy', Rev.Sci.Instrum. 62, 88-92.
11. v.Hulst, N.F., Moers, M.H.P., Noordman, O.F.J., Faulkner, T., Segerink, F.B.,

v.d.Werf, K.O., de Grooth, B.G. and Bölger, B. (1992) 'Operation of a scanning near field optical microscope in reflection in combination with a scanning force microscope', SPIE 1639, 36-43.

12. Park Scientific Instruments, Sunnyvale, California, USA.
13. Putman, C.A.J., v.d.Werf, K.O., de Grooth, B.G., v.Hulst, N.F., Segerink, F.B. and Greve, J. (1992) 'Atomic force microscope with integrated optical microscope for biological applications', Rev.Sci.Instrum. 63(3), 1914-1917.
14. Meyer, G. and Amer, N.M. (1988), 'Novel optical approach to atomic force microscopy', Appl.Phys.Lett. 53(24), 2400-2402.
15. Weisenhorn, A.L., Hansma, P.K., Albrecht, T.R. and Quate, C.F. (1989), 'Forces in atomic force microscopy in air and water', Appl.Phys.Lett 54, 2651-2653.
16. v.Hulst, N.F., Segerink, F.B., Achten, F. and Bölger, B. (1992) 'Evanescent-field optical microscopy: effects of polarization, tip shape and radiative waves', Utramicroscopy 42-44, 416-421.
17. Keller, D.J. and Chih-Chung, C. (1992) 'Imaging steep, high structures by scanning force microscopy with electron beam deposited tips', Surf.Sci. 268, 333-339.

# THE REFLECTION NEAR FIELD OPTICAL MICROSCOPE: AN ALTERNATIVE TO STOM

M. SPAJER, A. JALOCHA
*Laboratoire d'Optique P.M.Duffieux, URA 214 CNRS*
*Université de Franche-Comté, UFR Sciences, route de Gray*
*25030 Besançon cedex*
*France*

ABSTRACT. Reflection near field microscopy is potentially more accurate and efficient than STOM, its advantage beeing a localized illumination of the surface. Its working in far field and near field mode is described and the experimental z-curve is analyzed. A solution for partially solving the inherent problem of S/N ratio is presented. It uses electronic and numerical z-derivation of the signal.

## 1. Introduction: possible configurations.

The principle of a reflection microscope was demonstrated 6 years ago [1,2], but the fabrication of the suitable probe seemed the challenge, especially the separation of near field signal from stray light diffracted by the probe. The STOM/PSTM, with an evanescent illumination of the surface, was a smart solution to the problem of stray light. It has the advantage, and may be the necessity, of using a dielectric probe. In transmission microscopy, authors put the accent on the nanosource, and the advantage of local illumination, with leads to the choice of a metallized probe with a nanometric hole [3] or to "supertips" [4]. Other searchers remarked that the dielectric probe offers potentially more resolution, because there is no limit to its smallness. Figure 1 summarizes the main configurations proposed for reflection. The probe can be a tapered monomode fiber that insures, for the best efficiency, an adiabatic transition from the guided fundamental mode of the fiber to the extremity of the cone (a, b). On the contrary, it can be a conical tip with a larger angle, made by chemical etching (see § 3). The images presented in this contribution were obtained by a combination of pulling and etching.

Some results using the light reflected outside a metallized probe (a) were published [5]: the main difficulty is the masking effect of the metal layer, whose thickness is at least 100 nm. The external illumination of a vibrating probe combined with heterodyne detection, was also proposed [6]. The technique we have developed uses the light reflected by the object through the probe in the fiber core (b). A new solution uses an AFM cantilever where the probe is illuminated by a laser beam directly (d) [7] or through an optical fiber ended by a micro-optic (e) [8]. We shall now analyze the reflection microscope using dielectric tapers [9] in their two possible working modes: far field and near field.



*D.W. Pohl and D. Courjon (eds.), Near Field Optics, 87–96.*

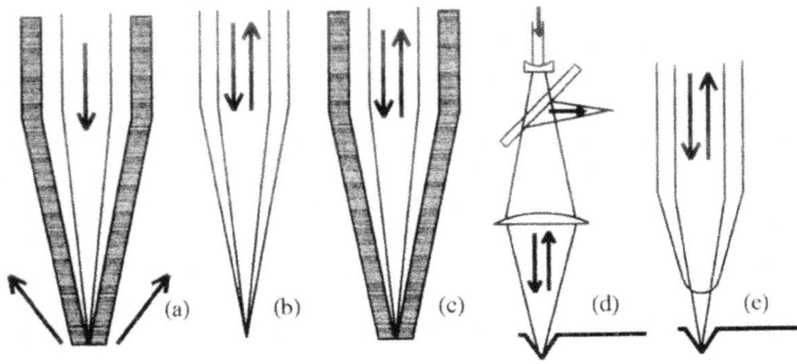


Figure 1. Different types of probe for reflection near field microscopy.

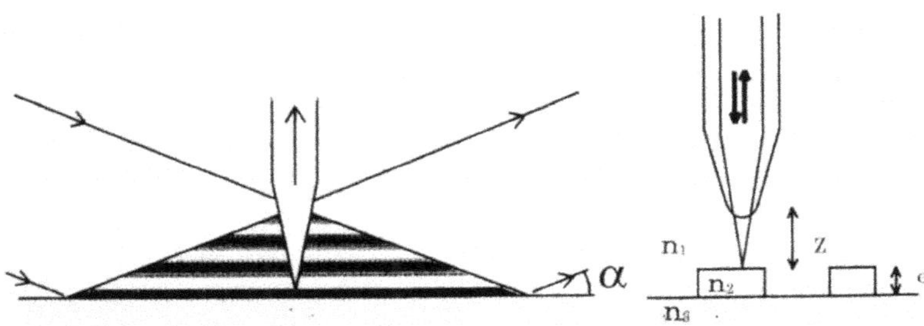


Figure 2. Two local probe profilometers: (a) using Wiener fringes (b) using a focusing probe.

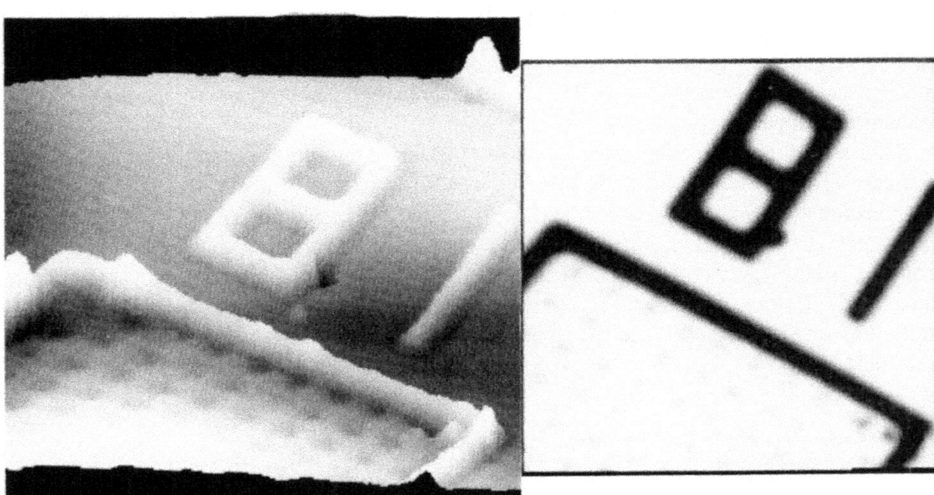

(a) topography (b) reflectance variations.

Figure 3. Images of a microcircuit $SIO_2$/SI obtained with the profilometer of figure 2b.

## 2. Local probe interferometric profilometer (far field).

The topography of a surface illuminated by an oblique wave can be studied with the help of a sub-wavelength probe in two detection modes, as it is explained in reference [10]. In reflection, the situation equivalent to the STOM/PSTM is represented by figure 2a. The fringes produced by interference of the incident and reflected beams near a surface allow distance control of a probe [11], in so far as the slope variations of the surface are small: they affect weakly the fringe spacing, and we can consider that a given fringe remains at the same distance from the surface. A nanometric probe diffracts the field near the apex without important perturbation of the fringes. The distance of the probe that allows the best sensitivity in distance control is $(\lambda/8)\sin\alpha$,i.e., 1/4 fringe. For an incidence of 45°, it corresponds to 110 nm: the same order as in PSTM configuration.

In the configuration we have chosen (object illuminated through the probe), the best classical (i.e., diffraction limited) image of the surface obtained will be given by a tapered fiber ended by a microlens, which focuses the beam in front of the probe (figure 2b). We shall consider the example of a microcircuit where lines and dots are made with a layer of $SiO_2$ on Si (thickness = 100 nm). The beam reflected by the lens is considered as the reference beam of two-wave interferometer. The second wave is reflected by the object. Getting a linear response from an interferometer requires a demodulation technique, or a heterodyne detection, or a displacement of the probe. We propose to use a vertical vibration of the probe, a synchronous detection which gives a signal proportional to the derivative of intensity dI/dz, and a feedback loop which maintains this signal at zero. On a pure metallic surface, the probe would follow its topography. On our sample, it will follow the optical thickness ne. In these scanning conditions, the whole reflected intensity I(z) depends on the reflectivity r(M) of the object. Hence we can measure simultaneously the topography and the reflectance of the surface, or its topography and index variations. Figure 3a shows the topographic information given by the feedback signal. Figure 3b is the complementary image of reflectance variations on the same zone. The field is 30 x 30 µm and the resolution is about 1.5 µm . It could be improved by optimizing the focusing properties of the microlens. The interest of this micro-optic profilometer is its ability to work at various wavelengths from the IR to the UV, without expensive optics.

## 3. Probe shape and approach curves.

To work in the near field domain, the probe used in the previous experiment has to be transformed in a sharp tip. As propagative waves are emitted, nanometric details diffract non homogeneous waves, with a greater wave vector $\mathbf{K}_i$ than a classical evanescent wave, and a smaller penetration depth (figure 4). These waves can be detected at small distance if the matching condition is fulfilled ($\pi/D<|\mathbf{K}_i|$, with D = diameter of the apex), which requires smaller probe. To describe more completely the tip/object interaction, we must consider the evanescent waves confined near the tip extremity, which are strongly modified by the presence of the object. This last effect is comparable to the working of PSTM, with this difference that the evanescent wave is bounded to the probe and not to the surface.

The same problems as with STOM/PSTM are encountered when large structures ($>\lambda/2$) or thick roughness is present in the object : they come from diffraction of homogeneous waves towards the probe, and of guiding or resonant field structures, which prevent a correct topographic working. The same consequence must be drawn: the best results in controlled mode will be obtained with extremely small roughness (a few nanometers). In these conditions only, the field can be correlated to the topography.

The tip modification is easily achieved by chemical etching, as schematized by figure 5. The initial shape (a) is obtained by pulling the fiber under the electric arc of a splicing device. The curvature radius of the microlens is typically 10 µm. In 50% hydrofluoric acid, the lens disappears after an etching time of 15 mn, and is replaced by a cone (b). The curvature radius of its extremity becomes typically 50 to 100 nm. A simpler method is the chemical etching of a clived fiber : the core is no more tapered and is only ended by a conical tip (c). The etching time is longer (> 1 hour), the tip roughness higher, and the result less reproducible. An angle of 30° can be obtained if the fiber is held in a ceramic connector or if the acid bath is covered by an oil film. We must mention the reproducible sharp tips obtained wether by a pure pulling technique [3] or by pure etching technique using the proper solution HF + $NH_4F$ [12, 13]. In this last case, great angles

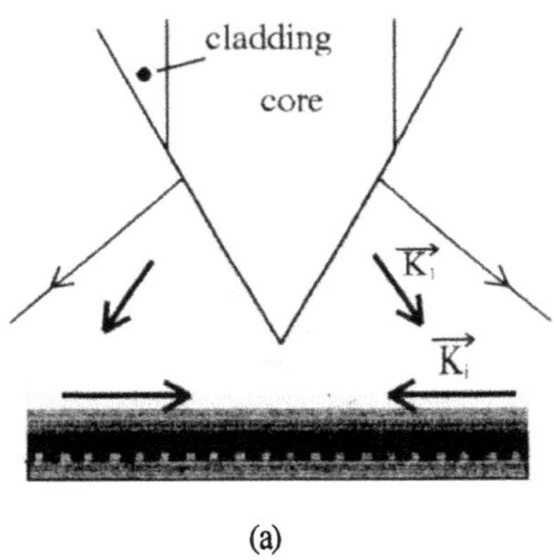


Figure 4. Schematic evolution of the field while the probe approaches the surface. (a) Probe in the far field: only the evanescent components are represented. (b) Probe in the near field. Propagative waves are emitted near the apex.

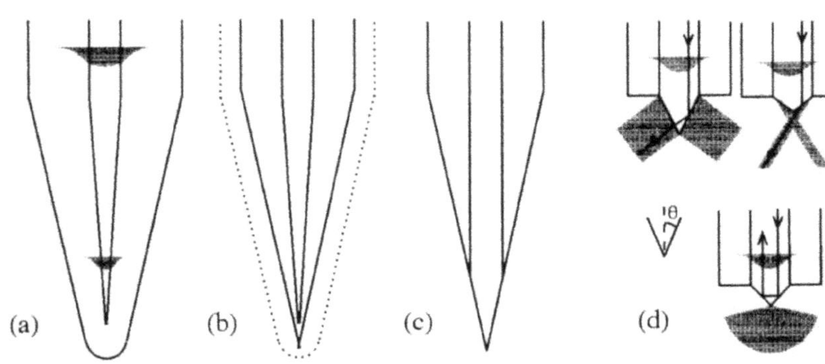


Figure 5. Different tip shapes. (a) taper with microlens (b) sharpened taper after chemical etching Tips obtained by etching: (c) with a standard fiber (d) with special fiber.

can be obtained more easily (figure 5d). Simple geometrical considerations lead to the hypothesis that for an unmetallized tip, the optimal angle can be 90°. The total reflection of the incident wave actually prevents emission of light by the probe, except the diffraction by the extremity. Tip angles we obtained with standard fibers were generally greater than 90°, which leads to pseudo-focusing effect .

For a real optimization of the geometry, the important questions are: what is the field structure around the conical probe and the intensity back reflected in the core (1) when the extremity is far from the surface (2) when its distance becomes $< \lambda$ ? We make the hypothesis that the penetration depth of the evanescent wave wrapping the probe, for a very small apex, depends mainly on the angle of the cone. What is the field intensity near the apex when it approaches close to contact ? The signal available for near-field detection depends on the answer. The resolution will depend on the dimension of the apex, which scans the surface in this modified field and generates near-field interactions. These hypothesis must be qualified by a model of the energy transfer from the fiber core to the apex, using a microscopic method [14, 15] or Maxwell equations [16].

A useful characteristic of the probe for near field imaging is the approach curve, i.e., the signal variation near contact, even if it gives no indication on lateral resolution. We shall analyze an example obtained with a metallized wafer and a probe made by chemical etching of a monomode fiber (figure 5c).

In the far field domain, i.e., when the distance of the apex to the surface is $> \lambda/2$, the light coupled into the guide after reflection can be expressed by the following formula:

$$\begin{aligned} I(X,Y,Z) &= | R + C_{//}(X,Y,Z) |^2 + | C_{\perp}(X,Y,Z) |^2 \\ &= R^2 + C^2(X,Y,Z) + 2\,R\,C_{//}(X,Y,Z) \cos [\Phi_{//}(X,Y,Z)] \end{aligned} \quad (1)$$

where $(X,Y,Z)$ are the coordinates of the extremity of the tip,

$R$ is the reflection coefficient of the probe in the absence of any object, with a phase arbitrarily chosen as zero,

$C_{//}(X,Y,Z) = | C_{//}(X,Y,Z) | \exp[\, j\Phi_{//}(X,Y,Z)]$ the amplitude of the light reflected back to the guide with the same polarization as the reflection by the probe,

$C_{\perp}(X,Y,Z)$ the amplitude of the reflected light perpendicularly polarized,

$C(X,Y,Z) = \{ | C_{//}(X,Y,Z) |^2 + | C_{\perp}(X,Y,Z) |^2 \}^{1/2}$ comprises both contributions of the near-field interaction at the end of the tip and of the propagating waves emitted by the whole probe and reflected by the object.

This description excludes multi-reflection between the tip and the object. We shall consider in this section that only Z is varying. In the far field, the phase varies with Z like:

$$\Phi_{//}(X,Y,Z) = \Phi_0(X,Y,Z) + 2\,k\,Z \qquad (k = 2\pi / \lambda) \quad (2)$$

under the assumption that the preponderant waves reentering the probe propagate in a direction close to the axis. This supposes that there is no strongly diffracting relief near the probe. We can expect that the near-field contribution is much smaller than the far-field one, so we proposed to enhance its visibility by a small vibration of the probe:

$$Z = z + \Delta z \cos \Omega t \quad (3)$$

and a synchronous detection. The signal at the frequency $\Omega$ of the vibration is:

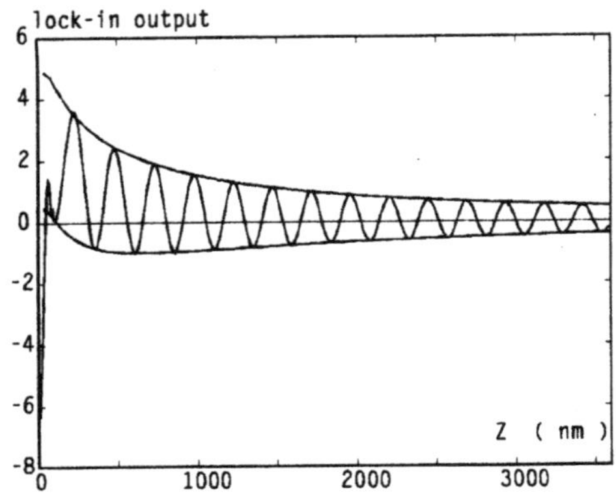


(a) Experimental curve. The envelopes $I_{max}(z)$ and $I_{min}(z)$ are calculated from the extrema (the first maximum, due to near-field interaction is excluded).

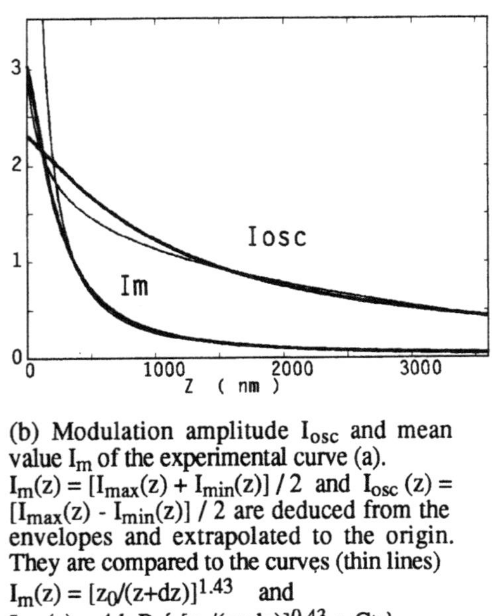


(b) Modulation amplitude $I_{osc}$ and mean value $I_m$ of the experimental curve (a). $I_m(z) = [I_{max}(z) + I_{min}(z)] / 2$ and $I_{osc}(z) = [I_{max}(z) - I_{min}(z)] / 2$ are deduced from the envelopes and extrapolated to the origin. They are compared to the curves (thin lines)

$I_m(z) = [z_0/(z+dz)]^{1.43}$ and

$I_{osc}(z) = 4\,k\,R\ \{\ [z_0/(z+dz)]^{0.43} + Ct\ \}$

with $z_0 = 4.18$ nm, $dz = 5$ nm, $Ct = -0.35$, $k = 2\pi / \lambda$ .

Figure 6. Approach curves of a vibrating dielectric probe near a metallic surface. The probe, obtained by chemical etching as represented by Fig.5c, has a cone angle $2\theta \approx 30°$.

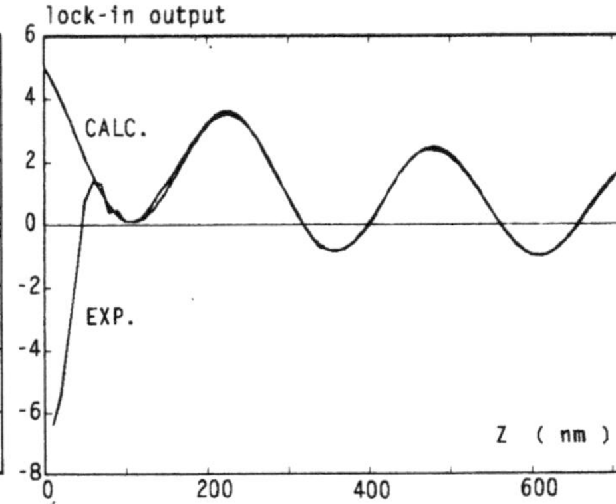


(c) Superimposition of the experimental curve to the "far field" curve reconstructed from $I_{osc}$ and $I_m$.

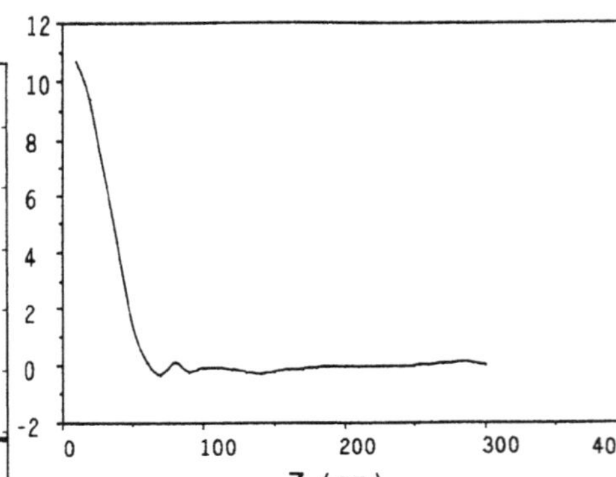


(d) Difference between experimental and "far-field" curve of (c).

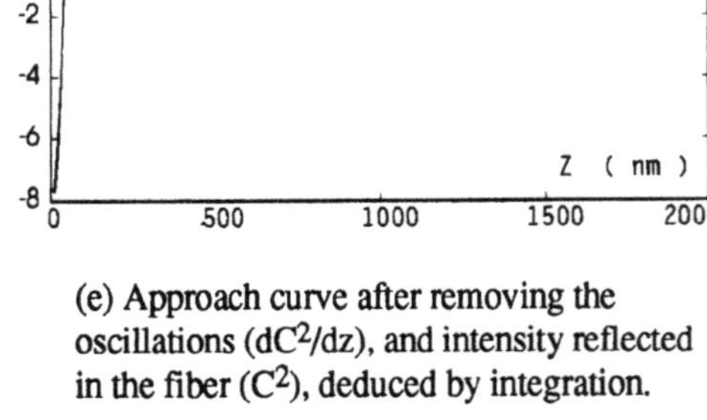


(e) Approach curve after removing the oscillations ($dC^2/dz$), and intensity reflected in the fiber ($C^2$), deduced by integration.

$$I_\Omega(z) = \Delta Z \left\{ \frac{d|C|^2}{dZ} + 2R\frac{d|C_{//}|}{dZ}\cos\Phi_{//} - 2Rk\,|C_{//}|\,\sin\Phi_{//} \right\} \tag{4}$$

The 2nd and 3rd term describe the oscillations due to the interference between the tip and the object. Their amplitude is given by:

$$I_{osc} = 2R\left\{ \left[\frac{d|C_{//}|}{dZ}\right]^2 + \left[2k\,|C_{//}|\right]^2 \right\}^{1/2} \tag{5}$$

The first term $I_m = d|C|^2/dZ$ is the derivative of the whole reflected light. $I_{osc}$ and $I_m$ represent the modulation amplitude and the mean value of the function $I_\Omega(z)$. Figure 6a gives the approach curve with its envelopes. Figure 6b shows the curves $I_{osc}(z)$ and $I_m(z)$ deduced from the envelopes. $I_m(z)$ is well fitted by a $z^{-1.43}$ curve for $\lambda < z < 6\lambda$. In this domain, if we suppose that $C_{//}$ and C are of the same order, the second terme of (5) is preponderant and $I_{osc}(z)$ might decrease like $(z/z_0)^{-0.43}$ + Ct. A constant Ct and a multiplying factor 2R can be calculated to fit the two curves with an error smaller than 15% on the same domain. The discrepancy can be attributed to the depolarizing effect of the conical probe. The value of R is about 2.5 greater than the maximum value of C(z).

Figure 6c superimposes the experimental curve with the curve reconstructed from the knowledge of $I_m(z)$, $I_{osc}(z)$, and of the phase of the oscillations. This last curve does not take into account the rapid variation near the origin: its value for z=0 is simply extrapolated. Figure 6d shows the difference between the experimental data and the reconstructed, or "far field" curve. This difference has significant values for $z < \lambda/8 = 80$ nm and its maximum is of the same order as the far field signal. Figure 6e shows the restored shape of $d|C|^2/dZ$ combining figure 6b and 6d, and the shape of |C(z)| deduced from it. The last one exhibit a maximum at a distance of 80 nm and a decrease of 20% as it approaches to contact. The first one show that the derivation effect of the vibrating probe increases the dynamic of the near field signal regarding the far field one.

The oscillations of the experimental curve (figure 6a) have an obvious drawback : they make the control of the distance more difficult, and the approach curve of figure 8e would be more advantageous. One solution for removing the oscillations is the heterodyne detection as used for vibration analysis. The reflected light guided in the fiber must interfere with a reference beam whose optical frequency is shifted by an acousto-optic modulator. Then the detected intensity becomes :

$$I(z,t) = |\,R_0 e^{j\omega t} + R_{//} + C_{//}(z)\,|^2 + |\,R_\perp + C_\perp(z)\,|^2 \tag{6}$$

where $R_0$ is the amplitude of the reference beam and $\omega$ its frequency shift. The indexes // and $\perp$ refer to the polarization of the reference beam. Replacing Z by its expression (3), formula (6) expanded to the first order in $\Delta Z$ and limited to the modulated terms in the neighbourhood of the carrier $\omega$ becomes :

$$\begin{aligned} I(z,t) = {} & 2\Delta Z\ R_0\ \{ R_{//}\cos(\omega t) + C_{//}(z)\cos[\omega t - \Phi_{//}(z)] \} \\ & + 2\Delta Z\ R_0 \frac{dC_{//}(z)}{dz} \{ \cos[(\omega+\Omega)t - \Phi_{//}(z)] + \cos[(\omega-\Omega)t - \Phi_{//}(z)] \} \\ & + 2\Delta Z\ R_0\ k\, C_{//}(z) \{ \sin[(\omega+\Omega)t - \Phi_{//}(z)] + \sin[(\omega-\Omega)t - \Phi_{//}(z)] \} \end{aligned} \tag{7}$$

Demodulation techniques use generally the first term, at frequency $\omega$, as the carrier, and gives the amplitude of the two sideband signals, at frequency $\omega+\Omega$ and $\omega-\Omega$. These amplitudes depend both on the phase modulation $k\ C_{//}(z)$ (3rd term) and on the amplitude modulation $d\ C_{//}(z)/dz$ (2nd term), which are in quadrature. The 2nd term is the only interesting one for near field detection. This involves a phase discrimination in the demodulation process, or the elimination of the 3rd term, which is preponderous. It can be achieved by introducing the same phase modulation

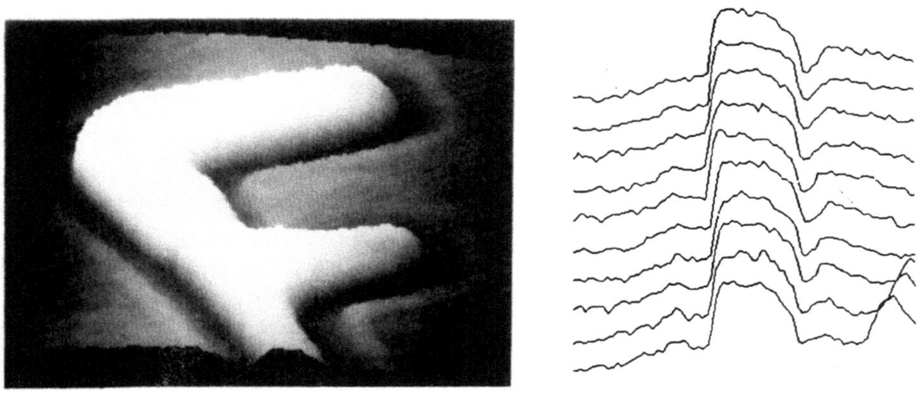

Figure 7. Image of a metallized microcircuit. Field = 10 μm x 10 μm.

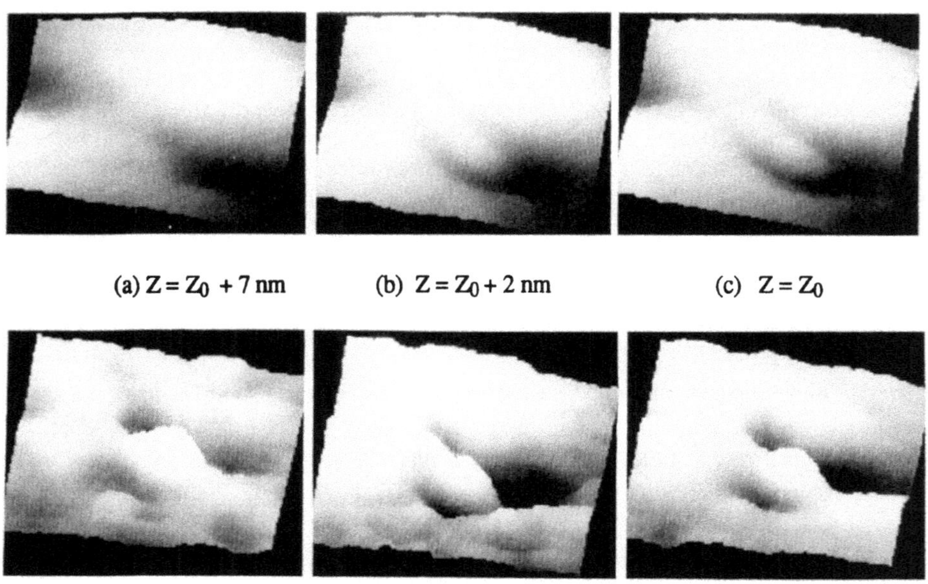

(a) $Z = Z_0 + 7$ nm (b) $Z = Z_0 + 2$ nm (c) $Z = Z_0$

(d) = (b) - (a) (e) = (c) - (b) (f) = (d) - (a)

Figure 8. Image of a poly cristalline silcon surface. Field = 200 nm x 200 nm.

on the reference beam.

The curves analyzed in this part were obtained on a flat surface. As they can be modified by the presence of diffracting structures, they can be used for distance control only in the case of small relief. In this case, the modification affects only the left extremity of the curve. On the objects studied until now, the reliefs were too high to allow this topographic mode. The slope inversion of the reflected intensity visible on figure 6e must be prudently interpreted. The conversion of the evanescent field wrapping the apex into propagating waves may have the paradoxical effect of reducing the intensity reflected in the core: these waves may be out of phase with the other waves emitted by the whole probe (and described by the far field diffraction of the probe) and reflected by the object. The effect of eventual water film on the surface must be also kept in mind.

## 4. Images in near field.

The following results illustrate two different situations with a common characteristic : the object is scanned at a constant height, without feedback. The experiences have been carried out with a 5 mW laser diode (l =790 nm), and a chemically etched taper.

Figure 7 is the image of a letter on a microcircuit. The line profile is supposed to be rectangular, with a height of 1.2 µm and a width of about 1.5 µm. The left edge seems sharper than the right one. This can be explained by the slope of the scanning plane, which is easily calculated from the two rectilinear fringes seen in the image, and by a dissymmetry of the probe. The image of the letter is surrounded by a fringe-like border, which can be attributed to the lateral diffracted beams of the probe.

The next result demonstrates the advantage of a double z-derivation to enhance the contrast of nanometric details. The second derivation is made by numerical subtraction between two images scanned at altitude z and z+dz. The object is a poly cristalline silicon surface, made of little pyramids randomly arranged : the roughness is typically 100 nm, and the mean width 150 nm. Figure 8a-c show 3 images of the same setting, obtained at 3 different distances (field = 0.2 µm). The smooth oscillations of figure 8a, which have a pseudo-period of $\lambda/2$, appear when the tip extremity "touches" the evanescent waves wrapping the object, i.e., for a distance < $\lambda/2$. Figure 8b and 8 c show some elliptic details, whose smaller dimension is about $\lambda/10$ (70 nm). It appears after a probe displacement of only a few nanometers. This result indicates unambiguously a near field interaction between the probe apex and a detail of the surface. Figure 8d-f are the different subtractions between the previous images.

## 5. Conclusion

The difference between near field microscope using a dielectric probe and STOM is the existence of a far field or interferential range (Z>100 nm) . Interference between a focusing probe and object can be turned to account for a profilometry of the surface with a classical lateral resolution (better than 1 µm if UV sources can be used). With a sharpened probe, a first analogy between the two devices appears in the "quasi near field" domain (Z<100 nm): we can compare the rapid variation which interrupts the interferential signal to the exponential Z-dependance observed with a STOM. In both cases, near field interaction of the probe apex induces slight perturbations of this approach curve. For near field imaging , a z-derivation of the signal improves the contrast of these interactions. It is achieved by a vertical dither motion of the probe and by numerical image subtraction. This procedure can be helpful for image interpretation even if the distance control is solved by force detection. Understanding of tip/surface interaction must be completed by a model of light propagation in a dielectric cone. Results obtained with a conical probe exhibit a resolution of about $\lambda/10$ (70 nm). The use of a greater wavelength would attenuate the influence of interferences and would not degrade the resolution, contrary to classical optics. Polarization control will give new mean of contrast enhancement.

## References


1 Fischer, U., Dürig, U., Pohl, D.W. (1988) 'Near-field optical scanning microscopy in reflection', Appl.Phys.Lett., 52, 249.

2 Fischer, U., Pohl, D.W. (1989) 'Observation of single-particle plasmons by near-field optical microscopy', Phys.Rev.Lett., 62, 458.

3 Betzig, E., Trautman, J.K., Harris, T.D., Wiener, J.S., Kostelak, R.L. (1991) 'Breaking the diffraction barrier: Optical microscopy on a nanometric scale', Science, 251 , 1468-1470.

4 Kopelman, R., Birnbaum, D., Shi, Z., Smith, S., Tan, W. 'Near field optical and exciton imaging, spectroscopy and chemical sensors', in this issue.

5 Trautman, J.K., Betzig, E., Wiener, J.S., DiGiovanni, D.J., Harris, T.D., Hellman, F. and Gyorgy, E.M. (1992) 'Image contrast in near-field optics', J.Appl.Phys., 71, 4659-4663.

6 Wickramasinghe, H.K., Williams, C.C. (1990) 'Apertureless near field optical microscope', U.S.Patent , n°4.947.034.

7 van Hulst, N.F., Moers, M.H.P., Noordman, O.F.J., Faulkner, T., Segerink, F.B., van der Werf, K.O., de Grooth, B.G., Bölger, B. (1992) 'Operation of a scanning near field optical microscope in reflection in combination with a scanning force microscope', SPIE vol. 1639, Scanning probe microscopies , 36-43.

8 Baida, F., Courjon, D., Tribillon, G., 'Combination of a fiber and a silicon nitride tip as a bifunctional detector; first results and perspectives', in this issue.

9 Spajer, M., Courjon, D., Sarayeddine, K., Jalocha, A., Vigoureux, J.-M. (1991).'Microscopie en champ proche par réflexion', J. Phys. III, 1, 1-12

10 Bainier, C., Leblanc, S., Courjon, D. 'Scanning tunneling optical microscopy: application to very low relief objects', in this issue.

11 Umeda, N., Hayashi, Y., Nagai, K., Takayanagi, A. (1992) 'Scanning Wiener-fringe microscope with an optical fiber tip', Appl.Optics, 31, 4515-4518.

12 Jalocha, A., Leblanc, S., Spajer, M.(1990) 'Etude des capteurs utilisés en microscopie tunnel optique', Conf. OPTO'90, Paris, may 1990, ESI Publications, Paris , 525-529.

13 Pangaribuan, T., Yamada, K., Jiang, S., Ohsawa, H., Ohtsu, M. (1992) 'Reproducible fabrication technique of nanometric tip diameter fibre probe for Photon Scanning Tunneling Microscopy', Jpn.J.Appl.Phys., 31, 1302-1304.

14 Girard, C., Spajer, M. (1990) 'Model for reflection near field optical microscopy', Appl. Optics, 29, 3726-3733

15 Dereux, A., Pohl, D.W.'Light scattering at a prism edge on approaching a surface', in this issue.

16 Hori, H. 'Quantum optical picture of photon STM and proposal of single atom manipulation', in this issue.

# SCANNING TUNNELING OPTICAL MICROSCOPY: APPLICATION TO VERY LOW RELIEF OBJECTS

C. BAINIER, S. LEBLANC and D. COURJON
*Laboratoire d'Optique P.M.Duffieux, URA214 CNRS*
*Université de Franche-Comté, UFR Sciences, route de Gray*
*25030 Besançon cedex*
*France*

ABSTRACT. Among the near field microscopes, one of the latest born, the so called Scanning Tunneling Optical Microscope (STOM or PSTM) presents a certain number of particularities due to its apparent similarity with the STM. In this paper we describe the mechanism of working of such a microscope and we show its actual ability to image very low relief objects with a subwavelength resolution.

## 1. Introduction

The scanning tunneling optical microscope has been proposed from 1988 by several groups independently [1-3]. Based on the use of an incident evanescent beam, its connection with the STM appeared obvious to the scientists who introduced the notion of optical tunneling microscopy. Many papers describing the field behavior both experimentally and theoretically, have been published in the last years [4-8].

From these works, it appears that the STOM/PSTM can be used in two different ways. The first way consists of working at about 50 nanometers or even more from the object surface. The interest of this solution is the ability of ensuring a simple control of the working distance by maintaining the detected intensity constant. This working mode can be exploited for measuring very small index variations [9]. The second way consists of working a few nanometers far from the surface in order to reach actual subwavelength resolutions. In that working mode, the microscope behaves exactly as a near field microscope lighted in total internal reflection. In this paper only that mode will be considered.

## 2. Mechanism of near field detection

In fact, these two ways hide a difference in the detection mechanisms. We know from the works of Newton that the detection of an evanescent field can be achieved by frustating the totally reflected beam over an interface by means of a suitable macroscopic piece of dielectric material. In this case, the conversion of non radiating fields into propagating ones is based on refractive phenomena as shown in figure 1. Since refractive phenomena imply macroscopic objects, such an approach is obviously not suitable for describing the interaction between the evanescent field and the detector in the case of subwavelength tips and samples. Fortunately, recent works have shown that a diffracting component (small hole, thin stylus, etc.) immersed in an evanescent beam is excited and can reemit a partially propagating field that can be then classically detected. In this case, the conversion of non radiating fields into propagating ones is based on diffraction rather than



*D.W. Pohl and D. Courjon (eds.), Near Field Optics*, 97–104.

refraction. Because diffracting objects can be exceedingly small, the use of diffraction for converting non radiating field into propagating field is not restricted to macroscopic analysis. It is thus suitable for reaching actual subwavelength resolutions. In the diffraction mode the STOM/PSTM behaves as a classical near field microscope and leads to similar results. However, some specific properties of such microscopes can justify the use of a specific acronym: the use of an incident evanescent beam has got some advantages such as the small amount of stray light. This point will be detailed further.

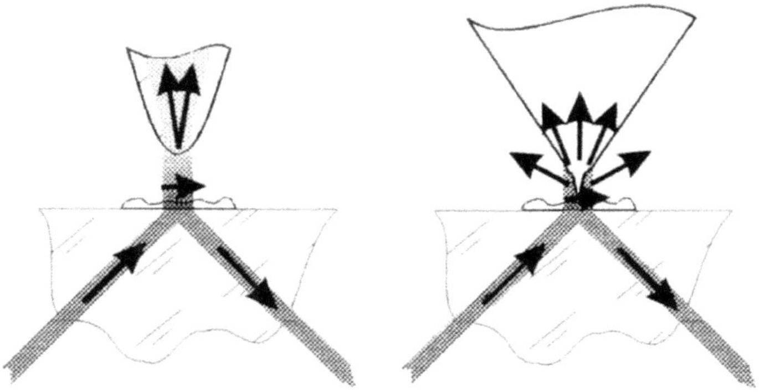

A B

Figure 1. Working modes of STOM:
A. detection of the evanescent field by refraction inside the fiber;
B. detection of the evanescent field by diffraction by the subwavelength extremity of the tip.

## 3. Decay length

In the following, we will compare the behavior of the detected field assuming first that the tip is a simple dipole (point-detector), second that the tip dimensions are larger than the wavelength (macroscopic description).

### 3.1. CASE OF A POINT DETECTOR AND A PERFECTLY FLAT OBJECT SURFACE

The behavior of the evanescent field generated by the reflection of a plane incident wave over a perfectly flat surface is described[10] by the following transmission coefficient:

$$T_1 = \exp\left(-2\pi \frac{z}{\lambda} C\right) \qquad \text{where} \quad C = \sqrt{n^2 \sin^2\theta - 1}\ ,$$

z is the distance between the surface and the observation point, $\lambda$ is the wavelength, $\theta$, the incidence angle and n the index of refraction of the object.

### 3.2. CASE OF A LARGE FLAT DETECTOR AND A PERFECTLY FLAT OBJECT SURFACE

If the detector is in dielectric material assumed to be flat and large enough in comparison with the wavelength, the dependence of the evanescent field versus the distance is somewhat more complex.

The transmittivity can be written:

$$T_2 = \{(A/C)\sinh^2(2\pi \frac{z}{\lambda} C) + 1\}^{-1} \qquad \text{where} \quad A = (n^2-1)^2 / 4n^2 \cos^2\theta,$$

we have assumed the same index for the tip and the sample.

We note in that case that the field varies slowly when the tip is very close to the object. This effect is due to the strong interaction between the tip and the object surface.

The associated curves A and B are shown in figure 2.

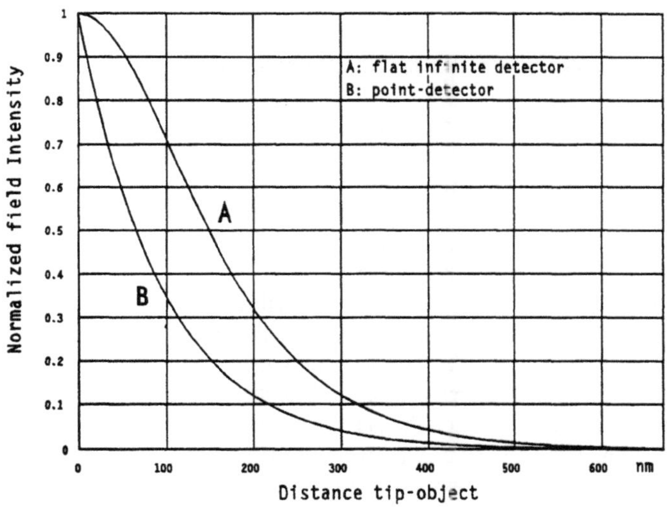


Figure 2. Detection of the evanescent field: in A the detector is assumed to be a flat and infinite piece of dielectric material; in B the detector is reduced to a point.

### 3.3. REAL CASE OF AN OBJECT AND A TIP WITH SUBWAVELENGTH PROTRUSIONS

In the real case, the object is not uniform and contains subwavelength structures. It is well-known now that such objects will diffract the light into propagating and non propagating fields [11]. The latter are non radiating and have got the same structure as the evanescent wave associated to a uniform and flat surface. It has been computed [12] that the penetration depth of such evanescent waves induced by the interaction of the incident evanescent beam with the object structure, decreases when the object frequencies increase. As an example, if the object is a grating of period 50 nm, the associated evanescent fields drop over distances smaller than 8 nm [13]. For detecting such fields, it is then necessary to place the detector a few nanometers far from the surface.

### 3.4. ROLE OF THE OBJECT MODULATION

Other works [7,8,14] have shown and demonstrated the difficulty of correlating the field profile with the object topography in the case of strong z-variation in the object structure. This character is mainly due to the interference of the light beam inside the object. Moreover, for such objects, the field versus the z coordinate is no longer monotonic, preventing the control of the distance by detection of the evanescent field.

On the contrary, if the object is flat enough, typically a few nanometers, the incident evanescent field will be slightly perturbed and the field variations will be correlated with the object structure. This behavior has been clearly demonstrated, both theoretically and experimentally, by various microscopic and macroscopic models. Moreover, in that case, the field near the surface is sufficiently monotonic to be used for controlling, by feedback, in a very simple way the distance between tip and sample.

## 4. Experiments

### 4.1. FIELD INTENSITY VERSUS THE DISTANCE TIP-SAMPLE

The electromagnetic field near an object depends strongly on the physical and chemical properties of the outer layers of the objects. Since one of the advantage of the SNOM is its ability to work in ambient conditions, the surface contaminants will play a key role in the resulting field mapping. In the following experiment, we have analyzed with a very thin probe (diameter around 50 nm) the field behavior at a given point of the surface of an optically flat glass plate. This glass has been cleaned up in a bath of acetone and handled in a clean room. One of the resulting z-curve is shown in figure 3. We note an excellent agreement with the theoretical curve (assuming point detector) at large distances from the sample. The difference between the two branches of the cycle is simply due to the hysteresis of the piezo actuator. On the contrary, at distances smaller than 100 nanometers, the field does not obey the theoretical predictions. Part B of the curve can be interpreted as an effect between the tip and a water film lying on the glass plate. This film is first attracted by Van der Waals forces and then redistributed by capillarity around the tip as shown in figure 4(b). During this phase, the field varies slightly because the detecting zone is no more the tip but the base of the liquid cylinder. The last step, (part C, Fig.3), which is certainly the most interesting, is when the tip enters the liquid film. The slope of the curve is not so sharp as in part A, because of small differences of index between the piece of glass ($n=1.5$) and the water ($n=1.3$). In this case, the critical angle being 60 degrees, the field is probably not evanescent.

When the phase of approach is achieved, the tip goes away from the surface. Because of non mechanical effect during the phase D (tip in the water film), the curve follows the part C of the approach phase faithfully. We note a strong hysteresis during the phase E, which can be easily explained by taking into account the capillarity forces (see figure 4(E)). Finally when the tip goes out of the water the classical evanescent regime is reproduced with a small shift due to the hysteresis of the piezo actuator. Mechanical effects due to the bending of the tip are probably very weak because of the stiffness of the free part of the tip. Moreover, such experiments have been faithfully reproduced several times proving the non destruction or modification of the tip.

This simple example points out the problems of interpretation one can meet even for flat objects. In the case of rough objects the problem becomes evidently more complex. The film layer depending on the object roughness, a 'large' distance analysis will essentially inform on the water film structure rather than on the actual object topography. A way would consist in preparing the samples carefully, by dehydratation or heating for example. Unfortunately, such preparations would strongly limit the interest of near field microscopy. Probably, as in the case of ambient AFM, the best solution consists of accepting to work in water that is at a very short distance from the object, where the mechanical artefacts due to capillarity forces are negligible.

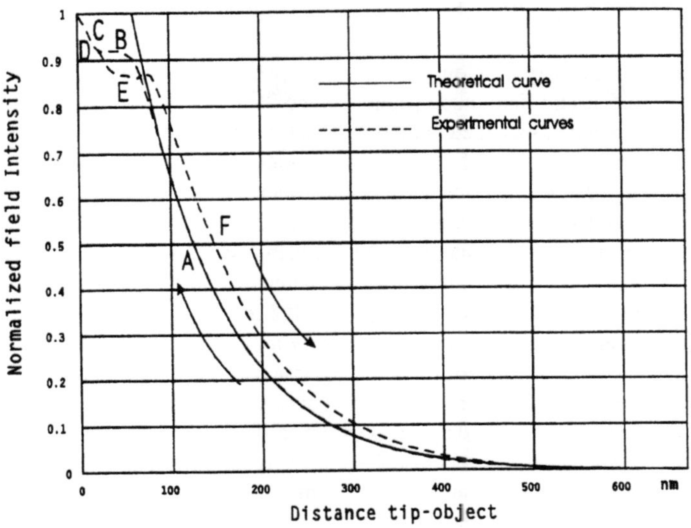


Figure 3. Detection intensity versus the distance between object and tip. The object is a piece of glass. The tip is the extremity of a monomode fiber chemically etched. The theoretical curve has been computed assuming a point detector.

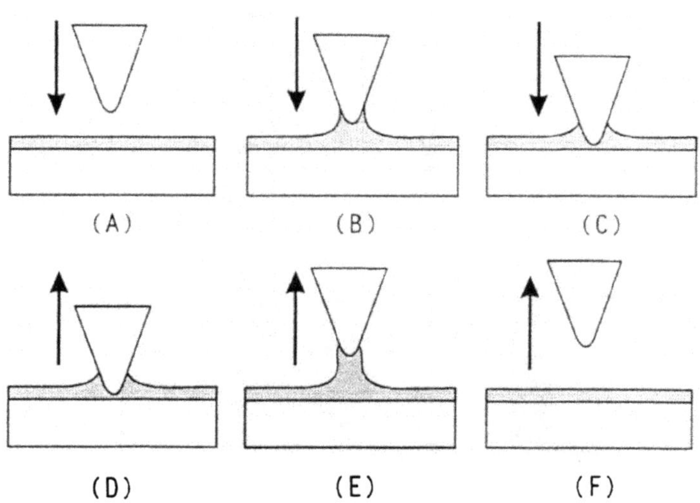


Figure 4. Possible scenario of the z-scanning of the tip. The object in dielectric material is covered with a thin film of water (or other liquid).

## 4.2. IMAGE OF LOW RELIEF GRATINGS

For completing the previous analysis, a low relief grating has been imaged both in constant altitude mode (figure 5) and in constant field mode (figures 6 and 7) with TM and TE polarization. For large distances (larger than 250 nanometers), the detected intensity is probably associated to the near field over the liquid. Whatever the polarization mode, it exhibits local strong intensity variations (fig. 6 a) which can be interpreted as complex interferences between object, liquid film and tip. When the tip is a few nanometers far from the surface (fig. 6 d and fig.7 a, b), the variations of the field are much weaker and the grating structure appears. At such distances we note some differences between TM and TE modes. In the latter case, the images seem to be better resolved (about 30 nanometers in the xy direction and 0.3 nanometer in the z direction) [15]. Several images have been recorded in constant intensity mode by decreasing the distance tip-sample, step by step (each increment corresponds to approaching the surface 6 nm by 6 nm). The following figure 7c, shows the difference between two successive images. We note that the edges of the line gratings are enhanced by this simple processing confirming the rapid variation of the near field in the vicinity of subwavelength structures. A simple way for getting an estimate of the image sharpness consists of making a spectral analysis of the image. Figure 8 shows the spectral density of Fig 7c. Since the energy in the high frequencies is very weak a log-scale representation has been chosen. The cutoff is defined as the signal on noise ratio $S/N \approx 1$. Concerning the distance between the tip and the sample, we note in the lower corner on the left (fig.7), a bright mark due to the contact with the sample. The average observation distance seems to be then a few nanometers. This rough estimate is confirmed by the theory, which predicts that a resolution x can be reached only for distances of the order of x/2p.

Figure 5. Image of a grating (period 417 nm, modulation 5nm) recorded in constant altitude and TM modes. Locally, the tip probably touches the surface (sharp and not reproducible variations of the field intensity from one line to the following one), (scanned area 3 x 3 μm).

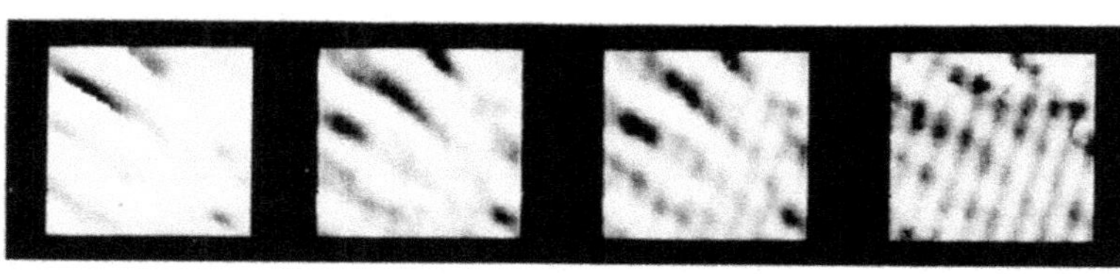

a b c d

Figure 6. Sequence of images of the same grating in constant intensity and TM modes. From a to d the tip comes near the surface. In a, the tip is far away from the surface (probably more than 250 nm), the observed high contrast variations could be due to a liquid film over the object, (scanned area 3 x 3 μm).

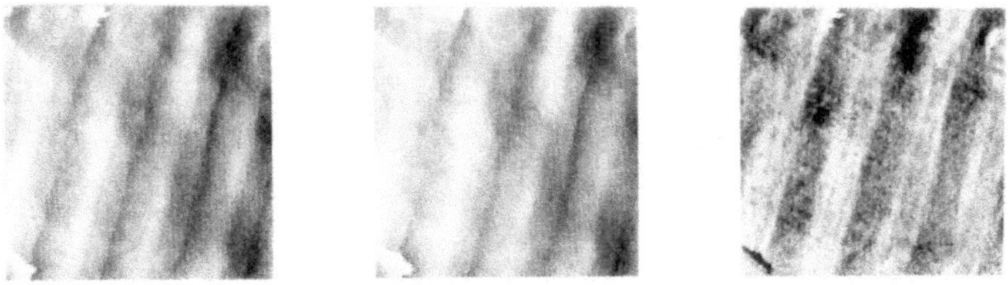

a b c

Figure 7. Images of the same grating, in TE mode. The tip is now a few nm far from the surface, the grating stucture is well reproduced. a and b have been obtained successively while the tip is coming near the surface (z-increment = 6nm). c is the subtraction of images a and b. This result experimentally confirms that the near field stuctures significantly vary over very short z-distances, (scanned area 1.5 x 1.5 mm).

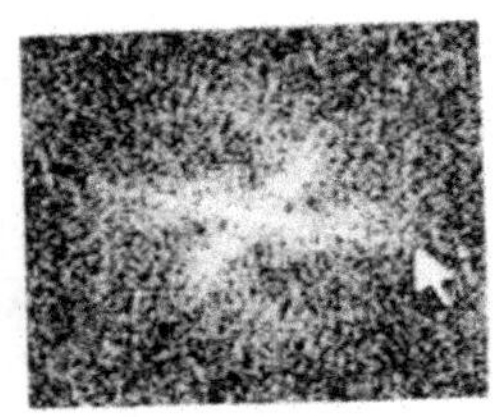

Figure 8. Fourier spectrum of image c fig.7 (log-scale intensity). The frequency cutoff is not far from the Shannon limit (see the arrow extremity). The corresponding resolution is estimated to 35 nm.

## 5. Role of the evanescent illumination mode

The ability of the STOM to image low relief objects with $\lambda/20$ spatial resolution raises a certain number of questions. First of all, the images are characterized by an unexpected good contrast in spite of the use of unmetallized tips. As an example, by using a 12 bit acquisition card (4096 levels), the average intensity in the image described above corresponds to about 3000 levels whereas the object variations range over 400 levels, i.e., about 13 % of the average intensity with a noise rms of 12 levels. By canceling the offset signal, the signal on noise ratio equals 33.

This good value can be partially explained by the use of a synchronous detection but the main reason is the use of an evanescent illumination mode. This technique, which is similar to a dark field illumination, reduces the amount of stray light entering the walls of the fiber. Let us recall that the evanescent beam is composed of two parts: a pure evanescent wave corresponding to the interaction of the light beam with the average surface of the object and a perturbed signal due to the object details. The first one is much more intense than the second one but the direction of its k vector is orthogonal to the direction of propagation inside the monomode fiber. It can be thus detected only by the very tip of the fiber that is multimode because of its exceedingly small size. The components entering the fiber by its walls are quickly absorbed by the cladding and then do not contribute to the detected signal. This property can be easily verified with a monomode fiber freshly cleaved. In this case, no signal is detected.

Finally, unlike classical and normal illumination, the use of an evanescent illumination introduces a drastic low frequency filtering allowing to image very small index or relief variations in dielectric materials. This explanation is only phenomenological. It must be verified and confirmed by a theoretical approach of the energy transfer from the incident beam to the transmitted one.

## 6. Conclusion

From the two examples described above, it appears that the STOM is certainly a powerful tool for analyzing low relief dielectric structures. However, the presence of pollutants or liquid films (water, oil, etc.) can play a key role in the interpretation of the images. Some supplementary studies and experiments have to be carried out for understanding more clearly the role of liquid over the object surface.

This work has been partly supported by the Ministère de la Recherche et de l'Espace, France.

References


1 Courjon, D., Sarayeddine K. and Spajer, M. (1989) 'Scanning tunneling optical microscopy', Opt.Commun., 71, 23-28.

2 Reddick, R.C., Warmack R.J. and Ferrell, T.L. (1989) 'New form of scanning optical microscopy' Phys. Rev. B, 39, 767-770.
de Fornel, F., Goudonnet, J.P., Salomon, L. and Lesniewska, E. (1989) 'An evanescent field optical microscope', ECO2, Opt. Storage Scanning Technol. 1139, 77-84.

3 Van Hults, N., de Boer, N.P. and Bölger, B. (1990) 'Scanning near field optical microscopy', Micro 90 London, 239-242.

4 Vigoureux, J.M., Girard, C. and Courjon, D. (1989) 'General principles of scanning tunneling optical microscopy', Optics Letters, 14, 1039-1041.

5 Courjon, D. (1990) 'Scanning tunneling optical microscopy' in R.J. Behm & al (eds.), Scanning Tunneling Microscopy and Related Methods, Kluwer Academic Publishers, 497-505.

6 Dereux, A., Vigneron, J.P., Lambin, P. and Lucas, A. (1991) 'Theory of near-field optics with applications to SNOM and optical binding', Physica B, 175, 65-67.

7 Nevière, M. and Vincent, P. (1992) 'Diffraction gratings as components for photon scanning tunneling microscope image interpretation', Workshop on Near Field Optics, Arc & Senans France, communication n°11.

8 Van Labeke, D. and Barchiesi, D. (1992) 'Scanning tunneling optical microscopy: a theoretical macroscopic approach', J. Opt. Soc. Am. , 9, 732-739.

9 Bourillot, E., de Fornel, F., Salomon L., Adam, P. and Goudonnet, J.P. (1992) 'Observation de structures guidantes en microscopie à effet tunnel photonique', J. Optics Paris, 23, 57-62.

10 Court, I.N. and von Willisen, F.K. (1964) 'Frustrated Total Internal Reflection and Application of Its Principle to Laser Cavity Design', Applied Optics, 3, 719-726.

11 Wolf, E., Nieto-Vesperinas, M. (1985) 'Analicity of the angular spectrum amplitude of scattered fields and some of its consequences', J. Opt. Soc. Am. A, 2, 886-890.

12 Bouju, X., Girard, C. and Courjon, D. (1992) 'Spectroscopie locale d'une surface par détection de champ proche : étude théorique comparative des métaux nobles', J. Phys. I France, 2, 1431-1444.

13 Courjon, D., Vigoureux, J.M., Spajer, M., Sarayeddine, K. and Leblanc, S. (1990) 'External and internal reflection near field microscopy: experiments and results', Applied Optics, 29, 3734-3740.

14 Girard, C. and Courjon, D. (1990) 'Model for scanning tunneling optical microscopy: a microscopic self-consistent approach', Phys.Rev. B, 42, 9340-9349.

15 Courjon, D., Bainier, C. and Spajer, M. (1992) 'Imaging of submicron index variations by scanning optical tunneling', J. Vac. Sci. Technol. B, 10(6), 1-3.

# QUANTUM OPTICAL PICTURE OF PHOTON STM AND PROPOSAL OF SINGLE ATOM MANIPULATION

Hirokazu Hori
*Faculty of Engineering, Yamanashi University,*
*4-3-11 Takeda, Kofu 400, Japan*

ABSTRACT. We study a quantum theoretical picture describing optical near-field and the photon scanning tunneling microscope (PSTM). We introduce an intuitive virtual photon model using Yukawa potential which is based on the fact that the fundamental process of PSTM is the quasi-static short-range interaction between microscopic objects. A similarity between particle tunneling and optical near-field problem is demonstrated, and the resolution of PSTM is analyzed by using the virtual photon picture. We investigate the atomic interaction with localized evanescent field and propose a novel technique of single atom manipulation using a nanometric optical fiber-probe of PSTM.

## 1. Introduction

The photon scanning tunneling microscope (PSTM) is considered as an optical analogue of STM using a localized object-probe interaction via evanescent photons [1,2]. The progress of sharp optical-fiber probes led to the achievement of nanometer resolution which was recently demonstrated on the PSTM image of bacteriophages T4 [2-4]. This confirms that the PSTM technique opened a new field of measurement and control of highly localized electromagnetic interaction between microscopic objects.

In this paper we focus our discussion to the theoretical analysis of localized evanescent field as a basis of the PSTM theory. The optical near field problems, in general, are concerned with mesoscopic phenomena in which electromagnetic fields are coupled strongly to sources, and a determination of the source distribution comes into question as well as solutions of the field equations. In fact, difficulties in solving Maxwell's equation as a boundary problem for a microscopic object-probe system of PSTM have been discussed in several articles [5-7], and electromagnetic theories based on self-consistent approach have been developed [6,7]. In general, models based on experimental results play an important role in the theory of mesoscopic phenomena, since complicated many-body interactions are involved. Instead of solving Maxwell's equation, we have proposed an intuitive model describing the localized evanescent field [4]. In this paper, we present a qualitative analysis of the optical near field problems and propose a semi-quantitative theory based on the intuitive model of the localized evanescent field. Discussions in this paper involve general descriptions of scanning near field optical microscopes (SNOM) including aperture type experiments. However, the term PSTM is used throughout this paper because this work is based on the recent results of PSTM experiments which are discussed by M. Ohtsu et. al. in the present proceedings volume and also on a similarity between PSTM and STM.

We begin our discussion with the assumption that the fundamental process of PSTM is described in terms of quasi-static short-range interaction between microscopic objects irradiated by propagating light. This assumption leads to a local theory describing the evanescent field, in which the complicated electromagnetic interaction of the whole system is reduced to a simplified model. This is in contrast to the boundary problem, in which one



*D.W. Pohl and D. Courjon (eds.), Near Field Optics*, 105–114.

has to solve, basically, the whole system. In order to construct the simplest model based on the assumption, we introduce virtual photons with a nanometric Compton wavelength described in terms of Yukawa potential [4]. The Compton wavelength, or the effective mass of virtual photons, is the unique parameter characterizing the interaction and determined by the size of the microscopic particle which is the source of the evanescent field. The validity of the model is tested by applying Yukawa potential to the small-aperture problem. Then we discuss the physical background of the model in terms of the coherent coupling of electromagnetic field with fundamental excitations of material surface. Based on the virtual photon model, the mechanism and resolution of PSTM is analyzed.

The virtual photons exhibit nature of quasi-particles, which helps understanding the PSTM process from the similarity with the electron scanning tunneling microscopes (STM), atomic force microscopes (AFM), and so on. Among them, the surface plasmon near field microscope (SPNM) [8] has been a clear example of the quasi-particle near field picture, in which the localized electromagnetic interaction is described in terms of surface plasmon. Contrasting with SPNM, in which real energy transfer to the plasmon mode takes place due to resonance, the photon-surface coupling in PSTM occurs instantaneously, and the energy is again transferred into reflected or transmitted photons. Therefore, the quasi-particle assumed is virtual in the sense that the coupling process is out of dispersion relations and no real resonance occurs.

The reversal process of the PSTM is also possible; we can control the localized electromagnetic interaction using an evanescent field produced in the narrow vicinity of the PSTM probe-tip by the laser light propagating inside of the fiber. It is especially interesting to study the atomic interaction with a localized evanescent field because of several remarkable features which can not be achieved in the usual atomic interaction with propagating photons. We present a possibility of trapping a single atom from a very cold source by resonance tunneling of virtual photons from the probe-tip to atom [4], which will be a manifestation of extraordinary large momentum of the localized evanescent field due to Heisenberg's uncertainty.

## 2. Virtual Photon Model of Optical Near Field

The nanometer-resolution PSTM experiment confirms that the object-probe interaction has a much smaller range of action than the optical wavelength. This implies that the interaction is considered to be mediated by a highly localized evanescent field, according to the Heisenberg's uncertainty principle: since the resolution of microscopes is determined by the shortest wavelength available in the measurement, an electromagnetic field with extraordinary large wavevector is involved in the PSTM measurement.

It is quite natural to describe the optical near field problem in terms of a quasi-static short-range interaction via evanescent photons. The term "quasi-static" is used in the sense that the propagation or retardation effects can be neglected because of the distance of action much smaller than the wavelength of the propagating light. As is well known, the quasi-static short-range interaction is simply described by using Yukawa potential,

$$\phi(r) \propto \frac{e^{-r/\lambda\!\!\!^{-}_C}}{r}, \tag{1}$$

where $\lambda\!\!\!^{-}_C$ is the Compton wavelength characterizing the interaction range. We propose to use the Yukawa potential to describe the localized evanescent field. This implies that the theory is based on the modified Helmholtz equation, $\{\nabla^2 - (1/\lambda\!\!\!^{-}_C)^2\}\,\phi = 0$ , which is the spatial reduction of the Klein-Gordon equation describing the field of particles with mass $m = \hbar/c \cdot \lambda\!\!\!^{-}_C$. This mass is considered here as the effective mass of virtual photons resulting from the coupling of electromagnetic field to the excitations on the material surface. The reason for using the modified Helmholtz equation may be simply explained from a consideration of dispersion relation of photons. In the half space outside of the material, let us consider the two dimensional dispersion relation, $-k_\perp^2 - k_\parallel^2 + (\omega/c)^2 = 0$, where $\omega$ is the temporal frequency, $k_\parallel$ and $k_\perp$ are the wavevectors parallel and perpendicular to the material surface, respectively. Because of locality assumed, the evanescent photons are on the section of the dispersion relation at $k_\parallel \sim 1/ƛ_C$, therefore the dispersion curve as a function of $k_\perp$ is parabolic, and photons on this section behave like massive particles. However the evanescent photons that we are considering on this section are virtual because the energy is below the vertex, i. e. $\omega/c \ll 1/ƛ_C$, as shown in Fig. 1.

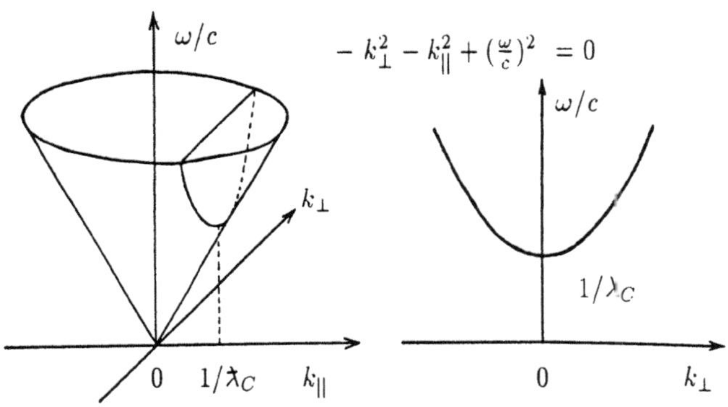


Fig. 1. The section of photon dispersion relation showing parabolic function in one dimension.

In order to demonstrate the validity of the virtual photon picture, the Yukawa potential model is applied for a calculation of evanescent field near a small circular aperture irradiated by propagating light with a wavelength, $\lambda$, much larger than the aperture radius. Assuming the Compton wavelength of virtual photons equal to the aperture radius, the intensity distribution in the near-field region is estimated in terms of a Hamiltonian obtained by integrating the Yukawa potential over the aperture, as shown in Fig. 2 (a),

$$\psi \propto \int_{aperture} \frac{e^{-r/ƛ_C}}{r} dS. \tag{2}$$

The Hamiltonian of the field described by Yukawa potential has a gradient term, $|\nabla\psi|^2$, and a mass term, $|\psi|^2/ƛ_C^2$. Figures 2 (b) and (c) show the numerical results for a $\lambda/50$ aperture radius. The results are in good agreement with the numerical results given by Leviatan [9] based on the boundary problem, when a Compton wavelength of virtual photons is assumed to be equal to the aperture radius. This provides a rule for the appropriate choice of the Compton wavelength. It should be noted that the gradient term showing spikes at the edge of the aperture and the mass term showing smooth decay at the aperture edge and taking its maximum at the aperture center, respectively, are very similar in their shape to the field distribution for TM and TE incident polarizations in the electromagnetic results. The reason is not clear at present, and only the total intensity is shown in Fig. 2. It is also a very interesting problem to consider polarization states of virtual photons since massive photons should exhibit a transformation property under spatial rotation, which is quite different from the property of free photons.

The origin of difficulty in handling optical near field problems is due to the fact that interaction, or source, of electromagnetic field is of fundamental importance. There are several ways to avoid explicit use of interaction in the electromagnetic theory, otherwise one has to solve complicated inhomogeneous differential equations. In the usual theory based on the macroscopic Maxwell's equation, interactions, except for magnetic interaction, are replaced by dielectric function in a material medium and by boundary conditions on a material surface. On the other hand, the idea of effective mass is popular in the field theory and simplifies complicated interactions of particles with material medium. In fact, the effective mass of virtual photons may be the counterpoint of magnetic current used for calculations of the evanescent field as a boundary problem [9] in the sense that the electromagnetic field is screened by magnetic current and photons acquire effective mass. This picture is analogous to the real screening current in the case of a conducting surface, especially in the case of a super-conductor, photons get real effective mass due to the direct proportionality of the screening current to the field described by vector potential.

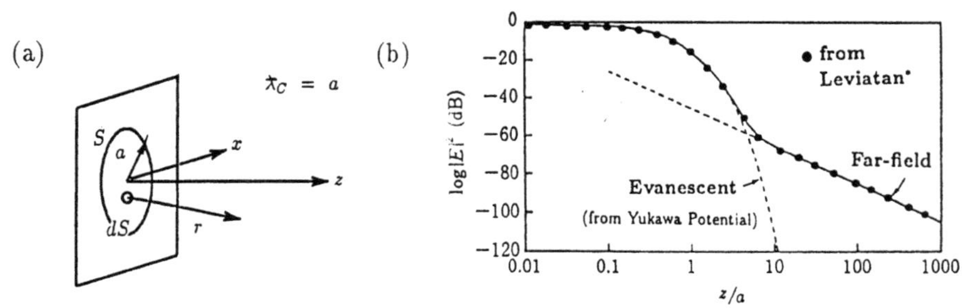


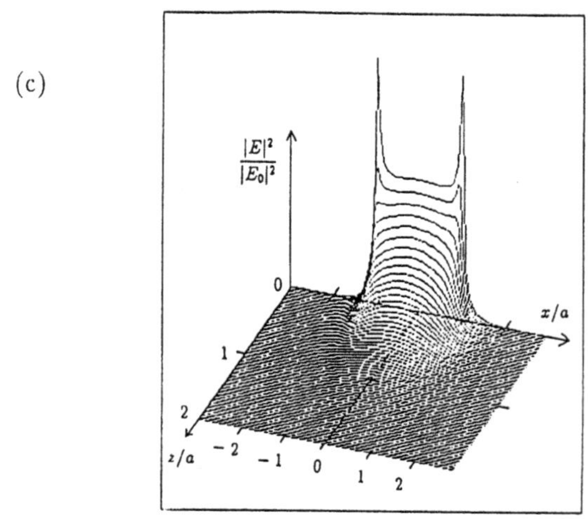


Fig. 2. The application of the Yukawa potential model for the intensity distribution of the near field of a small circular aperture of radius $a$ irradiated by propagating light with wavelength $\lambda = 50a$; (a) coordinate system, (b) intensity distribution on the symmetry axis, compared with the numerical results of Leviatan [9], and (c) three dimensional intensity distribution.

Compared with the surface plasmon near field microscope [8], the virtual photons employed may be described in terms of quasi-particles similar to polaritons or exciton-polaritons representing the electromagnetic field coupled to the mechanical vibrations of a dielectric medium via induced polarization. In contrast to usual fundamental excitations, no dispersion relation is assumed because the interaction occurs in a space-time volume much smaller than the periods of spatial and temporal oscillation. Basically it may be difficult to describe the electromagnetic field in a classical way in the microscopic problems where a highly confined space-time is considered. In this case, the use of the electromagnetic field defined at each point of the space time requires the non-locality of the dielectric constant or additional boundary conditions as a natural consequence. The use of the Feynman's picture of a point like photon propagating in any path might be the best way to treat mesoscopic problems, in which the evanescent field is described in terms of an electromagnetic propagator connecting matters placed in a space time volume smaller than $\lambda^4/c$. In this sense, no free photons are considered in the treatment of the evanescent electromagnetic field in this section. The microscopic substance of the evanescent field is considered to be a quasi-static or Coulomb-like interaction between electric polarizations induced on the material surface by incident and reflected lights. In this sense, the evanescent field correlates a pair of incident and reflected waves, $\phi_{in}(x_1)\phi_{ref}(x_1)$, at a position $x_1$, with that, $\phi_{in}(x_2)\phi_{ref}(x_2)$, at a closely spaced position $x_2$, on the material surface. This implies that the evanescent field represents the propagator similar to two particle Green's function, which integrates the many-body interaction on the material surface. This picture is consistent with the quantum theory of the evanescent electromagnetic field near a flat dielectric surface, developed by Carniglia and Mandel [10]. It should be noted that the Yukawa potential plays an important role in the Fourier expansion of Coulomb-like $1/r$ interaction propagator; Yukawa potential instead of Coulomb potential is Fourier transformed and afterward the Compton wavelength $\lambda\!\!\!^{-}_C$ is extended to infinity. In this procedure, if we consider a small interaction volume and dropout the long-range terms, the Coulomb-like potential remains as the Yukawa potential.

### 3. Theoretical Model of PSTM

Based on the considerations in the previous section, we assume that the fundamental process of PSTM is modeled as a localized interaction via evanescent photons between two closely spaced small apertures or dielectric spheres coupling to incoming and outgoing propagating photons, as schematically shown in Figs. 3 (a) and (b). This requires an extension of the optical near field propagator description to the case of closely spaced disconnected surfaces, i. e. the propagator connecting electric polarizations of two closely spaced dielectric objects. In this sense, the PSTM interaction has a close relation to the van der Waals force which is described by an exchange of two virtual photons connecting polarizations of two closely spaced dielectric objects. The range of the PSTM interaction is longer than the van der Waals force because the electric polarization is enhanced by incidence of propagating light. It should be noted that the solution of Maxwell equations with such disconnected boundary conditions might be very difficult to obtain.

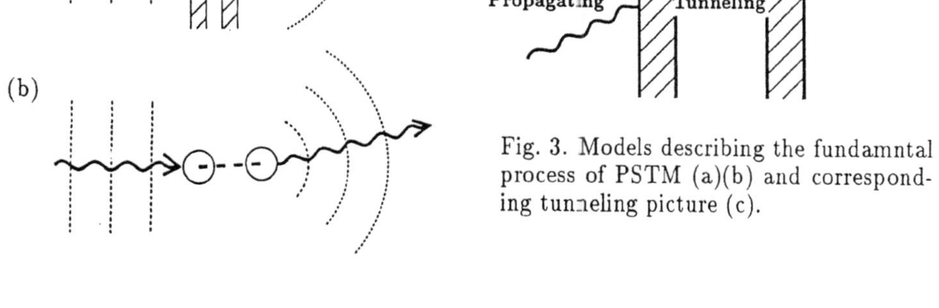


Fig. 3. Models describing the fundamntal process of PSTM (a)(b) and corresponding tunneling picture (c).

We focus our discussion to description of the interaction of microscopic object-probe system instead of calculation of the optical near field. The model together with the virtual photon picture makes it possible to interpret the PSTM process as a particle tunneling through an effective potential barrier. Similar to the resonant tnneling effect, multiple scattering processes between two microscopic objects are essential to produce a considerable increase of the tunneling probability. This implies that the PSTM process is, basically, a destructive measurement of the evanescent field. In the Yukawa potential model of the localized evanescent field, such many-body interaction is reduced to the effective mass or Compton wavelength of tunneling photons determined by the size of the interacting bodies. As a natural consequence of reduced many-body interaction, we assume an idea of resonance in the object-probe interaction, that is, the evanescent fields of two objects couple to each other if Compton wavelengths, $\lambda\!\!\!^{-}_C$, of the fields are equal. This is due to the momentum conservation in the exchange of evanescent photons between object and probe tip. Although evanescent fields having different Compton wavelengths may not be orthogonal, we can approximate the optical near field of the object-probe system as a sum of evanescent fields with several different Compton wavelength, which may correspond to Laplace expansion of optical near field. This is also equivalent to approximate the object-probe system by a set of small spheres or apertures having several different radii.

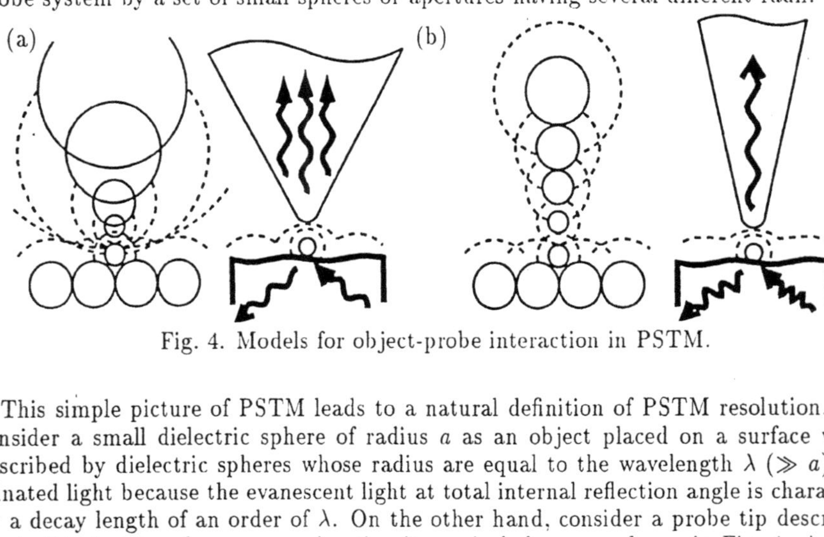


Fig. 4. Models for object-probe interaction in PSTM.

This simple picture of PSTM leads to a natural definition of PSTM resolution. Let us consider a small dielectric sphere of radius $a$ as an object placed on a surface which is described by dielectric spheres whose radius are equal to the wavelength $\lambda$ ($\gg a$) of illuminated light because the evanescent light at total internal reflection angle is characterized by a decay length of an order of $\lambda$. On the other hand, consider a probe tip described by a set of dielectric spheres approximating its conical shape, as shown in Fig. 4. According to the resonance assumption described above, the highest resolution of PSTM depends on the radius of the sphere at the extremity of the probe tip. The most important point to determine the PSTM resolution is considered to be the filtering character of evanescent background due to the surface on which the microscopic object is placed. Consider that the extremity of the probe tip is positiond at a distance equal to its curvature radius $a$ from the microscopic object. In this case, the tunneling probability of the evanescent field with Compton wavelength $\lambda\!\!\!^{-}_C = a$ is $\sim e^{-1}$, and the evanescent fields with shorter Compton wavelengths due to small bumps on the object are filtered out since their interaction ranges are much shorter than the probe-object distance. When the probe tip has an obtuse conical angle (Fig. 4(a)), the evanescent fields having Compton wavelengths larger than $a$ have a relatively large tunneling probability, and consequently the ratio of the pick-up powers from the object and background is less than unity, and the contrast of the PSTM picture will be rather low. In contrast, when the probe-tip has an acute conical angle (Fig. 4(b)), the evanescent fields with Compton wavelengths larger than $a$ are filtered out, since the corresponding spheres recede more rapidly than increase in their radius. As a consequence, the PSTM process can be viewed as a kind of spatial filtering of the optical near field, in which the resolution is dependent on the curvature radius and conical angle of the probe tip employed.

The validity of the resonance assumption is tested for the simplest case of evanescent field at total internal reflection with incident angle $\theta_{in}$, refractive index $n$, and optical wavelength $\lambda$. Assuming a Compton wavelength equal to the decay length of the evanescent field, $\lambda\!\!\!^{-}_C = \lambda/2\pi(n^2\sin^2\theta_{in} - 1)^{1/2}$, the evanescent field is considered to be coupled to the sphere of radius $\lambda\!\!\!^{-}_C$ which is a part of the probe tip. A simple calculation shows that, when the extremity of the probe-tip is held at a constant height from the surface, the pickup power from the background evanescent field characterized by the Compton wavelength $\lambda\!\!\!^{-}_C$ varies as $\exp[-2\alpha/\sin(\theta/2)]$ depending on the conical angle, $\theta$, of the probe tip. Here $\alpha$ is the form parameter depending on the shape of the probe tip, which is conical in the present case. Figure 5 shows a good agreement of this prediction with experiments presented by Ohtsu et. al. in the paper appearing in this volume. The experimental data showed a decay length of 350 nm which is in good agreement with the Compton wavelength $\lambda\!\!\!^{-}_C$ for $\theta_{in} = \pi/4$, $n = 1.5$, and $\lambda = 780$ nm. The solid line in Fig. 5 shows the theoretical prediction with a form parameter $\alpha = \pi$, which is given by the calculation of the overlap integral of the whole probe-object Yukawa field. Resulting from the spatial filtering character of the optical fiber probe, the pickup power $\sim 1$ pW due to interaction between the nanometric tail of Bacteriophage T4 and the nanometric sphere at the extremity of the probe tip becomes comparable to the evanescent background when the conical angle of the probe tip is decreased to $\sim 25°$.

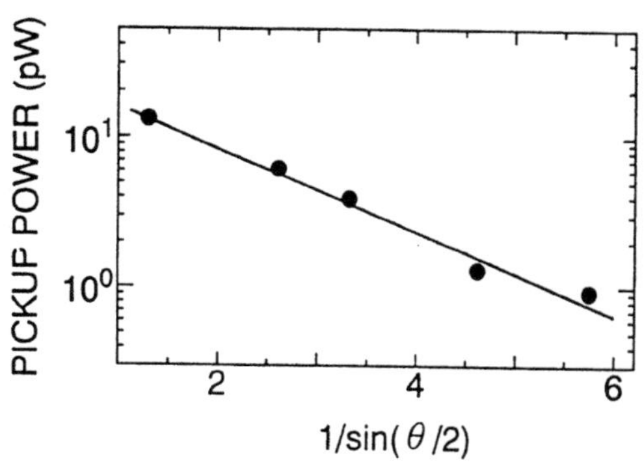


Fig. 5. Pickup power from the evanescent field at total internal reflection versus conical angle $\theta$ of the optical fiber probe held at a constant height from the surface. The experimental results (dots) taken from the paper of Ohtsu et. al. in the present volume show good agreement with the theoretical predictions giving $\exp[-2\alpha/\sin(\theta/2)]$ dependence (solid line) with $\alpha = \pi$.

The precise value of the tunneling photon current will be given by the overlap integral of the evanescent fields described by Yukawa potential, according to the resonance tunneling picture or the tunneling current description. Since the present model of PSTM is based on the modified Helmholtz equation, the calculation is almost similar to those in STM [11]. The magnitude of the overlap integral depends on the evanescent field momentum spectrum which is assumed to be Gaussian with a HWHM width $\hbar/a$ when the field is localized in $\sim a$ much smaller than the optical wavelength $\lambda$. The propagating photons within the cut-off, $\pm\hbar/\lambda$, are neglected since their contribution is on the order of $a^2/\lambda^2$ compared to the evanescent component. The tunneling probability or the transmissivity of the evanescent photon, has a maximum value when dielectric spheres of same radius interact, which is represented by virtual photons of Compton wavelength $\lambda\!\!\!^{-}_C = a$. This is the basis of the resonance assumption used in the intuitive consideration of the PSTM resolution.

The same discussion above can be used for an explanation of the mechanism which a conical tip of optical fiber probe transfers evanescent field into propagating field efficiently. An interaction between dielectric spheres in the conical probe tip is due to evanescent field (Fig. 4(b)), and the propagating field spread over wide space around the probe-tip gives small contribution. The transmission of field energy from one small sphere to its neighbour is described also by an overlap integral of uncertainty Gaussians. Because neighbouring spheres are similar in diameters and placed close, the overlap integral is close to unity, and the considerable amount of the field energy is transferred between them. The field energy transferred produces an evenescent field which, in turn, has a momentum spectrum described by an uncertainty Gaussian containing larger portion of the propagating component as the sphere diameter increases. In this way, a smoothly tapered probe-tip transforms the evanescent field into propagating field efficiently. This is considered as an impedance matching conditon for evanescent fields, which is a surface wave of electric polarization, in a similar way for such as an acoustic wave propagating in a horn.

As is shown in the example above, it is straight forward to calculate the tunneling photon current or pickup power since the model describes the interaction between microscopic objects characterized by a Compton wavelength determined by their size. This procedure allows to calculate the pickup power for an object and probe tip of any shape in the same manner, using expansion in spheres or apertures. This is an essintial feature of the local theory we requested. A direct quantitative comparison of the Yukawa potential model with the electromagnetic calculations based on the self consistent approach given by Girard et. al. [6] and Vigoureux et. al. [7] is underway.

## 4. Single Atom Manipulation using a PSTM Probe

The resonant interaction of an atom with a localized evanescent electromagnetic field will exhibit several remarkable features which cannot be obtained where using ordinary propagating light. An atom scattering an evanescent photon into a propagating photon via resonance absorption and spontaneous emission gets an extremely large recoil momentum parallel to the probe-tip surface. The large momentum of the evanescent electromagnetic field is attributed to the mechanical distortion associated with the induced polarization in the dielectric material [12]. The atom feels also a strong dipole force or gradient force [13] normal to the tip surface due to rapid decrease of the field intensity in the near-field region. These features altogether result in a mechanical interaction between the atom and the probe-tip, which is analogous to the inter-atomic van der Waals force. A novel technique to manipulate single atom can be realized by using the localized evanescent electromagnetic field produced in the vicinity of the nanometric probe-tip irradiated from inside of the fiber, as the reversal process of PSTM.

Let us assume that the fiber-probe tip is irradiated by laser light detuned to the red of atomic resonance, as shown in Fig. 6. An atom coming into the evanescent field resonantly absorbs the counterpropagating laser photon due to positive Doppler shift against the red detuning and spontaneously emits a propagating photon. By this process the atom is recoiled strongly in the direction parallel to the probe surface, since the absorbed momentum of the evanescent photon is much larger than the emitted momentum of the propagating photon due to locality of the evanescent field. This acts as a cooling force until the momentum of the atom becomes comparable to the recoil momentum. Because of the spatial modulation of the field momentum in its direction, the cooling force accompanies the centrifugal component, which will be compensated by the strong dipole attractive force in the near-field region where the field intensity decreases exponentially. The balance of the centrifugal component of scattering force and the dipole attractive force produces a potential valley along the probe-tip surface. The atom will move back and forth in this potential valley like a pinball trapping in the optical standing-wave.

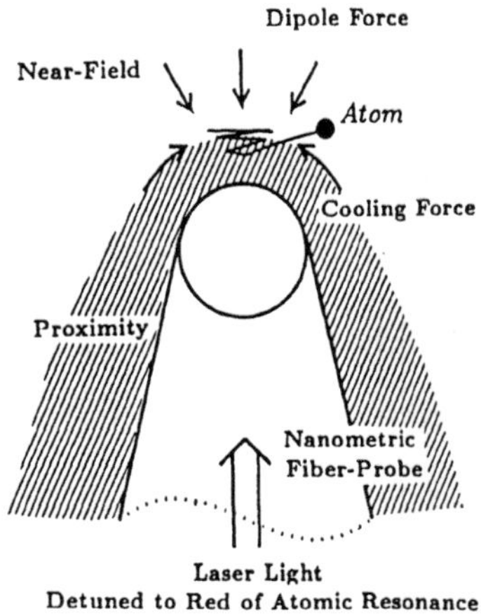


Fig. 6. The resonant interaction of an atom with the localized evanescent field produced at the extremity of the nanometric optical fiber-probe irradiated by laser light propagating inside of the fiber.

The trapping condition will be achieved when the cooled atom motion is reversed at each recoil and such scattering occurs several times per passage near the probe tip. This condition determines the probe-tip radius of curvature $a$ as

$$a \geq \sqrt{\frac{h\tau_{sp}}{2m_{atom}} N_{sp}}, \tag{3}$$

where $\tau_{sp}$ is the spontaneous lifetime of the atomic resonance, $N_{sp}$ the mean number of spontaneous emission events, $m_{atom}$ the atomic mass. The momentum of the trapped atom is assumed to be the half of the recoil momentum. As an example, in the case of the rubidium $D_2$-line, the trapping condition is satisfied by a fiber probe with a 15 nm tip radius, and the temperature of trapped atom will be $\sim$80 $\mu$K, if $N_{sp} = 4$ is assumed. The laser detuning is assumed to be 10 MHz to the red of the atomic resonance, being larger than the natural linewidth. Such atoms may be extracted from $\sim$300 $\mu$K sources realized by laser cooling techniques, such as optical molasses [14]. In order to discuss the feasibility of single atom manipulation in detail, the quantum optical analysis of the atomic interaction with localized evanescent field is required, because the process includes spontaneous emission as a basic ingredient.

As an experimental evidence of the large momentum transfer in the atomic interaction with evanescent field, we have observed the Doppler broadening of the Cs-$D_2$ line at 852.1 nm using evanescent field at total internal reflection on a glass surface with refractive index $n = 1.47$. A preliminary result shows a broadening of the Doppler width proportional to $n\sin\theta_{\rm in}$, where $\theta_{in}$ is the angle of laser incidence larger than the critical angle. This is in good agreement with theoretical prediction.

## 5. Conclusion

We have discussed the principle of the PSTM from the view point of the quasi-static short-range electromagnetic interaction, introducing the Yukawa potential model. This model is useful not only for the understanding of the PSTM process in terms of a generalized framework of the tunneling microscope but also for practical purposes, such as an estimation of the PSTM resolution. This is because the model dscribes the spatial frequency of microscopic samples in terms of the Compton wavelength of the localized field. The virtual photon model implies that the tunneling process is, in general, associated with a mechanical action between microscopic sample and probe tip due to the exchange of large photon momentum resulting from its locality.

Investigating the atomic interaction with localized evanescent field, we have proposed a novel technique of single atom manipulation. The study of the electromagnetic interaction localized in a nanometric region involves lots of interesting mesoscopic problems. concerned with such as photon localization or localization of fundamental excitations, cavity QED, tunneling-time, and so on. For example, the interaction of a single atom with a confined electromagnetic field in a high-Q cavity is the main concern of the cavity QED problems, in which the strongly restricted electromagnetic modes are due to a coherent interaction of the field with atoms forming the cavity wall. In this respect, the interaction of a single atom with confined evanescent electromagnetic field has a similar nature to the cavity QED problems.

ACKNOWLEDGEMENT. This work was done in collaboration with Dr. M. Ohtsu and his colleagues at Tokyo Institute of Technology. The author would like to thank Mr. M. Kozuma and Dr. S. Jiang for assistance and discussions, Dr. T. Yabuzaki and Dr. M. Kitano for many useful discussions.

## REFERENCES

[1] R. C. Reddick et al., Phys. Rev. B **39** (1989) 767, D. Courjon et al., Opt. Commun. **71** (1989) 23, and also papers by N. F. van Hulst, and by F. de Fornel in the present proceedings volume.
[2] S. Jiang et al., Jpn. J. Appl. Phys. **31** (1992) 2282.
[3] T. Pangaribuan et al., Jpn. J. Appl. Phys. **31** (1992) L1302.
[4] H. Hori et al., Tech. Digest of 18th Int. Quantum Electronics Conf. (1992, Vienna) 48.
[5] Theories of optical near field based on Maxwell's equation are given in the papers by J. P. Vigneron et. al., A. Dereux and D. W. Pohl, M. Nieto-Vesperinans, D. Van Lebke and D. Barchiesi, A. Roberts, in the present proceedings volume.
[6] C. Girard and D. Courjon, Phys. Rev. B **40** (1990) 9340.
[7] J. M. Vigoureux et al., Appl. Opt. **31** (1992) 3036.
[8] U. Ch. Fischer and D. W. Pohl, Phys. Rev. Lett. **62** (1989) 458, M. Specht et al., Phys. Rev. Lett. **68** (1992) 476.
[9] Y. Leviatan, J. Appl. Phys. **60** (1986) 1577.
[10] C. K. Carniglia and L. Mandel, Phys. Rev. D **3** (1971) 280.
[11] C. J. Chen, Phys. Rev. B **42** (1990) 8841.
[12] J. P. Gordon, Phys. Rev. A **8** (1973) 14.
[13] J. P. Gordon and A. Ashkin, Phys. Rev. A **21** (1980) 1606.
[14] E. L. Raab et al., Phys. Rev. Lett. **59** (1987) 2631.

# Low Temperature Photon Scanning Tunneling Microscope

C. L. JAHNCKE and M. A. PAESLER
*Department of Physics*
*North Carolina State University*
*Raleigh, North Carolina 27695*
*United States of America*

ABSTRACT. One of the advantages afforded by using near field optical techniques is the ability to obtain topographic and spectroscopic information simultaneously. For many samples -- in particular for semiconductors -- a detectable spectroscopic signal can only be obtained for specimens refrigerated to cryogenic temperatures. In order to spectroscopically study such materials a cryogenic photon scanning tunneling microscope (PSTM) has been developed. The design for this cryogenic compatible microscope is presented. Low temperature images and low temperature photoluminescence spectra are shown.

## 1 Introduction

Near field optical (NFO) probes have been used to demonstrate the capability of transducing images of surfaces with higher spatial resolution than those used in conventional far field optics. The techniques used to gather these images are varied and include transmission near field scanning optical microscopy (NSOM) [1], illumination NSOM [1], reflection NSOM [1], and photon scanning tunneling microscopy (PSTM) [2]. In addition there are now probes that combine force microscopy with NFO in order to gather images[3]. One reason for using these NFO techniques rather than higher resolution techniques such as scanning tunneling microscopy, atomic force microscopy or electron microscopy is that a wealth of optical information in addition to surface topography can be obtained [4]. For example, PSTM has been used to study stress by gathering photoluminescence data [5], NSOM has been used to write and image magneto-optic devices [6], and a force/reflection NSOM has been used to collect Raman data on nonlinear optical materials [7]. In order to further expand the spectroscopic applications of this technique, a low temperature PSTM has been developed.

In order to study the electronic structure of semiconductor materials and how this structure correlates to sample growth and topography, photoluminescence as a function of surface features can be observed. The photoluminescence effect in semiconductors is enhanced at low temperatures because the nonradiative recombination channels are no longer available for the electron-hole pairs. Therefore, a low temperature PSTM becomes a necessary tool for this type of study.

## 2. Design

The design of the microscope is based on Digital Instruments' Nanoscope D-head. The microscope fits in a Janis Super Varitemp which has a 1.5 inch diameter sample chamber and can be cooled below 2K. A Lake Shore Cryotronics DRC 80C temperature controller is used with a silicon diode temperature sensor. Some of the considerations made in



*D.W. Pohl and D. Courjon (eds.), Near Field Optics*, 115–120.

designing a low temperature PSTM are cryogenic compatibility, tip to sample approach, vibration isolation, and light access.

The microscope is made primarily of invar because of its low coefficient of thermal expansion. A long range (75 micron) piezoelectric tube scanner was chosen because of the reduced piezoelectric response at low temperatures. The tube scanner is attached to the invar using a macor screw and a macor nut which is glued [8] into the piezoelectric tube. The invar sample stage is attached to the invar mounting block with three 1/4-80 invar screws as legs, stainless steal ball bearings, and stainless steal compression springs. The sample stage is kinematically mounted to the three screws. The entire design is cylindrically symmetric with the optical fiber mounted up through the center of the tube scanner in order to minimize the effect of lateral contraction and expansion of the materials due to the changes in the temperature. [see figure 2.1]

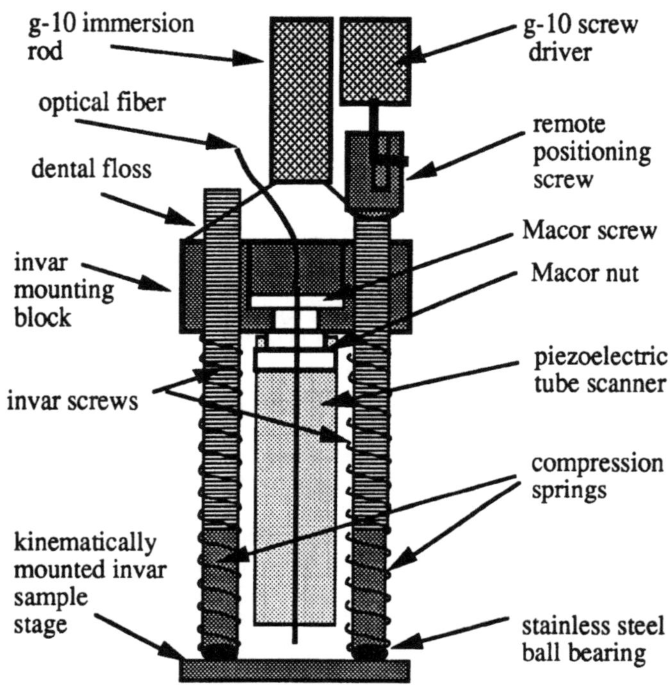


Figure 2.1. Key features of the low temperature photon scanning tunneling microscope.

The tip to sample approach can be accomplished external to the cryostat by adjusting the three 1/4-80 screws. Once inside the cryostat, a screwdriver engaged in one of the screws is remotely driven by a stepper motor that can produce tip to sample steps as small as 130 nm. In addition, further positioning is accomplished using the tube scanner which has a 5 micron range at room temperature. The tip of the screw driver is made of 1/16 inch stainless steel rod in the shape of a T. It fits in a 3/16 inch wide slot that is 3/4 inch deep so that it can be disengaged from the microscope by backing off several steps. This helps reduce vibrations.

In order to further minimize the effects of vibrations, the entire cryostat is placed on an optical table. The microscope itself is mounted to an immersion rod made of g-10 (chosen for its low thermal conductivity) with dental floss [9] in order to reduce the coupling of vibrations into the microscope.

The Janis Super Varitemp has sapphire windows that allow optical access to the microscope. A laser is directed through the windows and focused onto a sample where it is totally internally reflected providing the evanescent field necessary for PSTM. The evanescent field is then transduced with an optical fiber that is threaded through high pressure tubing and out the top of the cryostat to the detector.

## 3. Results

The microscope is designed to operate at temperatures below 2K, but all of the images shown were taken at liquid nitrogen temperatures. The light source used is the 514.5 line of an argon ion laser. The illumination is in a PSTM configuration. That is, the source field is totally internally reflected light. The evanescent field is used to aid in positioning the tip close to the sample, but it is not used as a feedback mechanism. All of the images shown are taken in a constant height mode. The probe is a tapered optical fiber made using a Sutter Instruments $Co_2$ laser puller and then coated with aluminum.

In figure 3.1 an image of a BK7 dove prism at 89.3 K is shown. In figure 3.2 an image of a Cr3+ ion implanted sapphire sample at 81.2 K is shown. The scale in each figure indicates that the scan range is 5000 nm, but the piezoelectric tube scanner has not been calibrated for cryogenic temperatures and the range is certainly less than that indicated due to the reduced piezoelectric effect at low temperatures. The sapphire sample was imaged in liquid nitrogen with helium blowing over the liquid in order to reduce bubbling. The BK7 sample was imaged dry with cold nitrogen blowing on it. Random streaking in the y-scan direction can be seen in the image taken in the liquid nitrogen. Similar streaking

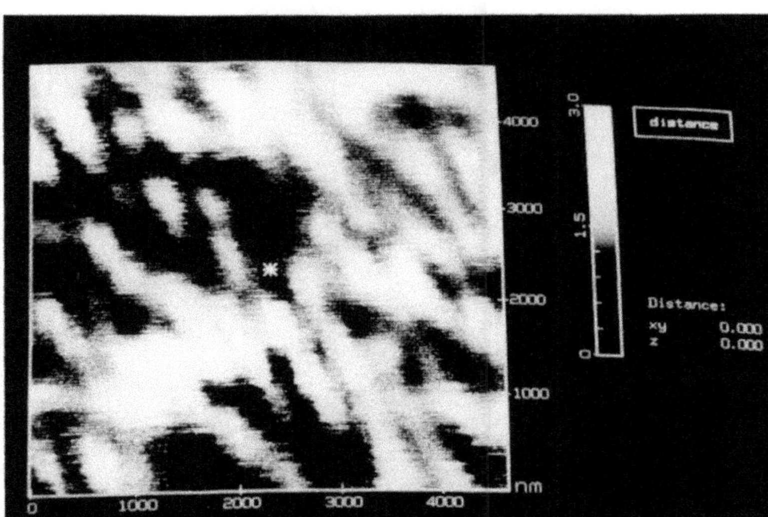


Figure 3.1. Cryogenic image of a BK7 dove prism at 89.3 K.

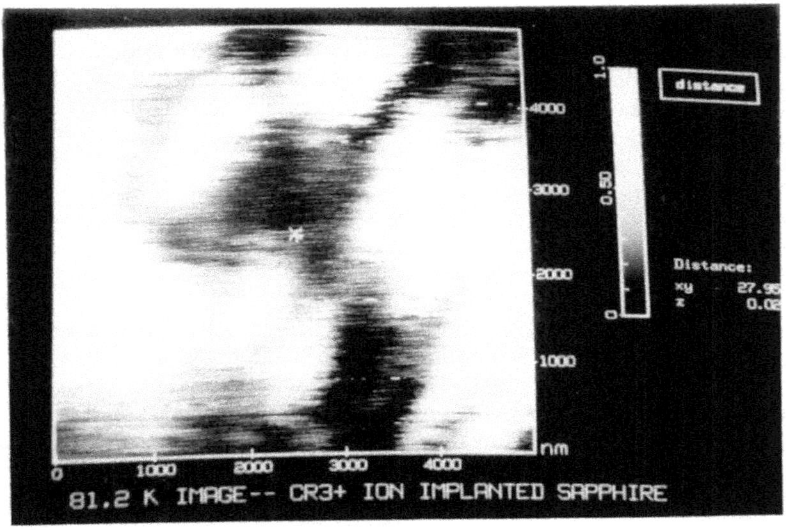


Figure 3.2 Cryogenic image of a Cr3+ ion implanted sapphire imaged in liquid nitrogen.

can be seen in room temperature images where the tip scrapes the sample surface. However, the streaking seen in the images taken in liquid nitrogen does occur in places besides the high points on the sample It is possible that the bubbling nitrogen perturbs either the tip or the sample in such a way that they come in contact with each other, or it could be some effect related to the viscosity of the liquid nitrogen.

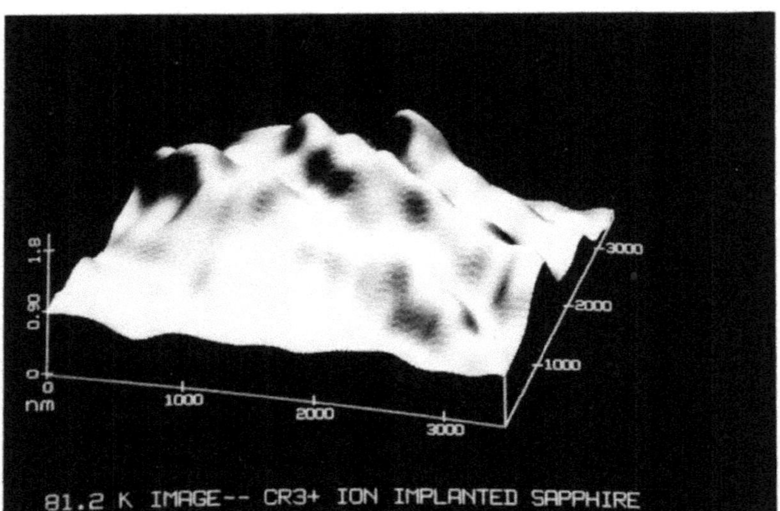


Figure 3.3. Cryogenic image of Cr3+ ion implanted sapphire.

In figure 3.3 an image of the surface of the Cr3+ ion implanted sample taken at 81.2 K is shown. Fluorescence spectra were taken on this sample at room temperature and at 81.2K, see figure 3.4. The purpose of this data is to demonstrate that optical spectra can be obtained with this instrument. The ruby fluorescence R lines shift towards the red and become sharper and stronger at low temperatures. A shift of 10 angstroms is observed between the low temperature and room temperature peaks and the full width half maximum of the low temperature peaks is 1.6 angstroms. The spectrometer used, SPEX 1403, has a theoretical resolution of 1.54 angstroms at 6920 angstrom wavelength and 400 micron slit widths. The spectra shown are consistent with previously reported low temperature ruby fluorescence given the resolution of the spectrometer [10].

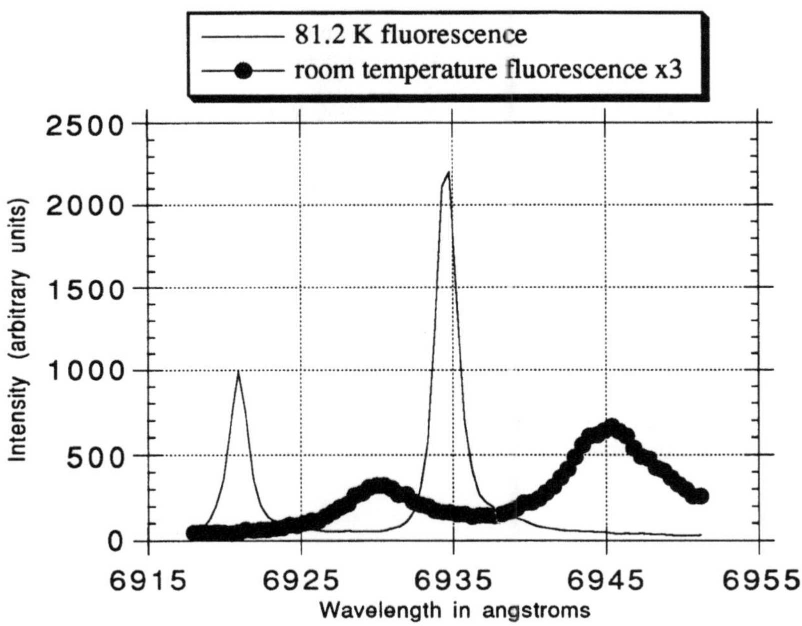


Figure 3.4. Fluorescence of Cr3+ ion implanted sapphire at room temperature and at 81.2 K

## 4. Conclusions

We have demonstrated a successful design of a cryogenic PSTM which allows for both imaging and spectroscopic sample analysis. Remaining problems center on mechanical decoupling of the microscope from the environment, which is apparently not complete due to the current design of the coarse z-control. The effects of this coupling may be minimized, however through decoupling elsewhere in the mechanical system or by elimination of the source of vibration. If boiling in the liquid nitrogen jacket is such a source, minimization of it (through the use of Teflon beads or gaseous helium in the $LN_2$ reservoir) may alleviate this current source of image noise.

Other improvements include modifications to the sample stage in order to allow the microscope to function in a transmission NSOM mode. This would simplify image interpretation as well as allow for further flexibility in the types of samples studied.

In this paper we have provided no discussion of the important issue concerning spatial resolution of the spectroscopic signal. This is a topic that is currently being studied.

## 5 References

1 Betzig E., Isaacson M., Barshatzky H., Lewis A., and Lin K. (1988) 'Near-field scanning optical microscopy (NSOM)', SPIE, 897,91-99.
2 Reddick R. C., Warmack R. J., and Ferrell T. L. (1989) 'New form of scanning optical microscopy', Physical Review B 39, 767-770.
3 Betzig E., Finn P. L., and Weiner J. S. (1992) 'Combined shear force and near-field scanning optical microscopy', Appl. Phys. Lett. 60, 2484-2466.
4 Trautman J. K., Betzig E., Weiner J. S., DiGiovanni D.J., Harris T. D., Hellman F., and Gyorgy E. M. (1992), 'Image contrast in near-field optics', J. Appl. Phys. 71, 4659-4663.
5 Moyer P.J., Jahncke C. L., Paesler M. A., Reddick R. C., and Warmack R. J. (1990) 'Spectroscopy in the evanescent field with an analytical photon scanning tunneling microscope', Phys. Lett. A 145, 343-347.
6 Betzig E., Trautman J. K., Wolfe R., Gyorgy E. M., and Finn P. L. (1992) 'Near-field magneto-optics and high density data storage' Appl. Phys. Lett. 61, 142-144.
7 Moyer P. J., and Paesler M. A., Workshop on Near Field Optics 92 proceedings.
8 Eccobond 285 by Emerson and Cuming Inc.
9 Adler J. G., Chen T. T., Gallagher M. C., Konkin M. K., and Mullin D. P. (1991) 'Comparison of various low temperature STM designs for use in spectroscopy and topography', J. Vac. Sci. and Technol 2, 992-995.
10 Schawlow, A. L. (1961) 'Fine structure and properites of chromium fluorescence in aluminum and magnesium oxide', J. R. Singer (ed), Advances in Quantum Electronics, Columbia University Press, New York, pp. 50-64.

# LIGHT FORCES ON DIELECTRIC PARTICLES AND ATOMS

OTHMAR MARTI AND VICTOR I. BALYKIN
*Fakultät für Physik, Universität Konstanz, D-W-7750 Konstanz, Germany*

ABSTRACT: The mechanical interaction of light with matter is subject to many ongoing research projects. This article gives an overview over basic principles of the forces light exerts on dielectric particles and on atoms. An easy to understand force is the light pressure that can be deduced from Maxwell's equations. Dielectric particles are also subject to the gradient force that accelerates them towards higher field intensities. A typical application of these forces is the single beam particle trap. We then discuss the interaction of light with atoms. The discussion is limited to the case of two-level atoms. Light forces arise because of a resonant interaction of the atom with the light field. Forces are therefore only substantial if the frequency of the light is in resonance or quasi-resonance with the atom. Two distinct forces are present, the scattering force and again a gradient force. The direction of the gradient force in this case depends on the sign of the detuning: red detuned light with respect to the atomic transition pulls atoms to the high intensity region whereas blue detuned light pushes the atom away from the high intensity region. Examples for the use of the scattering and gradient forces are the cooling of atoms to very low temperatures, to guidance of them in a light beam and mirrors for atoms.

## 1. Introduction

The idea that light might exert forces on matter dates back to Kepler, who proposed that the light of the sun would deflect the tails of the comets. The concept of forces due to electromagnetic radiation surfaced next in Maxwell's equations, which contain the ingredients to calculate the forces of light on continuous bodies. The theoretical predictions were then confirmed by Lebedev[1], Nichols, and Hull[2]. The force exerted by light on single Na-atoms was demonstrated by Frisch in 1933[3]. After a long delay Letokhov[4], Hänsch and Schawlow[5], and Wineland and Dehmelt[6] proposed the localization and cooling of single atoms by light. Ashkin (1971)[7] demonstrated the levitation of microscopic dielectric spheres. Bjorkholm *et al.*[8] achieved a focusing of an atomic beam, whereas Andreev *et al.* (1981)[9] and Phillips and Metcalf (1982)[10] were able to cool Na-atoms.
The purpose of this article is to outline and review some basic concepts of mechanical light interaction with matter. Besides the interaction with micrometer sized particles we also treat briefly the resonant interaction of light with atoms.

## 2. Light Forces on $\mu$m-Sized Particles

The interaction of light with small particles has been studied for a long time. Mie and Rayleigh calculated the scattering amplitudes as a function of the scattering angle. To understand light



*D.W. Pohl and D. Courjon (eds.), Near Field Optics,* 121–130.

forces on matter we first discuss the light pressure on a large body using plane waves. Then we investigate the interaction with light by smaller, spherical particles .

## 2.1. LIGHT PRESSURE

Maxwell's equations bear the effect of light pressure within. An electromagnetic wave has an associated energy density. The energy density has the same dimension as pressure and can be interpreted in this way.

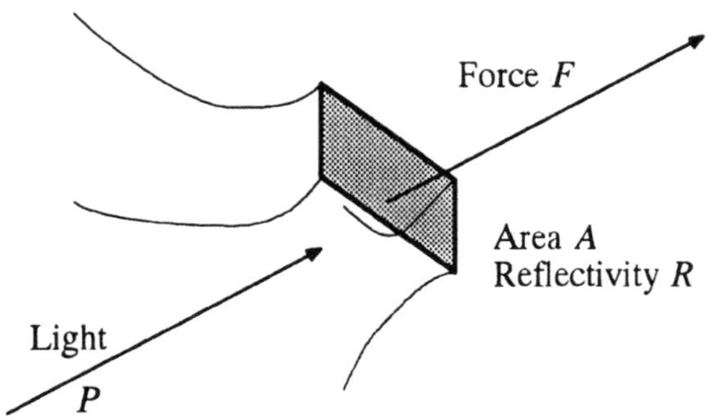


Fig. 1. Light pressure of a plane parallel wave incident on an area $A$.

In Fig. 1 we consider an area $A$, upon which a plane wave is incident. This plane wave is characterized by an optical power $P$. In vacuum we can write the energy density as

$$w = \varepsilon_0 E^2 \tag{1}$$

where $E$ is the magnitude of the electrical field in the electromagnetic wave and $\varepsilon_0$ is the dielectric constant of the vacuum. If we assume a reflectivity $R$ of the surface $A$, we obtain for the force

$$F = wA(1+R) \tag{2}$$

If we note that $wA = P_0/c$, where c is the speed of light in the medium in front of the surface, we obtain our final result

$$F = \frac{P_0}{c}(1+R) \tag{3}$$

The factor (1+$R$) says, that the force exerted on the area by the light doubles if we change from a perfectly absorbing surface to a specularly reflecting one. The same result can also be derived using quantum mechanics. Using $\hbar k = \hbar\omega/c$ for the photon momentum and $\hbar\omega$ for the energy of a single photon we can write for the force

$$F = (1+R)\frac{\partial}{\partial t}(\hbar k N) = (1+R)\hbar\frac{\omega}{c}\frac{\partial}{\partial t}N = (1+R)\hbar\frac{\omega}{c}n \tag{4}$$

$R$ is again the reflectivity of the surface, $N$ the total number of photons and $n$ the number of photons per unit time incident on the surface. A photon that is reflected transfers twice the momentum to the surface than a photon that is absorbed. $R$ is just the fraction of photons that are reflected. The number of photons n incident on the surface $A$ per unit time is calculated by

$$P_0 = \frac{\partial E}{\partial t} = \frac{\partial}{\partial t}(N\hbar\omega) = \hbar\omega n \qquad (5)$$

Combining equations (4) and (5) we arrive at exactly the same result as derived for the classical case. Assume an incident power $P_0$ of 1 mW, an interaction area $A$ of 100 $\mu m^2$, and a reflectivity $R$=0. The force on the area $A$ is then 3.3 pN, a magnitude that is detectable by force microscopes. Indeed it has been demonstrated that laser powers well below 1 mW are sufficient to stimulate the oscillation of cantilevers[11]. Using the signal to noise ratio given in paper [11] one can conclude that even powers of 1 nW would be detectable by measuring the oscillation of mechanical sensors.

2.2. FORCE OF REFRACTED LIGHT: GRADIENT FORCE

Light transfers momentum to bodies not only on reflection, but also when it is refracted. Figure 2 shows the momentum balance for a light beam incident from below. The incident light beam with an angle $\alpha$ to the interface normal is refracted and leaves with an angle $\beta$.

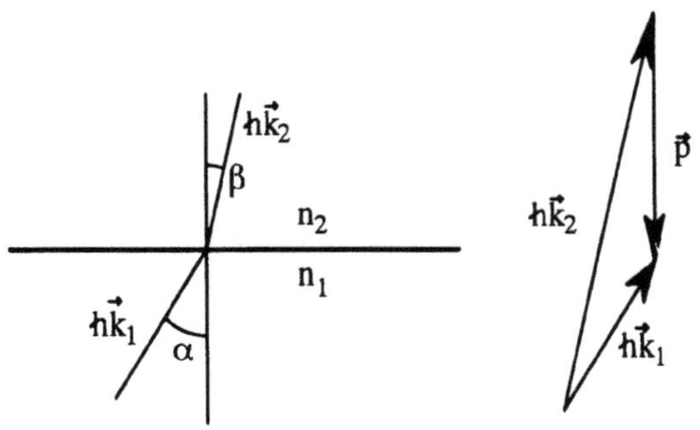


Figure 2. Momentum balance for refracted light. The light is incident from the lower left.

$\alpha$ and $\beta$ are related by Snell's law. Both of these laws are derived from the fact, that there is no momentum transfer from the light parallel to the interface. Hence a net momentum transfer perpendicular to the interface exists at the entry and the exit of the light beam. For plane parallel interfaces there is on a time average no momentum transfer through the refraction of light. However there is a time difference between the momentum transfer on entry and on exit. This means, that the time averaged displacement of a small plane parallel slab can be non-zero, if the intensity of the illumination is low enough so that only one or a few photons are within the slab at a time.

The forces arising from refraction of light at small spheres are always directed towards the higher intensity. Ashkin [12] therefore called this force the gradient force. We first consider a sphere large compared to the wavelength $\lambda$ of the light (Mie regime). In a sharply focused laser beam, as was considered by Ashkin [13], the gradient force tries to move particles towards the focus. At the same time light reflected (scattered) at the surface of the sphere pushes the sphere away from the high intensity region. Under certain conditions these two forces form a stable trap for particles.

The above analysis can be extended to wavelength-sized particles, in the so-called Rayleigh scattering regime. Again there exist two forces, the scattering force that is analogous to the light pressure and the gradient force. Ashkin[14] finds for the gradient force of light on small particles approximated as spheres

$$\vec{F}_{grad} \propto nr^3 \vec{\nabla} \left| \vec{E} \right|^2 \propto nr^3 \vec{\nabla} I_0 \tag{6}$$

where $n$ is the relative index of refraction of the particle (with respect to the environment), $r$ is the radius of the particle and $\left|\vec{E}\right|^2 \propto I_0$ is the intensity of the light. The scattering force for small particles can be written as

$$F_{scat} \propto \frac{I_0}{c} \frac{r^6}{\lambda^4} n \tag{7}$$

The force is again proportional to the intensity $I_0$ of the light, but also to the sixth power of the radius of the particles. This sixth power dependence of the scattering force on the radius is a sign, that the scattering force unlike the gradient force is a near-field effect in the Rayleigh regime [15].

### 2.3. THE SINGLE BEAM PARTICLE TRAP

The combined effect of the scattering force and the gradient force can be used in single beam optical traps for particles[14]. The general setup of such a trap is outlined in figure 3. A laser beam is sharply focused. Particles entering the volume of the focus are subject to three forces: gravitation pulling particles down, the scattering force pushing them up and the gradient force attracting them to the higher intensity regions.

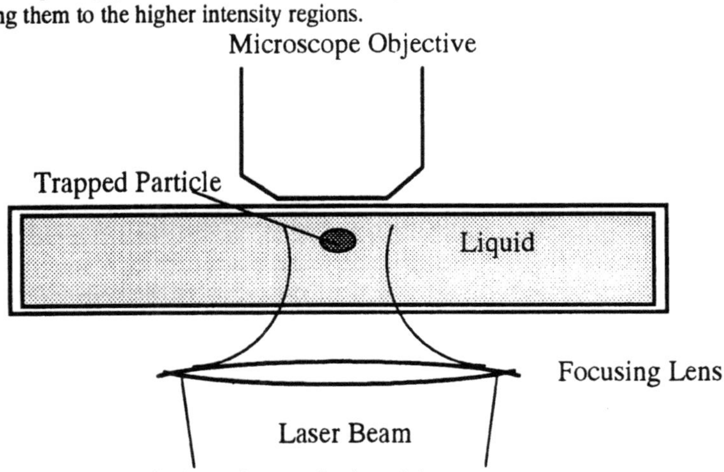


Fig. 3. Single beam optical trap for $\mu$m-sized particles.

The combined action of these forces for properly chosen laser intensity and laser wave-length traps the particle slightly away from the laser focus (seen from the laser). The trap created by the laser beam has a potential well of the depth

$$U = n\alpha \frac{|E|^2}{2} \propto n\alpha \frac{I_0}{2} \tag{8}$$

where n is the relative index of refraction, α the polarizability of the sphere and $|E|^2 \propto I_0$ the intensity of the laser beam at the focus. This trap is longitudinally stable. Axial stability is achieved when [14]

$$\frac{F_{grad}}{F_{scat}} > 1 \tag{9}$$

Typical applications for this kind of traps are the manipulation of small spheres and of biological matter [16].

## 2.4. PARTICLES IN THE EVANESCENT FIELD

Of particular interest for the near-field optics is the behavior of atoms and particles in the evanescent wave. A recent example is the movement of small latex or glass spheres (diameters ranging from 1 $\mu$m to 27 $\mu$m) in the evanescent wave at the surface[17,18] with speeds up to 8 $\mu$m/s.

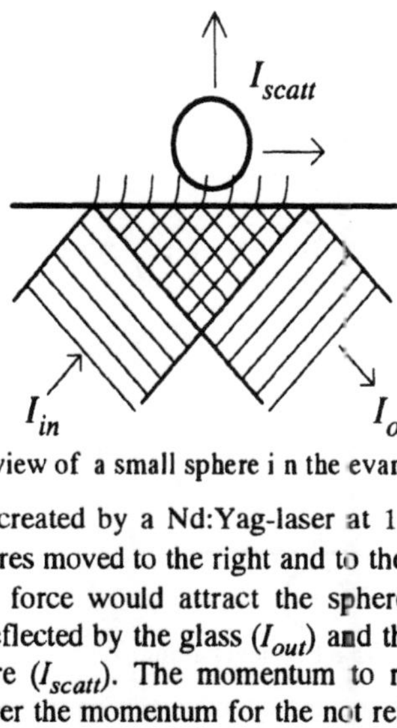


Fig. 4. Schematic view of a small sphere in the evanescent wave.

The evanescent wave was created by a Nd:Yag-laser at 1.06 $\mu$m with a power of 150 mW focused to 100 $\mu m^2$. The spheres moved to the right and to the top. The relevant force here is not the gradient force, since this force would attract the spheres to the surface. Of the incident intensity $I_{in}$ a fraction <1 is reflected by the glass ($I_{out}$) and the rest is scattered in a not yet well understood way by the sphere ($I_{scatt}$). The momentum to reflect a wave is furnished by the dielectric-air interface. However the momentum for the not reflected portion of the incident wave must be transferred to the particle, since the scattered wave has a random momentum distribution that averages to zero.

One must conclude from this experiment, that the scattering force dominates the interaction in the evanescent wave with $\mu$m-sized structures. Since the scattering force points in the direction of the incident laser beam a force microscope cantilever subject to the scattering force will measure a repulsive force.

## 3. Light Forces on Atoms

Unlike the light forces on small dielectric particles light forces on atoms are resonant. The exact behavior of the force is a complicated function of the frequency of the laser light, the frequency and width of the atomic transition, the spatial and temporal properties of the laser beam, and last but not least the speed of the atoms. We simplify in this paper the situation in two ways. First we will only consider two-level atoms. These are atoms prepared such that for the tested interaction they only consist of the ground level and one excited level. Examples of such atoms are for instance metastable helium or metastable argon. The second simplification will be that we only consider plane waves, which lead to the scattering force, and a focused Gaussian beam, which exerts in addition the gradient force on the atoms.

### 3.1. THE SCATTERING FORCE

The scattering force is the resonant analogy for atoms to the scattering force for small dielectric particles. The scattering force depends on the detuning $\Omega = \omega_L - \omega_0$ (frequency difference between the atomic transition and the frequency of the laser beam) of the laser, the width $\gamma$ of the atomic transition and the saturation parameter $G$

$$G = \frac{I}{I_S} = \frac{\pi hc}{3} \frac{\gamma}{\lambda^3} \tag{10}$$

Figure 5 depicts the atomic transition and the three possible processes, which are the

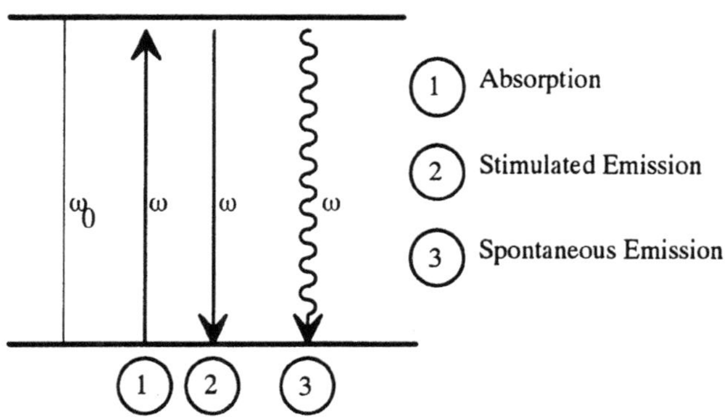


Fig. 5. Light interaction with a two-level atom.

absorption, the stimulated emission and the spontaneous emission. Absorbed photons with the energy $\hbar\omega$ transfer their momentum $\hbar\vec{k}$ to the atom. The stimulated emission process emits a photon with the same momentum $\hbar\vec{k}$ along the incident laser beam. Therefore a momentum $-\hbar\vec{k}$ is transferred to the atom. Hence on the average there is no momentum transfer to the atom in an absorption-stimulated emission process (However during each cycle there is a short time during which the momentum of the atom is not zero. On the average this time is the life time of the excited state of the atom. The displacement induced this way is minute and might only be observable as a beam splitting in a well-collimated monochromatic light beam.). The main part of the momentum transfer is due to the absorption-spontaneous emission process. The spontaneous emission process is random in the direction and hence does not give a net momentum transfer to the atom. Therefore we can write

$$\vec{p} = \vec{p}_0 + \hbar\vec{k}(N_+ - N_-) + \sum_s \hbar\vec{k}_s \tag{11}$$

where $N_+$ and $N_-$ are the number of stimulated absorption and emission processes, respectively. The last sum, over the spontaneously emitted photons, is zero, since the spontaneous emission is isotropic in space. The resulting force is then given by

$$\vec{F} = \hbar\vec{k}\frac{\langle N_i\rangle}{\Delta t} = \hbar\vec{k}\gamma\frac{G}{1+G+\left(\left(\Omega-\vec{k}\vec{v}\right)/\gamma\right)^2} \tag{12}$$

where the average number of stimulated transitions $<N_i>$ during the time $\Delta t$ has been calculated with the rate equations for the atomic transition. Also the Doppler shift for an atom moving with the velocity $v$ has been taken into account. Figure 6 shows plot of the scattering force as a function of the atom velocity.

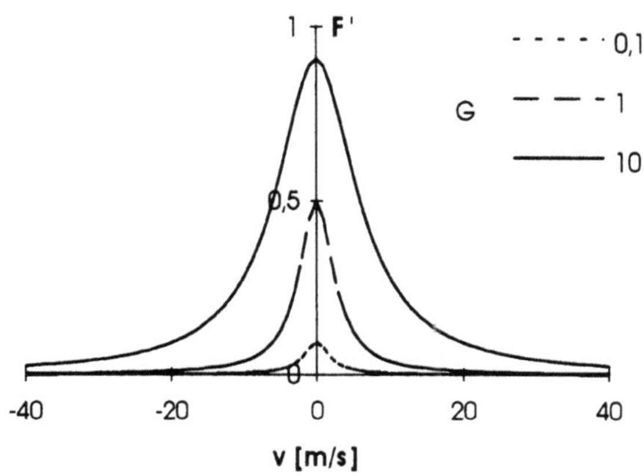


Fig. 6: Scattering force F as a function of the atomic velocity v for different saturation parameters.

It is obvious from equation (12) and from figure 6, that the scattering force saturates for large intensities, i.e., large saturation parameters $G$. The force has its maximum for $v$=0 with no detuning. Ashkin and Gordon [17] showed that, in analogy with the electrostatic Earnshaw theorem, a trap relying only on the scattering force would not be stable: this is the optical Earnshaw theorem. Later it was proposed to circumvent this theorem in the same way RF-traps overcome the electrostatic Earnshaw theorem [19]. However the scattering force has been extensively used to cool atoms to very low translational temperatures.

3.2. THE GRADIENT FORCE

If the light wave has a non-uniform intensity distribution, gradient forces along or opposite to the intensity gradient occur. Light interacting with an atom induces through absorption and emission processes a dipole in the atom. This dipole in turn interacts with the intensity gradient of the light. For a Gaussian beam the gradient force [20] is given by

$$F_{gr} = \frac{\hbar\left(\rho/\rho_0^2\right)(\Omega - kv_z)}{1 + G + \left((\Omega - kv_z)/\gamma\right)^2} \tag{13}$$

where $\rho_0$ is the radius of the laser beam, $\rho$ the distance of the atom from the laser beam axis, $2\gamma$ the width of the atomic transition, $\Omega$ is the detuning of the laser frequency from the atomic transition, and $v_z$ the velocity of the atom **along** the laser beam. Figure 7 depicts the shape of the gradient fore for different saturation parameters $G$, i.e., for different laser intensities.

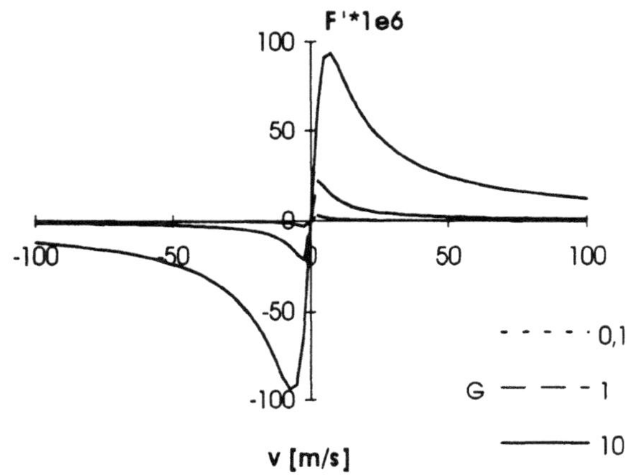


Fig. 7. The gradient force as a function of the detuning

It is obvious from equation (13) and from figure 7 that the gradient force on the atoms does not saturate. Moreover the gradient force has a dispersive shape that permits, depending on the detuning $\Omega$, to concentrate atoms to the center of the laser beam or to expel atoms. This feature can be used to guide atoms, to focus, or to reflect them[21].

3.3. APPLICATIONS OF THE GRADIENT FORCE: LENSES AND MIRRORS FOR ATOMS

The gradient force discussed exists for atoms interacting with with any light field of arbitrary intensity distribution. If an atomic beam propagates at the center of a laser beam and the laser beam is red detuned regarding the atomic transition, the beam will be guided by the light[8]. If we focus red detuned light to a diffraction limited spot, the atomic beam inside the laser beam will also be focused. Theoretical considerations expect a spot size for the atom beam of a few tenths of a nanometer[22].

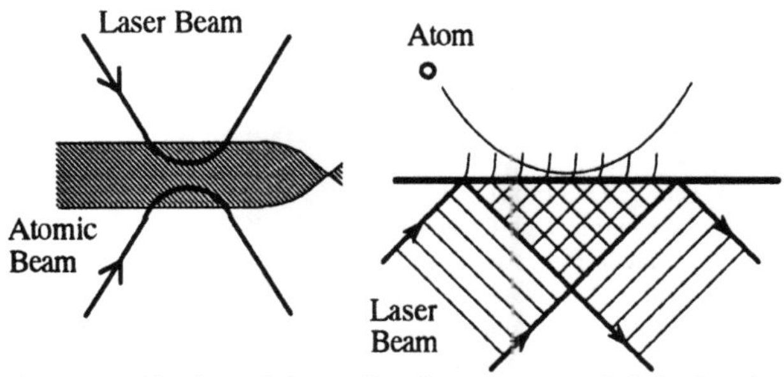


Fig. 8. Two applications of the gradient force on atoms: left the focusing of atoms, right the reflection of atoms by evanescent waves.

A second application of intensity gradients is the use of evanescent waves and plasmons to reflect an atomic beam. An atom approaching such a mirror under grazing incidence is deflected by the blue detuned light in the evanescent wave[22].

## 4. Summary

We have shown in this short review article the basic concepts of the mechanical interaction of light with atoms and small structures. It is found that small dielectric structures in most cases interact in a non-resonant way with the light. This interaction leads, for plane waves, to the scattering force (or light pressure) which is directed away from the direction of incidence. In light fields with non-uniform intensities there is an additional force towards the high intensity. The light forces have been used to build single beam traps for small particles and to move $\mu$m-sized spheres.

Atoms on the other hand interact in a resonant process with the light field. The two types of forces found for dielectric particles are also present. Again the scattering force always is directed away from the light source. Its magnitude depends on the detuning of the laser beam regarding the atomic transition in the moving frame of reference of the atom. The gradient force, however, exhibits for atoms the unique property that its direction can be chosen towards or away from the high intensity, depending on the detuning.

The applications of the gradient force for focusing and reflecting atomic beams are briefly discussed. The reflection geometry is very reminiscent of the near-field optical arrangement and might be relevant in the discussion of the manipulation of single atoms by near-field optical methods.

The theory of atom and particle motion in laser fields and the corresponding experimental techniques are quite well developed today. The interested reader may find reports on the main achievements in the field in three special issues[24-26].

ACKNOWLEDGMENTS

The authors would like to thank J. Mlynek, A. Ruf, H. Bielefeldt, A. Schnetz, M. Hipp, J. Colchero and their other colleagues at the University of Konstanz for many useful discussions. This work was supported by the DFG and the Land Baden-Württemberg.

## References

[1] P. Lebedev, *Ann. Phys. (Leipzig)* 6, 433 (1901).
[2] E.F. Nicholls, and G.F. Hull, *Phys. Rev.* 13, 207 (1901).
[3] R. Frisch, *Z. Phys.* 86, 42 (1933).
[4] V.S. Letokhov, *JETP Lett.* 7, 272 (1968).
[5] T.W. Hänsch and A.L. Schawlow, *Opt. Commun.* 13, 68 (1975).
[6] D.J. Wineland and H. Dehmelt, Bull. Am. Phys. Soc. 20, 637 (1975).
[7] A. Ashkin and J.M. Dziedzik, *Appl. Phys. Lett.* 19, 283 (1971).
[8] J.E. Bjorkholm, R.R. Freeman, A. Ashkin, and D.B. Pearson, *Phys. Rev. Lett* 41, 361 (1978).
[9] S. Andreev, V. Balykin, V. Letokhov, and V. Minogin, *JETP Lett.* 34, 442 (1981).
[10] W.D. Phillips and H.J. Metcalf, *Phys. Rev. Lett.* 48, 596 (1982).
[11] O. Marti, A. Ruf, M. Hipp, H. Bielefeldt, J. Colchero, and J. Mlynek, *Ultramicroscopy* 42-44, 345 (1992).
[12] A. Ashkin, Phys. Rev. Lett. 24, 156 (1970).
[13] A. Ashkin, *Science* 210, 1081 (1980).
[14] A. Ashkin, J.M. Dziedzic, J.E. Bjorkholm, and S. Chu, *Opt. Lett.* 11, 288 (1986).
[15] H.A. Bethe, *Phys. Rev.* 66, 163 (1944).
[16] A. Ashkin, J.M. Dziedzic, and T. Yamane, *Nature* 330, 769 (1987).
[17] A. Ashkin and J.P. Gordon, *Opt. Lett.* 8, 511 (1983).
[18] S. Kawata and T. Sugiura, *Opt. Lett.* 17, 772 (1992).
[19] A. Ashkin, *Opt. Lett.* 9, 454 (1984).
[20] V.G. Minogin and V.S. Letokhov, "Laser Light Pressure on Atoms", Gordon and Breach Science Publishers, 1987.
[21] V.I. Balykin and V.S. Letokhov, *Physics Today* 4, 23 (1989).
[22] V.I. Balykin and V.S. Letokhov, *Opt. Comm.* 64, 151 (1987).
[23] V.I. Balykin, V.S. Letokhov, Yu.B. Ovchinnikov, and A.S. Sidorov, *Phys. Rev. Lett.* 23, 2137 (1988).
[24] P. Meystre and S. Stenholm, eds., (1985) *J. Opt. Soc. Am.* Special Issue B2, N. 11, "Mechanical Effects of Light".
[25] S. Chu and C. Wieman, eds., (1989) *J. Opt. Soc. Am.* Special Issue B6, N. 11, "Laser Cooling and Trapping of Atoms".
[26] J. Mlynek , V.I. Balykin, and P. Meystre, eds., (1992) *J. Opt. Soc. Am.* Special Issue B54, N. 5, "Optics and Interferometry with Atoms".

# NANOMETER RESOLUTION PHOTON STM AND SINGLE ATOM MANIPULATION

M. OHTSU, S.JIANG, T. PANGARIBUAN, and M. KOZUMA
Interdisciplinary Graduate School
of Science and Engineering,
Tokyo Institute of Technology,
4259 Nagatsuta, Midori-ku, Yokohama 227, Japan

ABSTRACT. A photon STM was constructed by using a sharpened optical fiber with a radius of curvature of about 1 nm. Nanometer-resolution was achieved to measure biological and other kinds of samples with sub-micron structures. We also proposed a new method to trap a single neutral atom in the evanescent field localized near the fiber probe top.

## 1. INTRODUCTION

This paper reviews our recent activities on photon scanning tunneling microscopy ( PSTM ), and is composed of three parts. The first part describes the reliable fabrication technique of nanometric optical fiber probe. Several results of nanometer-resolution imaging are presented in the second part. The last part is to present a possibility of trapping a single neutral atom in the evanescent field emitted from the fiber probe.

## 2. FABRICATION TECHNIQUE OF FIBER PROBES

Based on the principle of the PSTM, the radius of curvature a and the cone angle $\theta$ of the probe, which is used to disturb and detect the evanescent light localized around small particles, must be as small as possible to obtain a high resolution and to avoid picking up low



*D.W. Pohl and D. Courjon (eds.), Near Field Optics*, 131–139.

spatial Fourier frequency components of the evanescent light generated from the substrate. We fabricated an optical fiber probe by using selective chemical etching in a buffered HF solution, whose volume composition is represented as $NH_4F$ : HF : $H_2O$ = X : 1 : 1. As shown in Figs.1(a) and (b), the minimum of $\theta$ realized so far is smaller than $15^o$, and a was estimated to be about 1 nm.

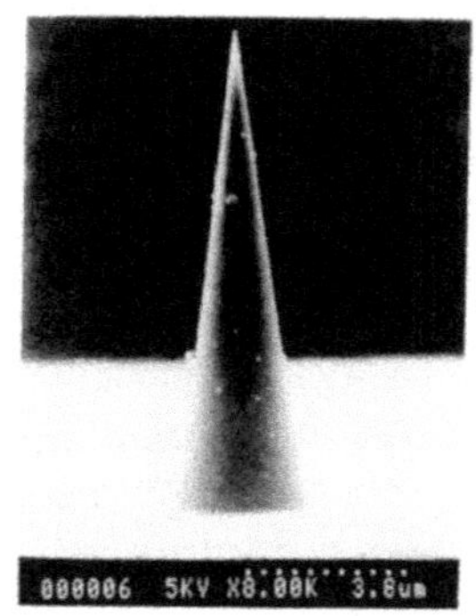


(a) (b)

Fig.1 SEM images of the sharpened fiber.(a)and(b) are the images with different magnifications.

The $\theta$ can be varied by controlling the compositions of the optical fiber core, cladding, and the buffered HF solution. It was found from the experimental results that $\theta$ was a monotonously decreasing function of the X for X < 6 and tends to a stationary value for X > 6. The stationary values of $\theta$ were $100^o$ and $20^o$ for the fibers composed of $SiO_2$ cladding and Ge-doped $SiO_2$ core with doping ratios of 3.6 mol% and 23 mol%, respectively [1]. For these fibers, the corresponding normalized refractive index differences between core and cladding were 0.36% and 2.3%, respectively. The probe with the minimum $\theta$ shown by Fig. 1 was realized with a special fiber composed of a Ge-doped core and a F-doped cladding, whose normalized refractive index difference was 2.9%. For all fibers, the standard deviation of the cone angle $\Delta\theta$ was kept within $0.5^o$, which means that the etching process is of high reproducibility. The high reproducibility of our fiber probe fabrication is also demonstrated by Fig.2, in which cone angle and height differences between two cores of a dual-core fiber are smaller than $0.2^{o}$ and 20 nm respectively.

Fig.2 The SEM picture of the sharpened dual-core fiber. The core diameter and separation between two cores are $6\mu m$ and $40\mu m$, respectively.

3. EXPERIMENTS OF NANOMETER-RESOLUTION IMAGING

A power-controlled 0.8 $\mu m$ wavelength AlGaAs laser was used as a light source in order to construct a reliable and compact system and to take advantage of its wide-band power modulation capability for the fast phase sensitive detection. Direct power modulation was employed for the phase sensitive detection in measuring the evanescent light power which was picked up by the fiber probe and detected by a photo-multiplier. The sample was fixed on an optically flat glass prism to which the laser beam was incident under total reflection condition. A PZT-scanner used for conventional STM or AFM was imported to the present system.

Figure 3(a) shows the exponential decreases of the picked up evanescent light power measured by increasing the probe-prism separation d. 0 nm-point of d was determind by the picked-up power variation as the probe is moved close to the prism. As the probe contacts with the prism, the picked-up power which increases with the increasing of the voltage applied to the PZT will not grow up further. We define such a point as the 0 nm-point for determination of the absolute d. Although the picked up power was on the order of pico-watts, it was high enough for phase sensitive detection. The relationship between $\theta$ and the picked up power is shown in Fig.3(b) measured by keeping d constant. From this figure, it can be seen that the picked up power is proportional to $\exp\{-1/\sin(\theta/2)\}$. This proportionality is in agreement with the theoretically estimated dependence based on the virtual photon model of the localized evanescent field.[2]

We also measured three dimensional profiles by imaging experiments in a constant height mode. The images were: (1) A 9 nm thick $SiO_2$ thin film[3], (2) a moth-eye optical disk with 300 nm diameter and 80 nm depth pits[4], (3) 80 nm diameter latex particles which were fixed on a glass substrate( see Fig.4 )[3], and (4) uncoated bacteriophages T4 ( see Figs.5(a) and (b) )[4].

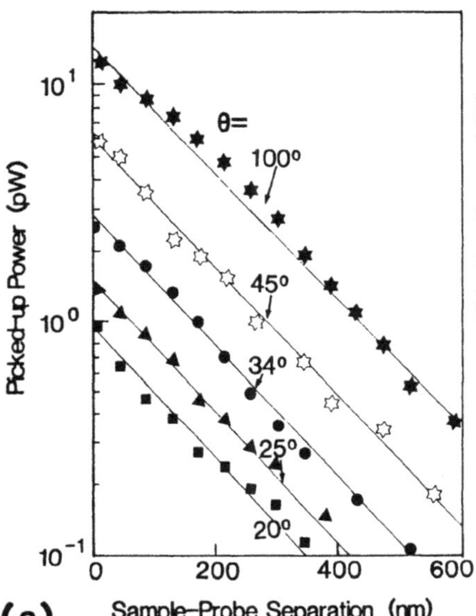


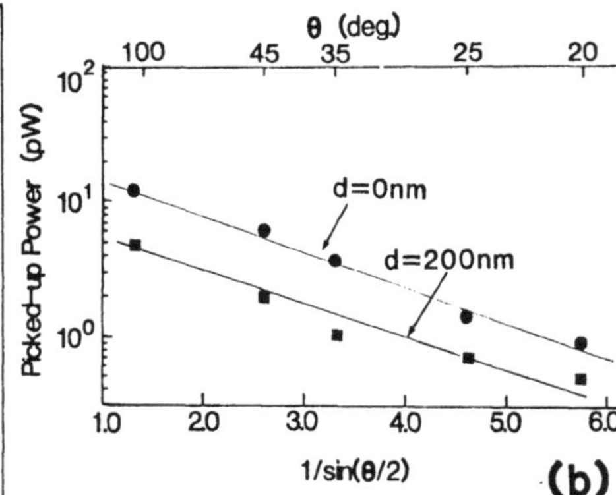


Fig.3 (a) Measured relationship between probe-prism separation d and picked up evanescent light power. The cone angle $\theta$ of the fiber probe was used as a parameter. (b)The relationship between $\theta$ and the picked up evanescent light power. d was used as a parameter.

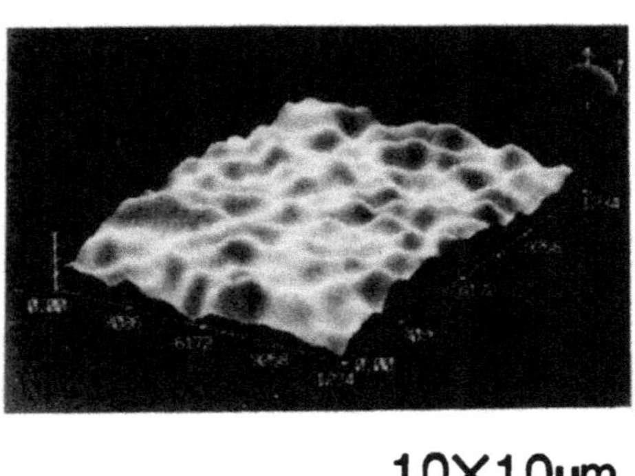


Fig.4 Images of latex particles with diameters of 80nm, which are densely fixed on a prism.The scanning range was 1.0 μm x 1.0μm[3].

Although the fully quantitative estimation of the resolution has not yet been completed, normal resolution limited by the signal to noise ratio of the PSTM system.( seeing the details in ref.4) was estimated to be 0.5 nm. The lateral resolution given in ref.4 was limited by the pixel size, and it could be improved by optimizing the probe scanning parameter. More fundamental lateral resolution limited by the radius of the curvature of the fiber probe was estimated to be 2 nm. As a quantitative measure to estimate the lateral resolution limited by the radius of the curvature of the fiber probe, we can use the 3-dB spatial frequency cut-off of the picked up evanescent power spectral density.

It should be noted that the signal-to-noise ratio of the evanescent power detection was still 40 dB lower than the detector-noise-limited value[4]. Careful acoustic shielding could improve the signal-to-noise ratio which was mainly limited by the acoustic vibration of the fiber probe, and thus, a resolution improvement up to the atomic level can be expected.

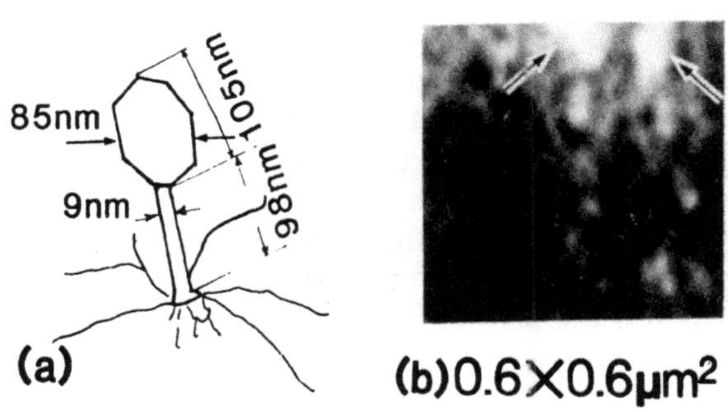


Fig.5 Image of bacteriophage T4[4].
(a) The structure of T4.
(b) The image taken by using the fiber probe with a cone angle of $25^0$. The scanning range was 0.6 $\mu$m x 0.6$\mu$m.

## 4. POSSIBILITY OF A SINGLE NEUTRAL ATOM TRAP

Considering a reversal process of the PSTM, we found a possibility of trapping a single neutral atom by a evanescent light which is localized around the nanometric top of a fiber probe[5], although it is impossible if one uses a conventional propagating light[6]. Based on our intuitive virtual photon model, this trapping is theoretically possible because of the intrinsic properties of the evanescent light localized around the top of the fiber probe, i.e., the evanescent light has a high power density, high spatial power variation, and a large tangential component of the wave vector( for a wavelength $\lambda$, it is $(\lambda/a)$ times larger than that of a conventional propagating light). In other words, it is possible because the photon tunneling process, in general, is associated with a mechanical action and mechanical force between the atom and probe due to the extraordinary large field momentum resulting from its locality.

Figure 6 shows the principle of this trapping. A fiber probe, coated with a metallic thin film except on the top whose radius is smaller than the laser wavelength, is used to emit the evanescent light. An atom coming into this evanescent field resonantly absorbs the counterpropagating laser photon. This is due to a large positive Doppler shift induced by a large tangential component of the evanescent wave vector. After this absorption, the atom spontaneously emits a propagating photon. Since the momentum absorbed by the evanescent photon is much larger than that of the emitted propagating photon, the atom looses its kinetic energy very quickly and is cooled. This large absorbed momentum is attributed to the mechanical distortion associated with the induced polarization in the nanometric fiber probe top.

As a result of this laser cooling process, the atom's momentum becomes comparable to that of the evanescent photon, i.e.,the final thermal velocity is inversely

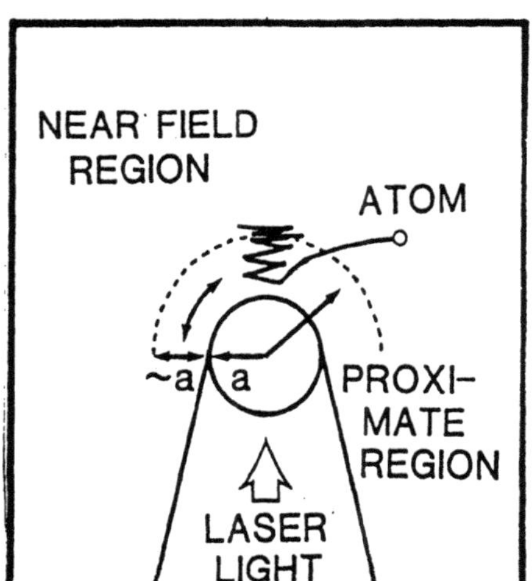


Fig.6 Illustration of the principle of a single neutral atom trapping.

proportional to the curvature radius of the fiber probe, and the atom exhibits a Brownian motion with this velocity along the probe surface. Although the atom is subjected to a centrifugal force and thus being pushed out from the field by this Brownian motion, it also feels an attractive force due to the dipole force induced by the rapidly decreasing power along the normal direction if the laser frequency is red-tuned to the atomic resonance frequency. These features together result in a mechanical interaction between the atom and the probe, which is analogous to an inter-atomic Van der Waals force. As a result of the balance between these two forces, the atom is trapped because it sees a potential minimum at the boundary between the proximate and near-field regions. In other words, this trapping is equivalent to the formation of a diatomic quasi Van der Waals molecule between this atom and an atom at the top of the fiber probe, where the exchange of virtual photons between the two atoms corresponds to the Van der Waals force. Only one atom will be trapped because of the velocity-selective cooling process and of the repulsive force between the atoms in the nanometric volume. Detection of trapped atom is possible by detecting the fluorescence emitted from the atom in the cooling process.

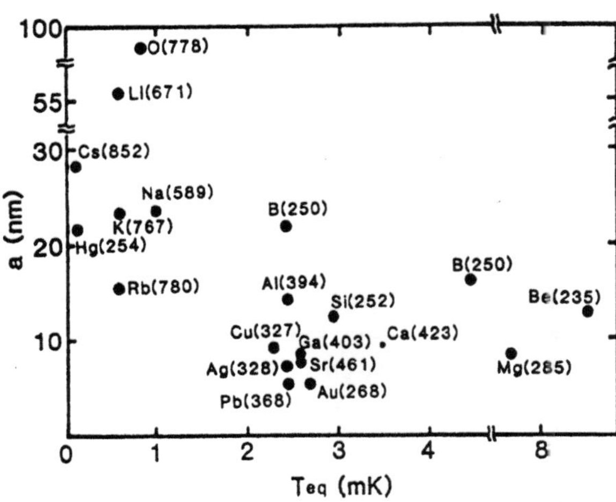


Fig.7 Calculated relationship between the equivalent temperature of the potential depth and the optimum radius of the curvature of the fiber probe. The number attached to each closed circle represents the resonance wavelength (in nanometer unit) of the atomic transition.

Based on the principle of the trapping mentioned above, the atom's momentum has to be equal to the recoil momentum for the purpose of trapping a single atom on the top of the fiber probe. By this requirement, the relation between the equivalent temperature of the potential depth of the trap and the optimum curvature radius of the fiber probe can be calculated. Figure 7 shows the relationship between the equivalent temperature of the potential depth and the optimum radius of curvature of the fiber probe for a variety of atoms. The number attached to each closed circle represents the resonance wavelength ( in nanometer unit ) of the atomic transition. It can be seen from this figure that the equivalent temperature corresponding to the potential depth is as shallow as several mK for the saturation power of each atomic transition. However, a conventional technique of optical molasses[7] can be employed as a pre-cooling technique to prepare an ultra-cold atomic source in a high vacuum by reducing the equivalent temperature of the molasses atoms to 1 $\mu$K or even lower. Therefore, it can be expected that an atom can be trapped by this fiber probe by catching the atom falling from this cold molasses. By using this trapping, an optical tweezer for single neutral atoms can be realized. Furthermore, if the trapped atom is brought to the surface of a cold substrate by translating the fiber probe and is accelerated by tuning the laser frequency to the blue, the atom can be fixed at a certain place on the cold substrate by Van der Waals force or other chemical binding forces. By this process, a single atom-level crystal growth is expected. For precise red- and blue-tunings within the accuracy of 1 MHz, coherent and tunable short-wavelength light sources are required. Such sources can be realized by using several semiconductor lasers, LD-pumped solid state lasers, and nonlinear optical crystals. For example, a Peta-Hz tunable range coherent optical frequency sweep generator has been developed by employing semiconductor lasers, high gain and wide bandwidth frequency control techniques, optical phase locking, and nonlinear optical frequency conversions[8].

## 5. CONCLUSION

By using a sharpened optical fiber with the radius of

curvature of about 1 nm, a photon STM was constructed and a nanometer-resolution imaging was achieved for the biological and other kinds of sub-micron samples. A novel method for trapping a single neutral atom was proposed by using the light pressure of the evanescent light localized around the fiber probe top. For further developments of photon scanning tunneling microscopy and atomic manipulation, careful studies on the concept of massive photon( i.e., the photon interacting with material, ), and quantum optical effects ( e.g., photon localization, cavity quantum electrodynamics ), should be necessary.

## REFERENCES

[1] Pangaribuan, T., Yamada K., Jiang, S., Ohsawa, H. and Ohtsu, M.(1992) 'Reproducible Fabrication Technique of Nanometric Tip Diameter Fiber Probe for Photon Scanning Tunneling Microscope', Jpn. J. Appl. Phys., 31, L1302-L1304

[2] Hori,H., 'Quantum optical picture of photon STM and proposal of single atom manipulation', the paper appearing in this proceeding

[3] Jiang, S., Tomita, N., Ohsawa, H. and Ohtsu, M. (1991) 'A Photon Scanning Tunneling Microscope Using an AlGaAs Laser', Jpn. J. Appl. Phys., 30, 2107-2111

[4] Jiang, S., Ohsawa, H., Yamada, K., Pangaribuan, T., Ohtsu, M., Imai, K. and Ikai, A. (1992) 'Nanometric Scale Biosample Observation Using a Photon Scanning Tunneling Microscope', Jpn. J. Appl. Phys., 31, 2282-2287

[5] Hori, H., Jiang, S., Ohtsu, M. and Ohsawa, H (1991) 'A Nanometer-Resolution Photon Scanning Tunneling Microscope and Proposal of Single Atom Manipulation', Tech. Digest of the 18th International Quantum Electronics Conference ( June 14-19, Vienna ) pp.48-49

[6] Lett, P.D., Phillips, W.D., Rolston, S.L., Tanner, C.E., Watts, R.N., and Westbrook, C.I., (1989), 'Optical molasses', J.Opt. Soc.Am. B, 6, 2084-2107

[7] Lett, P.D., Watts, R.N., Westbrook, C.I., Phillips, W.D., Gould, P.L. and Metcalf, H.J (1988), 'Observation of Atoms Laser Cooled Below the Doppler Limit', Phys. Rev. Lett., 61, 169-172

[8] Ohtsu, M. (1992) Highly Coherent Semiconductor Lasers, Artech House, Inc., Boston

# SCANNED PROBE OPTICAL MICROSCOPY USING A NON-INTRUSIVE PROBE

Lars Malmqvist and Hans M. Hertz
*Lund Institute of Technology*
*Department of Physics*
*P.O. Box 118*
*S-221 00 Lund*
*SWEDEN*

ABSTRACT. A scanned probe optical microscope allowing non-destructive studies of a wide range of objects and surfaces is described. The microscope utilizes a non-intrusive optical trap to position a microscopic probe light source in immediate proximity to the studied object. Recent theoretical and experimental work aiming at optical imaging with sub-diffraction-limited resolution of, e.g., living biological objects, is discussed.

## 1. Introduction

The non-intrusive character of classical optical microscopy is of great importance in the study of, e.g., living biological systems. Unfortunately, the resolution is limited by far-field diffraction. Near-field scanned probe optical methods[1,2] have demonstrated very high resolution by scanning a microscopic light source (e.g., an etched aperture, a protrusion[3] or a exciton capillary[4]) in close proximity to the studied object. However, the mechanical positioning of the probe restrict their use to mechanically accessible and smooth surfaces, limiting the methods' applicability for studies of, e.g., living biological material with intervening membranes.

In the present paper we discuss the use of different types of optically trapped particles[5] as microscopic light sources for non-intrusive scanned probe optical microscopy. The method, Trapped Particle Optical Microscopy (TPOM)[6], relies on the use of non-intrusive tweezers to position and scan the microscopic light source. In our case we use a single-beam gradient force optical trap [5,7] as tweezers although other traps, e.g., based on electrical fields[8], also might be useful. The optical trap consist of a strongly focused laser beam that traps dielectric particles just below the focus due to radiation forces. Due to the scattering of light from the trapping laser beam, the particle will act as a light source. These particles may be non-intrusively positioned and scanned very accurately in transparent media.



*D.W. Pohl and D. Courjon (eds.), Near Field Optics*, 141–146.

## 2. Experiments

Figure 1 shows the arrangement for the first experimental demonstration of TPOM. The optical trap is based on a $\lambda$=514 nm Argon ion laser. The beam is focused with a NA=1.25 water immersion objective into a water cell where 60 nm or 290 nm diameter $SiO_2$ particles are trapped. The scattered light from the particles is detected by a photomultiplier and a test object is scanned piezoelectrically close to the trapped particle.

It is evident that the stability of the trapped particle light source is very important for convenient microscopy work. Figure 2 shows measurements of the scattered intensity from 60 nm diameter $SiO_2$ particles as a function of time. Data was recorded with 10 samples/second and laser power was 150-200 mW. At t=3.1 minutes the particle was let out on purpose to record the zero level. Clearly the stability is sufficient for microscopy applications.

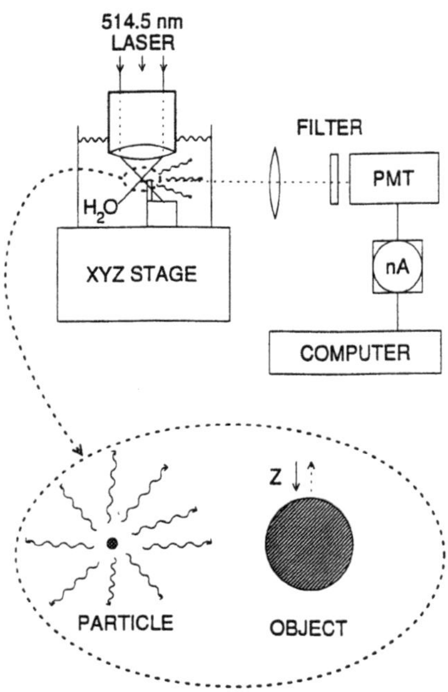


Fig. 1. Experimental arrangement for Trapped Particle Optical Microscopy.

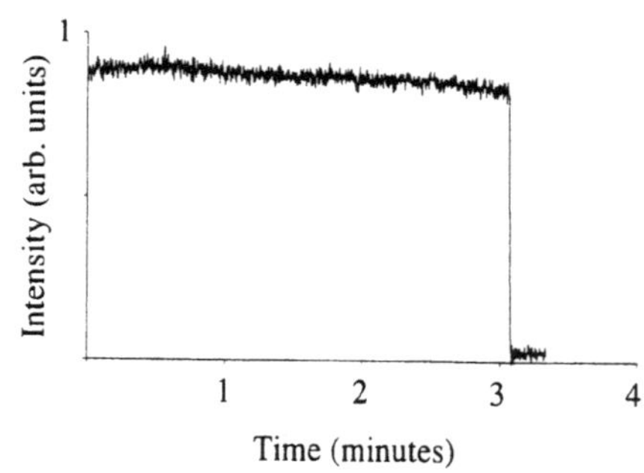


Fig. 2. Scattered intensity from 60 nm $SiO_2$ particle as a function of time.

Figure 3 shows the first image recorded with the system. Here 290 nm $SiO_2$ particles were used to image a test object consisting of two crossed 4 μm carbon wires by 2-D scanning. Each pixel is 1.3x1.3 μm. Evaluating similar 1-D scans we conclude that the resolution in the demonstration experiment is approximately 2 μm, which is significantly higher than the far-field diffraction limit. This is due to the difficulty to perform measurements with a sufficiently small particle-object distance when using the experimental arrangement of Fig. 1. When the test object is close to the trapped particle, laser light from the Gaussian tails of the laser beam will scatter off the

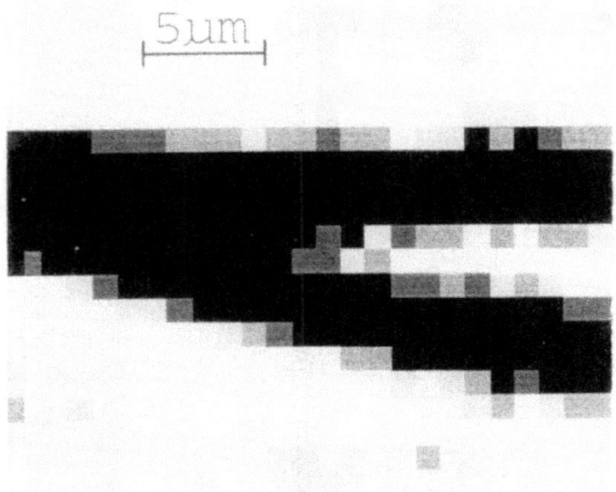


Fig. 3. TPOM image of two crossed carbon wires.

object. Since this scattered light has the same wavelength as the probe light scattered by the trapped particle, the signal-to-noise ratio deteriorates as the object approaches the source, making measurements with < 5 µm particle-object distance difficult.

To eliminate the restriction on particle-object distance we are now developing two-color TPOM, see Fig. 4. In this system the trapping and detection wavelengths are separated allowing the object to be very close to the particle. A cw Nd:YAG ($\lambda$=1.06 µm) laser is used to trap $LiNbO_3$ particles[9]. The particles were produced by grinding $LiNbO_3$ in a mortar before adding the powder to water. The size distribution is wide with a large majority of particles <1 µm. Due to the nonlinear properties of the crystal, green light ($\lambda$=532 nm) is generated when the particle is trapped, even when using a cw beam. By chopping the beam the average trapping intensity can be held below the biological damage threshold (approximately 100 mW at $\lambda$=1.06 µm)[10] while still increasing the green intensity due to the higher peak power. In Fig. 5 the stability of the light source is displayed using 100 mW trapping power. Microscopy experiments with this source is planned in the near future.

## 3. Discussion

Assuming that the particle-object distance can be made very small with the two-color TPOM system, the resolution of TPOM is ultimately determined by the effective size of the trapped particle, i.e., the particle's physical size convolved with its Brownian motion in the trap[6]. Assuming that the optical trap is a harmonic potential well, the particle mean square displacement along the vertical z-axis due to Brownian motion is[11]

$$<z^2>=\frac{kT}{M\omega^2}, \qquad (1)$$

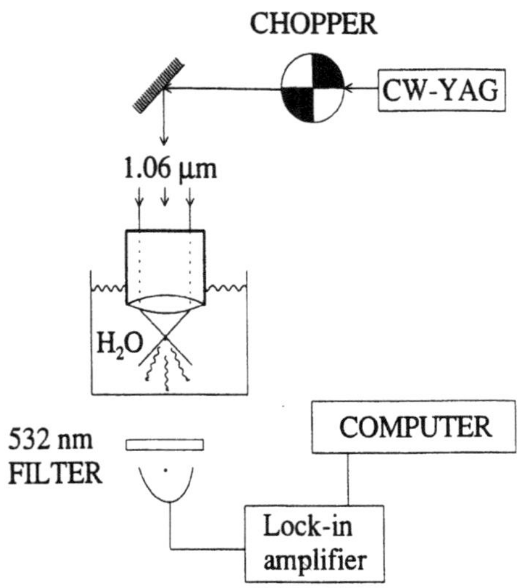


Fig. 4. Experimental arrangement for two-color $LiNbO_3$ TPOM.

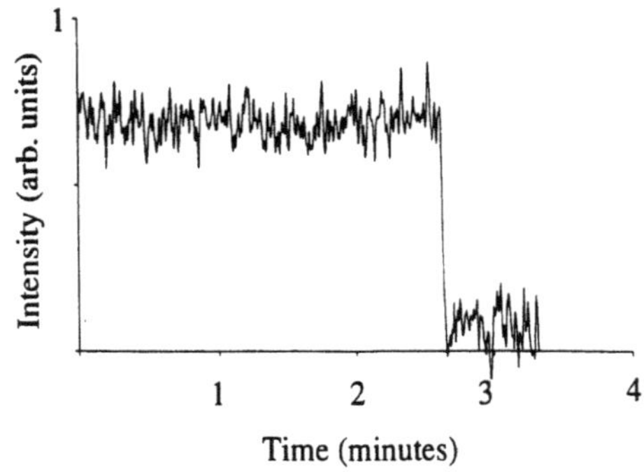


Fig. 5. λ=532 nm intensity from trapped $LiNbO_3$ particles.

where k is Boltzman's constant, T the temperature, and $M\omega^2$ the force constant of the harmonic oscillator. A trapped Rayleigh particle is axially effected by the scattering force

$$F_{scat} = \frac{n}{c}\frac{128\pi^5 a^6}{3\lambda^4}\left(\frac{m^2-1}{m^2+2}\right)^2 I \tag{2}$$

and the gradient force

$$F_{grad} = \frac{2\pi n\, a^3}{c}\left(\frac{m^2-1}{m^2+2}\right)\nabla I. \tag{3}$$

Here I is the intensity, c the speed of light, $a$ the particle radius, λ the wavelength in the medium and $m=n_p/n$, where $n_p$ is the refractive index of the particle and n the index of the media. The forces, and thus the force constant, may be calculated from Eq. (2,3) assuming Gaussian beams. Fig. 6 shows the axial force for a $a$=40 nm $LiNbO_3$ particle (n=2.2). Note that the trapping occurs just below z=0, the focus. The root mean square (rms) displacement is 44 nm for 100 mW of λ =1.06 μm radiation focused to $\omega_o$ = 370 nm (i.e., a slightly truncated Gaussian). Using higher-index particles (e.g, n=3), lower IR wavelengths (e.g., λ=800 nm) and strong focussing, result in rms displacements of approximately 25 nm for $a$=25 nm particles. This clearly indicates the potential for sub-diffraction limited imaging.

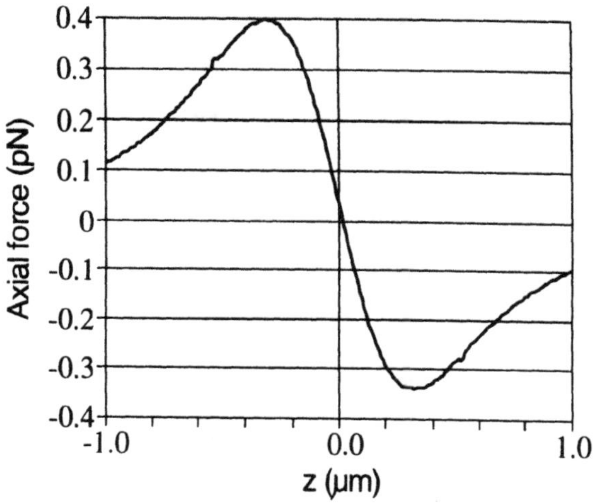


Fig. 6. Theoretical calculation of axial force on *a*=40 nm $LiNbO_3$ particle.

### 4. Conclusions

In summary, we have developed new forms of non-intrusive probes for scanned probe optical microscopy. By use of an optical tweezer the microscopic probe light source is non-intrusively scanned in proximity to the studied object. Potentially the method allows near-field optical microscopy to be performed on hitherto not accessible objects, such as living biological samples with intervening membranes. From the experimental and theoretical considerations in Sect. 2 and 3 it seems reasonable to expect a resolution considerably below the classical diffraction limit using the two-color TPOM.

### Acknowledgements

The support of and discussions with Bo Larsson, Lars Rymell, Sune Svanberg and Bengt Wesslén are gratefully acknowledged. This work was financed by the Swedish Natural Science Research Council, the Swedish Research Council for Engineering Sciences, the Swedish Board for Technical Development and the Crafoord Foundation.

### References

[1] D.W. Pohl, 'Scanning Near-field Optical Microscopy (SNOM)', in Advances in Optical and Electron Microscopy Vol. 12, pp. 243-312 (Academic Press, 1991).
[2] E. Betzig, M. Isaacson, H. Barshatzky, A. Lewis and K. Lin, 'Super-resolution imaging with near-field scanning optical microscopy (NSOM)', Ultramicroscopy **25**, 155 (1988).
[3] U. Ch. Fischer and D.W. Pohl, 'Observation of single-particle plasmons by near-field optical microscopy', Phys. Rev. Lett. **62**, 458 (1989).

[4] A. Lewis, 'The confluence of advances in light microscopy', in New Techniques in Optical Microscopy and Microspectroscopy, R.J. Cherry (Ed.), pp. 49-90 (Macmillan Press, London, 1991).
[5] A. Ashkin, J.M. Dziedzic, J.E. Bjorkholm, and Steven Chu, 'Observation of a single-beam gradient force optical trap for dielectric particles', Opt. Lett. **11**, 288 (1986); O. Marti, 'Light Forces', this volume.
[6] L. Malmqvist and H.M. Hertz, 'Trapped particle optical microscopy', Opt. Commun. **94**, 19 (1994).
[7] W.H. Wright, G.J. Sonek, Y. Tadir and M.W. Berns, 'Laser trapping in cell biology', IEEE J. Quant. El. **26,** 2148 (1990).
[8] G. Fuhr, P. Rösch, T. Mueller, V. Dressler and H. Göring, 'Dielectric spectroscopy of chloroplasts isolated from higher plants - characterization of double membrane system', Plant Cell Physiol. **31**, 975 (1990).
[9] S. Sato and H. Inaba, 'Observation of second harmonic generation from optically trapped microscopic $LiNbO_3$ particle using Nd:YAG laser', Electron. Lett. **28**, 286 (1992).

[10] S.M. Block, 'Optical tweezers: A new tool for biophysics', in Noninvasive Techniques in Cell Biology, pp. 375-402, Wiley (1990).
[11] S. Chandrasekhar, 'Stochastic problems in physics and astronomy', Rev. Mod. Phys. **15**, 1 (1943).

# THEORY OF EVANESCENT AND RADIATIVE FIELDS

J.P. VIGNERON
*Institute for Studies in Interface Sciences (ISIS)*
*Facultés Universitaires Notre-Dame de la Paix*
*Rue de Bruxelles 61*
*B-5000 Namur, Belgium*

ABSTRACT. Many local-probe experiments have in common the use of evanescent waves to circumvent the effects of diffraction on the imaging resolution. The presence of evanescent waves implies specific conditions, ranging from the highly dispersive behaviour of the traversed medium to abrupt geometry changes in the propagation boundary conditions. The scattering process which takes place at the interfaces between the tip and the object in a near-field scan, and at the interfaces within the sample structure can be viewed as a mechanism for conversion of radiative waves to and from evanescent modes. The conversion rate is strongly dependent on the proximity of resonant modes (surface plasmon or phonon polaritons) and it is important to determine the conditions of formation of these modes. A very general and practical scheme, based on the electromagnetic Green function is proposed as a tool to describe this conversion.

## 1. A simple example : pipelining wave guides

An electromagnetic wave guide carries electromagnetic energy along a one-dimensional path (in the $z$ direction, along the wave guide axis), with the advantage of minimising the geometric attenuation. If the wave frequency is higher than a specific threshold, determined by the mode polarisation and by the geometry of the guide normal section, the electromagnetic energy can be transported in the form of a one-dimensional propagating or "radiative" wave. If the frequency is lower than this threshold, there is no possibility of propagation and the field keeps localised at the entrance of the wave guide, in an "evanescent" wave : the field amplitude then decreases rapidly as it penetrates the hollow conductor. This is an example of the existence of radiative and evanescent fields in a simple, macroscopic object. The distinct behaviours (frequency admission or rejection) arise from the correlation of the axial wave number $k_z$ (corresponding to the wavelength measured along the pipe axis), and the length of the transverse wave vector $k_{//}$, lying within the normal section. This relation, arising from Maxwell's equations, involves the fixed frequency $\omega$ of the injected electromagnetic field,



*D.W. Pohl and D. Courjon (eds.), Near Field Optics*, 147–156.

$$k_z = \sqrt{\frac{\omega^2}{c^2} - k_{//}^2}\,. \tag{1}$$

In the case of a singly-connected wave guide, the boundary conditions on the border of the two-dimensional section enforce discrete values to the transverse wave numbers, classified as transverse electric or transverse magnetic. For a square section of side $a$, and for transverse magnetic ($H_Z = 0$) waves, (for which the electric field $E_Z$ has to vanish on the internal surface) the transverse wave numbers are given by

$$k_{//}^2 = \left(n^2 + m^2\right)\frac{\pi^2}{a^2}. \tag{2}$$

$n$ and $m$ are non-zero positive integers, used to label the mode and, for a fixed frequency, decide of the radiative or evanescent nature of the wave. The axial component of the electric field, for specific values of $n$ and $m$, takes the form

$$E_z = A_{nm}\ e^{\pm ik_z z}\ \sin\left(n\frac{\pi}{a}x\right)\ \sin\left(m\frac{\pi}{a}x\right) \tag{3}$$

with an obvious choice of the origin of the $(x,y)$ plane containing the square wave guide section. This expression represents travelling waves if the wave number $k_z$ assumes real values, *i.e.* if the quantity under the root sign in (1) is positive. For a fixed frequency, this limits the number of radiative modes, and prohibits any propagation for frequencies below the limit

$$\omega_c = \frac{\pi c}{a}. \tag{4}$$

For these "forbidden" frequencies in the interval $[0,\omega_c]$, a solutions of the form (3) to the wave equation still exist, but none can be tailored to match finite boundary conditions for $z \to \pm\infty$ : the factor $exp(\pm ik_Z z)$ is then a real exponential function of the co-ordinate $z$. Such evanescent solutions must be discarded on physical grounds for an infinite, translationally invariant wave guide, but are of tremendous importance when we come to the question, for instance, of connecting two different wave guides or introducing an optical defect in an otherwise translationally invariant wave guide.

When two wave guides ($a$) and ($b$) of different sections are connected in series, some frequencies may be radiative in one of them and evanescent in the other. For instance, if the length of the side of the square sections are $a$ and $b$, with $b$ smaller than $a$, a frequency in the interval

$$\frac{\pi c}{a} < \omega < \frac{\pi c}{b} \tag{5}$$

is not allowed to propagate in the small-size wave guide ($b$) but is easily transmitted into the wave guide ($a$). This situation gives rise to a total reflection of the wave incident in ($a$) at the junction with wave guide ($b$), a classical near-field effect in essence similar to the evanescent field appearing behind a small screen aperture [1]. The evanescent field in the rejecting wave guide is actually a superposition of modes which can be described by the infinite sum (we keep x=y=0 as a corner of the wave guide ($b$)) :

$$E_z^{(b)} = \sum_{r=1}^{\infty} \sum_{s=1}^{\infty} B_{rs}\; e^{-\left|k_z^{(r,s),(b)}\right| z} \sin\left(r\frac{\pi}{b}x\right) \sin\left(s\frac{\pi}{b}y\right) \tag{6}$$

The coefficients $B_{rs}$ must be chosen so that incident and reflected modes in wave guide ($a$) matches the transmitted field at the wave guide junction. In practical cases, these coefficients keep significant values for a very large number of modes ($r$,$s$), so that our summation (6) exhibits a very slow convergence rate. This means that the evanescent field will not be easily described by a pure exponential decay, and that a great flexibility remains in expression (6), with a large number of axial and transverse spatial frequencies, to build up a very localised distribution of the electromagnetic energy just past the wave guides junction. The total reflection of waves on moderately small aperture at the junction generates a large number of evanescent waves both in the emergence and in the incidence wave guides, and the energy stored in these waves can be extremely large, compared to the energy stored in the incident and reflected travelling fields. The ratio between these energies measures the efficiency of the "radiative" to "evanescent" conversion.

The change in the lateral boundary conditions imposed by the guided waves can be viewed as a three-dimensional corrugation of the propagating medium, which leads to a transformation of the ratio $\omega/k_z$ into an imaginary value. This transition to an imaginary value of the phase velocity can be obtained, in general, by a transition through a three-dimensional optically corrugated region, or by a refraction on a plane interface, in total-reflection situations. The efficiency of the "radiative" to "evanescent" conversion can be extremely high, if the field frequency is well chosen : it can easily be imagined how a wide wave guide, inserted between two narrower pipes can generate "resonant frequencies" which can be described as oscillatory modes (in fact, steady-state) connected to evanescent waves in the drain and source narrow pipes. Neglecting dissipation, such modes can be sustained without an incident travelling wave and constitute bound electromagnetic modes. These localised excitations of the electromagnetic fields can also, as the next section will show, be generated at flat interfaces separating a propagating ($\omega/k_z$ real) and a non-propagating medium ($\omega/k_z$ imaginary) [2].

## 2. Electromagnetic states guided by interfaces

Electromagnetic interface modes play a substantial role in problems as different as those involved in optical properties of stratified materials or in physisorption of highly polarisable molecules at surfaces. We will start by introducing a very simple description of the macroscopic excitation of a polarisable surface.

### 2.1 INTERFACE POLARITONS

Multilayer optics requires a full treatment of the electromagnetic excitations, including retardation effects. This requires solving more completely Maxwell's equations [3-6]. For stratified materials, which keep translational invariance with respect to any displacement perpendicular to the growth direction, two independent solutions of Maxwell's equations are the well-known $s$ and $p$ electromagnetic waves. In the former case ($s$), the radiation is polarised in a direction parallel to the surface and perpendicular to the direction of incidence ($x$). In the latter case ($p$), the electric field is kept within the plane of incidence ($y,z$). $s$ modes do not contribute to a charge accumulation corresponding to any interface polarisation, since the corresponding motion of the charges remains strictly parallel to the surface. No interface mode can develop for this polarisation. $p$ modes, by contrast, induce a superficial charge density which may react on the ions in the volume of the material and gives rise to interface excitations. Again, because of the translational symmetry parallel to the interface, the fields can all be described as a product of a wave travelling parallel to the interface and an envelope function varying in the normal ($z$) direction:

$$\vec{H}(r,t) = e^{iky-i\omega t}\ U_p(z)\ \vec{e}_x \tag{7}$$

If the wave travels in a medium with a stratified structure, described by a local dielectric function which varies only in the $z$ direction, $\varepsilon(z,\omega)$, Maxwell's equations reduce to a modified wave equation in one dimension:

$$\varepsilon(z,\omega)\frac{d}{dz}\left[\frac{1}{\varepsilon(z,\omega)}\frac{dU_p(z)}{dz}\right]+\left[\frac{\omega^2}{c^2}\varepsilon(z,\omega)-k^2\right]U_p(z)=0 \tag{8}$$

the other components of the electromagnetic fields being easily derived from this. The discovery of interface modes requires here the solution of the wave equation in both regions of space, again with dielectric constants $\varepsilon_1$ and $\varepsilon_2$, and matching of $U_p$ and $(1/\varepsilon)dU_p/dz$ at the boundary. In this geometry, the solutions behave as exponential distributions,

$$U_p(z) = A_1\ e^{-\alpha_1 z}\ \text{(for } z>0) \quad \text{or} \quad A_2\ e^{\alpha_2 z}\ \text{(for } z<0) \tag{9}$$

with

$$\alpha_i = \sqrt{k^2-\varepsilon_i\frac{\omega^2}{c^2}},\quad i=1,2 \tag{10}$$

The decays are different on both sides of the junction, a retardation effect due to the difference in the propagation speed of the field. Accounting for retardation effects, the distinction between evanescent and radiative waves requires the introduction of the wave number $k$ : the wave will be

$$\textit{evanescent}\quad \textit{if}\qquad k^2>\varepsilon(\omega)\frac{\omega^2}{c^2}\qquad \text{(always if } \varepsilon(\omega)<0) \tag{11}$$

$$\textit{radiative}\quad \textit{if}\qquad k^2<\varepsilon(\omega)\frac{\omega^2}{c^2} \tag{12}$$

Then, if $\alpha_i$ is imaginary, we have a radiative type of mode in the corresponding medium $i$. These modes cannot be sustained without an external source, as the radiation will drive away its energy. But we are more interested here in solutions where both $\alpha_i$ are real. The matching conditions lead to the following constraints:

$$A_1 = A_2 \quad \text{and} \quad \alpha_2\ \varepsilon_1(\omega)+\alpha_1\ \varepsilon_2(\omega)=0 \tag{13}$$

These conditions become

$$A_1 = A_2 \quad \text{and} \quad \varepsilon_1(\omega) + \varepsilon_2(\omega) = 0 \tag{14}$$

as $k$ becomes much larger than $\omega/c$ (non-retarded, or electrostatic limit) The basic difference introduced by retardation is dispersion (the frequency is different for each wavelength). For a surface polariton (one of the joining media is vacuum) one obtains the dispersion relation shown in Fig. 1, with a non-retarded limit set by $\varepsilon(\omega) = -1$, valid for large wave numbers.

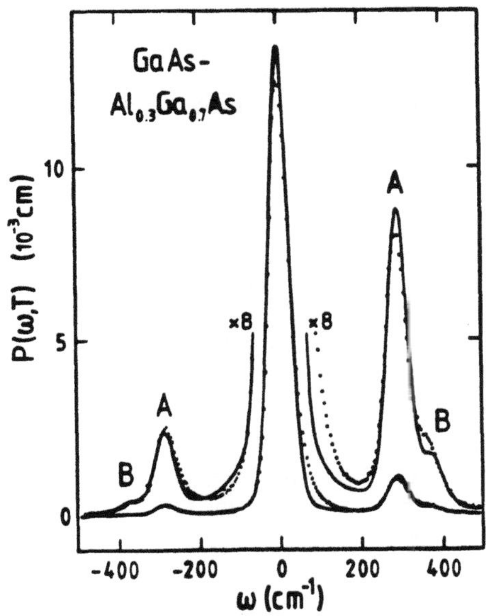


Figure 1. Surface polariton. Dispersion appears at wavelengths close to that of light at comparable frequencies. At short wavelengths (large $k$) the mode frequency approaches the value found for the electrostatic interface mode.

An important property of the surface and interface modes is that they determine an electric field outside the volume of the crystal. An electron in the vacuum, outside the crystal, will be able to induce and experience the effect of this evanescent field and modify its energy accordingly. This is the main idea pursued in using high resolution electron energy loss spectroscopy (HREELS) in the detection of these excitations, in the non-retarded limit [7]. Retarded interface modes are observed through light reflectance measurements (LR) and attenuated total reflection (ATR). The basic difference between the information released by these experiments lies in the very different phase velocities of the excitations preferentially excited. In HREELS, the "surfing effect" contributes to select excitations for which $\omega = kv_{//}$, where $v_{//}$ is the incident electron velocity along the interface plane; in light reflectivity, the (radiative) modes with $\omega = k\,(c/\sin\theta)$, where $\theta$ is the incidence angle, contribute to the absorption; in attenuated total reflection experiments, the relation is $\omega = kc/(n_{prism}\sin\theta)$, which means that the absorption now occurs in the region of evanescent modes.

## 2.2 COUPLING OF INTERFACE MODES

Interface modes are peaked at interfaces and decay into the bulk of the joining materials. When another interface is formed close to the first one, the interface modes couple to give new "slab" modes with different frequencies. This mechanism is, in essence, similar to that brought about by the tight-binding description of the molecular states starting from the atomic states: tunnelling takes place to couple atomic states. This is also true for electromagnetic waves. Tunnel effect is in fact a very general wave phenomenon, not necessarily connected to quantum-mechanical features of the described system. When a superlattice is considered, the large number of interfaces can give rise to the formation of bands, in a way very similar to the formation of bands in solids from the periodic repetition of atoms in the crystal structure.

To study the formation of multiple states by coupling of interface modes, we will again consider the simple example of the classical electromagnetic interface modes. The simplest compound state is that provided by the electromagnetic excitation of a thin plane slab, of thickness L. The retarded modes of a self-supported slab made of a material with dielectric constant $\varepsilon(\omega)$ and surrounded by vacuum ($\varepsilon$=1) can be classified as symmetric ($\omega_+$) and antisymmetric ($\omega_-$) with respect to the geometric symmetry plane. The field envelope is of the form

$$U_p(z) = A\ e^{\alpha z} + B\ e^{-\alpha z} \tag{15}$$

inside and outside the slab. Because of the mirror symmetry, the matching and boundary conditions may only be applied for z>0 and lead to the following algebraic equations to determine the slab mode frequencies

$$\varepsilon(\omega_\pm) = -\ \frac{\alpha_2}{\alpha_1}\ \mathrm{th}^{\pm 1}\left[\alpha_1(\omega_\pm)\frac{L}{2}\right]. \tag{16}$$

where

$$\alpha_1^2 = k^2 - \frac{\omega^2}{c^2} \qquad \text{and} \qquad \alpha_2^2 = k^2 - \ \varepsilon(\omega)\frac{\omega^2}{c^2} \tag{17}$$

These modes occur in very specific region of the ($\omega$,$k$) plane, which lie in the restrahl band ($\omega_{TO}<\omega<\omega_{LO}$) and verify $k>\omega/c$ . The modes can be described as a symmetric and antisymmetric superposition of surface modes which overlap inside the slab. The amount of overlap is determined by the ratio of the wavelength in the propagation direction along the interface ($2\pi/k$) and the thickness of the film $L$. At short wave lengths ($k$ large), both frequencies degenerate into a single non-retarded surface mode frequency [$\varepsilon(\omega)$ = -1]. For wavelengths of the order of the slab thickness, the interface modes influence each other and one is lead to two well-separated frequencies. Finally, if the wave number comes close to $\omega/c$, retardation effects force both frequencies to the value $\omega_{TO}$.

When we see more and more interfaces, the number of levels increases. In the case of an infinite or a semi-infinite superlattice, for instance, continuous bands are formed, defining propagating waves of a new kind. Each wave is formed by a "radiative" Bloch combination of evanescent interface modes. Figure 2 shows the result of a calculation of coupled interface polaritons in a GaAs-AlAs semi-infinite superlattice, that is a superlattice *terminated* by a free surface on vacuum. One observes the formation of three kinds of modes that can be labelled according to the short-wavelength limit of their frequencies. Modes of the first group converge, for large *k*, to the interface state located in the AlAs restrahl frequency. The isolated mode near 290 $cm^{-1}$ approaches the GaAs surface mode frequency, while the modes of the next group converging near 280 $cm^{-1}$ reproduce the independent interface modes located in the GaAs restrahl band. Some of these modes are part of a continuum, which shows that they can propagate energy, while some others are isolated, a sign that these modes are localised near the surface of the superlattice. These isolated surface modes are also built from a combination of surface and interface polaritons, but the corresponding potential amplitude is evanescent in both directions away from the free surface of the superlattice. Similar results where also described for plasmon-polariton excitations in metals [8-13]

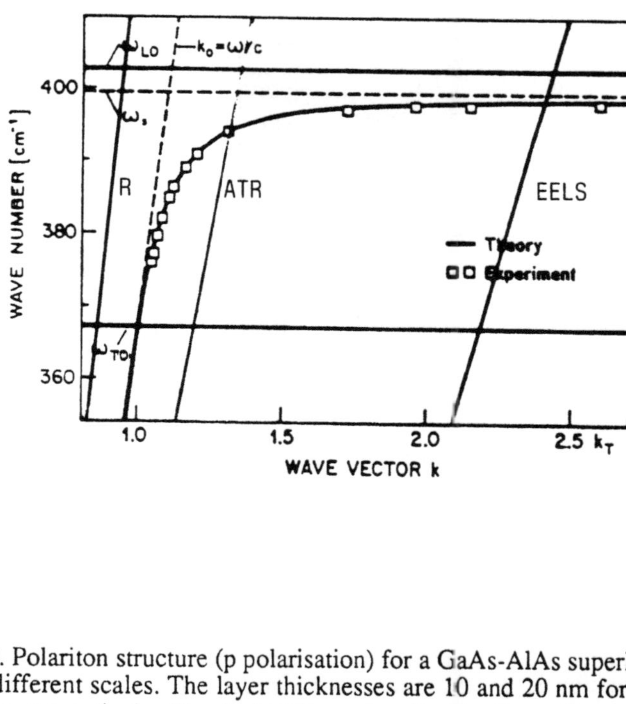


Figure 2. Polariton structure (p polarisation) for a GaAs-AlAs superlattice, represented on two different scales. The layer thicknesses are 10 and 20 nm for GaAs (first layer) and AlAs, respectively. The dashed lines visualise the lines of equation (a) $\omega=kc$, (b) $\omega=kc/n_{prism}$ and (c) $\omega=kv_{//}$ relevant for IR, ATR and HREELS spectroscopies, respectively.

The simple electromagnetic eigenmodes described above are also partially accessible to electron energy loss measurements [14-19], for which, in contrast to the case of optical reflectivity, retardation effects are negligible. When an electron with a few electron volts of kinetic energy is reflected by the external surface of the layer it will lose energy, coupling its own Coulomb field to the decaying field associated with the surface or shallow interface modes. Figure 3 shows the reflection energy-loss spectrum of a semi-infinite GaAs-$Al_{0.3}Ga_{0.7}As$ superlattice with about 200 Å layer thickness [20,21]. This spectrum can be interpreted as follows. The main peak, labelled A, on the right (loss) and left (gain) of the elastic peak is associated with an evanescent surface mode attached to the external GaAs surface, while the weaker structure, labelled B, is due to a propagating interface mode having primarily an Al-As stretching character.

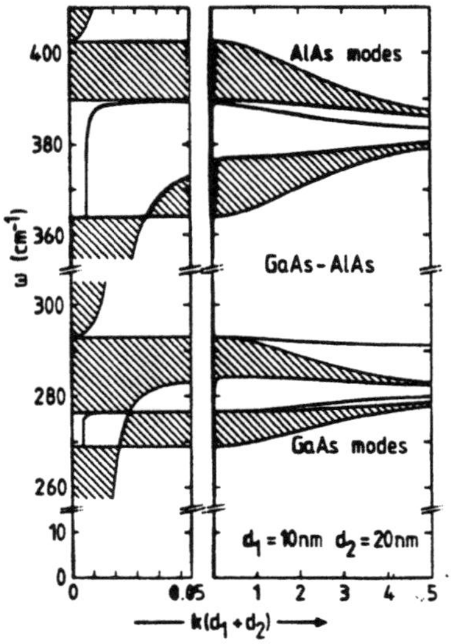


Figure 3 . Surface and interface electromagnetic modes, seen from electron energy loss experiment. These modes correspond to oscillating fields associated with the charge on a surface or an interface, evanescent in both directions away from the materials junction.

## 3. Computer Prediction of Near-Field Effects

Experiments such as attenuated total reflection or light tunnelling put to work the conversion of radiative to and from evanescent waves in the most direct way, but a detector placed in the far-field region cannot directly measure the evanescent field. The information concerning the interface between the sample and the medium in which the detector is placed, and the interfaces inside the sample is carried by the radiative part of the emerging field (usually only as a small disturbance, when compared to the incident intensity). The situation is quite similar for the electron tunnelling in a scanning tunnelling microscope, where the current in the external measuring circuitry is carried by purely travelling (would we say "radiative" ?) electronic waves but, on the other hand, is finely tuned by scattering mechanisms which take place at the tunnel junction. Here also, the electron scattering is merely a multiple conversion process of current-carrying waves to and from evanescent waves and the heart of a quantitative theoretical description of tunnelling is the understanding of the origin of this conversion.

A very instrumental approach of such a calculation is the localised Green function method, already heavily used for the description of point defects in solids. This method was introduced in the context of near-field effects to study the three-dimensional scattering of electrons at an STM tip [22] and later applied to light scattering [23]. The idea is actually quite simple and very effective to describe the effects of a local disturbance of the propagation medium on the far-field intensity distribution. The entire space is divided into two distinct regions : region A is the region where the perturbation $V_{AA}$ takes place (change of dielectric function, in a macroscopic view of the propagation, or change in the ion polarisability in a more detailed microscopic description), and region B, the undisturbed ($V_{BA} = V_{AB} = V_{BB} = 0$) far-field region where the field must be detected. The amplitude of the scattered field $\psi$ is then described in all space as a two-components quantity ($\psi_A$: near field, $\psi_B$ : far field) which verifies a Lippman-Schwinger system (The dyadic character of this equation, not emphasised here, should not be missed) :

$$\begin{bmatrix} \psi_A \\ \psi_B \end{bmatrix} = \begin{bmatrix} \psi_A^0 \\ \psi_B^0 \end{bmatrix} + \begin{bmatrix} G_{AA}^0 & G_{AB}^0 \\ G_{BA}^0 & G_{BB}^0 \end{bmatrix} \begin{bmatrix} V_{AA}^0 & 0_{AB} \\ 0_{BA} & 0_{BB} \end{bmatrix} \begin{bmatrix} \psi_A \\ \psi_B \end{bmatrix} \tag{18}$$

This quite naturally leads to the sequence of calculations

$$\psi_A = \psi_A^0 + G_{AA}^0 V_{AA} \psi_A \tag{19}$$

and

$$\psi_B = \psi_B^0 + G_{BA}^0 V_{AA} \psi_A \tag{20}$$

The first of this equation allows to compute the near-field scattered wave in region A, where the propagation is perturbed : this evaluation of the near field $\psi_A$ does not require any knowledge of the far-field wave $\psi_B$. This equation determines a kind of source field which determines the distribution of energy in the far field. This distribution of the far-field amplitude is then determined by the second prescription [Equ. (20)] which discloses the field at any point of region B, once the near-field is known in region A. This way to determine the prolongation of the near-field into the far-field region is not much more than a quantitative formulation of the Huyghens-Fresnel principle : many theories and simulations of the scanning near-field optical microscope and the photon scanning tunnelling microscope presented at the present workshop actually use this sound principle, in a way or another.

## REFERENCES


[1] Leviatan, (1986), J. Appl. Phys. 60, 1577 .

[2] Fuchs R. and Kliewer K. L., (1965), Phys. Rev. 140, 2076, Kliewer K. L. and Fuchs R., (1974), Adv. Chem. Phys. XXVII, 358 .

[3] Huang K., (1951), Proc. Roy. Soc. London A208, 352 .

[4] Economou E.N., (1969), Phys. Rev. 182, 539 .

[5] Borstel G. and Falge H. J., (1982), in *Electromagnetic Surface Modes* (Edited by A. D. Boardman), p. 219, John Wiley & Son Ltd., New-York .

[6] Lambin Ph., Vigneron J.P., Lucas A.A., and Dereux A., (1987), Physica Scripta, 35, 343 .

[7] see, for instance, Lucas A. A., (1986), Physica Scripta, T13, 150 .

[8] Economou E.N. and Ngai K. L., (1974), Adv. Chem. Phys. XXVII, 265 .

[9] Raether H., (1980), in *Excitation of plasmons and Interband Transitions by electrons* (Springer-Verlag, Berlin).
[10] Pokatilov E. P. and Beril S.L., (1983), Phys. Stat. Sol. (b) 118,567 .
[11] Giuliani G.F. and Quinn J.J., (1983), Phys. Rev. Lett. 51, 919 .
[12] Giuliani G.F., Quinn J.J., and Wallis R. F., (1984), J. Phys. (Paris) Colloq. 45, C5-285 .
[13] Quinn J.J., (1984), Solid State Commun. 52, 607 .
[14] Ibach H., (1970), Phys. Rev. Lett. 24, 1416 .
[15] Lucas A.A. and Sunjic M., (1971), Phys. Rev. Lett. 26, 229 .
[16] Ibach H. and Mills D. L., (1982), in *Electron Energy Loss Spectroscopy and Surface Vibrations*, pp. 63-126, Academic Press, New York .
[17] Lucas A.A., Vigneron J.P., Lambin Ph., Thiry P.A., Liehr M., Pireaux J.J., and Caudano R., (1986), Intern. J. Quantum Chem.: Quantum Chem. Symposium 19, 687 .
[18] Lucas A. A. and Vigneron J.P., (1984), Solid State Commun. 49, 327 .
[19] Liehr M., Thiry P. A., Pireaux J.J. and Caudano R., (1984), J. Vac. Sci. Technol. A2, 1079 .
[20] Lambin Ph., Vigneron J.P., Lucas A.A., Thiry P. A., Liehr M., Pireaux J.J., Caudano R., and Kuech T. J., (1986), Phys. Rev. Lett. 56, 1842 .
[21] Lambin Ph., Vigneron J.P., and Lucas A. A., (1985), Phys. Rev. B 32,8203 .
[22] Lucas A.A., Morawitz H., Henry G.R., Vigneron J.P., Lambin Ph., Cutler P.H., Feuchtwang T.E., (1988), Phys. Rev. B 37,10708-10720
[23] Dereux A., (1991), PhD Thesis .

# THEORETICAL PROBLEMS IN SCANNING NEAR-FIELD OPTICAL MICROSCOPY

D. VAN LABEKE and D. BARCHIESI
*Laboratoire d'Optique P.M. Duffieux*
*Centre National de Recherche Scientifique: U. R.A. 214*
*Université de Besançon. U.F.R. Sciences et Techniques*
*Route de Gray 25030 Besançon Cedex FRANCE*

ABSTRACT. The rigorous calculation of Scanning Near-field Optical Microscope (SNOM) images is a difficult task which needs the resolution of a complicated problem of diffraction in unusual conditions: two diffracting objects are present, they contain details smaller than the wavelength and are very close. Various approaches have been proposed to attack this problem and the purpose of this paper is to present, classify and compare these theoretical works.
First we rapidly present the pioneer theoretical papers which give the principle of SNOM or calculate the near-field above various small apertures. After the various models of tip detection are discussed: energy flow or scattering theory.
Then we present the simpler model where the coupling between the sample and the tip is neglected but which give analytical solutions and thus can lead to a real discussion of the role of the various parameters. Finally we describe more rigorous models where the tip-sample coupling is taken into account.

## 1. Introduction

Last year in Interlaken was celebrated the 10 years of Scanning Tunneling Electronic Microscopy (STM) [1]. The Scanning near-field Optical Microscopy (SNOM) and the Scanning Tunneling Optical Microscopy (STOM) are younger and cannot already present images of atoms or molecules. But since their birth in 1984 and 1989 they have experienced an important development. Beautiful images and results have been shown in Interlaken [1] and now in Besançon new and interesting images are presented with high resolution and contrast on various materials.

Descriptions of the beginning of the evolution of SNOM, both experimentally and theoretically, are well exposed in a paper by Pohl [2]. The purpose of this paper is to present the recent theoretical works in SNOM and to discuss the most important theoretical problems.

We first rapidly present, in chapter 2, the first theoretical papers which describe the principle of SNOM or which calculate near-fields. After that (chapter 3) we present a classification of the various theoretical approaches. It is found on two criteria: global or non-global theories and macroscopic or microscopic theories. The non-global theories need a model for tip detection. These models are analyses in chapter (4). Then (chapter 5) a macroscopic non-global theory is described. Finally, in the last chapters main faitures of global macroscopic (6) and microscopic (7) theories are presented.



*D.W. Pohl and D. Courjon (eds.), Near Field Optics*, 157–178.

## 2. First theoretical papers

The first theoretical papers on SNOM can be classified into papers which presents the general principle of this new microscopy and papers devoted to the calculation of near-field above small apertures.

### 2.1. SNOM PRINCIPLE

The theoretical principle of SNOM was perfectly described in 1928 prophetic paper of Synge [3].

Then the first papers on SNOM are principally devoted to descriptions ofexperimental set-up and results but however some theoretical discussions can be found.

The problem of the resolution limit was discussed in terms of spatial frequencies and evanescent waves in 1967 by Mc Cutchen [4] and more recently by Massey [5] and Fischer et al [6].

To model a transmission SNOM Leviatan [7] calculated the coupling between two conducting screens each of which having a small slot. We have to note the very interesting papers of Vigoureux et al [8, 9, 10]. The first one [8] published in 1989 presents the general principle of detection in near-field Microscopy in terms of a conversion by scattering of evanescent waves into propagating modes. The second [9] discusses the super-resolution obtained in SNOM in connection with properties of confined electromagnetic modes. The last one [10] analyses the resolution obtained in SNOM in light of the Rayleigh criterion and Heisenberg principle.

### 2.2. NEAR FIELD CALCULATIONS

From the beginning of the century many papers were consecrated to near-field calculation in diffraction problems of Electromagnetism. We only present here some papers which can be useful in Near-field Microscopy. For modeling the tip detection calculations of the near-field above small apertures and around small protrusions are necessary. The Near-field above many kinds of samples are available.

Following the suggestion of Synge, the first SNOM apparatus uses a small aperture in a metallic screen as a probe for scanning the near-field above the sample [2] and most of the first theoretical works were devoted to calculations of the near-field above small apertures.

The starting point of these calculations was the famous paper of Bethe [11] who calculated the transmission of a small aperture in a thin perfectly conducting screen illuminated by a plane wave. This work was improved by Bouwkamp [12] who provides a correct solution for the near-field above the aperture. For transmission properties these authors have shown that a small aperture can be described as a superposition of an electric dipole and a magnetic one. A recent bibliography on diffraction by small apertures in thin screen can be found in [13] where the near-field above a square aperture is calculated.

The investigation of electromagnetic diffraction theory for an aperture in a thick screen is a more difficult problem. The simplest case is a one dimensional aperture, an infinite slit. The first paper was by M. Wirgin [14] and various methods have been used to attack this problem [15, 16, 17]. In 1988 E. Betzig et al [18] made a careful study of the near-field diffraction by a thin slit in a thick perfectly conducting screen with implications for SNOM For a circular aperture into a thick screen we have to quote a paper by Roberts [19] who used a modal formulation to obtain a rigorous expression for the diffracted field and for the transmission coefficient for plane incident wave.

By using a similar method, the same author [20] has recently proposed an interesting model for a small-hole tip used in SNOM experiments. She calculated the transmission efficiency of a plane wave into a conical waveguide.

To model the dielectric tips, which are now currently used in SNOM, other theoretical works have to be used. The scattered field of a dipole is the first pedagogical example of a non radiative near-field [21]. Mie scattering of a plane wave by a sphere or a cylinder is known since 1909 [22, 23] but the properties of the near-field around a sphere illuminated by a plane wave has been recently discussed [24]. For spheroidal or conical tips no rigorous theory seems available except in the non-retarded case [25].

In order to interpret the experimental results we do not only need the theoretical properties of probes but we have to know the near-field above various samples.

Diffraction of Electromagnetic field by simple smooth objects can be found in many text books [25, 26] but most of works are for the metallic case and only few papers are consecrated to the dielectric one.

In microscopy, the resolution limit is generally tested by looking on periodic structures. Many papers are available on grating theory and the various rigorous methods are described and compared in review papers of Petit [27] and Maystre [28]. For a random surface a bibliography can be found in [29].

When the height of the sample is small compared to the wavelength a perturbation theory can be used for calculating the diffracted field [30 - 32].

## 3. Classification of theories

The true aim of this paper is to present and compare the recent theoretical works which go further into the interpretation of SNOM experiments and give, or will be able to give, images of a realistic object with a realistic model for the detection tip.

To simplify the following discussions we limit ourselves to a tunneling microscope working in transmission. Of course all the theories and results can be apply to other kinds of near-field microscopes.

For calculating the detected signal we have to solve Maxwell equations into the sample, into the vacuum, into the tip and into the fiber and to match the various boundary conditions. But it is a tremendous task which cannot be solved rigorously because the problem has no symmetry: the interface has plane-symmetry, the fiber a cylindrical one. Moreover the sample has roughness and the apex of the tip is spherical or conical.

Approximations are thus necessary to calculate the SNOM images. Before to present the various theoretical approaches we first try to classify them. This classification is build upon two kinds of criteria: macroscopic or microscopic formalism and global or non-global theory.

In a **macroscopic** theory all the components of the microscope, sample and tip, are supposed to be continuous solids. Their optical properties are described by macroscopic parameters, for instance the dielectric constant $\varepsilon(\vec{r})$.

In **microscopic** method the system is no longer continuous. The tip and the sample are composed of a set of discrete scattering centers (atoms or small polarizable systems).

In a **non-global** approach of SNOM theory we first determine the diffracted field above the sample. A model for the detection of this field by the tip is then employed for calculating the detected signal. In this context it is supposed that the probe does not modify the sample near-field. This is an approximation, the tip is not a passive probe, it absorbs a part of the sample field energy and modifies the field distribution. Moreover a part of this field is certainly reflected back to the sample which can cause a deeper modification of the near-field distribution.

For small tip-surface separation there is a coupling between the tip and the sample and the most rigorous theories of SNOM need a **global treatment** of all the system: tip and sample cannot be separated.

Global macroscopic theories are presented in chapter (6): we investigate first the exactly solvable model of tip-sample coupling of Denk and Pohl which lead to a localized plasmon between the tip and the sample. Then we describe the principle of 3 methods which are able to furnish theoretical images: Green function, grating theory and macroscopic self-consistent equation.

The microscopic method is exclusively used by Girard et al and has been improved along all the published papers [33-38]. The coupling between the tip and the sample is described by a self-consistent equation which is numerically solved. The principle of this method is presented in chapter (7). C. Girard presents new results in another lecture of this workshop [39].

## 4. Models for tip detection

In non-global approach the near-field above the sample is first determined alone and for calculating the detected intensity a model for the tip detection is necessary. The true question is "what is detected by the probe?"

### 4.1. ELECTROMAGNETISM APPROACH

In Electromagnetism the energy flow of a wave is represented by the Poynting vector $\vec{P} = \frac{c}{4\pi}\vec{E} \wedge \vec{H}$ [21]. For one monochromatic plane homogeneous wave the time averaged flux of energy can be written:

$$< \vec{P} > = \frac{c}{8\pi} \, |\vec{E}|^2 \sqrt{\varepsilon}\, \vec{n} \qquad (1)$$

Where $\varepsilon$ is the dielectric constant of the medium and $\vec{n}$ the direction of the wave.
This result is well known by opticians: the intensity of a wave is proportional to the square modulus of the electric field. But it supposes one and only one plane homogeneous wave and (1) is no longer valid for two or many plane waves, for a general wave and certainly not in the near-field which contains many evanescent modes. Even for one evanescent wave equation (1) cannot be used without care.

The detected intensity in SNOM must be calculated by integrating the flux of the Poynting vector along the tip surface (Figure 1a). If the tip is composed of a dielectric without absorption the detected signal is the flux of the Poynting vector through the tip section S which is equal to the flux along the lateral surface $\Sigma$. However, for calculating the detected intensity some papers integrated $|\vec{E}|^2$ and not $<\vec{P}>$ along the tip surface. Salomon et al [40] used such an approximation for studying efficiencies of a cleaved fiber and of a parabolic tip immersed in one evanescent wave above a tip surface.

With the same model for tip detection Cites et al [41] have calculated images of a dielectric grating in STOM. The probe is a conical tip which is modeled as a dielectric paraboloid.

## 4.2. SCATTERING MODEL FOR TIP DETECTION

Another approach of tip detection has been first proposed by Vigoureux et al [8].

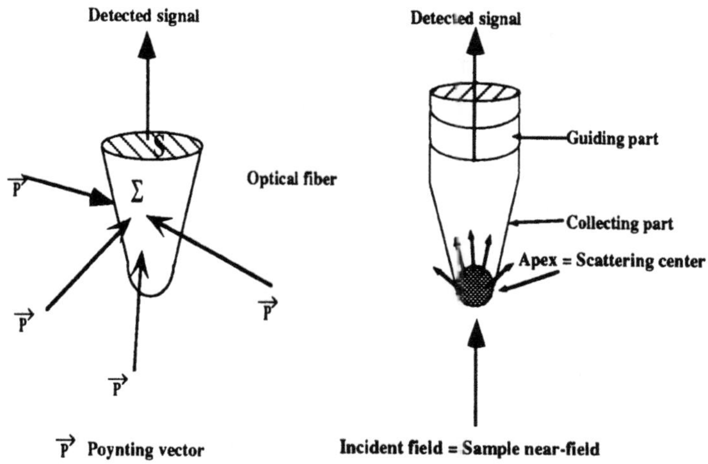


Figure 2a. Figure 2b.

Figure 1. Two models for tip detection.
(1a). The electromagnetic model (1b).The scattering model

The near-field above the sample is composed of many non-propagating evanescent waves and the tip transforms this waves by scattering into propagating modes which can be guided to the detector. The probe is thus decomposed into 3 parts a scattering apex, a collecting part and the guiding part (Fig. 1). The collecting part only integrates the scattered intensity over a define solid angle. The shape and the width of this apex are the key for the resolution limit which could be obtained in STOM.

4.2.1. *Small Electric Dipolar Polarizable Center.* Many models can be used for the scattering center. The simplest is to consider the tip as a small electric polarizable system characterized by its polarizability $\alpha$. In this case the near-field $\vec{E}(x,y,z)$ induces on the tip an electric dipole $\alpha\vec{E}(x,y,z)$ and the detected intensity is:

$$I_d(x, y, z) = \frac{\omega^4}{96c^3}|\alpha|^2\Big[\left(|E_x(x,y,z)|^2 + |E_y(x,y,z)|^2\right)\left(16 - 15\cos\delta - \cos 3\delta\right) + |E_z(x,y,z)|^2\left(16 - 18\cos\delta + 2\cos 3\delta\right)\Big] \quad (2)$$

where $\delta$ is the angle of the collecting angle (figure 1b).

For a small electric dipolar tip the detected signal is the electric field but the detection is not isotropic. Even in this simple case the tip induces a polarization effect. The 3 components of

$\vec{E}(x,y,z)$ are not detected in the same way. When the aperture angle becomes larger, the detection becomes more isotropic:

$$\lim_{\delta \to \pi/2} I_d(x, y, z) = \frac{\omega^4}{6c^3} | \alpha \vec{E}(x,y,z) |^2 \tag{3}$$

For a dielectric probe, obtained by etching an optical fiber, the apex of the tip can be considered as a small sphere of radius R and of dielectric constant $\varepsilon_{tip}$. In this case the polarizability can be written [21-23]:

$$\alpha = \frac{\varepsilon_{tip} - 1}{\varepsilon_{tip} + 2} R^3 \tag{4}$$

This electric dipolar model is often used as a first attempt for theoretical calculations of SNOM images [8, 33, 42].

We have used [42] such a tip model to calculate STOM images of a dielectric grating with a period smaller than the wavelength. In that paper we discuss the influence of various parameters on STOM signal: sample period, laser wavelength and polarization...We show that for a small sample period ($a<\lambda$) the characteristic tip-sample distance is neither the wavelength, neither the evanescent decay length but rather the sample period. We also discuss the relation between the image and the sample profile in terms of spatial filtering by diffraction and propagation.

For a small metallic sphere, formula (4) remains valid. The polarizability can become very large when the laser frequency is such that $\varepsilon \simeq -2$. It has been suggested to use this plasmon resonance to enhance the detected signal in STOM [2].

4.2.2. *Small Sphere.* In fact previous formulas are valid for a negligible sphere radius. In Electromagnetism, the problem of the scattering of a plane wave by a sphere of finite radius has been solved in 1905 by Mie and is presented in various books [21-23].

The sphere has for index N and radius R and is centered at point $\vec{R}_c$. The formalism is based upon a multipolar expansion of a plane wave into spherical ones. We consider a more general plane wave of wave vector $\vec{k}$. The electric and magnetic fields of this wave are noted: $\vec{E} = \vec{E}_0 \exp(i\vec{k}.\vec{r})$, $\vec{B} = \vec{B}_0 \exp(i\vec{k}.\vec{r})$. The direction of the wave vector is labeled by the angles ($\theta_k$, $\psi_k$). In this case the scattered intensity becomes, in Gauss units [45]:

$$I = \frac{c}{8\pi k^2} \sum_{L,\,M} [\,| a_L(\xi) f_{L,M} e^{i\vec{k}.\vec{R}_c} |^2 + | b_L(\xi) g_{L,M} e^{i\vec{k}.\vec{R}_c} |^2)] \tag{5}$$

$$( L = 0, +\infty \,;\, M = -L, L )$$

In this formula, $k=2\pi/\lambda$, where $\lambda$ is the wavelength in vacuum. $a_L(\xi)$ and $b_L(\xi)$ are respectively the electric and magnetic Mie coefficients of order L [22, 23]. L=1 corresponds to a dipolar term,

$L = 2$ to a quadrupolar one... These coefficients are complicated combinations of spherical Bessel and Hankel functions and derivatives [43, 44]. They depend upon the size parameter $\xi = 2\pi R/\lambda$.

When the sphere is small compare to the wavelength ($\xi \ll 1$), $a_L(\xi)$ and $b_L(\xi)$ are rapidly decreasing function of L. Asymptotic expression of $a_L(\xi)$ and $b_L(\xi)$ can be found in Kerker's book [23]. In this case the electric terms are much more important than the magnetic ones.

The $f_{LM}$ and $g_{LM}$ are rotational coefficients which depend on the direction of vector $\vec{k}$ and on the polarization of the field [45 ]:

$$f_{LM} = 4\pi i^L\, \vec{E}_0 \cdot \vec{X}_{LM}(\theta_k, \psi_k)^* \;;\quad g_{LM} = 4\pi i^L\, \vec{B}_0 \cdot \vec{X}_{LM}(\theta_k, \psi_k)^* \qquad (6)$$

$\vec{X}_{LM}(\theta, \phi)$ is the vectorial spherical harmonic of order L [43].
An exponentially decreasing function can also be expanded into a multipolar expansion [44] and it can be easily verify that formulae (5-6) remain valid for an evanescent wave characterized by a complex wave vector. Such a wave can be easily obtained in internal total reflection or in the near field of a grating with a small period. It can also be defined by a complex angle $\theta_k$ or a $\sin(\theta_k)>1$.

For a propagating incident wave the scattered intensity does not depend on the incident angle and polarization. For an evanescent wave, the vector $\vec{k}$ being complex, the scattering problem has no more symmetry $f_{LM}$ and $g_{LM}$ depend on the complex angle $\theta_k$, but the physical parameter is in fact $\sin(\theta_k)$. $f_{LM}$ and $g_{LM}$ are rapidly increasing function of $\sin(\theta_k)$ and L. These functions are bound for a propagating mode ($\theta_k$ real and $\sin(\theta_k)<1$) but are unbound polynomials for evanescent waves ($\mathrm{Sin}(\theta_k)>1$).

With such variations the multipolar expansion can be strongly modified when the incident wave on the sphere is evanescent. When a small sphere is excited by a propagating wave, the Mie coefficients decrease rapidly with L and the magnetic terms are negligible compared to the electric ones. The leading term is the electric dipolar one. But when the wave is evanescent the multipolar series, even for a small sphere, is no longer a monotonous decreasing function of L. The quadrupolar term can become stronger than the electric one and the magnetic terms are enhanced.

We have applied this result to study theoretically the influence of the tip radius on the resolution and the quality of images of a dielectric grating [45, 46]. A general near-field can always be expressed as a plane wave series, composed of both evanescent or propagating waves. The scattered intensity by the sphere immersed in the near-field is already given by formula (5) but $f_{LM}$ and $g_{LM}$ coefficients have to be summed over the plane wave structure of the field.

4.2.3. *Small Hole Tip* Another kind of tip is currently used In SNOM It is a small aperture probe, build by drawing a micropipette or a fiber which is then metallized. The resulting probe has a flat end face with a small adjustable aperture from 20 nm to 500 nm [47]. This kind of probe is very close to the suggestion of Synge [3].

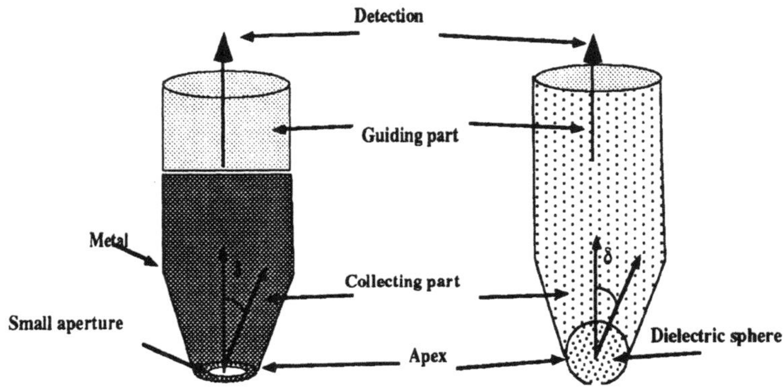


(2a) (2b)

Figure 2. Comparison of two models of tip 2a: small aperture tip. 2b: Dielectric tip

The intensity detected by this probe can be modeled by using the diffraction theory of Bethe and Bouwkamp [11, 12]. For a small aperture in a thin perfectly conducting metal can be describe as a system with effective electric ($\alpha_{hole}$) and magnetic ($\beta_{hole}$) dipolar polarizabilities. As a small aperture is an anisotropic system the two polarizabilities are anisotropic tensor. For a circular aperture of radius R, if the z axis is perpendicular to the aperture plane, the only non vanishing components of these tensors are:

$$(\alpha_{hole})_{zz} = -\frac{R^3}{3\pi} \quad \text{and} \quad (\beta_{hole})_{xx} = (\beta_{hole})_{yy} = +\frac{2R^3}{3\pi} \tag{7}$$

In this model the detected signal is obtained by integrating the scattered intensity over the collecting angle δ, which leads to:

$$\begin{aligned} I(x,y,z) = \frac{\omega^4}{96c^3} \Big[ & \left( |\mu_x|^2 + |\mu_y|^2 + |m_y|^2 + |m_y|^2 \right)\left(16 - 15\cos\delta - \cos 3\delta\right) \\ & + \left( |\mu_z|^2 + |m_z|^2 \right)\left(16 - 18\cos\delta + 2\cos 3\delta\right) \\ & + 6\left(\cos 2\delta - 1\right)\left(m_x d_y^* + d_y m_x^* - m_y d_x^* - d_x m_y^*\right) \Big] \end{aligned} \tag{8}$$

Where $\vec{\mu}$ and $\vec{m}$ are respectively the induced electric and magnetic dipoles.

The small hole tip detects both electric and magnetic near-fields. In this case the detection process also leads to strong polarization effects.

By using models presented above we have compare [48] the theoretical images of the same sample (a grating and a letter E) obtained with a spherical dielectric tip and with a small hole tip of the same radius.

In fact formulas (7) are the leading terms of a multipolar series. The method used by Roberts [19-20] for calculating the efficiencies of a conical tip excited by a plane wave can lead to the following terms of this development.

## 5. Macroscopic Global theories

This chapter is devoted to the presentation of a non-global macroscopic theory applied to calculate STOM images [42, 45, 46, 48]. In this review paper we only present the main features of the model for a linear periodic sample imaged with a dielectric tip. The formalism is more completely described and applied to a true two-dimensional sample in another part of this workshop [48]. In that paper a theoretical comparison of images of a same sample obtained by two kinds of tips is exposed.

### 5.1. NEAR-FIELD EXPRESSION

The sample is the surface of a glass of index n and profile z = f(x). The profile is a periodic function of period a and height h which can be expanded as a Fourier series:

$$f(x) = \sum_{p=-\infty}^{+\infty} F(p)\, e^{2i\pi px/a} \tag{9}$$

The p=0 harmonic corresponds to the mean height of the grating. Information on the sample profile is in fact contained in the other harmonics $p \neq 0$.

This sample is illuminated, in total internal reflection, by a laser of wavelength $\lambda$ in vacuum and amplitude $E_0$. The angle of incidence is i, the plane of incidence is the principal plane. Conical incidence can be described by formulas of [48]. The field above the sample is composed of a transmitted field plus a diffracted field:

$$\vec{E_2}(x, z) = \vec{E_t}(x, z) + \vec{E_d}(x, z) \tag{10}$$

The transmitted field is alone if the sample surface is perfectly plane, it is evanescent when the angle of incidence is over the critical one. The amplitude of this field, given by Fresnel formulas, has about the same magnitude than the incident field. $\vec{E_t}(x, z)$ is independent of grating height and profile. It leads to a constant signal (independent on x) when the probe scans the surface at a constant z distance.

True information on the sample is contained in the diffracted field. For calculating $\vec{E_d}(x, z)$ the usual Optical-Physics approximation cannot be used as the sample period is too small. To determine the diffracted field we have to use Maxwell equations in the sample and in vacuum and to match the boundary conditions on the sample surface. For this purpose the rigorous grating theories are available. The various methods are presented and compared in [27, 28]. Recently Neviere [49] has apply a differential method for calculating the near-field above a grating in a theoretical study of STOM.

However in STOM experiment the height h of the grating grooves must be small compared to the wavelength and to get $\vec{E_d}(x, z)$ a perturbative method is enough. Such an approach has been used by Toigo [30] and Agarwal [31] for studying in far-field the diffracted light by a rough surface. When $h \ll \lambda$ the diffracted field above the grating can be expressed as a perturbation expansion that we limit to the first order [42]:

$$\vec{E_d}(x,z) = E_0 e^{2i\pi x \sin(i)/\lambda} \sum_{p=-\infty}^{+\infty} e^{2i\pi px/a} \frac{2\pi F(p)}{\lambda} \vec{D}(p)\, e^{iw_p z} \tag{11}$$

Apart a phase factor, the diffracted field can be considered as a Fourier series which contains the sample spatial spectrum. Each diffracted order p is characterized by its wave vector: $\vec{k_p} = (u_p, 0, w_p)$ where the x component (parallel to the z=0 plane) is equal to the x component of the incident wave vector plus a wave vector of the reciprocal lattice:

$$u_p = \frac{2\pi}{\lambda} n\sin(i) + p\frac{2\pi}{a} \quad \text{and} \quad w_p = \sqrt{\frac{\omega^2}{c^2} - u_p^2} \tag{12}$$

Whatever is the value of the ratio $a/\lambda$ the development (11) contains evanescent waves: for high diffracted orders ($|p| \gg a/\lambda$) the component $k_{px} = u_p$ is too large and the z component $w_p$ becomes purely imaginary.

## 5.2. DISCUSSION

The formalism presented above leads to analytical expressions so the physical discussion is very easy.

First, let us notice that the diffracted term is proportional to the ratio $h/\lambda$. Thus, as it is supposed that the grating has a small height ($h \ll \lambda$), the true signal $\vec{E_d}(x, z)$ is weak and embedded in a very intense continuum $I_t(z) = |\vec{E_t}(x, z)|^2$. This continuum could be easily subtracted but, its noise still remains. This problem is a serious limit to quality and resolution of experimental STOM images.

Each term of the series (11) is proportional to the corresponding Fourier coefficient F(p) of the grating profile but it is multiplied by two filtering coefficients: $\exp(iw_p)$ and $\vec{D}(p)$. The exponential is a low-pass filter and probe-sample distance is the key parameter of this filtering. Let us consider a grating with a very small period. In fact, for n=1.5, $a < \lambda/3$ is enough. In this case, with an angle of incidence above the critical angle all the diffracted orders are evanescent. No signal is detected in Far Field, Near-Field methods have to be used and we have to determine what is the probe-surface distance to get image of the sample with enough spatial harmonics.

For the p diffracted order, the characteristic decay length is the inverse of the z component of its wave vector:

$$l_p = \frac{1}{iw_p} = \frac{\lambda}{2\pi\sqrt{(n\sin(i)+p\lambda/a)^2-1}} \tag{13}$$

For p=0 formula (13) coincides with the decay length of the evanescent transmitted wave. But this length has no importance because the p=0 order contains no information on the grating shape, but only on its mean height.

The other orders have to be considered. Moreover, as nsin(i) is always near one, the most important term in (13) is $p\lambda/a$ when $\lambda/a > 1$, which is the case considered here. So the decay lengths are (practically) independent on the wavelength and only depend on the grating period:

$$l_p \cong \frac{a}{2\pi|p|} \quad \text{for } p \neq 0 \text{ and } a < \lambda/3 \tag{14}$$

The 3 components of $\vec{D}(p)$ have not the same value, so this filtering is called a polarization filtering. They do not depend on z so, even at small distances when the exponential filtering is negligible, the Fourier spatial spectrum of the sample is modified by the polarization filtering. Thus, generally the profile of the near field does not reproduce the profile of the sample.

The properties of this filtering effect have been fully discussed in [42]. For a large period ($a >> \lambda$), in normal incidence, many propagating plane waves are produced. In this case the coefficients $\vec{D}(p)$ are rather constant versus p and do not depend on the polarization The approximation used in Optical Physics is recovered in these conditions.

For a small period ($a << \lambda$ but $a < \lambda/3$ is enough) all the modes are evanescent (except the p=0 one for normal incidence) and the coefficients $\vec{D}(p)$ vary with p, which leads to polarization and filtering effects. The main result is that for a small grating period, the polarization filtering strongly enhanced high harmonics of the diffracted field. For the electric field the T.E. diffracted field is weaker than the T.M. one. For the magnetic field these polarization properties are inverted.

### 5.3. APPLICATIONS

When the near-field above the sample is known (Eqs. 10 and 11) a model of tip detection is used for calculating the detected signal. In [41] the tip is conical and the signal is obtained by integrating the field square modulus along the tip surface. In [42] the probe is modelled by a very small dipolar sphere and a discussion of resolution, of tip-surface distance and of images filtering is presented. In [45] the Mie Scattering of evanescent waves discussed above is used to study the influence of tip radius on resolution of STOM images.

The tip-sample distance is the critical parameter to get high resolution. To illustrate this point we present on figure (3) the calculated intensity of a sample wearing details with two lateral scales. The sample is a dielectric rectangular grating of period 200 nm and height 20 nm. On the top of the grooves there is a superstructure of tracks with a smaller period and height. The tip is a dielectric tip of radius R = 10 nm. In T.E. polarization we cannot resolve the sample profile (Fig. 3b, 3d). With T.M. polarization, for a tip-sample distance equal to 28 nm we resolve the larger profile. To resolve the smaller details we have to go closer to the sample: for z = 15,5 nm higher harmonics can be detected. But when the tip radius is increased the lateral resolution is quickly degraded.

With a tip radius R = 50 nm, the sample profile is not well reproduced and only the grating period is visible [45]. Similar results are obtained by global microscopic models of Girard.

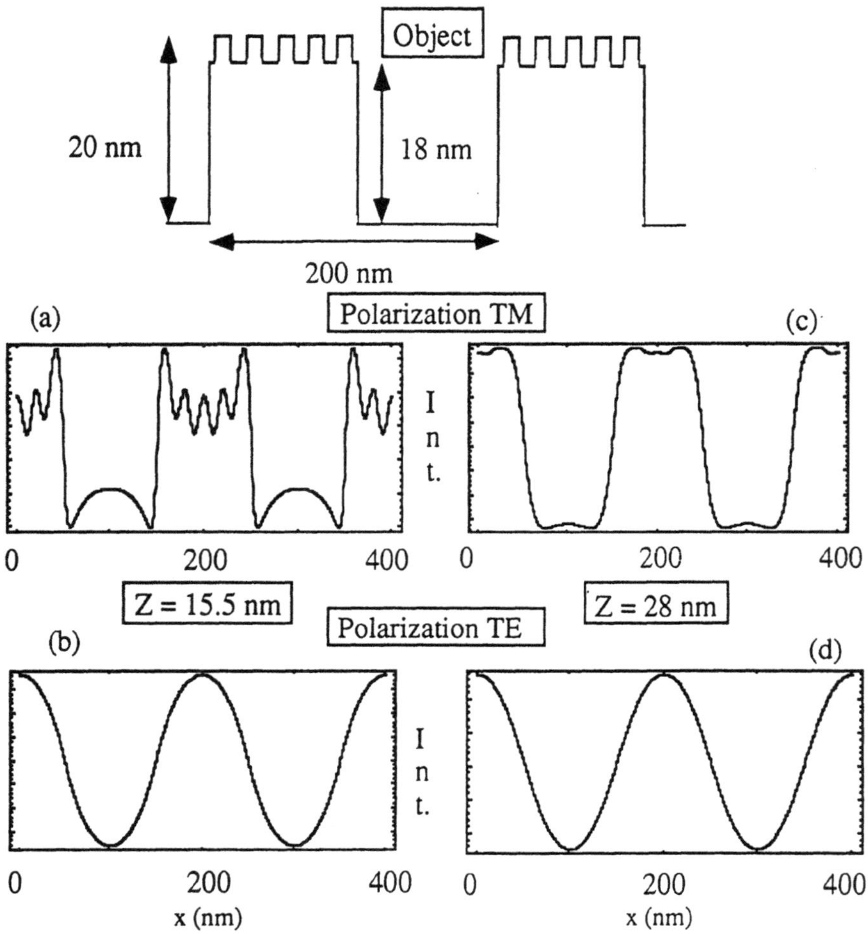


Figure 3: Calculated intensity of a sample with two scales of details.
dielectric tip R = 10 nm
3top: Sample profile. 3a. 3c. T.M. polarization. 3b. 3d. T.E. polarization.

In [48] theoretical images of a same sample are calculated with two models of tip: a dielectric tip and the small hole tip. Images show very important polarization variations. In the studied configuration the small hole tip leads to the best images as it detects both electric and magnetic near-field.

## 6. Macroscopic Global theories

### 6.1. A NON-RETARDED EXACTLY SOLUBLE MODEL

The first example of macroscopic global approach is described in a paper by Denk and Pohl [50]. It is a very interesting case of strong coupling between the tip and the sample.
The tip-sample geometry is modeled by a hyperboloid for the tip separated from the plane by a gap. Prolate spheroidal coordinates allow separation of the Laplace's equation for this geometry. The non-retarded solution can thus be found. For suitable media the paper presents plasmon solutions which are highly localized in the gap region. This exact solution is possible in this case because of the high symmetry of the model. Moreover only two parameters can be chosen freely: for a given tip radius and semi-angle the tip distance is prescribed. The extension to other geometry and to retarded solution seems very difficult. However this localized plasmon between metallic tip and sample are very important for the understanding of light emission in STM [51].

### 6.2. GREEN FUNCTION

For really calculating images the global macroscopic approach proposed by Dereux et al [52, 53] from the University of Namur in Belgium seems promising. It is found upon a Green's function method. This method had already been used in diffraction theory. For instance recently Greffet used it to study the scattering by a 2D obstacle on an interface or a deep inhomogeneous grating [54-55]. A Similar method is used for electronic tunneling.

They start from the knowledge of the dyadic Green's function $\overleftrightarrow{G}_0(\vec{r}, \vec{r}')$ of a reference system. For application to SNOM the reference system can be a plane interface, a multilayered structure or eventually a grating. The Green function verifies the equation:

$$- \mathrm{Curl}[\overleftrightarrow{G}_0(\vec{r},\vec{r}')] + \frac{\omega^2}{c^2}\varepsilon(\vec{r})\overleftrightarrow{G}_0(\vec{r},\vec{r}') = \overleftrightarrow{I}\delta(\vec{r} - \vec{r}') \qquad (15)$$

For a flat interface this Green function is for instance given in [53-56 and references there in].

The geometrical and optical properties of the system are contained in the local dielectric constant $\varepsilon(\vec{r})$. Then, the system is perturbed (for instance by a roughness or a tip) and the perturbation is characterized by a dyadic $\overleftrightarrow{V}(\vec{r})$.

The solution $\vec{E}(\vec{r})$ is deduced from a Lippmann-Schwinger equation:

$$\vec{E}(\vec{r}) = \vec{E}^0(\vec{r}) + \int d\vec{r}\,\overleftrightarrow{G}_0(\vec{r},\vec{r}')\,\overleftrightarrow{V}(\vec{r}')\vec{E}(\vec{r}') \qquad (16)$$

Where $\vec{E}^0(\vec{r})$ is the solution without perturbation ($\overleftrightarrow{V}(\vec{r}') = 0$).

When the perturbation is localized into a bounded region P the scattered field $\vec{C}(\vec{r}')$ inside the perturbation is solution of a linear integral equation:

$$\vec{E}^0(\vec{r}\,'') = \int_P d\vec{r}\,'\{\overleftrightarrow{I}\delta(\vec{r}\,''-\vec{r}\,') - \overleftrightarrow{G}_0(\vec{r}\,'',\vec{r}\,')\overleftrightarrow{V}(\vec{r}\,')\}\vec{C}(\vec{r}\,') \tag{17}$$

Where $\vec{r}\,'$ and $\vec{r}\,''$ are in the perturbed region P. When $\vec{C}(\vec{r}\,')$ is known, the field anywhere in the space is given by the equation:

$$\vec{E}(\vec{r}) = \vec{E}^0(\vec{r}) + \int_P d\vec{r}\,'\overleftrightarrow{G}_0(\vec{r},\vec{r}\,')\overleftrightarrow{V}(\vec{r}\,')\vec{C}(\vec{r}\,') \tag{18}$$

The previous equations are rigorous but they have no analytic solution and computation is necessary. For this purpose the perturbed area is discretized into a grid of N points and the integration over P becomes a weighted summation over the grid. Equation (17) becomes a linear system of equations with unknown the scattered field $\vec{C}(\vec{r}_j)$ on the points of the grid $\vec{r}_j$:

$$\vec{E}^0(\vec{r}_k) = \sum_{j=1,N} w_j\,[\overleftrightarrow{I}\delta_{kj} - \overleftrightarrow{G}_0(\vec{r}_k,\vec{r}_j)\overleftrightarrow{V}(\vec{r}_k)]\,\vec{C}(\vec{r}_j) \tag{19}$$

Where $w_j$ is the weight of the integration point j.

The quality of the approximation depends on the density of the discretization grid and on the weighting method.

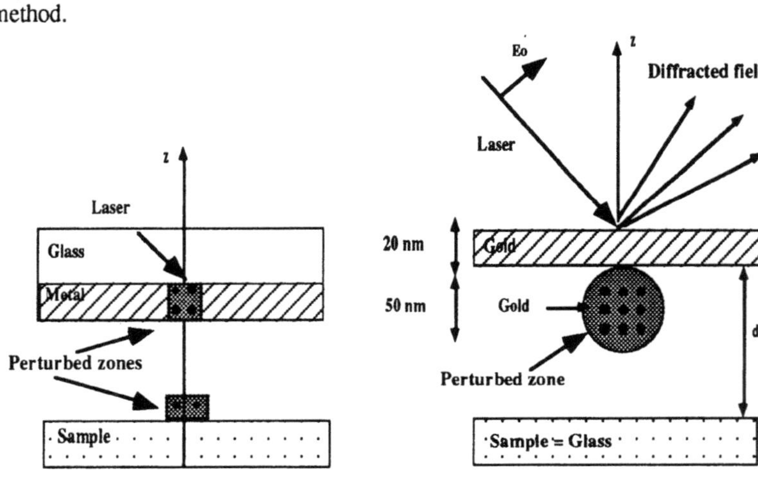


(4a) (4b)

figure 4. Two examples of application of Green's function method.
4a. Small aperture SNOM 4b. Small protrusion SNOM

Dereux has tested his algorithms by calculating Mie scattering cross section and by comparing his results with rigorous ones. He also studied the optical binding between two spheres.

The Green's function method could easily be applied to model a small aperture SNOM where the perturbation is a roughness on the sample and the small aperture (Figure 4a).
Dereux has succeeded in reproducing quite well the results of a SNOM experiment realized by Fischer and Pohl [57]. The probe is a small gold sphere (Fig. 4b) In the model $\overleftrightarrow{G}_0$ corresponds to the multilayer sample-air-metal and the perturbation zone is the gold sphere. As in experiments a resonance is obtained when varying the tip-sample distance. For this last example of calculation the time of computation is very long. For a simple grid of 27 points (3x3x3) in the sphere, the calculation of the scattered field for one tip-sample distance needs one hour of CPU on a 11 Megaflops computer.

## 6.3. RIGOROUS GRATING THEORY

Nowadays various numerical algorithms are available for calculating with a high precision the diffracted field by a periodic structure. The grating can have any shape and can be a multilayered structure. The various methods are well described in two review papers [24, 25].

In order to interpret the experiments of STOM realized in Dijon by Goudonnet and de Fornel, Maystre [49] has calculated the field above a rectangular Dielectric grating. In this first paper there is no tip in the theoretical study and the authors only study the properties of the Poynting vector above the grooves.

By principle a SNOM with a tip is a non-periodic structure and grating formalism is not valid in this case. However, a truck can be used to avoid this difficulty. It consists to model a SNOM with an artificial periodic array of identical tips. Grating theory can thus be applied to this periodic multilayered structure: layer 1-sample, layer 2-vacuum, layer 3-tips (Fig. 5).

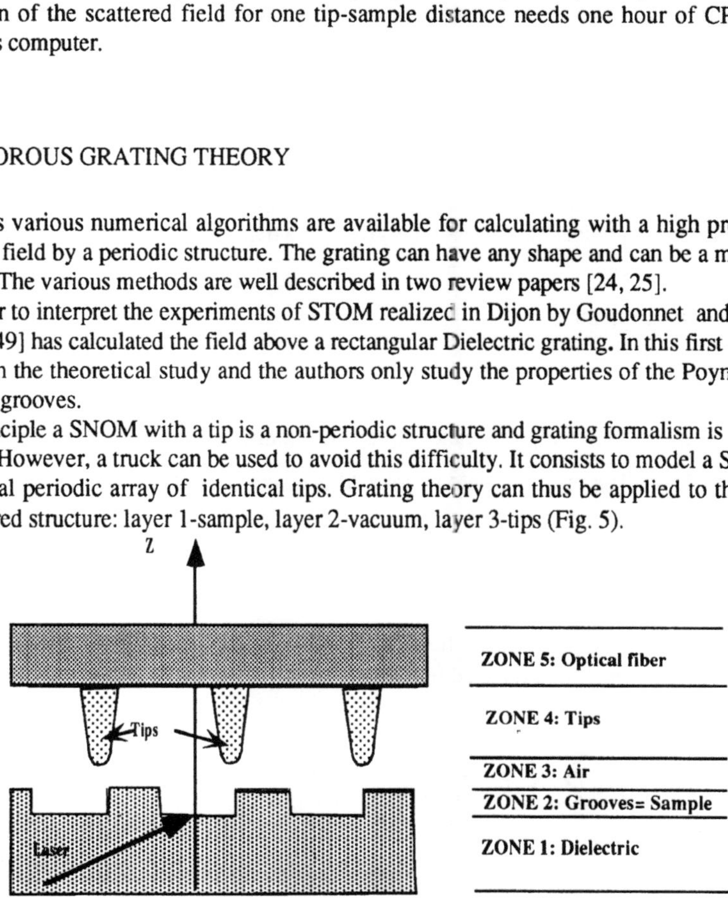


Figure 5. Principle of grating theory applied to STOM

In homogeneous zones (1, 3, 4) the field are expanded in Rayleigh series of plane waves. Then if the differential method is used Maxwell equations are solved in the grooves zone (zone 2) and in the tips zone (zone 4) where the dielectric constant is inhomogeneous. Matching of the boundary conditions lead to an infinite linear system of equations which is solved after truncation.

A transmitted matrix is then constructed which connect the fields in various regions in terms of the incident field. The method is detailed and many examples of calculations are presented in the lecture of Neviere [58]

The period of the tips-grating must be equal or be a multiple to the period of the sample grating. The coupling between the various fictitious tips is an artifact of this method. It can be avoided by increasing the distance between the tips without changing the sample period. But this constraint strongly enlarges the computing memory and time.

### 6.4. MACROSCOPIC SELF-CONSISTENT EQUATION

Recently S. Bernsten et al [59] have proposed a self-consistent macroscopic approach based on Maxwell equations to describe the tip-surface interaction in the external reflection near-field microscopy. For a one-dimensional flat sample the problem is solved and leads to a single integral equation, in terms of the Fourier components of the field at the interface. This equation has been extended to take into account a point-like scattering center on the sample which is the simplest model for roughness. A numerical discussion is given for a trapezium shaped tip.

## 7. Microscopic approach

### 7.1 PRINCIPLE

We consider a discrete set of dipolar scattering centers (atoms, molecules, small spheres) located at points $\vec{R_i}$ (i=1, N). The resulting field at point M is a sum over all the fields radiated by the dipoles $\vec{d_i}$. Every scattering center is characterized by its polarizability $\alpha_i$ (supposed isotropic):

$$\vec{d_i} = \alpha_i \, \vec{E}(\vec{R_i}) \tag{20}$$

So for calculating the dipole $\vec{d_i}$. the field at the point $\vec{R_i}$ is needed. This local field can be divided into the "incident" field (created by sources outside the set of dipoles) and the scattered fields by the other dipoles:

$$\vec{E}(\vec{R_i}) = \vec{E_0}(\vec{R_i}) + \sum_{j=1}^{N} \left[ \overleftrightarrow{P}(\vec{R_i}, \vec{R_j}) \, \alpha_j \, \vec{E}(\vec{R_j}) \right] \tag{21}$$

where $\overleftrightarrow{P}(\vec{r}, \vec{r}')$ is a tensor, or a dyadic, which describes the dipolar propagation of the field between points $\vec{r}$ and $\vec{r}'$. This dyadic is often named the dipolar propagator. In vacuum, for a field of pulsation ω, it can be expressed as:

$$\overleftrightarrow{P}_0(\vec{r}, \vec{r}') = e^{(iR\omega/c)} \left[ \frac{[3 \vec{R} \vec{R} - R^2 \overleftrightarrow{I}]}{R^5} + \frac{i\omega}{R^4 c} [3 \vec{R} \vec{R} - R^2 \overleftrightarrow{I}] + \right.$$
$$\left. - \frac{\omega^2}{R^3 c^2} [\vec{R} \vec{R} - R^2 \overleftrightarrow{I}] \right] \tag{22}$$

In (22) we have introduced the relative vector $\vec{R} = \vec{r} - \vec{r'}$ and its modulus R.

The equation (36) is **self-consistent** as the wanted solution, $\vec{E}(\vec{R_i})$ is in the right member and is needed for solving the problem.

For a continuous set of dipoles, the resolution of (36) leads for a whole space to the Lorentz-Lorenz relation and to the Ewald-Oseen extinction theorem for an interface problem [21].

For a discrete system, the self-consistent equation (21) can be easily formally solved. For this purpose two "super-vectors of dimension 3N are introduced:

$$\mathbb{F} = [E_x(\vec{R_1}), E_y(\vec{R_1}), E_z(\vec{R_1}), E_x(\vec{R_2}),...E_\alpha(\vec{R_i}).....E_Z(\vec{R_N})]$$

$$\mathbb{F}_0=[E^0{}_x(\vec{R_1}), E^0{}_y(\vec{R_1}), E^0{}_z(\vec{R_1}), E^0{}_x(\vec{R_2}),...E^0{}_\alpha(\vec{R_i}).....E^0{}_Z(\vec{R_N})] \tag{23}$$

$\mathbb{F}_0$ and $\mathbb{F}$ represent respectively the incident field and the local field on every dipole. Thus, (21) can be re-written as a linear system of equations between $\mathbb{F}$ and $\mathbb{F}_0$:

$$\mathbb{F} = \mathbb{F}_0 + \overleftrightarrow{\mathbb{B}}\ \mathbb{F} \tag{24}$$

where $\overleftrightarrow{\mathbb{B}}$ is a 3Nx3N matrix whose coefficients are determined by using (34-36) and depend on polarizabilities, on propagators and on the positions of every center.

The linear system (24) is easily inverted and the incident field $\mathbb{F}$ can be expressed in terms of the incident field $\mathbb{F}_0$:

$$\mathbb{F} = \frac{1}{1-\overleftrightarrow{\mathbb{B}}}\ \mathbb{F}_0 \tag{25}$$

If the incident field is known on each dipole, the field at every point of space can be calculated. Formula (25) is the real solution of the scattering problem but it is a formal solution because, even for a system with a few scattering centers (2 or 3) analytical expressions from (25) are difficult to manage. A numerical inversion of the matrix $1-\overleftrightarrow{\mathbb{B}}$ is needed for all the realistic problems. Moreover this computing problem can become extremely difficult when the number of centers is increased.

However, when the matrix elements of $\overleftrightarrow{\mathbb{B}}$ are small (weak coupling case) a perturbation expansion of (68) can be used for calculation purpose:

$$\mathbb{F} = \mathbb{F}_0 + \overleftrightarrow{\mathbb{B}}\ \mathbb{F}_0 - \overleftrightarrow{\mathbb{B}}\,\overleftrightarrow{\mathbb{B}}\ \mathbb{F}_0 + ....... \tag{26}$$

Restricted to the first order this expansion leads to a linear summation of each dipole contribution. The other terms of (26) can be interpreted as multiple scattering of the field of one dipole by the others.

If for the applications the dipolar approximation is no longer valid, a similar formalism can be developed with multipolar propagators and polarizabilities [38].

## 7.2. APPLICATIONS

The previous method has been 50used for studying the Mie scattering by small particles of any shape [60, 61].

In the context of Local Probe Microscopy, Girard and al have published papers that use this microscopic approach for calculating images in Atomic Force Microscopy [38 and references therein] and in near -field microscopy.

In STOM the incident field is the evanescent transmitted wave beyond the surface z=0. The sample is fixed on the flat surface of this interface and it is composed of m scattering centers. The probe is also a discrete set of n centers. These n+m dipoles are coupled by an self-consistent equation like (62). But in this interface problem, the surface appears in the propagator: $\overleftrightarrow{P}(\vec{r},\vec{r}')$ is composed of two terms, the vacuum propagator $\overleftrightarrow{P}_0(\vec{r},\vec{r}')$ and a surface term $\overleftrightarrow{S}(\vec{r},\vec{r}')$ [34-38]. In this case the matrix $\overleftrightarrow{\mathbb{B}}$ is composed of 4 blocks:

$$\overleftrightarrow{\mathbb{B}} = \begin{Bmatrix} \mathbb{B}_S & \mathbb{B}_{SP} \\ \mathbb{B}_{PS} & \mathbb{B}_P \end{Bmatrix} \tag{27}$$

Where the matrix $\mathbb{B}_S$ contains the electromagnetic couplings between centers of the sample. $\mathbb{B}_P$ contains the coupling between dipoles of the probe. The interactions between the probe and the sample are in $\mathbb{B}_{SP}$ and $\mathbb{B}_{PS}$

In first papers [33-35] the perturbation method based on formula (26) is used for calculating the images of a monolayer grating of dipoles [33] and of 5 parallels dielectric rectangular tracks [34-35]. They are composed of a large number of atoms with a polarizability $\alpha$=1,5 $Å^3$. The images of the tracks appear with a contrast inversion: a hole on the sample corresponds to a large intensity and a high relief to a weaker signal. A correct, image of the sample (apart the inversion of the contrast) is obtained for a very small tip-surface separation d=0.25nm. For a distance just a bit larger, we can resolve the 5 tracks but their shape is strongly degraded.

In [36] the method is apply to a SNOM working in reflection where the fiber is used both for illuminating the sample and for collecting the near-field.

Recently [37] a very interesting theoretical spectroscopic study has been obtained by Girard and Bouju. The sample is a layer of 25 gold spheres which are imaged in STOM. When the laser frequency is far from the localized plasmon frequency the spheres have small polarizabilities and are seen in reversal contrast like dielectric spheres. But when the laser can excite the plasmon ($\omega_0$ = 1.5 eV and T.M. polarization) the polarizabilities of the spheres are strongly enhanced, the diffracted field by the sphere is very strong and the spheres appear with a normal contrast.

## 8. Conclusion

We have described the evolution of the theoretical works in SNOM and described the principle of the theories which can lead to calculated images of realistic samples.

The most promising theories are of course global theories which in principle can lead to near-field intensity calculations where coupling between the tip and the sample is taken into account. Grating theory can only be applied to periodic samples but it is a very powerful method which has been applied to sample and tip of any shape. Moreover it needs an artificial periodic array of tips which could lead to non real coupling effects. The solution of this problem is in progress.

The difference between the macroscopic theory of Dereux and the microscopic theory of Girard is almost a question of pedagogy. The propagator $\overleftrightarrow{P}$ is proportional to $\overleftrightarrow{G}$. The Green's function method is a macroscopic theory and must be applied to a continuous sample and tip. But for computing purpose a discretization is necessary and points of the discretization grid of Dereux can be considered as the discrete scattering centers of the microscopic method.

The global methods (grating theory, Green's function or microscopic theory) could be applied to a large variety of sample and tips and are particularly well suited to study the influence of tip shape on image quality. However they lead to only numerical algorithms and need very long computing times.

By using a perturbation method the non-global models can calculate the near-field above a large variety of samples. But they are restricted to tips with a very simple shape (dipole, sphere, circular hole...). The models of tips can be easily improved. For instance, by using the formalism of Roberts [19, 20] the efficiency of an aperture or of a conical tip could be extended further than the dipolar terms of Bethe.

Non-global theories are not rigorous as they do not take the tip-sample coupling into account. But non-global theories are useful and will be always useful because they need small computing times and lead to analytical formulas. So they are a very precious tool for discussing the influence of the various parameters on the detected intensity, on images quality and resolution.

Moreover the important experimental parameters (tip-sample distance, tip shape, tip radius...) are generally badly known or unknown and the precision of non-global theories is todays enough for describing the main features of SNOM. For a given SNOM experiment, by using a non-global theory to calculate the detected intensity we obtain in a short computing time the leading term of a "rigorous" result eventually given by a global theory and this leading term can be very useful for testing the convergence of numerical algorithms of global theories.

The previous theoretical results suggest experiments: a study of tip radius on the resolution will be useful, a comparison of STOM images with different kind of tips will be very interesting for testing the various models. But such a comparison between theories and experiments supposes a precise knowledge of experimental parameters.

## References

[1] Descouts, P. and Siegenthaler, H. (1991) '10 Years of STM', Proceedings of the Sixth International Conference on Scanning Tunneling Microscopy Interlaken North Holland .

[2] Pohl D. W. , (1990) 'Scanning near-field optical microscopy', in C .J. R. Sheppard, and T. Mulvey (eds), Advances in optical and electron microscopy, Academic Press, London, pp. 243-312 .

[3] Synge E. A. , (1928) 'A suggested method for extending microscopic resolution into the ultra-microscopic region', Phil. Mag. C, 356-362.

[4] Mc Cutchen, C. W. (1967) 'Superresolution in Microscopy and the Abbe resolution limit', J. Opt. Soc. Amer. 57, 1190-1192.

[5] Massey, G. A. (1984) 'microscopy and pattern generation with scanned evanescent waves', Appl. Opt. 23, 658-660.

[6] Ch. Fischer, U., Dürig, U. and Pohl, D. W. (1987)' Near-field optical scanning microscopy and enhanced spectroscopy with submicron aperture', Scanning Microscopy Supplement 1, 47-52.

[7] Leviatan, Y. (1988) 'Electromagnetic coupling between two half-space regions separated by two slot-perforated parallel conducting screens', IEEE trans. Microw. Theory Tech. MTT-36, 44-50.

[8] Vigoureux, J. M., Girard, C. and Courjon, D. (1989) 'General principle of scanning tunneling optical microscopy,' Opt. Lett. 14, 1039-1041.

[9] Vigoureux, J. M., Depasse, F. and Girard, C. (1992) 'superresolution of near-field optical microscopy from properties of confined electromagnetic waves', Appl. Opt. 16, 3036-3045.

[10] Vigoureux, J. M. and Courjon, D. (1992) 'Detection of nonradiative fields in light of the Heisenberg uncertainly principle and the Rayleigh criteriom', Appl. Opt. 31 p 3170-3177

[11] Bethe, H.A. (1944) 'Theory of diffraction by small holes', Phys. Rev. 66, 163-182.

[12] Bouwkamp, C. J. (1954) 'Diffraction theory', Reports in Physics, 27, 35-100.

[13] English, R. E. and George, N. (1988) 'Diffraction from a small square aperture: approximate aperture fields', J. Opt. Soc. Am. A5 , 192-199.

[14] Wirgin, M. (1970) 'Influence de l'épaisseur de l'écran sur la diffraction par une fente', C. R. Acad. Sci. Paris 270 , 1457-1460.

[15] Neerhoff, F. L. and Mur, G. (1973) 'Diffraction of a plane electromagnetic wave by a slit in a thick screen placed between two different medìa', Appl. Sci. Res. 28 , 73-88.

[16] Harrington, R. F. and Auckland, D. T. (1980) 'Electromagnetic transmission through narrow slots in thick conducting screens', IEEE Trans. Antennas Propagat. AP 28, 616-622.

[17] Mata Mendez, O., Cadihac, O. and Petit, R. (1983) 'Diffraction of a two-dimensional electromagnetic beam wave by a thick slit pierced in a perfectly conducting screen', J. Opt. Soc. Am. 73 , 328-331.

[18] Betzig, E., Harootunian, A., Lewis, A. and Isaacson, M. (1986) 'Near-field diffraction by a slit: implications for supperresolution microscopy', Appl. Opt. 25, 1890-1900.

[19] Roberts, A. (1987) 'Electromagnetic theory of diffraction by a circular aperture in a thick, perfectly conducting screen,' J. Opt. Soc. Am. A4, 1970-1983.

[20] Roberts, A. (1992) ' Small hole coupling of radiation into a near-field probe,' J. Appl. Phys. 70 , 4045-4049.

[21] Born, M., Wolf, E. (1970) 'Principle of Optics', Pergamon press, New York .

[22] Van de Hulst, H.C. (1957) 'Light scattering by small particles', Wiley, New York.

[23] Kerker, M. (1983) ' The scattering of light and other electromagnetic radiation', Academic Press, New York.

[24] Tirosh, E. and Cohen, A. (1989) 'Near-field scattering by passive and active small spherical particles', J. Opt. Soc. Am. 6, 523-524 .

[25] Jones, D. S. (1986) 'Acoustic and Electromagnetic waves', Clarendon Press, Oxford.

[26] Au Kong, J. (1984) 'Electromagnetic wave Theory', John Wiley, New York.

[27] Petit, R. (1980), 'Electromagnetic Theory of gratings' Vol. 22 of Topics in current Physics, Springer-Verlag, Berlin.

[28] Maystre, D. (1984) 'Rigorous vector theories of diffraction gratings', in E. Wolf (ed) Progress in Optics XXI, Elsevier, Amsterdam, pp 1-66.

[29] Soto-Crespo, J. M. and Nieto-Vesperinas, M. (1989) 'Electromagnetic scattering from very rough random surfaces and deep reflection gratings', J. Opt. Soc. Am. A 6 , 367-384.

[30] Toigo, F., Marvin, A., Celli, V. and Hill, N. R. (1977) 'Optical properties of rough surfaces: General theory and the small roughness limit', Phys. Rev. B13, 5618-5626.
[31] Agarwal, G. S. (1977) 'interaction of electromagnetic waves at rough dielectric surfaces', Phys. Rev. B15, 2371-2383.
[32] Maradudin, A. and Mills, D. (1975) 'Scattering and absorption of electromagnetic radiation by a semi-infinite medium in the presence of surface roughness', Phys. Rev. B 11, 1392-1415.
[33] Labani, B., Girard, C., Courjon, D. and Van labeke, D. (1990) 'Optical interaction between a dielectric tip and a nanometric lattice: implications for near-field microscopy', J. Opt. Soc. Amer. A 7, 936-943.
[34] Girard, C. and Courjon, D. (1990) 'Model for scanning tunneling optical microscopy: A microscopic self-consistent approach', Phys. Rev. B 42, 9340-9349.
[35] Girard, C. and Spajer, M. (1990) 'Model for reflection near-field optical microscopy' Appl. Opt. 29, 3726-3733.
[36] Girard, C. and Bouju, X. (1991) 'Self-consistent study of dynamical and polarization effects in near-field optical microscopy', J. Opt. Soc. Amer. B 9 , 298-305.
[37] Girard, C. and Bouju, X. (1991) 'Coupled electromagnetic modes between a corrugated surface and a thin probe tip', J. Chem. Phys. 95 , 2056-2064.
[38] Girard, C. (1992) 'Multipolar propagators near a corrugated surface: Implication for local-probe microscopy', Phys. Rev. B 45, 1800-1810.
[39] Girard, C. (1993) 'Near-field detection and local spectroscopy of a surface: a self consistent theoretical study,' N.F.O. Besançon to be published in this review.
[40] Salomon, L., De fornel, F. and Goudonnet, J.P. (1991) ' Sample-tip coupling efficiencies of the photon scanning tunneling microscope', J. Opt. Soc. Amer. 8, 2009-2015.
[41] Cites, J., Sanghadasa, M. F. M., Sung, C. C., Reddick, R. C., Warmack, R. J. and Ferrell,T. L. (1992) 'Analysis of photon scanning tunneling microscope images', J. Appl. Phys. 71, 7-10 .
[42] Van Labeke, D and Barchiesi, D. (1992) ' Scanning-tunneling optical microscopy: a theoretical macroscopic approach,' J. Opt. Soc. Am. A 9, 732-739.
[43] Jackson,J. D. (1975) ' Classical electrodynamics', John Wiley, New York.
[44] Abramowitz, M. and Stegun, I. A. (1964) 'Hanbook of mathematical functions,' Dover, New-York.
[45] Barchiesi, D. and Van Labeke, D. (1993) 'Application of Mie scattering of evanescent waves to scanning optical tunneling microscopy theory', to be published in Modern Optics.
[46] Van Labeke, D. (1992) 'Near-field optical microscopy: theoretical concepts and models', in S. Gauthier and C. Joachim ( eds.) 'Scanning Probe microscopy: beyond the images' Editions de Physique, les Hulis, France, pp. 228-275.
[47] Betzig, E., Trautman, J. K., Harris, T. D., Weiner, J. S. and Kostelar, R. L (1991) 'Breaking the diffraction barrier: optical microscopy on a nanometric scale', Science 251, 1468-1470.
[48] Barchiesi, D. and Van Labeke, D. (1993) 'Scanning tunneling optical microscopy: theoretical study of polarization effects with two models of tip', N.F.O. Besançon to be published in this review.
[49] Goudonnet, J. P., Salomon, L., De Fornel, F., Bouillot, E., Adam, P. and Neviere, M. (1992) 'Recent progress in photon scanning tunneling microscopy', Conference SPIE, Los Angeles.

[50] Denk, W. and Pohl, D. W. (1991) 'Near-field optics: Microscopy with nanometer-size fields', J. Vac. Sci. Technol. B 9, 510-513.

[51] Bernt, R., Gimzewski, J. K. and Schlittler, R.R. (1992) 'Enhanced photon emission from the STM: a general property of metal surfaces', Ultramicroscopy 42-44, 355-359.

[52] Dereux, A., Vigneron, J. P., Lambin, Ph. and Lucas, A. A. (1991) 'Theory of near-field optics with applications to SNOM and optical binding', Physica B 175, 65-67.

[53] Dereux, A. (1991) 'Théorie de l'optique du champ proche,' thesis Namur , Belgium.

[54] Greffet, J. J. and Ladan, F. R. (1992) 'Comparison between theoretical and experimental scattering of an s-polarized electromagnetic wave by a two-dimensional obstacle on a surface', J. Opt. Soc. Am. A 8, 1261-1269.

[55] Santenac, A. and Greffet, J. J. (1992) 'Scattering by deep inhomogeneous gratings,' J. Opt. Soc. Am. A 9, 996-1005.

[56] Efrima, S. and Metiu, H. (1902) 'Classical theory of light scattering by an adsorbed molecule. I. Theory' J. Chem. Phys. 70, 1602-1613.

[57] Fischer, U. Ch. and Pohl, D. W. (1989) ' Observation of single-particle plasmon by near-field optical microscopy', Phys. Rev. Lett. 62, 458-461.

[58] Neviere, M. (1993) 'Diffraction gratings as components for photon scanning tunneling microscope image interpretation', N.F.O. Besançon to be published in this review.

[59] Bernstsen, S., Bozhevolnaya, E. and Bozhevolnyi, S. (1993) 'Macroscopic self consistent model for external reflection near-field microscopy,' to be published in J. Opt. Soc. Am.

[60] Singham, S. B. and Saltzman, G. C. (1986) 'Evaluation of the scattering matrix of an arbitrary particle using the coupled dipole approximation,' J. Chem. Phys. 84, 2658-2667.

[61] Singham, S. B. and Bohren, C. F. (1992) 'Light scattering by an arbitrary particle: the scattering-order formulation of the coupled-dipole method,' J. Opt.Soc. Am. A 5, 1867-1872.

# SCANNING TUNNELING OPTICAL MICROSCOPY (STOM): THEORETICAL STUDY OF POLARIZATION EFFECTS WITH TWO MODELS OF TIP

D. BARCHIESI , D. VAN LABEKE
*Laboratoire d'Optique P.M. Duffieux U.R.A. 214 C.N.R.S,*
*U. F. R. des Sciences et des Techniques,*
*Université de Franche-Comté,*
*16 route de Gray, 25030 BESANCON Cedex .*
*Tél: (33) 81 66 64 00. Fax:(33) 81 66 64 23.*
*FRANCE*

ABSTRACT. Two kinds of probe are actually used in scanning near-field optical microscopy: dielectric tip and small aperture probe. The purpose of this paper is to compare theoretically the two different probes by calculating and comparing the images of a same sample. We choose to describe a Scanning Tunneling Optical Microscope (S.T.O.M.) where the sample is illuminated by total internal reflection. The dielectric tip is modeled as a small scattering dipolar center. The intensity detected by the other probe is calculated by using the diffraction theory of Bethe and Bouwkamp. It is shown that the two probes do not detect the same things: the dielectric tip picks up the square modulus of the electric near-field. The other probe is sensible both to the electric and magnetic fields. The models are used to calculate images of a two-dimensional object (a letter) of dimension smaller than the wavelength. The images are quite different and are very sensitive to polarization of the incident field.

## 1. Introduction

Scanning Near-field Optical Microscopes (S.N.O.M.) exist both in reflection and transmission and the technical solutions are now very similar: to get an image of the sample a probe smaller than the wavelength is scanned very close to the surface of the sample. However two kinds of probes are used:

- The first one is a small aperture probe. It is build by drawing a micropipette or a fiber which is then metallized. The resulting probe has a flat end face with a small adjustable aperture from 20 nm to 500 nm [6].
- The second kind of probe is a dielectric tip [3,4,5] build from an optical fiber chemically etched.

The lateral resolutions obtained by microscopes which use the two kinds of probes are very similar and are now around 20 nm [6, 7].

The purpose of this paper is to compare theoretically the two different probes by calculating and comparing the images of a same sample. We choose to describe a Scanning Tunneling Optical Microscope (S.T.O.M.) where the sample is illuminated by total internal reflection. For these calculations we apply a macroscopic formalism, based on Maxwell equations, diffraction and scattering theories. We consider a general two-dimensional object. The two kinds of probe are modeled in terms of polarizabilities.The dielectric tip is modeled as a small scattering



*D.W. Pohl and D. Courjon (eds.), Near Field Optics,* 179–188.

dipolar center. The intensity detected by the other probe is calculated by using the diffraction theory of Bethe and Bouwkamp.

It is shown that they do not detect the same things: the dielectric tip picks up the square modulus of the electric near-field. The other probe is sensible both to the electric and magnetic fields. Then we show that the different behaviors of the probes lead to very different polarization properties of the calculated images.

**2. Near-field**

The principle of S.T.O.M. with two types of probe is presented on fig. 1.

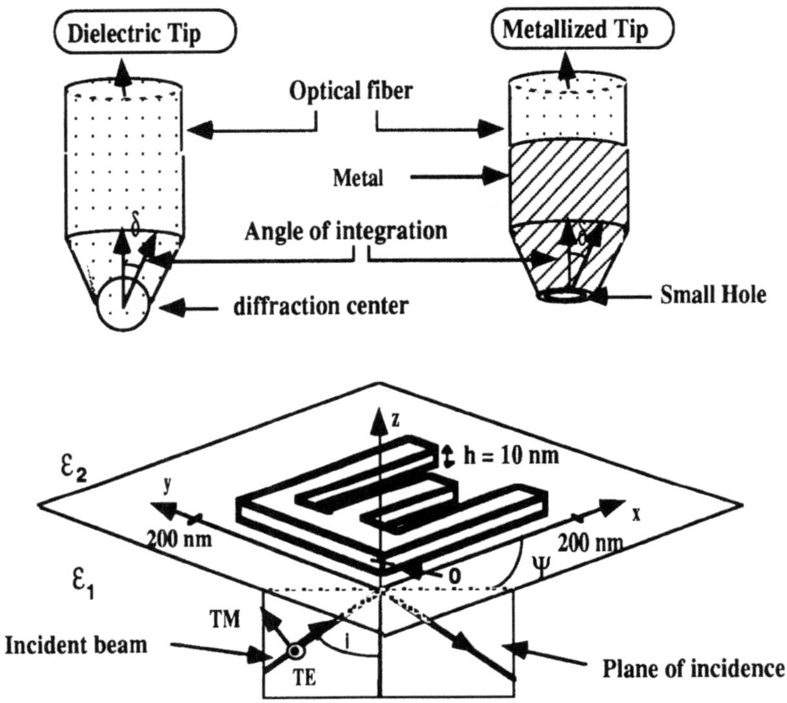


Figure 1 : Schematic of S.T.O.M. device and parameters of incident field, object and probe. Here the sample (letter "E") is in glass of index $n_1$= 1.5. The pattern is on a square of 200 nm tall and is 10 nm height.

In our model the sample is the surface of a glass of dielectric constant $\varepsilon_1 = (n_1)^2$ and profile z = f(x,y). The mean plane of this profile is the z = 0 plane, the z-axis being perpendicular. The probe is immersed in a medium of dielectric constant $\varepsilon_2 = (n_2)^2$ (generally air). This sample is illuminated in total internal reflection by a laser of wavelength $\lambda$ in vacuum. The angle of incidence is i, the plane of incidence is labeled by the angle $\psi$ with the x-axis.

The wave vector $\vec{k_i}$ of the incident light has for components $\vec{k_i} = (u^0, v^0, w^0{}_1)$ which verify relation:

$$(u^0)^2 + (v^0)^2 + (w^0{}_1)^2 = \varepsilon_1 (2\pi/\lambda)^2 \qquad (1)$$

The incident field has for amplitude $E_0$, it is plane polarized and the direction of polarization is noticed by the angle $\phi$ ($\phi = 0$ corresponds to a Transversal Electric polarization T.E., $\phi = \pi/2$ to a Transversal Magnetic one T.M.). This field can thus be written:

$$\vec{E_i}(x, y, z) = \vec{E_i} \exp[i\vec{k_i}.\vec{r}\,] \qquad (2)$$

The field in the medium $\varepsilon_2$ is the sum of the transmitted field which exist even if the glass surface is a perfect plane and the diffracted field which can be expressed as a plane wave spectrum (Rayleigh hypothesis) and is related to the glass corrugations. They are noted respectively:

$$\vec{E_t}(x, y, z) = \vec{E_t} \exp[i\vec{k_t}.\vec{r}\,] \qquad (3)$$

$$\vec{E_d}(x, y, z) = \int_{-\infty}^{+\infty} \vec{E_d}(u,v) \exp[i\vec{k_2}.\vec{r}\,]\, du\, dv \qquad (4)$$

where the wave vectors $\vec{k_t}$ and $\vec{k_2}$ verify the relations:

$$\vec{k_t} = (u^0, v^0, w^0{}_2) \text{ with } (u^0)^2 + (v^0)^2 + (w^0{}_2)^2 = \varepsilon_2 (2\pi/\lambda)^2$$

$$\vec{k_2} = (u, v, w_2) \text{ with } (u)^2 + (v)^2 + (w_2)^2 = \varepsilon_2 (2\pi/\lambda)^2. \qquad (5)$$

When, the angle of incidence is over the critical one ($n_1 \sin(i) > n_2$), the wave vector $\vec{k_t}$ is complex, $w^0{}_2$ is pure imaginary and the transmitted wave is evanescent. The amplitudes $\vec{E_t}$ are linearly connected to the incident one $\vec{E_i}$ (Fresnel relations in cartesian coordinates).

We also introduce the diffracted magnetic field and the wave vector of the diffracted field in the glass for later purpose:

$$\vec{B_d}(x, y, z) = \int_{-\infty}^{+\infty} \vec{B_d}(u,v) \exp[i\vec{k_2}.\vec{r}\,]\, du\, dv \qquad (6)$$

$$\vec{k_1} = (u, v, w_1) \text{ with } (u)^2 + (v)^2 + (w_1)^2 = \varepsilon_1 (2\pi/\lambda)^2. \qquad (7)$$

Integrals (4-6) are 2D Fourier integrals over the spatial frequencies u, v. They are angular spectrum of plane waves [8] but they generally contain both homogeneous and evanescent waves.

The plane wave is homogeneous when $w_2$ is real ($(u)^2 + (v)^2 \leq \varepsilon_2(2\pi/\lambda)^2$ ) and evanescent when $w_2$ is pure imaginary ($(u)^2 + (v)^2 > \varepsilon_2(2\pi/\lambda)^2$ ).

The calculation of the diffracted field amplitudes $\vec{E_d}(u,v)$ is a very difficult task and many papers are considering this subject. For a sample with details of lateral dimension (a) smaller than the wavelength, the significant spatial frequencies of the object are high and verify $(u)^2 + (v)^2 >> \varepsilon_2(2\pi/\lambda)^2$. In this case the decay lengths become practically independent of the exciting wavelength, are rather determined by "a" and can be very small [8-9].

For a sample with details smaller than $\lambda$, the diffracted field rapidly decay with z and only the sample with low corrugation ($f(x,y) << \lambda$) can be imaged. A perturbation method can thus be used for calculating $\vec{E_d}(u,v)$. We have also used such a method in our theoretical work on S.T.O.M. [8, 9 and references there in]. For a general surface shape, we introduce the 2D Fourier spatial spectrum of the profile:

$$f(x,y) = \int_{-\infty}^{+\infty} F(u,v) \exp[i(ux+vy)]\, du dv \tag{8}$$

By limiting the perturbation series to the first order ($f(x,y) << \lambda$), the diffracted amplitude is proportional to the profile Fourier coefficient $F(u-u^0, v-v^0)$ and is linearly connected to the transmitted field :

$$E_{dx}(u,v) = i(\varepsilon_1 - \varepsilon_2)F(u-u^0, v-v^0)\left[\frac{\omega^2/c^2}{w_1 + w_2}E_{tx} - \frac{u}{\varepsilon_2 w_1 + \varepsilon_1 w_2}(uE_{tx} + vE_{ty} + w_2 E_{tz})\right]$$

$$E_{dy}(u,v) = i(\varepsilon_1 - \varepsilon_2)F(u-u^0, v-v^0)\left[\frac{\omega^2/c^2}{w_1 + w_2}E_{ty} - \frac{v}{\varepsilon_2 w_1 + \varepsilon_1 w_2}(uE_{tx} + vE_{ty} + w_2 E_{tz})\right] \tag{9}$$

$$E_{dz}(u,v) = i(\varepsilon_1 - \varepsilon_2)F(u-u^0, v-v^0)\left[\frac{\omega^2/c^2}{w_1 + w_2}E_{tz} - \frac{w_1}{\varepsilon_2 w_1 + \varepsilon_1 w_2}(uE_{tx} + vE_{ty} + w_2 E_{tz})\right]$$

For a periodic sample integrals (4-6) are in fact series. For a grating with grooves parallel to the y-axis $F(u,v)$ is independent of v. When this grating is illuminated in the principal plane (x-z plane: $\psi = 0°$) the diffracted waves have $v = 0$ and in this case formula (9) coincide with those given in our previous papers [8, 9].

Now, using Maxwell equations, the magnetic field for the transmitted and diffracted waves can be easily calculated:

$$\vec{B_t} = \frac{c}{\omega}\vec{k_t} \wedge \vec{E_t} \tag{10}$$

$$\vec{B_d} = \frac{c}{\omega}\vec{k_2} \wedge \vec{E_d} \tag{11}$$

## 3. Intensity detected by the probe

### 3. 1 DIELECTRIC PROBE

The first kind of tip which is used in S.N.O.M. experiments is a chemically etched optical fiber. Some theoretical papers use the Poynting's vector for calculating the intensity captured by the tip [10-11]. Other studies [12 and references there in] suppose that only the end part of the tip is active. This apex is modelized by a small sphere which scatters the near-field above the sample into the fiber within a solid angle defined by angle $\delta$ (Fig. 1).

This apex acts as a scattering center which transforms the evanescent near-field into propagating mode which are guided by the fiber to the detector. It is characterized by its radius R and its dielectric constant $\varepsilon$. The center of the apex has for coordinates $\vec{R} = (x, y, z)$. Moreover we suppose that the sphere is so small that it can be considered as a small electric dipolar polarizable system with a polarizability:

$$\alpha = \frac{(\varepsilon - 1)}{(\varepsilon + 2)} R^3 \tag{12}$$

The near-field above the grating induces an electric dipole at the sphere center:

$$\vec{d} = \alpha \vec{E}_2(x,y,z) = \alpha\, [\vec{E}_t(x,y,z) + \vec{E}_d(x,y,z)]. \tag{13}$$

After integration over solid angle the detected intensity can be written (14):

$$I_d(x, y, z) = \frac{\omega^4}{96c^3}\left[ (|d_x|^2 + |d_y|^2)(16 - 15\cos\delta - \cos 3\delta) + |d_z|^2 (16 - 18\cos\delta + 2\cos 3\delta)\right]$$

In the present model the dielectric probe only detects the electric near-field at its center. For a larger sphere, for calculating the scattered intensity by the apex, other multipolar terms have to be taken into account. For this purpose we have applied Mie scattering of an evanescent wave to study the influence of tip radius on S.T.O.M images [9].

The polarizability of the sphere is isotropic. But the scattered intensity is collected into a finite solid angle centred around the z-axis. Thus when $\delta \neq \pi$ formula (14) is anisotropic, the 3 components of the dipole are no longer equivalent.

Moreover the 3 components of the field $\vec{E}_2(x,y,z)$ are not equal and their relative values depend on the polarization of the laser and on the sample width by way of formulas (3 and 9). So the polarization properties of the detected signal can be modify by laser, sample and tip geometry.

### 3.2 SMALL APERTURE PROBE

The second kind of probe is a small aperture tip. It is made from a micropipette [13] or an optical fiber [3] and can be described (Fig.1) as a conical metallized tip, with a flat end face but leaving a transmissive aperture of radius R. This kind of tip is used as a small source or as a probe in SNOM working in the collection mode.

In this paper, in order to have a model similar to that used for the dielectric tip, we simply apply the "small hole" theory of Bethe and Bouwkamp [14,15].
The hole of radius R and is centered at point (x, y, z).

When illuminated by the near-field above the grating, $\vec{B}_2(x,y,z)$ and $\vec{E}_2(x,y,z)$, the aperture radiates as an effective electric dipole moment $\vec{\mu}$ plus an effective magnetic dipole moment $\vec{m}$. The first one is proportional to the normal component of the electric field:

$$\mu_x = \mu_y = 0 \quad , \quad \mu_z = -\frac{2R^3}{3\pi} E_{2z}(x,y,z) \tag{15}$$

The magnetic dipole is proportional to the tangential component of the magnetic field:

$$m_x = \frac{4R^3}{3\pi} B_{2z}(x,y,z) \, , \quad m_y = \frac{4R^3}{3\pi} B_{2y}(x,y,z) \quad , \quad m_z = 0 \tag{16}$$

The detected signal is proportional to the radiated intensity by the two dipoles into the finite solid angle defined by $\delta$. Generally speaking the intensity radiated by an electric dipole and a magnetic dipole is not the sum of the intensities of the two dipoles alone. It contains interference terms because the two dipoles are excited coherently by the same field :

$$\begin{aligned} I_{sh}(x,y,z) = \frac{\omega^4}{96c^3} [ ( |\mu_x|^2 + |\mu_y|^2 + |m_y|^2 + |m_y|^2 ) (16 - 15\cos\delta - \cos 3\delta) \\ + ( |\mu_z|^2 + |m_z|^2 )( 16 - 18\cos\delta + 2\cos 3\delta ) \\ + 6(\cos 2\delta - 1)( m_x \mu_y^* + \mu_y m_x^* - m_y \mu_x^* - d_x \mu_x^* ) ] \end{aligned} \tag{17}$$

But in the case of the small hole these interference terms vanish as the two dipoles are perpendicular. We can consider that the two dipoles radiate independently.
The polarization discussion remains valid in the case of this tip. Moreover the dielectric probe (with a small radius) is only sensitive to the electric field. On the contrary the small aperture probe detects both the electric and magnetic fields. So the two probes lead to different detected intensities and polarization properties. There are differences of sign in (15-16) between the various papers on the subject [14-16], they have no influence on our results as we calculate scattered intensity and not the field.

## 4. Some results

In order to compare the two kinds of tips we have calculated the images of some sample with formulas (14) and (17) for various polarizations of the incident field. In the following the samples are in glass of index $n_1 = 1.5$. They are illuminated in total internal reflection by a laser of wavelength $\lambda = 670$ nm, the angle of incidence being i= 45 degrees. The probe is in the air ($\varepsilon_1 = 1$).

Betzick et al. [6] have produced S.T.O.M. images of letters with a small aperture probe. So we calculate with the two probes images of a two-dimensional object which is a letter "E".The pattern is on a square of 200 nm tall and is 10 nm high (fig. 1). The plane of incidence is the x-z plane ($\psi = 0$). the detected intensities are calculated for the two tips, with formulaes (9) and (14-17) and for T.E. and T.M. polarization. The dielectric sphere has the same radius as the aperture R= 20 nm. In order to compare the intensities under identical conditions we take the same distance between bottom of the tip and top of the sample: d= 5 nm.

Resulting intensities I(x,y,z), z being fixed, are presented in figures. (2-3) in 2D grey-levels images.

First we can verify that images have a really good contrast, even for a sample made of glass with the same index as the substrate. We choose letter "E", because it is made of perpendicular lines which can show interesting polarization effects which are easily seen by comparing Figs. 2(a) and 2(b), Figs. 3(a) and 3(b).

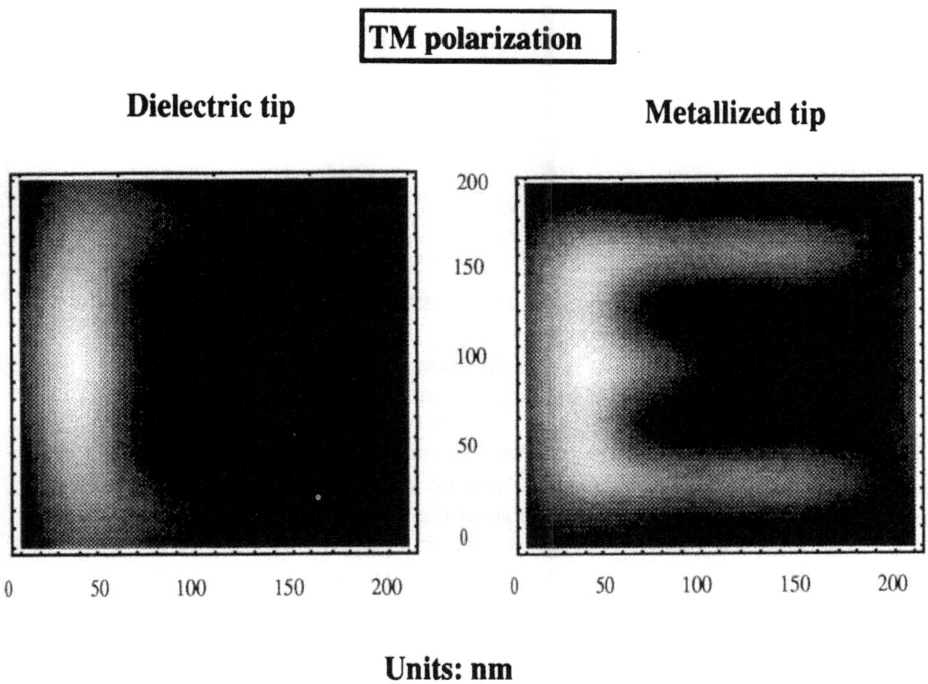


Figure 2. S.N.O.M. images with a dielectric tip (a) and metallized tip (b).
The sample is described on figure 1. The tip radius is R= 20 nm. The distance from the bottom of the tip to the top of the letter is d = 5 nm. The laser wavelength is $\lambda = 670$ nm. The angle of incidence is beyond the critical angle (i = 45°). The plane of incidence is the x-z plane. The polarization is T.M.

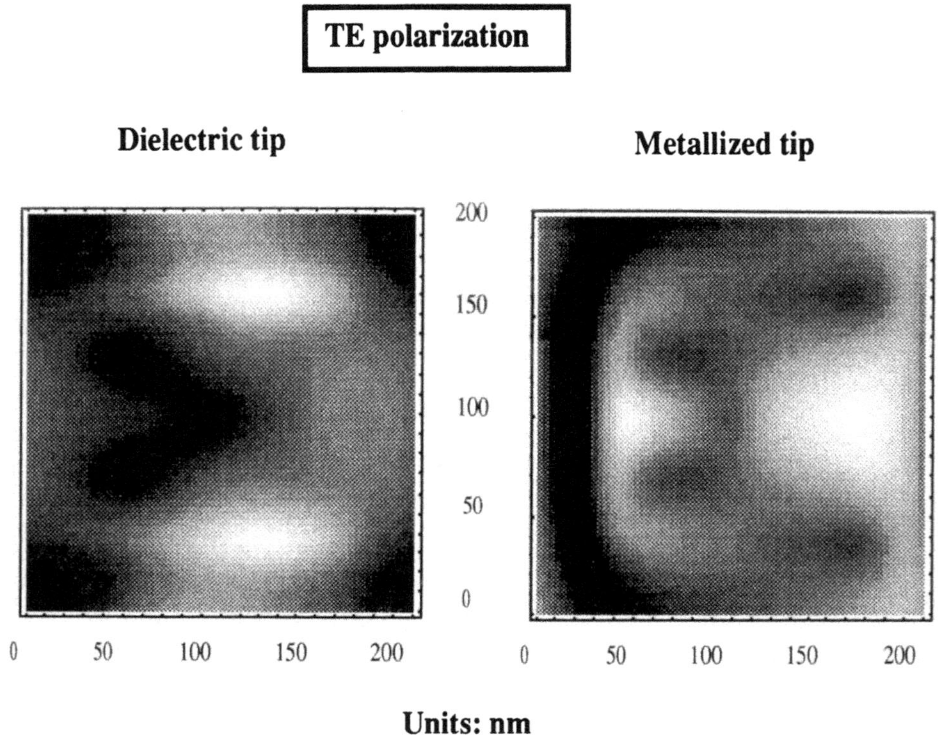


Figure 3. S.N.O.M. images with a dielectric tip (a) and metallized tip (b).
Parameters are the same as in Figure 2. except the polarization which is T.E.

In T.E. polarization, the dielectric probe only detects lines parallel to the x-axis (Fig. 2(a)), in T.M. polarization the lines parallel to the y-axis are more intense (Fig. 2(b)). This property is perfectly coherent with the polarization behavior that we have found for a grating, where the grooves parallel to the y-axis are difficult to detect in T.E. polarization, for $\psi = 0$. In the case of the small aperture probe, the polarization properties are inverted when compared to images obtained by the other tip (compare Figs. 2(a) and 3(b)). For instance lines parallel to the y-axis can be seen in T.E. polarization (Fig. 3(a)). Moreover the small aperture tip leads to an image well correlated to the original object (Fig. 3(b)) because this probe detects both the electic and magnetic field.

## 5. Conclusions

This paper is a first attempt to compare theoretically images obtained by the two kinds of probes used in near-field microscopy. It would be useful to experimentally perform this comparison which has not yet be realized. However a realistic interpretation of experimental images needs parameters which is difficult to determine. In [6] it is claimed that the radius of the small aperture is adjustable from 20 nm to 500 nm. It is more difficult to know the true radius of the dielectric tip, 50 nm seems to be the smaller radius [19]. The conical angle δ is not exactly known. Finally it is difficult to have a good knowledge of the tip-surface distance.

When images with well known parameters will be available, our crude models will be easily improved. A better approximation for the near-field above the sample can be obtained using rigorous electromagnetic theory, or an iterative method. For the dielectric tip, Mie theory can be extended to the scattering of the near-field [9]. For the small aperture tip, more rigorous theories are available but they suppose an homogeneous incident wave. It would be important to extend these theories to evanescent incident field. Coupling between the sample and the tip have been neglected in this paper. More rigourous models where the tip-sample interaction is taken into account are now available: microscopic model of Girard et al., Green function method of Dereux and grating theory of Nevière [21, references therein]. It would be interesting to compare results of such calculations to our crude model.

The models presented here describe main features of the phenomenom occuring in S.T.O.M and can be easily applied to other types of near-field microscopy.

## 6. References

[1] Pohl, D. W. Denk , W. Dürig, U. (1985) ' Optical stethoscopy : imaging with λ/20', in K.M. Monahan (ed.), Micron and Submicron Integrated circuit metrology', Proc. SPIE 565, pp. 56-61.

[2] Betzig, E. Lewis, A. Harootunian, A. Isaacson, M. Kratschmer, E. (1986) 'Near-field scanning optical microscopy', Biophys. J. 49, 269-279.

[3] Reddick, R. C. Warmack, R. J. Ferrel, T. L. (1989) 'New form of scanning optical microscopy', Phys. Rev. B 39, 767-770.

[4] Courjon, D. Sarayedine, K. Spajer, M. (1989) 'Scanning tunneling optical microscopy', Opt. Comm. 71, 23-28.

[5] de Fornel, F. Goudonnet, J. P. Salomon, L. Lesniewska, E. (1989) 'An evanescent field optical microscope', Proc SPIE 1139, pp. 77-84.

[6] Betzig, E. Trautman, J. K. Harris, T. D. Weiner, J. S. Kostelar, R. L (1991) 'Breaking the diffraction barrier: optical microscopy on a nanometric scale', Science 251, 1468-1470.

[7] Courjon, D. Bainier, C. Spajer, M. 'Imaging of submicron index variations by scanning optical tunneling microscopy', submitted to J. Vac. Sci. Technol.

[8] Van labeke, D. Barchiesi, D. (1992) 'Scanning-tunneling optical microscopy: a theoretical macroscopic approach', J. Opt. Soc. Am. A 9, 732-739 .

[9] Barchiesi, D. Van Labeke, D. 'Application of Mie scattering of evanescent waves to scanning optical tunneling microscopy theory', to be published.

[10] Salomon, L. de Fornel, F. Goudonnet, J. P. (1991) 'Sample-tip coupling efficiencies of the photon-scanning tunneling microscope', J. Opt. Soc. Am. A 8, 2009-2015 .

[11] Cites, J. Sanghadasa, M. F. M. Sung, C. C. Reddick, R. C. Warmack, R. J. Ferrell, T.L. (1992) 'Analysis of photon scanning tunneling microscope images', J. Appl. Phys. 71, 7-10 .

[12] Labani, B. Girard, C. Courjon, D. Van Labeke,D. (1990) 'Optical interaction between a dielectric tip and a nanometric lattice: implications for near-field optical microscopy', J. Opt. Soc. Am B 7, 936-942 .

[13] Betzig, E. Isaacson, M. Lewis, A. (1987) 'Collection mode near-field scanning optical microscopy', Appl. Phys. Lett. 51, 2088-2090 .

[14] Bethe, H.A. (1944) 'Theory of diffraction by small holes,' Phys. Rev. 66,163-182.

[15] Bouwkamp, C. J. (1954), Diffraction theory', Reports in Physics, 27, 35-100 .

[16] Betzig, E. Harootunian, A. Lewis, A. Isaacson, M. (1986) ' Near-field diffraction by a slit: implications for supperresolution microscopy', Appl. Opt. 25, 1890-1900.

[17] Roberts, A. (1991) 'Small hole coupling of radiation into a near-field probe', J. Appl. Phys. 70, 4045-4049 .

[18] Jackson, J. D. (1975) Classical Electrodynamics Wiley, New York, chap. 9.

[19] Jaloçha, A. Leblanc, S. Spajer, M. (1990) 'Etudes des capteurs utilisés en microscopie tunnel optique', Conference 'Opto 90', E.S.I. publications, Paris.

[20] Van Labeke, D. Barchiesi, D. 'Theoretical problems in scanning near-field optical microscopy', published in this revue.,and authors papers in this revue.

# THE 90° PRISM EDGE AS A MODEL SNOM PROBE: NEAR–FIELD, PHOTON TUNNELING, AND FAR–FIELD PROPERTIES

A. DEREUX [(a)] and D. W. POHL
*IBM Research Division*
*Zurich Research Laboratory*
*CH–8803 Rüschlikon*
*Switzerland*

ABSTRACT: A two–dimensional model SNOM is studied with the localized Green's dyadic method, allowing the determination of the relevant near and far–fields as well as the tunneling, radiation and imaging characteristics. The model consists of an illuminated 90° prism edge that approaches a planar object with a small protrusion at its surface. The edge allows coupling to the object via evanescent waves (photon tunneling), while the main part of the radiation is totally reflected at the prism side walls. The model comprises all the features that are instrumental for a realistic SNOM device. The far–field providing the signal is emitted to the sides rather than along the symmetry line, and the lateral resolution does not exceed the diffraction limit significantly.

## 1. Introduction

In scanning near–field optical microscopy (SNOM), a localized probe together with the substrate surface creates a near–field that is characteristic for the particular combination. For detection, the far–field emerging from the near–field is monitored as a function of probe position. Hence, understanding the probe–sample near–field interaction and how it influences the detectable far–field is important.

The description of the coupling between near– and far–fields requires the inclusion of retardation effects; hence the electrostatic approximation is not sufficient. On the other hand, the objects of interest generally have subwavelength sizes, so Kirchhoff theory is not applicable either. It is necessary to solve the complete set of Maxwell's equations, which is traditionally done by separation techniques. Unfortunately, analytical solutions can only be found for a few specific geometries, since separability of the vector wave equation is a necessary but not sufficient condition for this purpose. Low symmetry systems such as the arrangement of a pointed probe tip in front of an arbitrary surface profile cannot be easily studied in this way.

The Green's dyadic method associated with perturbation theory [1–3] is a convenient tool for achieving this purpose for arbitrary shapes and optical constants. This method explicitly avoids the matching of boundary conditions and provides general solutions of Maxwell's equations through a self–consistent integral equation. It is applicable whenever the perturbation is restricted to a small volume in space.

In order to get some insight into the optical interaction between tip and sample, this report will present theoretical investigations and computations of optical near– and far–field distributions in a primitive, though realistic, two–dimensional SNOM model. It consists of a 90° glass prism approaching a glass substrate. In this system, the incident light is totally



*D.W. Pohl and D. Courjon (eds.), Near Field Optics*, 189–198.

reflected at the prism faces so that only the edge allows coupling to the substrate. For computational convenience, the prism is assumed to be small, its entrance plane extending over 440 nm only. Light arrives from the top in the form of a plane wave with a wavelength of $\lambda = 488$ nm, but is prevented from entering the region beneath the prism by a metallic screen and total reflection at the prism walls. The only way to transmit radiation into the lower half plane is by frustration of the evanescent wave near the prism edge, i.e. photon tunneling (cf. Fig. 1). It will be seen that this arrangement allows the relevant fields and the imaging process to be studied in great detail

We will only discuss the results for the TE polarization (electric field along $x$, perpendicular to the plane of the paper), for which the present computational procedure is well suited. In the TM case (electric field in the $y$–$z$ plane) charges are induced at the rims of a slit [4]. These charges travel along the screen and generate a very strong near–field (divergent for a perfect conductor). This unwanted effect tends to mask the relevant near–field effects. In order to describe the coupling to the far–field of SNOM probes for both polarizations, our programs still have to be refined.

Within the present model, we discuss the TE field distributions as the edge comes close to the substrate. We will also investigate the changes of the near– and far–field patterns when the prism edge scans above a protrusion on the substrate. They turn out to influence the far–field in a sensitive way.

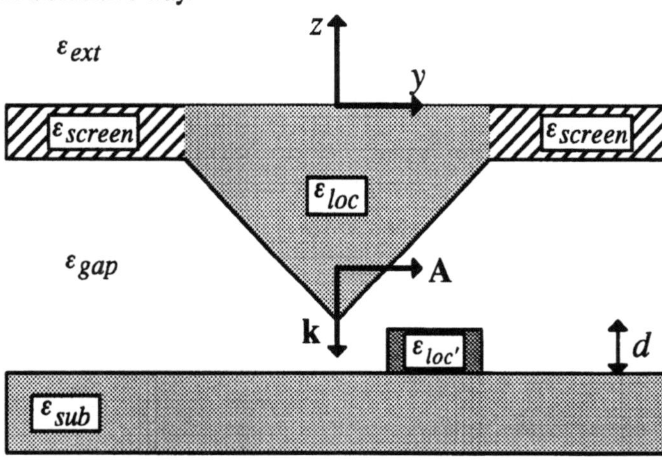


Figure 1: Multilayer reference with local perturbation by a prism $\varepsilon_{loc} = 2.25$ (glass) and a rectangular protrusion on the substrate. The different layers from top to bottom and their dielectric constants are: the base of the glass prism ($\varepsilon_{ext} = 2.25$), the metallic (silver) screen ($\varepsilon_{screen} = -9.2+0.35i$ , thickness = 50 nm), the vacuum gap ($\varepsilon_{gap} = 1$) and the glass substrate ($\varepsilon_{sub} = 2.25$).The rectangular protrusion measures 50 nm by 25 nm with $\varepsilon_{loc'} = 2.25$ or 5. The vector **A** represents either the incident electric field (TM) or magnetic field (TE).

## 2. Vector Lippmann–Schwinger Equation with Two–Dimensional Perturbation

In this section, we briefly describe the analytical framework underlying the numerical study. Assuming the $\exp(-i\omega t)$ time dependence in Maxwell's equations, the electric field solution $\mathbf{E}(\mathbf{r})$ of the perturbed vector wave equation

$$- \nabla \times \nabla \times \mathbf{E}(\mathbf{r}) + \frac{\omega^2}{c^2} \varepsilon(z) \mathbf{E}(\mathbf{r}) = V(\mathbf{r}) \mathbf{E}(\mathbf{r}) \tag{1}$$

is given by the vector Lippmann–Schwinger equation [5]:

$$\mathbf{E}(\mathbf{r}) = \mathbf{E}^0(\mathbf{r}) + \int d\mathbf{r}' \, G^0(\mathbf{r},\mathbf{r}') \, V(\mathbf{r}') \, \mathbf{E}(\mathbf{r}') \, . \tag{2}$$

With $\mathbf{\mathit{1}}$ as the unit dyadic, the perturbation $V(\mathbf{r})$ describes a spatially localized dielectric function change relative to the multilayer reference system characterized by $\varepsilon(z)$ (Fig. 1):

$$V(\mathbf{r}) = V(y,z) = \mathbf{\mathit{1}} \, [\, \varepsilon(z) - \varepsilon_{loc}(y,z) - \varepsilon_{loc'}(y,z)] \, \frac{\omega^2}{c^2} \, . \tag{3}$$

$\mathbf{E}^0(\mathbf{r})$ is the solution of Eq. (1) for $V(\mathbf{r}) = 0$: In our case, it is the analytically known solution of an electromagnetic field propagating in a multilayered medium. The Green's dyadic $G(\mathbf{r},\mathbf{r}')$ pertaining to the multilayered reference system is defined by

$$- \nabla \times \nabla \times G^0(\mathbf{r},\mathbf{r}') + \frac{\omega^2}{c^2} \varepsilon(z) \, G^0(\mathbf{r},\mathbf{r}') = \mathbf{\mathit{1}} \, \delta(\mathbf{r} - \mathbf{r}') \, . \tag{4}$$

Due to the planar symmetry of the reference system, both $\mathbf{E}^0(\mathbf{r})$ and $G^0(\mathbf{r},\mathbf{r}')$ can be formulated as two–dimensional Fourier integrals:

$$\mathbf{E}^0(\mathbf{r}) = \frac{1}{4\pi^2} \int dk_x \int dk_y \, \hat{\mathbf{E}}^0(z; k_x, k_y) \, e^{ik_x x + ik_y y} \tag{5}$$

$$G^0(\mathbf{r},\mathbf{r}') = \frac{1}{4\pi^2} \int dk_x \int dk_y \;\; \mathbf{g}^0(z,z'; k_x, k_y) \, e^{ik_x(x-x') + ik_y(y-y')} \, . \tag{6}$$

The particular form of the perturbation defined in equation (3) implies that the $x$ component of the wave vector is conserved. To take advantage of this symmetry along the $x$ direction, we Fourier analyze the electric field solution as follows:

$$\mathbf{E}(\mathbf{r}) = \frac{1}{2\pi} \int dk_x \;\; \hat{\mathbf{E}}(y,z;k_x) \, e^{ik_x x} \, . \tag{7}$$

We introduce this Fourier decomposition in the vector Lippmann–Schwinger equation. Using the well–known representation of the Dirac delta function

$$\delta(k_x - k_x') = \frac{1}{2\pi} \int dx' \;\; e^{i(k_x - k_x')x'} \tag{8}$$

we clearly see that the vector Lippmann–Schwinger equation can be solved for each $k_x$ component separately and that the discretization in direct space is limited to the $y$–$z$ plane. If we further assume an incident plane wave, the $x$–coordinate can be eliminated from (2):

$$\hat{\mathbf{E}}(y,z;k_x) = \frac{1}{2\pi}\, e^{ik_y y}\, \hat{\mathbf{E}}^0(z;k_x,k_y) + \int dy' \int dz'\, \mathbf{g}^0(y,y',z,z';k_x)\, V(y',z')\, \hat{\mathbf{E}}(y',z';k_x) \tag{9}$$

where $\mathbf{g}^0(y,y',z,z';k_x)$ is the $k_x$–resolved Green's dyadic resulting from the integration of (6) over $k_y$ alone:

$$\mathbf{g}^0(y,y',z,z';k_x) = \frac{1}{2\pi} \int dk_y\, \mathbf{g}^0(z,z';k_x,k_y)\, e^{ik_y(y-y')} . \tag{10}$$

If the incident wave vector is perpendicular to the $x$ direction, we have $k_x = 0$ and the $k_x$ – resolved Green's dyadic has a simplified structure [$Y = (y-y')$]:

$$\mathbf{g}^0(y,y',z,z';k_x = 0) = \int_0^\infty \frac{dk_y}{\pi} \begin{vmatrix} \cos(k_yY)\, F_{yy} & 0 & 0 \\ 0 & \cos(k_yY)\, F_{xx} & i\, \sin(k_yY)\, F_{xz} \\ 0 & i\, \sin(k_yY)\, F_{zx} & \cos(k_yY)\, F_{zz} \end{vmatrix} . \tag{11}$$

In Eq. (11), the functions $F_{ij}=F_{ij}(z,z';k_y)$ [with $i,j= x,y,z$] depend only on the magnitude of $k_y$. Their detailed mathematical descriptions for the case of a planar surface and that of a slab were given by Maradudin and Mills [6,7]. Their generalization to an arbitrary stratified medium can be found in [5]. The TE polarization involves the functions $F_{yy}$ only, while the TM polarization requires the four functions $F_{xx}$, $F_{xz}$, $F_{zx}$ and $F_{zz}$. Thus, at normal incidence the two polarizations can be treated independently.

Thanks to the spatial localization of the perturbations $\varepsilon_{loc}$ and $\varepsilon_{loc'}$, the integral in (9) is restricted to a domain area $S$ (prism plus the optional protrusion on the substrate in Fig. 1). As a consequence, the electric field inside $S$ can be computed independently of the electric field outside $S$. For this purpose, Eq. (9), reshaped to

$$\iint_S dy'dz' \Big[\, \mathit{1}\, \delta(y''-y')\delta(z''-z') - \mathbf{g}^0(y'',y',z'',z';k_x)\, V(y',z') \Big]\, \hat{\mathbf{E}}(y',z';k_x) = \frac{1}{2\pi}\, e^{ik_y y''}\, \hat{\mathbf{E}}^0(z'';k_x,k_y) , \tag{12}$$

is solved numerically by discretization of the area $S$ of the integration. All space coordinates $y''$, $y'$, $z''$ and $z'$ belong to $S$. Particular attention must be paid to the diagonal elements $y' = y''$ and $z' = z''$ of the resulting matrix [8–10]. Once the electric field inside $S$ is known, it can be inserted into Eq. (9) in order to find the field values anywhere else outside the perturbation. To determine the far–fields, we assume $z/\lambda \rightarrow \pm\infty$ and substitute $\mathbf{g}^0(y,y',z,z';k_x)$ by its analytical asymptotic expression [3,4,6]. The integrand of Eq. (10) can now be factorized.

$$\lim_{z \to \pm\infty} \mathbf{g}^0(y,y',z,z';k_x) = \frac{1}{2\pi} \int dk_y\, f(z';k_x,k_y)\, e^{-ik_y y'}\, e^{ik_y y + ik_\perp z} \tag{13}$$

with $k_\perp{}^2 = k_s{}^2 - k_y{}^2$, where $k_s = \omega\sqrt{\varepsilon_{far}}/c$, $\varepsilon_{far} = \varepsilon_{ext}$ if $z \to +\infty$ and $\varepsilon_{far} = \varepsilon_{sub}$ if $z \to -\infty$. The far–field emerges thus as a plane wave superposition:

$$\lim_{z \to \pm\infty} \hat{\mathbf{E}}(y, z; k_x) = \frac{1}{2\pi} \int dk_y \; \mathbf{F}(k_y; k_x) \; e^{ik_y y + ik_\perp z} \tag{14a}$$

$$= \frac{1}{4\pi} \sqrt{\frac{2k_s}{\pi}} \; \frac{e^{i(k_s r - \pi/4)}}{\sqrt{r}} \cos\theta \; \mathbf{F}(k_s \sin\theta; k_x) \; . \tag{14b}$$

In going from (14a) to (14b), the transition from cartesian to cylindrical coordinates was made with $\theta = \arcsin k_y/k_s$. The function $\mathbf{F}(k_y,k_x)$ is the result of an integral restricted to the localized perturbation

$$\mathbf{F}(k_y; k_x) = \int\!\!\int_S dy' dz' \; \mathbf{f}(y', z'; k_x, k_y) \; e^{-ik_y y'} \; V(y', z') \; \hat{\mathbf{E}}(y', z'; k_x) \; . \tag{15}$$

It is obvious from Eq. (14b) that $\mathbf{F}(k_s \sin\theta; k_x)$ represents the far–field characteristics of the light source at $S$, which appears point–like ($\mathbf{E} \propto 1/\sqrt{r}$) at large distance. The differential cross section $\sigma(\theta)$ in the planes $z = \pm\infty$ hence is proportional to $\cos^2\theta \, |\mathbf{F}(k_s \sin\theta; k_x)|^2$.

## 3. Numerical Results

The computational results discussed in this section were performed using the parameters described above in Fig. 1 and for TE polarization. In the near–field intensity (defined as $|\hat{\mathbf{E}}(y,z,;k_x = 0)|^2$ ) patterns to be displayed below, the contour lines progress by 5% steps of the maximum intensity located near the prism corner. The normalization was such that $|\hat{\mathbf{E}}^0(z{=}0;k_x = 0,k_y)|^2 = 4\pi^2$ so that the unperturbed field intensity is $|\hat{\mathbf{E}}^0|^2 = 1$ at $z = 0$ where, for the multilayered reference system considered here, the unperturbed field at $z = 0$ is the sum of the incident and the reflected field.

### 3.1. NEAR–FIELD AROUND THE EDGE

Figure 2 displays the $|\hat{\mathbf{E}}(y,z;k_x = 0)|^2$ pattern proportional to the electric field intensity [see Eq. (7)] around the edge when no substrate is facing the prism. A large lobe is found around the prism edge. Its width and length equal about $\lambda/2$, which corresponds approximately to the diffraction limit in air. Notice that the maximum intensity is not centered at the edge but slightly displaced to the interior of the prism. The field extending outside the prism is practically non–radiative since the transmitted far–field cross sections $\sigma(\theta)$ associated with this geometry are vanishingly small, as is to be expected.

### 3.2. NEAR–FIELD AROUND THE EDGE APPROACHING A FLAT SURFACE

Figure 3 shows the modification of the TE near–field intensity $|\hat{\mathbf{E}}(y,z;k_x = 0)|^2$ when the substrate approaches the edge. At large distances, the presence of the sample compresses the highest isointensity lines into the vacuum gap. As the gap is reduced, tunneling sets in,

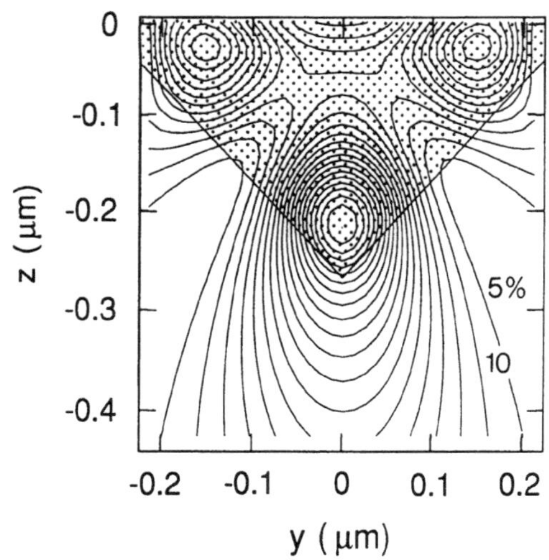


Figure 2: Calculated TE near–field intensity distribution around the edge when $d \rightarrow \infty$.

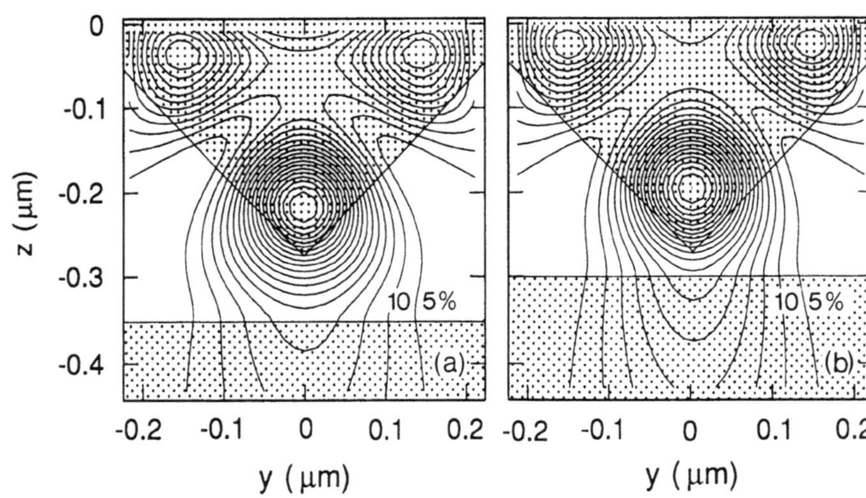


Figure 3: Calculated TE near–field intensity distribution around the edge when (a) $d = 80$ nm and (b) $d = 30$ nm.

characterized by penetration of the electric field into the sample.Notice the narrower iso-intensity lines inside the sample. The lateral FWHM now equals $\simeq$ 180 μm, i.e. half the wavelength inside the substrate. This pattern is identified as an optical tunneling channel because part of the field that has reached the sample is converted into radiative waves, giving rise to an increasing far–field total cross section as the prism comes closer to the sample surface. It is interesting to note that the far–field is mainly radiated into two lobes

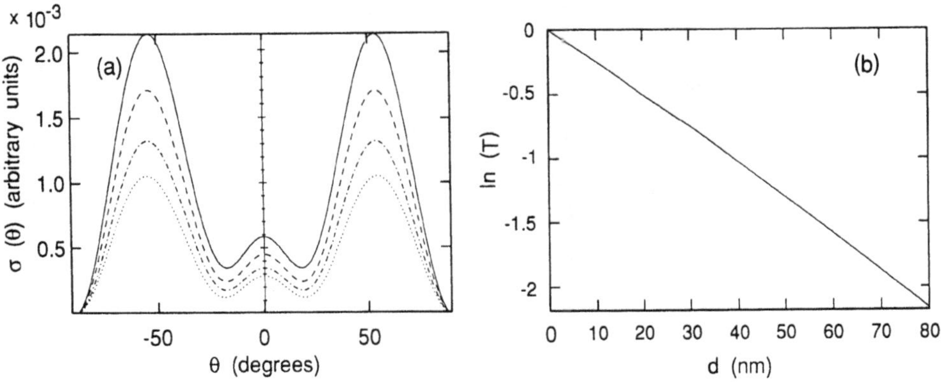


Figure 4: (a) Evolution of $\sigma(\theta)$ as a function of sample approach: $d = 0$ nm (solid line), $d = 10$ nm (dashed line), $d = 20$ nm (dot–dashed line), $d = 30$ nm (dotted line). (b) Logarithm of the total transmission $T$ [Eq. (16)] as a function of the gap $d$. The normalization is such that $T = 1$ for $d = 0$.

located symmetrically to the $z$ direction. The almost exponentially decreasing total transmitted far–field intensity

$$T = \int_{-\pi/2}^{+\pi/2} \sigma(\theta)\, d\theta \tag{16}$$

as a function of the gap width $d$ is a clear indication of the critical role played by evanescent waves (Fig. 4). The decay of the transmitted intensity is strongly dependent on the gap material. In other calculations reported elsewhere in these proceedings [11], we found a much steeper decay when the gap was filled with metal.

### 3.3. NEAR–FIELD AROUND THE EDGE NEXT TO A SURFACE PROTRUSION

For a fixed sample–to–prism edge distance $d$ in the tunneling range, we introduce a small glass protrusion with rectangular cross section (50 nm × 25 nm) on the sample surface. The distortion of the near–field pattern as the prism moves towards the protrusion is demonstrated in Fig. 5. The strongest alteration of the near–field pattern occurs on a scale fixed by the protrusion size. It causes a significant variation of the far–field differential cross section (Fig. 6). Noteworthy is the strong asymmetry of the main lobes, both with regard to amplitude and angular orientation of the maxima. The calculated $\sigma(\theta)$ provides thus a theoretical description of the SNOM imaging capabilities.

When the protrusion has the same dielectric constant as the glass substrate (Fig. 6a), the total transmitted intensity $T$ increases as the prism scans to the position of the protrusion. Additional information is contained in the far–field differential cross section. As already mentioned, the cross section as a function of $\theta$ is asymmetric if the prism edge is not exactly positioned above the defect: a higher peak occurs in the direction in which the protrusion is located (solid line in Fig. 6a). If the prism edge is exactly above the protrusion

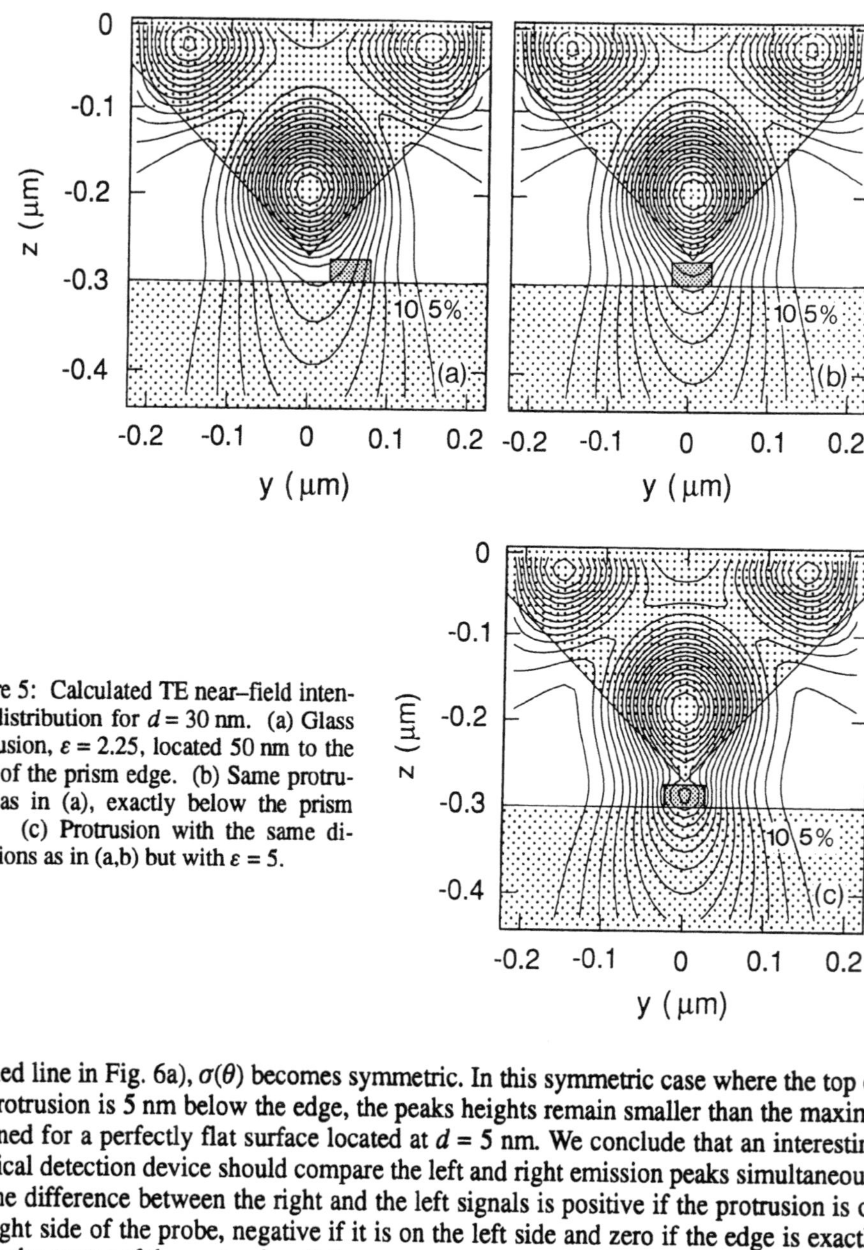


Figure 5: Calculated TE near–field intensity distribution for $d = 30$ nm. (a) Glass protrusion, $\varepsilon = 2.25$, located 50 nm to the right of the prism edge. (b) Same protrusion as in (a), exactly below the prism edge. (c) Protrusion with the same dimensions as in (a,b) but with $\varepsilon = 5$.

(dashed line in Fig. 6a), $\sigma(\theta)$ becomes symmetric. In this symmetric case where the top of the protrusion is 5 nm below the edge, the peaks heights remain smaller than the maxima obtained for a perfectly flat surface located at $d = 5$ nm. We conclude that an interesting practical detection device should compare the left and right emission peaks simultaneously. The difference between the right and the left signals is positive if the protrusion is on the right side of the probe, negative if it is on the left side and zero if the edge is exactly above the center of the protrusion. It thus represents a sort of differential image of the protrusion object; the sum of the two, on the other hand, may be considered its normal near–field optical image.

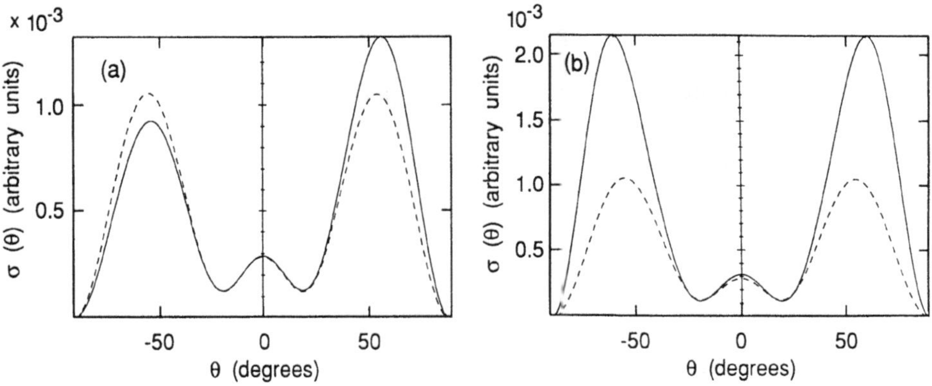


Figure 6: Calculated far–field patterns resulting from the near–fields of Fig. 5: (a) Radiation from the distributions of Figs. 5a (solid line) and 5b (dashed line). (b) Same for the patterns of Figs. 5c (solid line) and 5b (dashed line).

### 3.4. FAR–FIELD PATTERNS

In order to test the sensitivity to a variation of the dielectric function, we keep the prism edge exactly above the rectangle center but increase the protrusion dielectric constant to 5. The resulting transmitted intensity (solid line in Fig. 6b) is found to be considerably higher than for the glass protrusion with $\varepsilon$=2.25 (dashed line in Fig. 6b).

Since the amplitudes of the peaks found in the far–field contain such rich information, it is important to understand their physical origin. The peaks arise at an absolute value of $\theta$ and are such that $(\omega/c)\sin\theta\sqrt{\varepsilon_{sub}} > (\omega/c)\sqrt{\varepsilon_{gap}}$, i.e. these angles are larger than the critical angle for total reflection $\theta_{crit}$ at the vacuum–substrate interface. Indeed, the maxima arise around 55°, while for the vacuum–glass interface considered here $\theta_{crit} = 42°$. This means that the detected intensities at these large angles are related to nonradiative wave components in the gap. Thus, in the geometry considered here, the large angle scattering cross sections are directly related to evanescent waves in the gap. The evanescent waves play the main role in establishing the subwavelength near–field pattern. Therefore, the fact that our calculated scattering cross sections at large angles dominate the far–field signal explains the success of SNOM devices for achieving subwavelength resolution. It shows that the far–field emitted by the SNOM is strongly dependent on the gap nonradiative wave vectors responsible for the near–field optical interaction.

## 4. Conclusions

In this paper, we studied the light scattering at a prism edge upon approach of a sample surface using numerical solutions of the perturbed Green's dyadic method. The description of the scattering phenomenon consists of computing the near–field patterns from which far–field cross sections can be calculated. This procedure allows the observation of the transition to tunneling when the sample comes sufficiently close to the surface. In the tunneling condition, the near–field pattern is sensitively altered by the presence of a protrusion lying on the surface. This near–field modification was found to induce significant changes in the computed far–field differential cross sections. For the considered geometry, the far–field is emitted preferentially at angles larger than the critical angle for total reflection. This feature clearly identifies the evanescent wave origin of the radiation. Still, the lateral resolution of this model SNOM cannot be expected to exceed the diffraction limit by a sizeable factor: The extension of the near–fields is comparable to $\lambda/2$, hence components with a very large wave vector are not excited. We have begun a more detailed study of the imaging characteristics.

## Acknowledgments

We wish to thank A. Baratoff, Ch. Girard, A.A. Lucas, O. Martin, L. Novotny and J.P. Vigneron for fruitful discussions. One of us (A.D.) gratefully acknowledges both the Walloon Ministry for Research and Technology (Brussels) and IBM Belgium (Brussels).

## References

(a) Visiting Scientist from the Institute for Studies in Interface Sciences, Facultés Universitaires Notre Dames de la Paix, Namur, Belgium.

[1] Girard, C., Bouju, X. and Dereux, A., these proceedings.

[2] Greffet, J.J., these proceedings.

[3] Bozhevolnyi, S. and Keller, O., these proceedings.

[4] Sommerfeld, A. *Optics*, Academic Press, New York (1954).

[5] Dereux, A. Ph. D. Thesis, Facultés Universitaires N.D. de la Paix, Namur (1991).

[6] Maradudin, A.A. and Mills, D.L. Phys. Rev. B **11**, 1392 (1975).

[7] Mills, D.L and Maradudin, A.A. Phys. Rev. B **12**, 2943 (1975).

[8] Born, M. and Wolf, E., *Principles of Optics*, Pergamon, Oxford (1964), appendix 5.

[9] Van Kranendonk, J. and Sipe, J.E., Progress in Optics **XV**, 245 (1977), appendix 4.

[10] Yaghjian, A.D. Proceedings of the IEEE **68**, 248 (1980).

[11] Pohl, D.W., Courjon, D., Bainier, C., Dereux, A. and Heinzelmann, H., these proceedings.

# OPTICAL NEAR FIELD DETECTION AND LOCAL SPECTROSCOPY OF A SURFACE: A SELF-CONSISTENT THEORETICAL STUDY

**Christian Girard and Xavier Bouju**
*Laboratoire de Physique Moléculaire UA CNRS 772*
*Université de Franche-Comté*
*25030 Besançon Cedex, FRANCE .*

**Alain Dereux**[(a)]
*IBM Research Division, Zurich Research Laboratory*
*Säumerstrasse 4, CH-8803 Rüschlikon*
*SWITZERLAND .*

**Abstract.** *Optical microscopy by near field detection belongs to the new born family of local probes. This promising technique theoretically allows to image structures with typical sizes smaller than the radiation wavelength (C. Girard and D. Courjon, Phys. Rev.* **B42**, *(1990) 9340). The present work generalizes this concept to the local optical spectroscopy of surfaces. Starting from a Lippmann- Schwinger self-consistent equation, we study the near field behavior near a monolayer of metallic nanoparticles deposited on a transparent sample. In this context, we discuss the influence of different external parameters (approach distance, probe size, ...) on the spectral resolution.*

## 1 Introduction

A great variety of surface profiles appears in nanometer scale defined materials (lamellar lattices, semi-conductor superlattices, stepped surfaces, metal island films, ...). With the advent of local probe microscopies such substrates can be imaged in real space by controlling the scan of a nanometer-size detector in the immediate proximity of the sample [1]. The potential of these scanning probe techniques for nanometer-scale resolution optical microscopy has been convincingly demonstrated by many groups over the last years [2-7]. Such experiments carried out both in constant distance mode and in constant intensity mode allows one to obtain subwavelength ($\simeq \lambda/40$) resolved images. The individual structures lying on the surface induce some distorsion of the near field established by the self-consistent optical interaction between probe and sample. These distorsions are confined around the individual structures. Experimental studies clearly indicate that such a confined field strongly varies along the surface reproducing a certain relief of the object under test. Recently, by exploiting confined fields associated with resonance phenomena, the scanning plasmon near field microscopy reaches the super-resolution of $\lambda/200$ on silver samples [8].

In this work, we will address questions raised by such confined field associated with resonance (plasmon) effect occuring on the sample surface as a function of the incident



*D.W. Pohl and D. Courjon (eds.), Near Field Optics, 199–208.*

wavelength. Our study is pertinent with respect to the spectroscopic potential of the SNOM. Indeed, if a spectroscopic event occurs at a given site of the substrate, the signal detected by the probe can be used to perform a local spectroscopy of the surface. Recently, ruby photoluminescence spectra has been monitored with such a SNOM device [9]. In a similar way, the SNOM spectroscopy offers an interesting opportunity for studying localized surface plasmons near a metal island film. For such metallic objects the incident field frequency $\omega_0$ is an important parameter: the efficiency of the tip-sample coupling depends on the relative magnitude of the probe-object system plasmons with respect to $\omega_0$ [10].

This paper applies a general framework for solving Maxwell's equations to the local spectroscopy of metallic nano-particles deposited on a transparent sample. The traditionnal method of solving the Maxwell's equations by matching boundary conditions is restricted to a few specific geometries where the vector wave equation is separable and is not efficient for studying low symmetry systems such as the tip-sample system to be studied here. The method used to solve Maxwell's equations should avoid boundary conditions on both object and detector. The field susceptibility method [10] associated with perturbation theory [11-13] to be discussed in this paper offers this feature as we explain in the next section. Once the analytical framework will be settled, individual response properties of metal particles will be described in section (3) and simulations of local spectroscopy experiments near monolayers of metal particles will be presented in section (4).

## 2 The self-consistent problem

### 2-1 GENERAL

Let $\mathbf{E}_0(\mathbf{r},\omega)$ be the known Fourier component of the incident field on a highly symmetrical system (for example the perfectly planar surface that we will consider in our application).

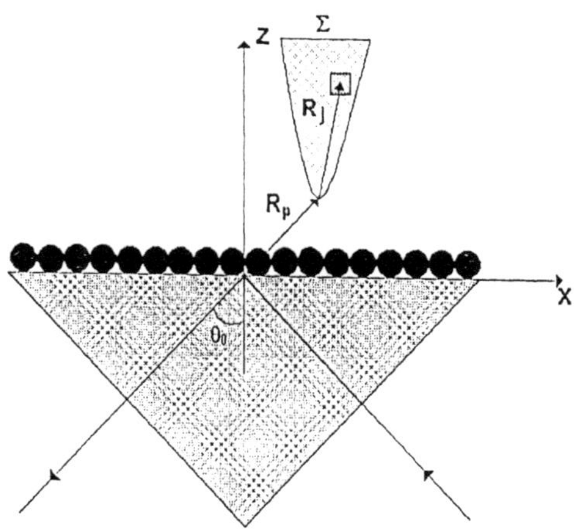


**Fig.1:** *Illustration of an experimental device working in total reflection. The light beam of wave vector* $\mathbf{k}_0$ *is totally reflected on the plane surface of a transparent medium. The object is composed by (m) metallic spheres deposited on the surface* ($z = 0$). *The vector* $\mathbf{R}_p = (X_p, Y_p, Z_p)$ *defines the detector apex position and* $\mathbf{R}_j$ *characterizes the location of a given volume element inside the tip.*

In presence of both a localized defect and the SNOM detector, the general solution of the perturbed field $\mathbf{E}(\mathbf{r},\omega)$ verifies the following implicit Lippmann-Schwinger equation [14]

$$\mathbf{E}(\mathbf{r},\omega) = \mathbf{E}_0(\mathbf{r},\omega) + \int\int \mathbf{S}(\mathbf{r},\mathbf{r}',\omega).\chi(\mathbf{r}',\mathbf{r}'',\omega).\mathbf{E}(\mathbf{r}'',\omega)d\mathbf{r}'d\mathbf{r}'' \quad (1)$$

where $\mathbf{S}(\mathbf{r},\mathbf{r}',\omega)$ represents the field susceptibility of the reference system (cf figure (1)) and $\chi(\mathbf{r}',\mathbf{r}'',\omega)$ is the linear susceptibility of the perturbation (localized defect + probe tip). An essential advantage of this description is provided by the spatial localization of the perturbations (detector + spheres) which allows to solve exactly the self-consistent equation (1) as will be shown in the next section. Moreover working in direct space avoids difficulties due to the poor convergence of the Fourier series arising in standard reciprocal space descriptions.

In the upper half space ($z \geq 0$), the second rank tensor $\mathbf{S}(\mathbf{r},\mathbf{r}',\omega)$ is the sum of two contributions

$$\mathbf{S}(\mathbf{r},\mathbf{r}',\omega) = \mathbf{S_0}(\mathbf{r},\mathbf{r}',\omega) + \mathbf{S_s}(\mathbf{r},\mathbf{r}',\omega) \quad (2)$$

where the first term represents the field propagator in vacuum. The presence of a surface at $z = 0$ is described by the field susceptibility $\mathbf{S_s}(\mathbf{r},\mathbf{r}',\omega)$. In general, this response function can be expressed in term of quantum states $|r>$ of the reference system [15]

$$\mathbf{S_s}(\mathbf{r},\mathbf{r}',\omega) = \frac{1}{\hbar}\sum_r\{\frac{\mathbf{E}_s^{0r}(\mathbf{r}')\mathbf{E}_s^{r0}(\mathbf{r})}{\omega-\omega_{r0}} - \frac{\mathbf{E}_s^{0r}(\mathbf{r})\mathbf{E}_s^{r0}(\mathbf{r}')}{\omega+\omega_{r0}}\} \quad (3)$$

In this expression the terms $\mathbf{E}_s^{0r}(\mathbf{r})$ represent the matrix elements of the field operator of the solid surface (reference system) between the fundamental state and the $r^{th}$ excited state. The field susceptibility $\mathbf{S_s}$ can also be built by calculating the response field of the reference system to an arbitrary fluctuating dipole lying in its neighbourhood. In the particular case of a solid (dielectric or metallic) limited by a perfectly planar surface, various theoretical methods have been described in the literature to obtain this response function [16].

## 2-2 EFFECTIVE FIELD INSIDE THE PROBE-SAMPLE SYSTEM

In this subsection we consider the problem of a pointed dielectric stylus placed in front of spherical nanoparticles finite size monolayer deposited upon a plane dielectric surface. For such a system, the general electric susceptibility $\chi(\mathbf{r},\mathbf{r}',\omega)$ describing the linear non local response properties of the perturbation is the sum of two contributions associated to the probe and the object respectively (cf. Fig.1)

$$\chi(\mathbf{r},\mathbf{r}',\omega) = \chi_{pr}(\mathbf{r},\mathbf{r}',\omega) + \chi_{ob}(\mathbf{r},\mathbf{r}',\omega) \quad (4)$$

We define that the glass substrate occupies the lower half space ($z < 0$), the upper half space being vacuum. The spheres centers lie at a constant distance $Z_0$ above the surface. We can write $\chi_{ob}$ as a sum of individual susceptibilities characterizing the dynamical response of each spheres of the monolayer

$$\chi_{ob}(\mathbf{r},\mathbf{r}',\omega) = \sum_{i=1}^{m}\chi_i(\mathbf{r},\mathbf{r}',\omega) \quad (5)$$

where the subscript $(i)$ is used to define the location of a sphere in a plane parallel to the surface and $(m)$ is the number of spheres. For metallic particles of subwavelength size, the nonlocal susceptibility $\chi_i$ can be expanded as a series of dynamical multipolar polarizabilities

$$\chi_i(\mathbf{r},\mathbf{r}',\omega) = \sum_p \frac{1}{[(2p-1)!!]^2}\alpha^{(p)}(\omega)[2p]\nabla_{\mathbf{r}}^{(p-1)}\nabla_{\mathbf{r}'}^{(p-1)}\delta(\mathbf{r}-\mathbf{r_i})\delta(\mathbf{r}'-\mathbf{r_i}) \tag{6}$$

where $\nabla_{\mathbf{r}}$ is the gradient operator, $\mathbf{r}_i$ represents the position vectors of the spheres and $\alpha^{(p)}(\omega)$ defines their multipolar polarizabilities. Moreover, the symbol $[p]$ indicates a total contraction of two $p^{th}$ order tensors. In the case of noble metal particles, the charge-density susceptibility approach provides a good framework to describe the response functions $\alpha^{(p)}$ in the optical range.

From the relations (4),(5) and (6) the integral equation (1) can be expressed as

$$\mathbf{E}(\mathbf{r},\omega) = \mathbf{E}_o(\mathbf{r},\omega) + \mathcal{E}_{pr}(\mathbf{r},\omega) + \mathcal{E}_{ob}(\mathbf{r},\omega) \tag{7}$$

where the second term is proportional to the perturbation induced by the detector extremity

$$\mathcal{E}_{pr}(\mathbf{r},\omega) = \int\int_{pr} \mathbf{S}(\mathbf{r},\mathbf{r}',\omega).\chi_{pr}(\mathbf{r}',\mathbf{r}'',\omega).\mathbf{E}(\mathbf{r}'',\omega)d\mathbf{r}'d\mathbf{r}'' \tag{8}$$

The response of this pointed dielectric stylus is local. It is then possible to express $\mathcal{E}_{pr}$ from the dielectric constant $\epsilon_{pr}(\omega)$ of the detector

$$\mathcal{E}_{pr}(\mathbf{r},\omega) = \frac{\epsilon_{pr}(\omega)-1}{4\pi}\int_{pr} \mathbf{S}(\mathbf{r},\mathbf{r}',\omega).\mathbf{E}(\mathbf{r}',\omega)d\mathbf{r}' \tag{9}$$

Let us notice that quantum and non local effects in the probe tip have been introduced in ref.[17]. The integral in equation (9) can be optimized by a tridimensional Gauss discretization method [12] which leads to

$$\mathcal{E}_{pr}(\mathbf{r},\omega) = \frac{\epsilon_{pr}(\omega)-1}{4\pi}\sum_{j=1}^{n} W_j \mathbf{S}(\mathbf{r},\mathbf{r_j},\omega).\mathbf{E}(\mathbf{r_j},\omega) \tag{10}$$

where $W_j$ represents the weight of the $j^{th}$ discretized element in the tip apex.

The term proportional to the perturbation induced by the metal particles is given by

$$\mathcal{E}_{ob}(\mathbf{r},\omega) = \sum_p\sum_{i=1}^{m} \frac{1}{[(2p-1)!!]^2}\{\nabla_{\mathbf{r}'}^{(p-1)}\mathbf{S}(\mathbf{r},\mathbf{r}',\omega)\}_{\mathbf{r}'=\mathbf{r_i}}[p]\alpha^{(p)}[p]\mathbf{F}^{(p)}(\mathbf{r_i},\omega) \tag{11}$$

where the tensors $\mathbf{F}^{(p)}$ describe the successive field-gradients experienced by each metal particles

$$\mathbf{F}^{(p)}(\mathbf{r_i},\omega) = \{\nabla_{\mathbf{r}}^{(p-1)}(\mathbf{E}_o(\mathbf{r},\omega) + \mathcal{E}_{pr}(\mathbf{r},\omega) + \mathcal{E}_{ob}(\mathbf{r},\omega))\}_{\mathbf{r}=\mathbf{r_i}} \tag{12}$$

For metallic particles sizes smaller than the incident wavelength, multipolar contributions higher than the dipolar one in Eq. (11) can be neglected resulting in a single summation on the sphere positions. In this approximation the number of discretization points is then identical to the number of metallic particles. The equation (11) reduces to

$$\mathcal{E}_{ob}(\mathbf{r},\omega) = \sum_{i=1}^{m} \mathbf{S}(\mathbf{r},\mathbf{r_i},\omega).\alpha^{(1)}(\omega).\mathbf{E}(\mathbf{r_i},\omega) \quad (13)$$

An essential strength of the present description is the spatial localization of the defect which allows to solve the self-consistent equation (7) exactly. Indeed, both indexes $n$ and $m$ remain finite so that the resulting matrix equations can be solved by standard linear algebra procedures

$$\mathbf{E}(\mathbf{r_k},\omega) = \mathbf{E_0}(\mathbf{r_k},\omega) + \sum_{l=1}^{n+m} \mathbf{S}(\mathbf{r_k},\mathbf{r_l},\omega).\alpha_l(\omega).\mathbf{E}(\mathbf{r_l},\omega) \quad (14)$$

where the quantity $\alpha_l(\omega)$ has the dimension of a dynamical polarizability and changes its value according to the index $l$:

$$\alpha_l(\omega) = \alpha^{(1)}(\omega), \qquad if(1 \leq l \leq m) \quad (15)$$

and

$$\alpha_l(\omega) = \{\frac{\epsilon_{pr}(\omega) - 1}{4\pi}\}W_l, \qquad if(m+1 \leq l \leq n+m) \quad (16)$$

We note that the procedure avoids matching boundary conditions. The only approximation of the technique lies in the density of the discretization grid which is arbitrarily adjustable.

2-3 SIGNAL DETECTED BY THE PROBE TIP

The conversion process of evanescent waves present at immediate proximity of the object into propagating waves in the detector can be described from the above results. The near field solution between the tip and the monolayer is not required for the far field computation. The knowledge of the effective field distribution inside the tip and the monolayer is sufficient to describe the far-field $\mathbf{E}_{far}$ crossing a surface located in the wave zone above the dielectric stylus (cf Fig.1):

$$\mathbf{E}_{far}(\mathbf{R_p} + \mathbf{r_0},\omega) = \sum_{k=1}^{n+m} \mathbf{S}(\mathbf{R_p} + \mathbf{r_0},\mathbf{r_k},\omega).\alpha_k(\omega).\mathbf{E}(\mathbf{r_k},\omega) \quad (17)$$

where $\mathbf{R_p}$ defines the location of the tip apex with respect to an absolute frame and the vector $\mathbf{r_0}$ represents the position of a point of the $\Sigma$ surface with respect to the apex of the detector. Thus, by using an asymptotic form of the propagator $\mathbf{S}$ in the far field range, it is possible to describe regions far from the perturbation. Finally, integration of the Poynting vector associated to $\mathbf{E}_{far}(\mathbf{R_p} + \mathbf{r_0},\omega)$ on the $\Sigma$ surface leads to the detected intensity.

## 3 Individual response properties of metallic nano-particles

The study of the optical linear response properties of small metallic aggregates has been extensively investigated for the last twenty years [18-24]. In the case of metal spheres embedded in a transparent dielectric, the interface between the two materials introduces a surface plasmon resonance whose the frequency depends on the optical properties of the metal and on the dielectric constant of the surrounding. Recently, a theoretical approach based on the charge-density susceptibility formalism has been applied to a spherical jellium [25]. In this work the main aspects of the optical properties of the particles are introduced by generalizing the method of Newns [26] to an electron gas confined in a spherical well. From this model we have calculated the dipolar polarizabilities of gold and silver particles by introducing the interband electronic contributions of the metal from standard experimental data (cf Fig.2) [27]

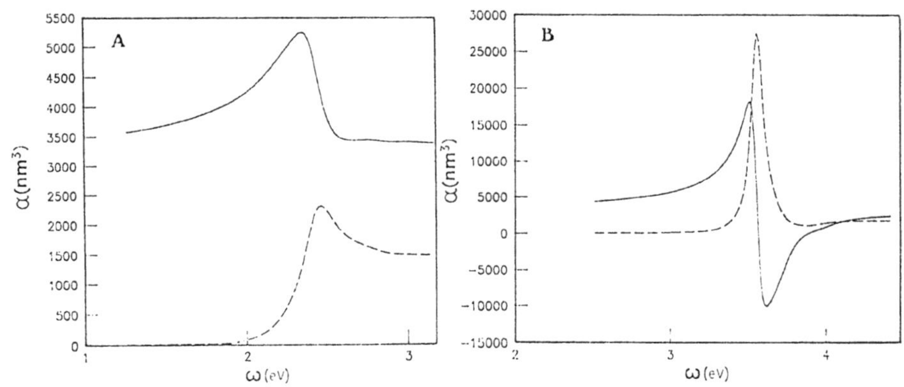


**Fig.2:** *Real and imaginary part of the dynamical polarizability of 15nm radius spherical metallic particles: (a)gold sphere; (b)silver sphere.*

Let us notice that more general approaches, including exchange and correlation effects and working in the framework of a microscopic finite-barrier model have been discussed in the literature [21,22]. Nevertheless, such calculations are computationally expensive and therefore limited to metallic particles of very small sizes. In this paper we prefer a simpler approach [25] leading to an easy evaluation of non-local effects for the metallic particles which nevertheless retains the main features of the physical mechanisms occuring in a local spectroscopy experiment.

## 4 Local spectroscopy of metal particles deposited on a transparent sample

The two relations (14) and (17) defined in the previous sections are quite general since they account for the real profile of the tip-apex and they include, through $\alpha(\omega)$, the dynamical properties of the metallic particles forming the object. From these equations it is thus possible to calculate the scattered signal $I(\mathbf{R_p},\omega)$ and to simulate for a fixed location of the detector a local spectroscopy experiment. For this study, we will use the values of the dynamical polarizabilities of gold and silver particles defined in section 3 and the discussion will be presented in the total reflection configuration described in Fig.1.

For metallic objects, the excitation frequency $\omega$ of the incident field becomes an important parameter. The influence of $\omega$ on the magnitude of the detected intensity is numerically studied on Figs. 3, 4 and 5 where we present the frequency dependent intensity $I(\mathbf{R_p}, \omega)$ for particular detector positions. We have placed the tip-apex just above the sphere located at the origin of the absolute frame, i.e., for $\mathbf{R_p} = (0, 0, Z_p)$ (see Fig.1).

Figure 3 gives the spectrum recorded at the immediate proximity of a 15 nm radius silver particle. This simulation has been performed by introducing 36 grid points in the tip-apex. The approach distance ($Z_p$ in Fig.1) between the detector edge and the plane support varies between 1nm and 16nm by step of 5nm. The curvature of the tip end is 20 nm.

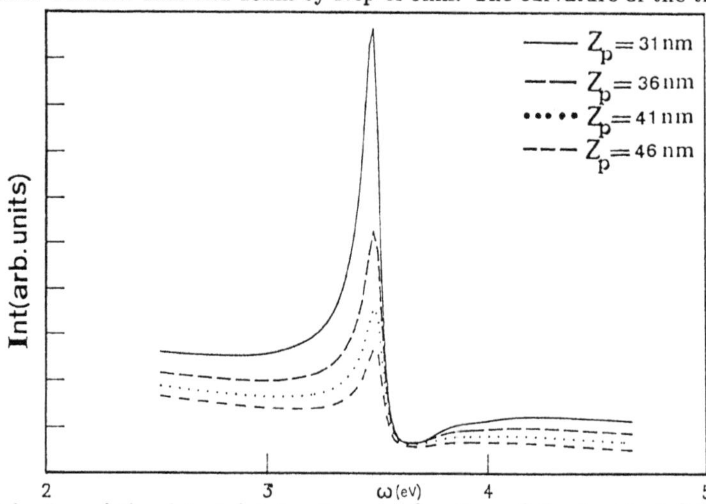


**Fig.3:** *Behavior of the detected intensity as a function of the incident field frequency detected above a single 15 nm radius silver particle.*

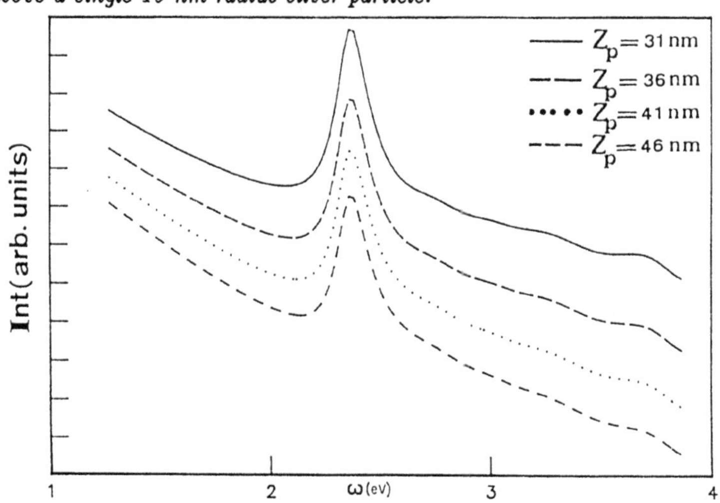


**Fig.4:** *Representation of the detected intensity above a set of 25 gold spheres of 15 nm in radius. The particles are placed at the nodes of a square array. The tip-apex is the same as the one used for the calculation described in figure 3.*

One observes on the curves of figure 3 that a slight red shift of the plasmon resonance occurs when $Z_p$ decreases. This shift is due to the presence of the detector approaching the metal particle. It is similar to the shift observed when noble metal particles are embedded in a transparent dielectric medium.

Figure 4 represents a spectrum recorded above a set of 25 gold spheres of 15 nm radius. In the optical range, this system displays a single resonance at the frequency 2.35 eV. For this system the red shift caused by the interaction with the dielectric stylus is too weak to be observed. This behavior is consistent with individual response properties of the gold particles described in Fig. (2-a). Moreover it may be seen on the figure 4 that the shape of the near field spectra and the peak plasmon width at half maximun are weakly modified by the detection device as $Z_p$ decreases.

Figures (5a) to (5d) describe the evolution of the spectrum of silver particles when the number N of spheres increases from 1 to 49. The 20 nm radius spheres are organized on the surface as a hexagonal lattice and the detector is located at the origin of the reference system.

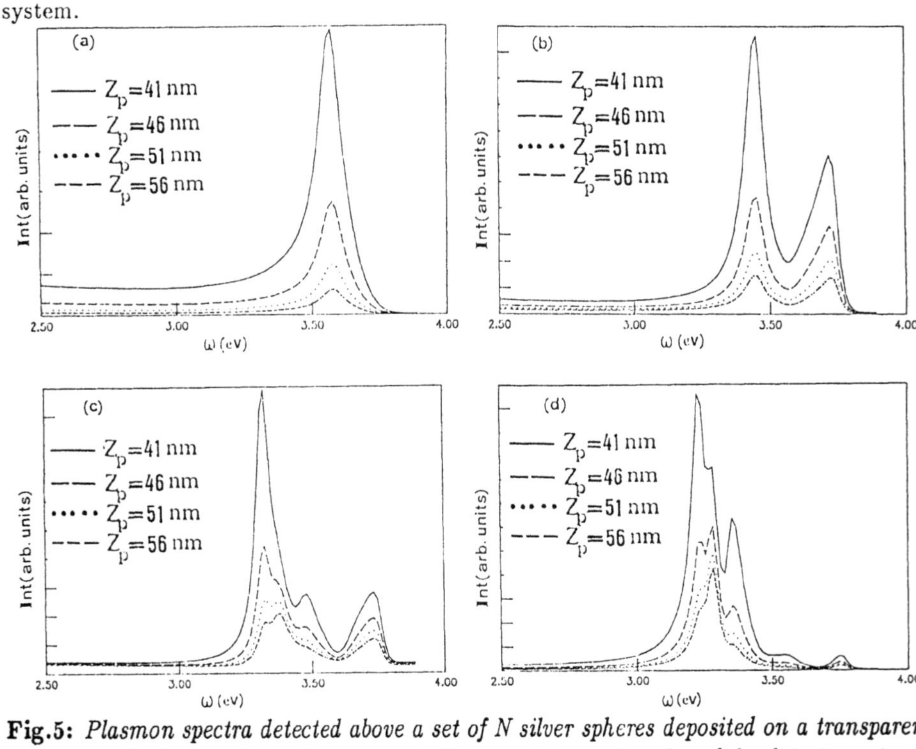


**Fig.5:** *Plasmon spectra detected above a set of N silver spheres deposited on a transparent substrate of index* $n = 1.5$. *The approach distance between the edge of the detector extremity varies between 1 nm and 16 nm by step of 5 nm: (a)* $N = 1$;*(b)* $N = 9$;*(c)* $N = 25$;*(d)* $N = 49$.

The main peak plasmon is shifted toward low frequency when the number N increases. This effect is due to the many-body interactions inside the monolayer. Moreover in the case of silver the detected intensity displays secondary resonances when the number of particles increases. Their physical origin lies in the fact that the monolayer outer metallic spheres have a smaller number of nearest neighbours than the spheres located just below the detector. This boundary effect gives rise to secondary peaks.

## Conclusion

We have theoretically analyzed in two systems of experimental interest the possibility to perform a surface local spectroscopy with a scanning near field optical microscope. We have studied the configuration of a dielectric probe-tip approaching a monolayer of metallic objects as the external frequency is varied in the optical range. The direct space discretization procedure used to solve this coupled tip-sample system appears to be well suited to realistic situations where dielectric function profile are spatially localized. We found that the approach distance of the dielectric probe weakly perturbs the own spectral characteristic of the metallic monolayer. Moreover, our simulations clearly show the interest of such a local spectroscopy method for the morphologic studies and for the characterization of mesoscopic structures.

## Acknowledgments

The authors express their sincere thanks to Drs. D. W. Pohl and D. Courjon for a number of stimulating and interesting discussions. One of us (A. D.) acknowledges both the Walloon Ministery for Sciences and Technology (Brussels) and IBM Belgium (Brussels) for financial support.

## References

(a) Visiting scientist from the Institute for Studies in Interface Sciences, Facultés Universitaires N. D. de la Paix, Namur, Belgium.

[1] H. ROHRER, in "*Scanning Tunneling Microscopy and Related Techniques*" **NATO** serie **E**,Vol. **184**,edited by J.R. Behm, N. Garcia and H. Rohrer, page 1.

[2] D. W. POHL, Adv. Opt. Electron. Micros. **12**,243(1991) and references therein.

[3] E. BETZIG, J. K. TRAUTMAN, T. D. HARRIS, J. S. WEINER and R. L. KOSTELAK, Science **251**, 1468(1991).

[4] R. REDDICK, R. J. WARMACK and T.J. FERRELL, Phys. Rev. B**39**, 767(1989).

[5] D. COURJON, K. SARAYEDDINE and M. SPAJER, Opt. Com. **71**,23(1989).

[6] J. A. CLINE, H. BARSHATZKY and M. ISAACSON, Ultramicr. **38**,299(1991).

[7] N. F. VAN HULST, F. B. SEGERINK and B. BOLGER, Opt. Com. **87**,212(1992).

[8] M. SPECHT, J. D. PEDARNIG, W. M. HECKL and T. W. HANSCH, Phys. Rev. Lett. **68**,496(1992).

[9] M. A. PAESLER, P. J. MOYER, C. J. JAHNCKE, C. E. JOHNSON, R. C. REDDICK, R. J. WARMACK and T. L. FERRELL, Phys. Rev. B**42**,6750(1990).

[10] C. GIRARD and X. BOUJU, J. Chem. Phys. **95**, 2056(1991).

[11] J. J. GREFFET, Opt. Com. **72**, 274(1989).

[12] A. DEREUX, "*Théorie de l'optique de champ proche*", PhD Thesis, Faculté Universitaire N. D. de la Paix, NAMUR, Belgium(1991).

[13] A. DEREUX, J. P. VIGNERON, Ph. LAMBIN and A.A. LUCAS, Physica B**175**,65(1991).

[14] E. N. ECONOMOU, *"Green Functions in Quantum Physics"*, 2d ed. Springer, Berlin (1983).

[15] G. S. AGARWAL, Phys. Rev. A**11**,230(1970).

[16] C. GIRARD, Phys. Rev. B**45**,1800(1992).

[17] O. KELLER, in these proceedings.

[18] A. KAWABATA and R. KUBO, J. Phys. Soc. Japan, **21**,1765(1966).

[19] L. GENZEL, T. P. MARTIN and U. KREIBIG, Z. Phys. **B 21**,339(1975).

[20] B. B. DASGUPTA and R. FUCHS, Phys. Rev. **B 24**, 554(1981).

[21] E. ZAREMBA and B. N. J. PERSSON, Phys. Rev. **B 35**, 596,(1987).

[22] W. EKARDT, Phys. Rev. **B 34**, 526(1986).

[23] T. TAKEMORI, M. INOUE and K. OHTAKA, J. Phys. Soc. Japan, **56**,1587(1987).

[24] M. C. BUNCICK, R. J. WARMACK and T. L. FERRELL, J. Opt. Soc. Am. **B 4**, 927(1987).

[25] C.GIRARD and F. HACHE, Chem. Phys. **118**, 249(1987).

[26] D. M. NEWNS, Phys. Rev. B**1**,3304,(1970).

[27] P. B. JOHNSON and R. W. CHRISTY, Phys. Rev. B**6** ,4370(1972).

# Near field scattered by subsurface particles

F. PINCEMIN, A. SENTENAC AND J.-J. GREFFET
*Laboratoire d'Energétique Moléculaire et Macroscopique, Combustion ;*
*Ecole Centrale Paris, Centre National de la Recherche Scientifique,*
*Grande voie des Vignes, 92295 Châtenay-Malabry Cedex, France*

Abstract. A volume integral method is developped in order to study the scattering of a plane wave by a small particle placed below a metallic surface. The scattering patterns produced in the near field are analysed. For p-polarization, a plasmon is excited producing interferences along the surface. It is shown that this interference pattern might be used to measure the dielectric constant of the metal. The scattered field has a pronounced maxima for p-polarization if the particle is close to the interface. This effect is due to a resonant interaction between the surface and the dipole already identified in the context of SERS as an image enhancement effect. Finally, the resolution of a near field detection of a subsurface scatterer is discussed on the basis of the transverse extension of the scattered field.

## 1. Introduction

The development of near field optics has recently stimulated the development of theoretical models of scattering of illuminating beams by given features of the surface such as dielectric contrast, microroughness or the presence of a tip. See for instance the papers by Dereux and Pohl and by Girard, Bouju and Dereux in these proceedings. A review is given in the paper by Van Labeke. Such problems have already been investigated in different contexts such as radiowave propagation along the earth, modelling of the interaction of an antenna with a structure, scattering of acoustic waves by the bottom of the ocean, scattering of radar waves by the sea, scattering of optical waves or atoms by surfaces for instance.

The elementary problem to solve is the scattering by a dipole near a surface. The first expression of the field scattered by a dipole in the vicinity of an interface has been given by Sommerfeld [1]. A lot of work has been done on this subject in the context of the propagation of radio waves on earth. In particular, the existence of surface waves known as Zenneck modes has been widely discussed. Banos has summarized the abundant literature on the subject and has provided a unified and detailed description of the subject [2]. Well known asymptotic expressions of the near field are given in [3, 4].

In the last twenty years, several numerical methods have been developped to deal with the problem of scattering by a structure. Integral techniques have proven to be very useful to achieve this goal.



*D.W. Pohl and D. Courjon (eds.), Near Field Optics, 209–220.*

An overview of the integral formulation of the scattering problems as well as a list of references of techniques developped among the electrical engineering community can be found in ref [5-7]. The difficult problem of the numerical evaluation of the Green's tensor is discussed at length in ref. [5, 8] . In the context of the studies of scattering by surfaces, a lot of numerical techniques have been developped in the optics community and a number of reviews of the field are available both for periodic [9, 10] and random surfaces [11-14].

Most of the techniques described above are restricted to the problem of a surface separating two homogeneous media. The first attempt to compute the scattering by a large particle of arbitrary shape was based on a representation by a set of dipoles [15]. Later, several volume integral approaches based on the macroscopic Maxwell equations have been developped. They allow to deal with the scattering by inhomogeneities [16, 17] . The relationship between both approaches has been studied in detail in ref. [18] . In the present paper, we study the near field scattered by a particle embedded in a metallic half space. We assume that it is close to the interface so that it can be illuminated by the transmitted field. We will consider a particle with a finite size rather than a point like dipole.

## 2. Theory

In this section, we briefly outline the derivation of the numerical technique that allows to compute the field scattered by a particle embedded in a medium. Details of the approach will be reported elsewhere.

### 2.1. GEOMETRY OF THE PROBLEM

A typical system in near field optics involves a surface separating two isotropic linear media denoted by $\Omega_1$ (z>0) and $\Omega_3$ (z<0). Their complex dielectric constants are $\varepsilon_1=n_1^2$ and $\varepsilon_3=n_3^2$, respectively. In what follows, the subscript f refers to this reference system. In particular we call $\varepsilon_f$ the dielectric constant that is defined by

$$\begin{aligned} \varepsilon_f(z) &= \varepsilon_1 \text{ for } z>0, \\ \varepsilon_f(z) &= \varepsilon_3 \text{ for } z<0. \end{aligned} \tag{1}$$

We will consider in what follows that medium 1 is a metal whereas medium 3 is vacuum($\varepsilon_3=1$). This reference system is modified because there might be some local modification of the dielectric constant or of the shape of the surface. These features will behave as scattering sources. In addition, the presence of a tip will also scatters light. More generally, any region in which the dielectric constant is different from $\varepsilon_f$ will break the invariance of the system and diffract the light. This region is denoted by $\Omega_2$, and characterized by a dielectric constant $\varepsilon_2(x,z)$. In what follows, we will consider only particles embedded in medium 1 so that $\Omega_2$ is entirely included in the half plane defined by z>0. In this work , we do not account for the presence of a tip. The geometry and the notations are shown in Fig. 1. The dielectric constant of the entire system is denoted by the space dependent dielectric constant $\varepsilon$, with $\varepsilon(x,z)=\varepsilon_i$ for (x,z) lying in $\Omega_i$.

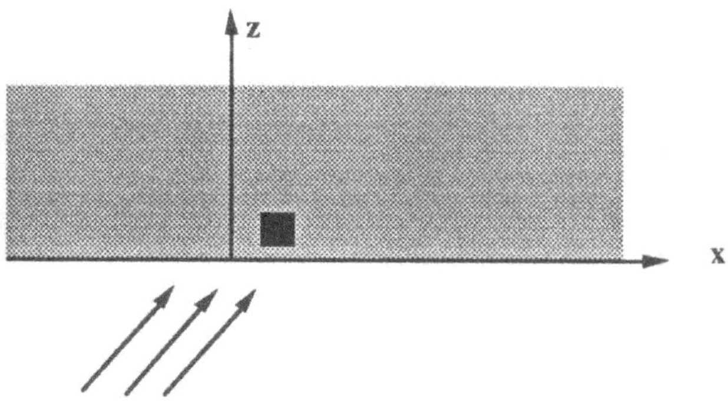


Figure 1. Schematic diagram of the geometry of the system.

2.2.EVALUATION OF THE FIELD IN THE SOURCE REGION $\Omega_2$

Starting from macroscopic Maxwell equations, we now show explicitly that the field scattered is completely known as soon as the field within the scattering sources is determined.

*2.2.1 Differential equation.* Let

$$\mathcal{E}_i(x,z,t)=\mathrm{Re}\{\mathbf{E}_i(x,z)\exp(-i\omega t)\}=\mathrm{Re}\{(E_{incx}\mathbf{u}_x+E_{incz}\mathbf{u}_z)\exp(i\kappa_{inc}x+i\gamma_1^{inc}z-i\omega t)\} \quad (2)$$

$$\mathcal{E}_f(x,z,t)=\mathrm{Re}\{\mathbf{E}_f(x,z)\exp(-i\omega t)\} \quad (3)$$

represent the incident field and the total electric field in the case of a flat interface, respectively (i.e. either incident and reflected field or transmitted field). The diffracted field $\mathbf{E}_d$ is defined by

$$\mathbf{E}_d = \mathbf{E}-\mathbf{E}_f\,, \quad (4)$$

where $\mathbf{E}$ denotes the total field. The Helmholtz equation, valid in the sense of distributions, satisfied by $\mathbf{E}$ is

$$\mathbf{rot\ rot\ E}-\varepsilon(x,z)k_0^2\mathbf{E}=0 \quad (5)$$

where $k_0^2=\omega/c$. The field $E_f$ satisfies a similar equation :

$$\mathbf{rot\ rot\ E}_f-\varepsilon_f(z)k_0^2\mathbf{E}_f=0\,. \quad (6)$$

By substracting (6) to (5), we obtain the differential equation satisfied by the diffracted field $\mathbf{E}_d$ :

$$\mathbf{rot\ rot\ E}_d - \varepsilon_f(z) k_0^2 \mathbf{E}_d = (\varepsilon(x,z) - \varepsilon_f(z)) k_0^2 \mathbf{E}\,. \tag{7}$$

The source term, $(\varepsilon(x,z)-\varepsilon_f(z))\mathbf{E}$ is null every where outside the volume of the scatterer $\Omega_2$. Note that the source term is known as soon as the total field $\mathbf{E}$ is known within $\Omega_2$

2.2.2 *Integral equation.* In order to derive the expression of the field everywhere as a function of the source term, we introduce the Green dyadic $\overleftrightarrow{G}$ which is the solution of the equation

$$\mathbf{rot\ rot}\ \overleftrightarrow{G}(x-x',z,z') - \varepsilon_f(z) k_0^2 \overleftrightarrow{G}(x-x',z,z') = \overleftrightarrow{\mathbf{I}}\delta(x-x')\delta(z-z') \tag{8}$$

subjected to a radiation condition at infinity. $\overleftrightarrow{\mathbf{I}}$ is the unit dyad. The exact expression of is given in ref[17] . Note that $\overleftrightarrow{G}.\mathbf{p}$ is simply the electric field created by a point like dipole $\mathbf{p}$. Its tensor nature simply expresses the fact that the electric field radiated by a dipole is not parallel to the dipole. The Green's dyadic is also known as Green's tensor or as propagator. Note also that all the information on the surface (i.e. reflection, excitation of surface waves for instance) is included in $\overleftrightarrow{G}$. Using Eqs. (4), (7), (8), and the properties of the source term, we may write $\mathbf{E}$ in the form :

$$\mathbf{E}(x,z) = \mathbf{E}_f(x,z) + \iint_{\Omega_2} \overleftrightarrow{G}(x-x',z,z') k_0^2 (\varepsilon(x',z') - \varepsilon_f(z')) \mathbf{E}(x',z') dx'dz'\ , \tag{9}$$

where integration is performed over $\Omega_2$ which means that the scattered field is only due to the inhomogeneous regions. Here, we can attribute a new physical meaning to $\overleftrightarrow{G}$. As before, $\overleftrightarrow{G}.(\varepsilon(x,z)-\varepsilon_f(z))k_0^2\mathbf{E}$ is the field created at (x',z') by a point like dipole $(\varepsilon(x,z)-\varepsilon_f(z))k_0^2\mathbf{E}$ located at (x,z). Hence, in eq. (9) $\overleftrightarrow{G}$ appears as a potential of interaction between two points. To compute $\mathbf{E}$ we restrict eq. (9) to the domain $\Omega_2$. Then, ~~we~~ assume that $\Omega_2$ can be mapped in a set of rectangular cells and that the field can be considered constant throughout each cell. To solve eq. (9) we use a standard moment method [17] which leads to the linear system :

$$E_x^j = E_{fx}^j + M_{xx}^{ji} E_x^i + M_{xz}^{ji} E_z^i \tag{10 a}$$

$$E_z^j = E_{fz}^j + M_{zx}^{ji} E_x^i + M_{zz}^{ji} E_z^i \tag{10 b}$$

where $E_x{}^i$ and $E_{fx}{}^i$ are the fields at the center of the cells $(X_i,Z_i)$ :

$$E_x{}^i = E(X_i,Z_i) \tag{11 a}$$

$$E_{fx}{}^i = E_f(X_i,Z_i) \tag{11 b}$$

and with

$$M_{uv}^{ji} = (\varepsilon(X_i,Z_i) - \varepsilon_f(Z_i)) k_0^2 G_{uv}^{ji} \tag{12}$$

and

$$G_{uv}^{ji} = \int_{X_i-\Delta X/2}^{X_i+\Delta X/2} dx' \int_{Z_i-\Delta Z/2}^{Z_i+\Delta Z/2} G_{uv}(X_j-x',Z_j,z')dz'\ . \tag{13}$$

where $\Delta x$ and $\Delta z$ are the side of the square cell. All the difficulties lies in the correct evaluation of the Green's tensor. Its expression and details on its numerical evaluation are given in ref.[17] .

### 2.3 NEAR FIELD

The system (10) is solved by standard routines of matrix inversion. Thus, the field in $\Omega_2$ is deduced by multiplying the inverse matrix by the field existing for the flat surface under illumination. Note that this can be done for any incident field, either a surface plasmon, a plane wave or a bounded beam. In order to obtain the near field at a given point out of $\Omega_2$, we use eq. (9) again or in numerical form, eq.(10) but with $(X_j,Z_j)$ being the coordinates of the point of observation, located outside of $\Omega_2$.

### 2.4 CONTRIBUTION OF SURFACE WAVES

When studying the near field over a metal surface, we have to pay attention to the excitation of surface plasmons. The surface plasmon can be viewed as an electromagnetic solution of a homogeneous problem . Its analytical form is given by

$$z>0 \quad E^{sp} = E_x{}^{sp}\,(1, 0, -\kappa_{sp}/\gamma_1{}^{sp})\,\exp[i(\kappa_{sp}x+\gamma_1{}^{sp}z] \qquad (14)$$

$$z<0 \quad E^{sp} = E_x{}^{sp}\,(1, 0, \kappa_{sp}/\gamma_3{}^{sp})\,\exp[i(\kappa_{sp}x-\gamma_3{}^{sp}z)] \qquad (15)$$

In the above theory, the surface plasmon appear though the presence of poles in the expression of the Green's tensor. These poles come from the Fresnel factors for p-polarization whereas no poles appears in the Fresnel factor for s-polarization. Indeed, the denominator of the Fresnel factors $\varepsilon_1\gamma_3 +\varepsilon_3\,\gamma_1$ have zeros for a value of the x-component of the wave vector given by :

$$\kappa_{sp}^2 = \frac{\omega^2}{c^2}\frac{\varepsilon_1\varepsilon_3}{\varepsilon_1+\varepsilon_3} \qquad (16)$$

The numerical treatment of the poles has to be done with some care. This point will be examined in detail elsewhere.

## 3. Near field

Let us first consider a square particle placed in the metal near the vacuum-metal interface illuminated by a s-polarized plane wave with an angle of incidence $\theta_i$ of 20°. We assume that the particle has a refractive index of 1 and its side is $\lambda/40$ so that the field can be considered constant over the cell. The dielectric constant of the metal is -3.47–i 0.18 which corresponds to the index of silver at 382 nm [19]. Note that if the distance between the square particle and the surface is reduced, this system becomes a rectangular groove.

We display in Fig. 2 a the square modulus of the scattered electric field. Note that the dark point is the groove. It is seen that the scattered field is localized in the metal in a sphere centered on the particle with a radius of a few skin depths. On the other hand, above the surface, the field is not so localized. The width of the scattering pattern gives an idea of the resolution that can be expected for such a system. It also illustrates the behavior of this width with the distance above the surface. The physical quantity that can be measured is the square modulus of the total field, that is the sum of the illuminating field and the scattered field. Note that, since the particle is very small, the scattered field would be the same with an incident surface plasmon of same amplitude. By contrast, the total field would be different. In Fig. 2 b, we display the total field observed when we illuminate with an incident plane wave. It appears that the scattered field is rather small and does not modify the near field.

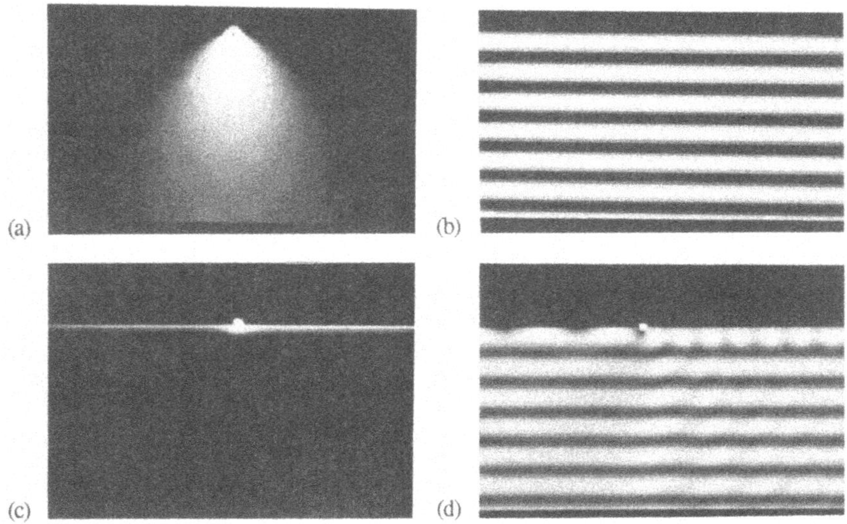


Figure 2. Square modulus of the field scattered by a groove $\Delta x = \Delta z = \lambda/16$ illuminated in s or p-polarization with an angle of incidence of 30°. (a) scattered field, polarization s. (b) total field, polarization p, (c) scattered field, polarization p, (d) total field, polarization p.

Figure 2c shows the square modulus of the scattered field when we illuminate the particle with a p-polarized plane wave with an angle of incidence of 30°. The unique feature of this system is the presence of a strong field along the interface due to the excitation of a surface plasmon. Its decay due to ohmic losses is seen in Fig. 2c. Figure 2d shows the square modulus of the total field. It is seen that a fringe pattern appears due to the interference between the surface plasmon and the transmitted field below the surface or the incident plus reflected field above the surface.
The x dependence of the surface plasmon is characterized by the wave vector $\pm\kappa_{sp}$ whereas the transmitted, incident and reflected fields have a wave vector $k_0\sin\theta_i$ along the surface. Hence, on the right of the particle, the interference system of fringes has a periodicity $|2\pi/(k_0\sin\theta_i - \kappa'_{sp})|$ whereas it is $|2\pi/(k_0\sin\theta_i + \kappa'_{sp})|$ on the left of the particle. We have used the notation $x = x' + ix''$ for the real and imaginary part of a complex. These interference effects have been observed experimentally by Keilman [20] coupling the light through a groove. A complete analysis of this pattern will be reported elsewhere.

The visibility of the fringes decays away from the scatterer, this is due to the decay of thé surface plasmon as seen in Fig. 2c. From this remark, it appears that the measurement of the interference structure provides a means for the determination of the dielectric constant of the metal. Indeed, the visibility of the fringes decays as $\exp(-\kappa''_{sp}x)$ where x is the distance from the scatterer. The periodicity of the fringes gives the real part of $\kappa_{sp}$. Hence, one might obtain ε from eq. (16).

## 4 Image enhancement effect.

In the above section, it has been shown that the scattered field is larger in p-polarization. In order to explain this phenomena, we represent in Fig. 3 the decay of the modulus of the electric field in the particle as a function of its depth. It is seen that for a depth larger than λ/100, the decay follows an exponential law $\exp[-\gamma_1''^{inc} z]$. By contrast, the decay for very small depths is extremely higher for p-polarization. This effect is not observed for s-polarization. This feature is characteristic of the so-called image enhancement already described in the context of studies on SERS. The root of the effect observed in Fig. 3 is the interaction between the scatterer and its image through the interface. Note that if we compare the results of Fig. 3 obtained with a particle of finite size, we do not recover the very fine peak of resonance that is produced with a point like dipole as in ref.[21, 22]. Indeed, this fine peak is due to the idealized point like behaviour of the model. It is also of interest to mention that it has been reported that accounting for quantum effects due to the spatial extension of the wave functions also leads to a reduced enhancement [23].

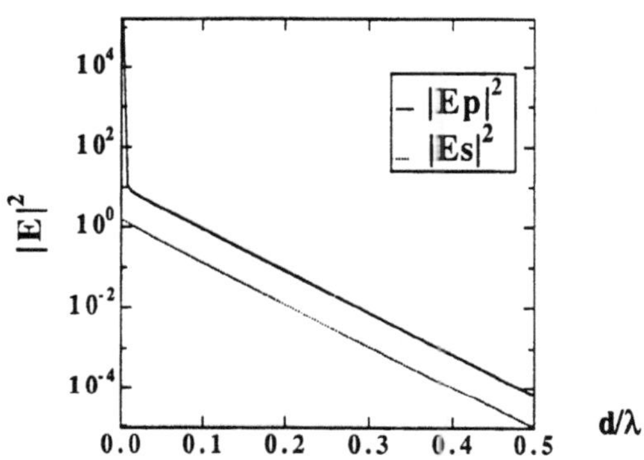


Figure 3. Square modulus of the field in the particle as a function of the distance d particle-surface.

Since the distance 2d between the scatterer and its image is very small as compared to the wavelength, retardation effects are negligible and an electrostatic model is justified. This can be obtained either by a direct calculation or by taking the limit $k_0 \rightarrow 0$ in the expression of the exact Green's tensor.

The calculation given below shows that as a result of the interaction, an effective polarizability may be introduced. The main feature of this modification is that the effective polarisability diverges for a given scatterer-surface distance generally of the order of a few nanometers.

We now turn to the derivation of the expression of the dipole within two approximations : we work with a point like dipole and we neglect retardation effects. The value of the dipole satisfies the folowing expression

$$\mathbf{p} = \alpha\left(\mathbf{E}_{ext} + \mathbf{E}(A'\to A)\right) \tag{17}$$

where $\mathbf{E}(A'\to A)$ is the field created by the image A' of the dipole (i.e. by the induced charges on the surface) at the location A of the dipole. It can be shown (see e.g. Jackson) that this field is given by

$$\mathbf{E}(A'\to A) = \frac{1}{2\pi\varepsilon_1}\frac{\varepsilon_3 - \varepsilon_1}{\varepsilon_3 + \varepsilon_1}\frac{1}{4d^2}\left(p_x\hat{\mathbf{x}} - p_z\hat{\mathbf{z}}\right) \tag{18}$$

in two dimensions and by

$$\mathbf{E}(A'\to A) = \frac{1}{4\pi\varepsilon_1}\frac{\varepsilon_3 - \varepsilon_1}{\varepsilon_3 + \varepsilon_1}\frac{1}{8d^3}\left(p_x\hat{\mathbf{x}} + p_y\hat{\mathbf{y}} - 2p_z\hat{\mathbf{z}}\right) \tag{19}$$

in three dimensions. It clearly appears that this field due to the interface is a linear operator L acting on **p**. Hence we may write (17) in the form

$$\mathbf{p} = \frac{\alpha}{1-\alpha L}\mathbf{E}_{ext} \tag{20}$$

which shows that there might be a resonance for a given value of d at a given frequency. This resonance results from the fact that a dipole at A induces surface charges on the interface. These charges produce a field at A that induces a dipole. Hence, for a given distance, this dipole may exist without an exterior field.

Note that the same basic mechanism can be invoked to recover the electrostatic resonance in a metallic sphere in vacuum at a frequency $\omega$ such that $\varepsilon_r(\omega)=-2$. Let us look for a resonance of this system. Hence a polarization may exist without any exterior field. Let us assume that there is a uniform polarization **P** in the sphere. It produces a uniform field $-\mathbf{P}/3\varepsilon_0$. In turn this field creates an induced polarization $\mathbf{P}_{in}= \varepsilon_0(\varepsilon_r -1)\mathbf{E}= -\varepsilon_0(\varepsilon_r -1)\mathbf{P}/3\varepsilon_0$. The self consistent relation $\mathbf{P}_{in}=\mathbf{P}$ leads to the condition $\varepsilon_r=-2$. In the presence of a surface, there might be a shift of the resonance frequency for a sphere. Indeed, assuming that a small sphere with a radius a and a dielectric constant $\varepsilon_p$ is located near an interface separating two media of dielectric constants $\varepsilon_1$ and $\varepsilon_3$, we can obtain the effective polarisability from eq. (20). The value of the polarisability $\alpha$ without interface is $4\pi a^3 (\varepsilon_p-\varepsilon_1)/(\varepsilon_p+2\varepsilon_1)$. The resonant condition without interface is $\varepsilon_p+2\varepsilon_1=0$. In the presence of an interface, we obtain

$$\varepsilon_p + 2\varepsilon_1 - \left(\varepsilon_p - \varepsilon_1\right)\frac{\varepsilon_3 - \varepsilon_1}{\varepsilon_3 + \varepsilon_1}\frac{a^3}{8d^3} = 0 \tag{21}$$

for the x and y component of the field and

$$\varepsilon_p + 2\varepsilon_1 + 2\left(\varepsilon_p - \varepsilon_1\right)\frac{\varepsilon_3 - \varepsilon_1}{\varepsilon_3 + \varepsilon_1}\frac{a^3}{8d^3} = 0 \qquad (22)$$

for the z component of the field. Analog shifts are found in 2D in p-polarization. Finally, we note that for s-polarization, the concept of a dipole has no meaning in 2D since the system is invariant along the y axis. Indeed, by looking at the electrostatic limit of the Green's function in s-polarization [24] , we find that the field due to the interface tends to zero. We also note that a treatment including the retardation effects for a spere below a metal surface can be found in ref.[25]

## 5 Field scattered by two particles

We now consider a system of two particles. The first question to adress is what is the possible relationship between the presence of a groove on the surface and the field along the surface. The second question to adress is the relationship between the structure of the field and the shape of the groove. The results that will be given in this section does not account for the presence of a probe. This presence might modify the structure of the field. However, even without a detection instrument, the relationship between a scatterer and the square modulus of the field is not so simple. In order to try to develop some physical insight in this problem we consider the simplest problem with characterized structure : two small square grooves separated by a distance $\lambda/10$.

In Fig. 4 a, we show the square modulus of the near field scattered by two square grooves whose depth and width are $\lambda/20$. In this case, it is seen that the near field pattern clearly allows a detection of two particles in terms of resolution and relationship between the square modulus of the field and the shape of the surface. However, the scattered field is small as compared with the interference pattern produced by the incident and reflected field as seen in Fig. 2b.

Figure 4 . Square modulus of the scattered field for s-polarization (a) and p-polarization (b).

From section 4, we know that the scattered field is much more important in the case of an illumination with a p-polarized beam. Fig. 4b shows the square modulus of the scattered field and the total field is seen in Fig.5. In this case, we obtain a pattern that could be measured. Again, it clearly appears that the scattered field has basically the same structure as the shape of the surface very close of the interface but the pattern is totally different at $\lambda/5$ above the surface where a dark region appears. Note that the scale of the figure have been modified because the field is enhanced on the particle. For an incident field of square modulus equal to 1, the interference pattern produces maxima around 3. The square modulus of the field in the particles is around 1 for s-polarization and 70 for p-polarization in the case of Fig. 5.

Let us now examine the interference effects between the surface plasmons scattered by two different grooves separated by a distance d. Each groove scatters the incident beam and the plasmon scattered by the other particle. The difference of phase between the two surface plasmons scattered is simply, $(k_0 \sin\theta_i \pm \kappa_{sp})$ d. By choosing the distance d such that this difference of phase is a multiple of $\pi$, either $2p\pi$ or $(2p+1)\pi$, it is possible to create constructive or destructive interferences. This is illustrated in Figure 6 where we have shown the square modulus of the scattered field only for normal incidence. It is seen that a localization of the surface plasmons between the two scatterers is produced by destructive interferences between the two scattered plasmons. Between the two scatterers, we observe a typical interference pattern.

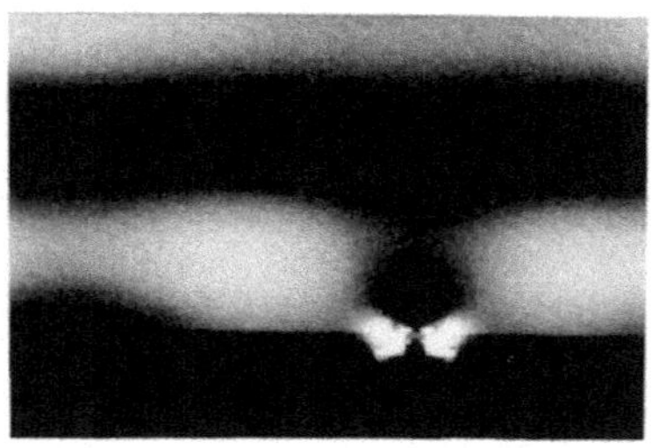

Figure 5. Square modulus of the total field scattered by two grooves. The grooves have a square section of side $\lambda/20$ and the size of the numerical cell is $\lambda/80$. Note the enhancement of the field in the grooves.

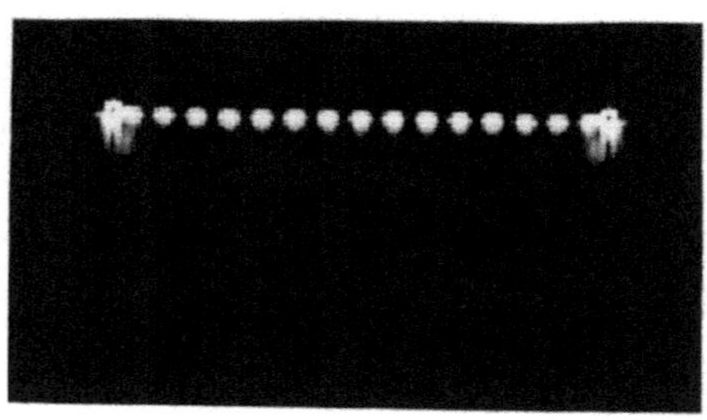

Figure 6. Square modulus of the scattered field showing the localization of the surface plasmon.

## 6. Conclusion

We have developped a theoretical approach of the scattering by a particle embedded in a metal by means of a volume integral equation based on the Green's dyadic. It has been shown that the field scattered in p-polarization may be resonantly enhanced by an image effect. The field due to the surface plasmon interferes with the illuminating beam producing fringes.

## Acknowledgements

This work was supported in part by a grant from Saint Gobain and by a fellowship to F.P. from the ANRT. A. S. acknowledges a fellowship from the CNRS and SNECMA.

## References

1. A. Sommerfeld, ( 1909), "Über die Ausbreitung der wellen in der drahtlosen Telegraphie.," Ann. Phys. (Leipzig), **28,** 665-737 .
2. A. Banos (1966)*Dipole radiation in the presence of a conducting half-space* (Pergamon, New-York).
3. L. M. Brekhovskikh (1980), in *Waves in Layered Media* (Academic Press, San Diego)
4. L. B. Felsen and N. Marcuwitz, (1973)*Radiation and scattering of waves* (Prentice Hall, Englewood Cliffs N.J.).
5. J. V. Bladel, (1991)*Singular electromagnetic fields and sources* (Clarendon Press, Oxford).
6. J.J. Wang (1991)*Generalized Moment Methods in Electromagnetics* (John Wiley, New York).
7. C. Hafner, (1990)*The Generalized Multiple Technique for Computational Electromagnetics* (Artech, Boston).
8. A. D. Yaghjian, (1980), "Electric Dyadic Green's Functions in the Source Region," Proc. IEEE, **68,** 248-263 .
9. R. Petit, (1980)*Electromagnetic theory of gratings* 1 (Springer Verlag, Berlin).
10. D. Maystre ( 1984) in *Progress in Optics* (E. Wolf ed.) (North Holland, Amsterdam), pp 3-67
11. J. A. Ogilvy, (1991)*Theory of wave scattering from random rough surfaces* 1 (Adam Hilger, Bristol, Philadelphia and New-York).
12. J. A. DeSanto and G. S. Brown, ( 1986), in *Progress in Optics* (E. Wolf ed.) (North Holland, Amsterdam), pp 3-64

13. A. A. Maradudin, T. Michel, A. R. McGurn and E. R. Mendez, (1990), "Enhanced backscattering of light from a random grating," Ann. Phys. (N.Y.), **203,** 255-307 .

14. M. Nieto-Vesperinas, (1991)*Scattering and Diffraction in Physical Optics* (John Wiley & Sons, Inc., New York).

15. E. M. Purcell and C. R. Pennypacker, (1973),"Scattering and absorption of light by nonspherical dielectric grains", Astrophys. J., **186,** 705 .

16. J.-J. Greffet and F. R. Ladan, (1991), "Comparison between theoretical and experimental scattering of an s-polarized electromagnetic wave by a two-dimensional obstacle near an interface," J.Opt.Soc.Am.A, **8,** 1261-1269 .

17. A. Sentenac and J.-J. Greffet, (1992), "Scattering by deep inhomogeneous gratings," J.Opt.Soc.Am.A, **9,** 996-1006 .

18. A. Lakhtakia, (1990), "Macroscopic theory of the coupled dipole approximation method," Opt. Commun., **79,** 1-5 .

19. P. B. Johnson and R. W. Christy (1972), "Optical constants of the noble metals," Phys. Rev. B **6,** 4370-4379 .

20. F. Keilmann, K. W. Kussmaul and Z. Szentirmay, (1988), "Imaging of optical wavetrains," Appl. Phys. B, **47,** 169-176 .

21. C. Girard, (1992), "Plasmon resonance and near-field optical microscopy: a self consistent theoretical model," Appl. Opt., **31,** 5380-5387 .

22. G. W. Ford and W. H. Weber, (1984), "Electromagnetic interactions of molecules with metal surfaces," Physics Reports, **113,** 195-287 .

23. P. R. Hilton and D.W.Oxtoby, (1980), "Surface enhanced Raman spectra: a critical review of the image dipole description," **72,** 6346-6348 .

24. J.-J. Greffet, (1989), "Scattering of s-polarized electromagnetic waves by a 2 D obstacle near an interface," Opt. Commun., **72,** 274-278 .

25. T.Takemori and M. Inoue (1988),"Field enhancement in a spherical cavity beneath a metal surface", J. Phys. Soc. Jpn **57,** 3188-3197.

# THEORETICAL CONSIDERATIONS ASSOCIATED WITH NEAR-FIELD SCANNING BY SUBWAVELENGTH APERTURES

A. ROBERTS
*School of Physics*
*University of Melbourne*
*Parkville, Victoria, 3052*
*Australia*

ABSTRACT. Various theoretical aspects of image formation and positioning constraints in near-field scanning aperture microscopies are considered in this paper. As well as a study of the way evanescent radiation fields behave immediately behind a subwavelength aperture, it includes a preliminary study of the role that the polarization of the illuminating radiation plays in the quality of the image obtained. In particular, three different models are used to explore specific aspects of SNOM: coupling of light through a subwavelength circular aperture, coupling of light into a tapered conical structure (to model commonly used probes), the effect on images produced by the illumination parameters and interactions between the detector itself and the structure under observation.

## 1. Introduction

Technical aspects of scanning near-field microscopy (SNOM) have undergone rapid advancement over the last few years [1]. Various theorietical studies (see the review by van Labeke and Barchiesi [2] in this volume) have led to a greater understanding of both the advantages and limitations of SNOM over other forms of high resolution imaging. Along with the limitations come severe constraints of the positioning of the detector and the amount of light which can be coupled into very small detecting or illuminating apertures [3]. In this paper some of these constraints and polarization and detector effects are explored theoretically.

Although there are many different types of near-field optical scanning microscopies, the studies presented here are those which involve imaging the object under study through an aperture with characteristic dimensions very much less than the wavelength of the light being used to obtain the image. The models presented here apply to the 'collection mode' where the light produced by the sample is collected through the subwavelength aperture. Images can also be obtained in 'illumination mode' by illuminating the sample through such an aperture. The results obtained here can be extended to similar configurations operating in such an illumination mode.

To understand the way light behaves in the near-field of various structures it is necessary to consider rigorous electromagnetic theories of diffraction. Although there are many applications which can be analysed using Fresnel and Fraunhofer theories, they are not appropriate for the study of near-field behaviour as they do not adequately account for the behaviour of evanescent fields. Since evanescent fields decay over a distance of the order a wavelength they do not necessarily play an important role in the far-field behaviour of diffraction patterns. One way of viewing near-field microscopy is that it allows the direct detection of the high-spatial frequency components that do not appear in standard models of far-field diffraction. Some authors [4] have used modified Fresnel theories that include these high spatial frequency components and have found reasonable agreement with other, more rigorous models for the geometries they considered. In this paper rigorous electromagnetic theories are used. These include the effect of the presence evanescent fields in a natural way and ensure that the appropriate boundary



*D.W. Pohl and D. Courjon (eds.), Near Field Optics*, 221–228.

conditions are satisfied. Most of the models presented here are based on such rigorous solutions of Maxwell's equations. The theoretical problems are generally extremely complex, so simplifying assumptions are still made here, usually in the type and shape of objects being considered. For example, the structures studied here are considered to be perfectly conducting.

## 2. Propagation of light through a circular aperture

### 2.1 BETHES' SMALL-HOLE THEORY

One of the most important considerations in near-field imaging is the total amount of power that can be coupled through an extremely small aperture. Although rigorous electromagnetic models are used extensively here to explore the near-zone fields of various diffracting structures, an approximation developed by Bethe [5] is used in two examples to calculate the total power transmitted through a circular aperture having dimensions very much less than the wavelength of the imaging radiation. This approximation has been widely used elsewhere (see, for example, [6]) in calculations relevant to scanning near-field optical microscopy. Here, this small-hole approximation is used to calculate the power transmitted through 'detectors' assumed to be perfectly conducting planes. The relevant fields at the surface of the 'detector' plane are calculated using rigorous theory and Bethe's model used to calculate the power transmitted through a subwavelength detecting aperture.

Bethe's model is based on the assumption that the fields in the vicinity of a small aperture are approximately those that would exist in the absence of the aperture. Using this he showed that if radiation with wavenumber $k_0$ is normally incident on a circular aperture of radius R the total transmitted power is given by

$$P = \frac{4k_0^4R^6}{27\pi Z_0}(4Z_0^2H_0^2 + E_0^2)$$

where $E_0$ is the component of the incident electric field perpendicular to the plane containing the aperture, and $H_0$ the component of the incident magnetic field parallel to the plane.

### 2.2 RIGOROUS ELECTROMAGNETIC THEORY

The Bethe theory outlined above assumes that the screen containing the aperture is infinitesimally thin. To obtain sufficient contrast between light that propagates through the aperture and that which leaks through the metal surrounding the hole, however, the metal must be sufficiently thick. For apertures in thick metallic screens the light propagating through the aperture is attenuated strongly with increasing screen thickness. So, although the screen must have sufficient thickness to block the light leaking through the metal, it cannot be so thick that the light propagating through the aperture is attenuated to an unacceptably low level. Thus, there is a finite range of metal thicknesses that provide sufficient contrast between the light coming through the aperture and that leaking through the surrounding metal while maintaining a reasonable intensity of the light propagating through the aperture.

Another characteristic of SNOM that can be explored by looking at models of diffraction through apertures is the constraint on positioning of the detector with respect to the sample under investigation. Since evanescent fields decay rapidly with distance away from the diffracting structure, variations in the detector-sample distance (eg. due to surface roughness) could lead to the erroneous interpretation of images. Studying the theoretical variation of fields in the near-field of a circular aperture (Figure 1) can provide an indication of the constraints that must be placed on detector-sample positioning to obtain accurate images.

In the model used here [7,8], the fields above and below the screen containing the aperture are written as integrals over a continuum of plane waves - both propagating and evanescent. The electromagnetic fields within the circular aperture itself are written as a sum over circular waveguide modes. The amplitudes of the plane waves and the waveguide modes are found by

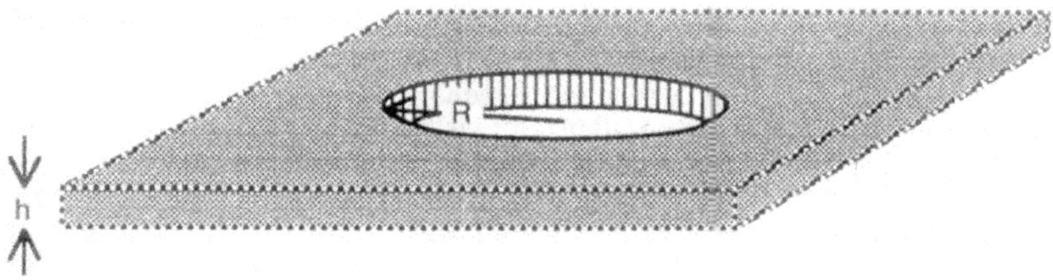


**Figure 1:** Parameters used in describing diffraction by a circular aperture in a thick, perfectly conducting screen.

applying the appropriate boundary conditions at the upper and lower surfaces of the screen. These continuity equations are then projected onto either the waveguide modes or plane waves, and after rearranging the expressions, an infinite set of linear equations in the amplitudes of the waveguide modes is obtained. The mode sum is truncated by ignoring higher order modes that make a negligible contribution to the total field and the amplitudes found using numerical methods. Once the mode amplitudes have been found, the plane wave amplitudes can be calculated. Thus the total power transmitted through the aperture can be computed as can the near-field diffraction patterns. The solutions for the electric and magnetic fields above and below the screen obtained using this technique satisfy the Helmholtz equation and the appropriate boundary conditions.

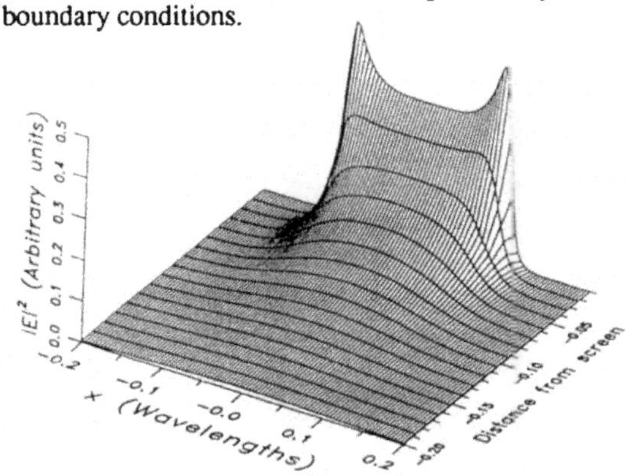


**Figure 2:** Electric field intensity behind a circular aperture with a radius of $\lambda/5$. The field is plotted in a plane containing the axis of the aperture and the electric field vector of the incident wave.

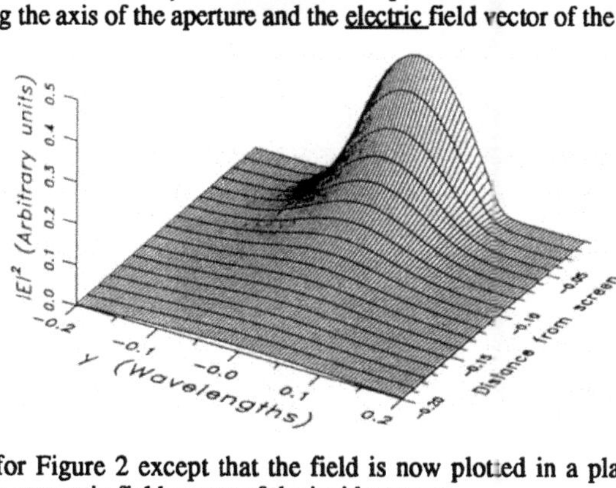


**Figure 3:** As for Figure 2 except that the field is now plotted in a plane containing the axis of the aperture and the magnetic field vector of the incident wave.

Estimates of the leakage of light through the surrounding metal can be obtained by ignoring the presence of the aperture and comparing this result with the transmittance of the light through the aperture in the perfectly conducting screen. These studies have shown that apertures with a diameter of approximately λ/10 (eg. for visible light a diameter around 50nm) in a screen of thickness λ/40 (eg. 12.5nm at 500nm) have a transmittance of approximately 11 percent compared to the transmission of light through an aluminium screen of the same thickness of around 2 percent. This provides reasonable contrast between the aperture and the screen. Although the transmission through apertures in thinner screens is greater, this increase in intensity is offset by an increase in the leakage of light through the surrounding screen.

This technique has also been used to compute near-field diffraction patterns produced by a circular aperture. In Figures 2 and 3, the electric field intensity $|E|^2$ in two orthogonal planes containing the axis of the aperture are plotted as a function of the distance behind the aperture. The aperture has a radius of λ/10 and is assumed to have infinitesimal thickness. Figure 4 contains a an 'image' of the beam immediately behind such an aperture. Figure 5 shows this spot a distance of 0.08λ behind the aperture (i.e. approximately 40nm for visible light) These figures show that the collimation of the light immediately behind the circular aperture although very good immediately behind the aperture, deteriorates rapidly over a distance of approximately 0.1λ (50nm )to a point where it is extremely poor. The rapid falloff in intensity with distance from the aperture gives an indication of the constraints on the position of a detector with respect to a rough surface. If a uniform distance is not maintained, small variations in the sample

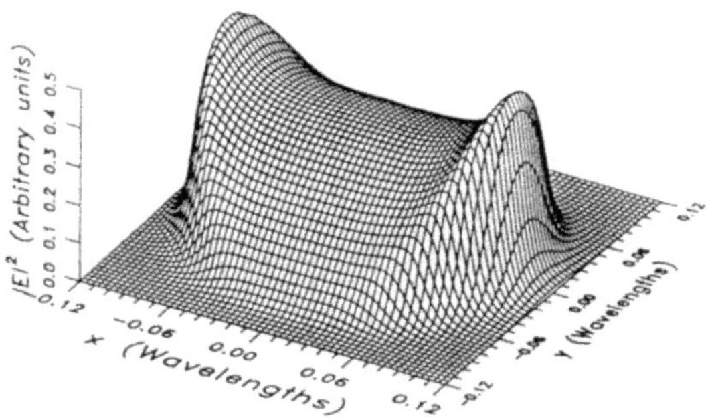


**Figure 4** : Two dimensional electric field intensity pattern a distance of 0.01λ behind a circular aperture with a radius of λ/5.

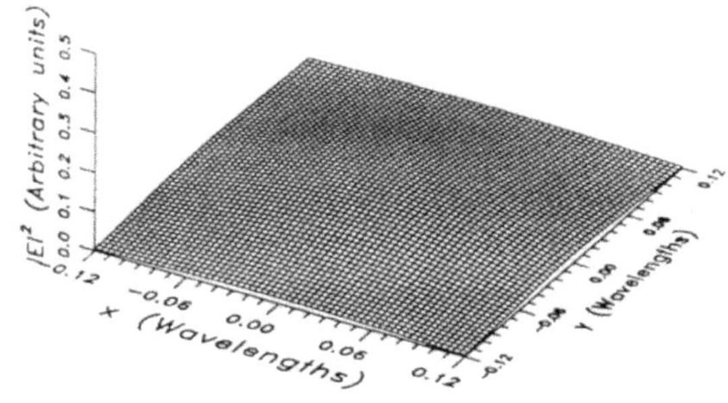


**Figure 5:** Two dimensional electric field intensity pattern a distance of 0.08λ behind a circular aperture with a radius of λ/5.

detector position will produce a large increase in measured intensity and possibly an incorrect interpretation of the image produced. Note that Figures 2 and 3 suggest that the detector-sample distance should be maintained with an accuracy of approximately 0.01λ (5nm) to avoid this.

Figure 6 is similar to Figure 2 although the screen thickness has been increased to λ/40 (12.5nm). The electric field scale is the same as in Figure 2. The most striking difference introduced by increasing the screen thickness is the decrease in the transmitted field. The collimation and spreading of the beam behind the aperture is not substantially affected by increasing the screen thickness. In Figure 7, the effect of decreasing the aperture radius by a factor of 2 is considered. Only values in the plane corresponding to that shown in Figure 2 is presented here for the sake of brevity. The horizontal and vertical scales are the same in both figures. It can be seen that the intensity transmitted through the smaller aperture in Figure 7 is substantially less than that shown in Figure 2. The collimation of the beam immediately behind the aperture is approximately equal to the size of the smaller aperture corresponding to a higher resolution in the image obtained It can be seen, however, that the deterioration in the beam collimation is much more rapid in this case, thus increasing the constraints on detector-sample positioning outlined above.

Another feature of the diffraction patterns shown above is their asymmetry compared to the symmetry (circular) of the structure itself. This arises because the polarization of the incident wave imposes a preferred direction. The peaks in the two-dimensional diffraction patterns are a

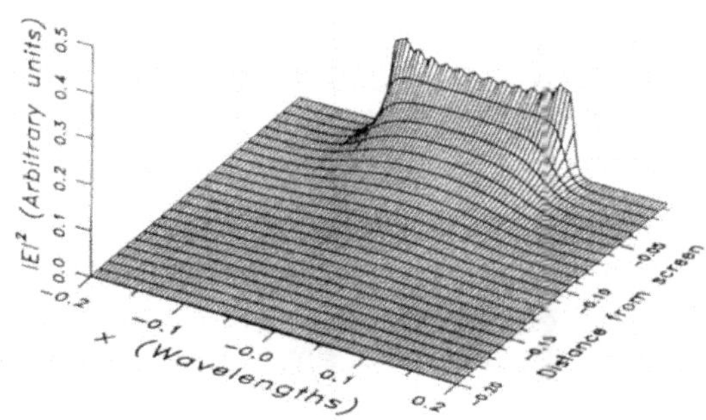


**Figure 6:** Electric field intensity behind a circular aperture with a radius of λ/5 in a screen of thickness λ/40. The field is plotted in a plane containing the axis of the aperture and the electric field vector of the incident wave.

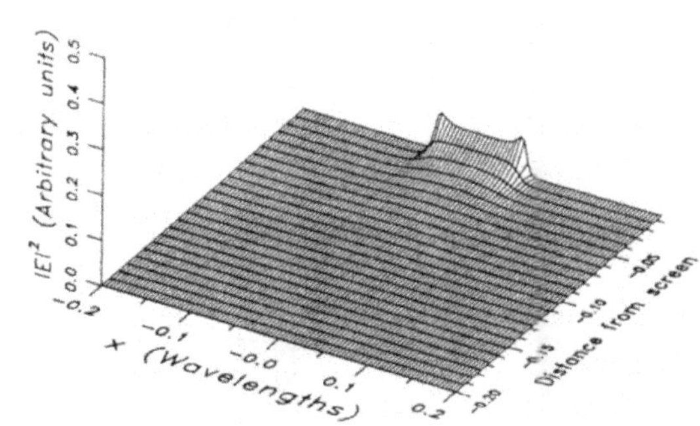


**Figure 7:** Electric field intensity behind a circular aperture with a radius of λ/10 in a screen of zero thickness. The field is plotted in a plane containing the axis of the aperture and the electric field vector of the incident wave.

result of the allowed discontinuity in the component of the electric field normal to a perfectly conducting surface. Note, however, that the component of the field parallel to the perfectly conducting surfaces approaches zero as required. Polarization effects are explored further in section 4.

## 3. Coupling of light into tapered lightguides

Many optical scanning probe techniques rely on metal coated tapered guides. Examples of these include micropipettes and optical fibres drawn down to a tip size of the order of one micron and containing a subwavelength aperture. In order to gain some estimate of the effect of this geometry on the total power transmitted into such a probe, a simple model of radiation coupling into a conical waveguide through a subwavelength aperture was considered. [9] Parameters considered were the size of the coupling aperture, the probe taper and the size of the probe itself at the point where the coupling aperture is located. Along with the strong $R^6$ dependence on the coupling aperture radius, the most important factor influencing the amount of radiation that could be coupled into such a probe is the size of the probe at the coupling point. If the radius of the probe at the coupling point is less than the cutoff radius of a circular waveguide at the wavelength of the radiation being used, then very little power can propagate along the probe to a detector. This power increases gradually with increasing taper angle, but there is a more striking increase in the power as the radius of the probe at the point of coupling increases above the cutoff radius of a circular waveguide.

As yet, no attempt has been made to take into account the material filling the probe, whether it be a tapered optical fibre or a micropipette.

## 4. Probe-sample interactions and polarization effects

One aspect of SNOM that has been considered experimentally is the affect of polarization on the 'image' that is obtained by scanning a structure with a subwavelength aperture. [1] Looking at the near-field diffraction patterns of a symmetrical structure, such as a circular aperture, it can be seen that the 'image' obtained is not symmetrical. There are field enhancements at the sides of the circular aperture that are perpendicular to the electric field of the incident light. This is a result of the boundary conditions which must be satisfied at the interface between two material media. Although the component of the electric field that is tangential to such an interface must be continuous, there can be a discontinuity in the component of the field which is normal to this interface. Hence, the electric field intensity varies with position.

To obtain a clearer relationship between polarization and the orientation of various components of a metallic structure being investigated, the fields in the near-zone behind a perfectly conducting diffraction grating were considered. (Figure 8)

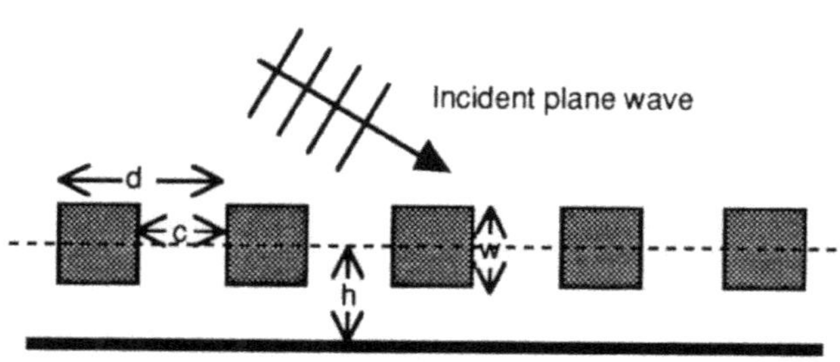


**Figure 8:** A lamellar diffraction grating of period d, thickness w, and wire separation c is backed by a perfectly conducting plane, a distance h below the centre of the grating.

As in the case of the circular aperture a modal theory was used. The main differences lie in the fact that the continuum of plane waves now becomes a discrete set as a result of the uniform periodicity of the grating and the waveguide modes now become those of a parallel plate guide. To model the presence of a detector, a perfectly conducting plane was placed below the grating. By calculating the electric field normal to, and the magnetic field parallel to this plane, estimates of the power transmitted through a subwavelength aperture situated within this plane can be obtained. The plane of the wavevector was assumed to be perpendicular to the grating wires and two orthogonal polarizations were considered. If the wavevector of the incident plane wave is taken to be perpendicular to the grating 'wires', then any incident plane wave of arbitrary polarization can be regarded as a superposition of two orthogonal polarizations: one where the incident electric field is parallel to the wires (P-polarization) and one where the incident magnetic field is parallel to the wires (S-polarization).

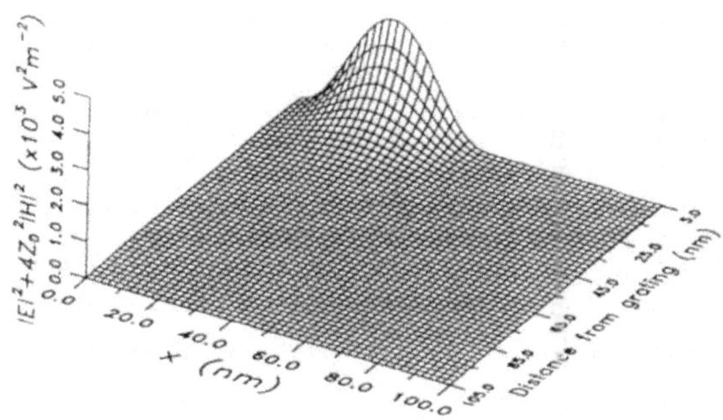


**Figure 9:** Variation in the quantity $4Z_0^2H_0^2 + E_0^2$ behind a grating with d=100nm, c=50nm, w= 50nm subjected to light incident at 45° with a wavelength $\lambda$=633nm polarized with its <u>electric</u> field parallel to the grating wires.

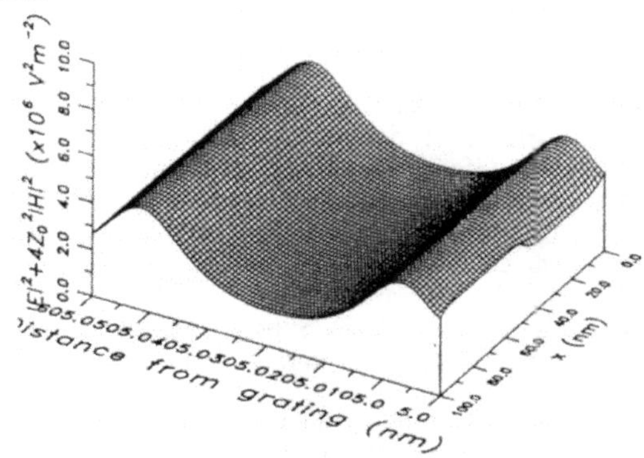


**Figure 10:** As for Figure 9, but with the <u>magnetic</u> field of the incident light parallel to the wires.

Without the plane approximating the 'detector', an analysis of the electric field patterns behind the grating reveals that behind the wires, the 'image' obtained appears dark and behind the air-gap between the wires appears bright. The 'image' thus obtained provides a reasonable representation of the original object for <u>both</u> polarizations. If we look at the relevant combination of fields $|E|^2 + 4Z_0|H|^2$ at the plane (detector) for the two polarizations, very different images are obtained. For P-polarization, the image is very similar to that obtained if only the electric field was measured and no interaction was taken into account. That is, the image was 'bright' behind the air-gap and 'dark' behind the wires. For S-polarization, however, the image appears about 20 per cent brighter behind the metal wires than behind the air-gap. In all images, the contrast deteriorates rapidly with distance behind the grating. This is consistent with the result obtained for the circular aperture. With the presence of the perfectly conducting plane modelling the detector, interference maxima appeared at distances approximately $\lambda/2$ apart behind the grating. Given the stringent requirements of positioning inherent in obtaining meaningful images using SNOM, these peaks should not adversely affect near-field images.

## 5. Conclusions and future work

This paper has outlined only some of the aspects of scanning near-field optical microscopy using a subwavelength apertures. The two main areas of investigation have involved either studying the amount of light which can be collected through such an aperture, or the interaction of the illuminating light, the structure under investigation and the probe itself. By studying the light propagating though apertures having dimensions of $\lambda/10$ - $\lambda/20$, it can be seen that the most severe constraints on near-field imaging are on the total amount of power which can be coupled through such an aperture and on the positioning of the probe with respect to the aperture. The latter arises due to the rapid deterioration of the collimation of the beam with distance from the collimating aperture. The distance over which the beam spreads decreases as the size of the aperture decreases and, thus, the positioning constraints become more severe as the characteristic dimensions of the structure being investigated become smaller.

By investigating the interaction between a diffracting structure (a metallic diffraction grating) and a 'probe' modelled as an infinite plane, the influence on the final image of the polarization of the incident light with respect to the structure was explored. It was found that the polarization of the incident light had a profound effect on the image obtained, with the best contrast obtained when the electric field of the illumination radiation was parallel to the wires of the grating being investigated.

## 6. Acknowledgments

The author would like to thank Keith Nugent, Duncan Butler and Michael Hrynevych for useful discussions.

## 7. References

1. Betzig E. and Trautmann J.K. (1992) 'Near-field optics: microscopy, spectroscopy, and surface modification beyond the diffraction limit', Science 257, 189-195.
2. Van Labeke D. and Barchiesi D. (1992) 'Theoretical problems in scanning near-field optical microscopy', NFO Besancon, to be published.
3. Pohl D.W. (1990) 'Scanning near-field optical microscopy (SNOM)', in C.J.R. Sheppard and Mulvey T. (eds), Advances in optical and electron microscopy, Academic, Orlando, pp. 243-312.
4. Vigoureux J.M, Depasse and Girard C. (1992) 'Superresolution of near-field optical microscopy defined from properties of confined electromagnetic waves', Applied Optics 31, 3036-3044.
5. Bethe H.A. (1944) 'Theory of diffraction by small holes', Physical Review 66, 163-182.
6. Leviatan Y. (1986) 'Study of near-zone fields of a small aperture', Journal of Applied Physics 60, 1577-1583.
7. Roberts A. (1987) 'Electromagnetic theory of diffraction by a circular aperture in a thick, perfectly conducting screen', Journal of the Optical Society of America A 4, 1970-1983.
8. Roberts A. (1989) 'Near-zone fields behind circular apertures in thick, perfectly conducting screens', Journal of Applied Physics 65, 2896-2899.
9. Roberts A. (1991) 'Small-hole coupling of radiation into a near-field probe', Journal of Applied Physics 70, 4045-4049.

# ON THE RESOLUTION LIMIT OF NEAR-FIELD OPTICAL MICROSCOPY

O. Keller, S. Bozhevolnyi and M. Xiao
Institute of Physics, University of Aalborg,
DK-9220 Aalborg, Denmark.

ABSTRACT. Using the near-field approximation to describe the interaction between two electric dipoles (a probe and an object dipole) placed near a surface we show that the resolution of the system can be substantially increased by scanning the probe dipole along the surface at a specific (resonance) distance from the surface. In such an approach the resolution of the near-field optical microscope is found to depend only upon the system damping and the resonance distance.

## 1 . Introduction

A central question to be considered in scanning near-field optical microscopy (SNOM) is that of the resolution of the system. Usually the resolution is expressed in terms of the radiation wavelength [1-3]. For a quantitative description of the SNOM a microscopic approach is widely exploited [4-6]. In such an approach the fiber tip is assumed to be small sphere and the tip-surface interaction is considered in the electric dipole approximation. The general conclusion is that the resolution of the SNOM is restricted by the size of the tip and decreases with increase of the tip-surface distance [6,7]. Considering a lossless system of point-like dipoles placed near the bulk we have recently shown [8] that the resolution can be substantially increased by scanning the probe dipole along the surface at a specific (resonance) distance from the surface. The resonances in the dipole-surface system and in the system of two identical dipoles recently have been studied by Girard [5] and by Markel' [9] accordingly, but without implications to the resolution of the SNOM. Here we discuss the resolution of the system in the presence of damping taking into account the interaction of the probe dipole only with one (object) dipole placed near the surface. In this case the resolution of the SNOM is found to be strictly related neither to the wavelength nor to the size of the tip or to the tip-surface distance, but to the system damping and the resonance distance. We also present a quantitative explanation of the lateral superresolution ($\sim 3nm$) recently demonstrated with the scanning plasmon near-field microscope by Specht et al. [10].



*D.W. Pohl and D. Courjon (eds.), Near Field Optics, 229–237.*

## 2 . System of two dipoles

Using a Green's function formalism in the near-field approximation we have previously established a set of self-consistent algebraic equations to describe the local field at the sites of point-like dipoles placed near the bulk [8]. For two dipoles in free space (Fig.1) the local field at the site of the probe dipole immediately can be written down in an explicit form, viz.

$$E_{x(y)} = E^0_{x(y)}(1 - \frac{\alpha_2(\omega)}{4\pi\epsilon_0 R^3})/(1 - \frac{\alpha_1(\omega)\alpha_2(\omega)}{(4\pi\epsilon_0 R^3)^2}) \quad , \tag{1}$$

$$E_z = E^0_z(1 + \frac{\alpha_2(\omega)}{2\pi\epsilon_0 R^3})/(1 - \frac{\alpha_1(\omega)\alpha_2(\omega)}{(2\pi\epsilon_0 R^3)^2}) \quad , \tag{2}$$

where $E^0_{x(y,z)}$ are the components of the field which would prevail in space in the absence of the dipoles, $R$ is the distance between the dipoles and $\alpha_1(\omega)$ and $\alpha_2(\omega)$ are their (isotropic) polarizabilities cf. Fig.1.

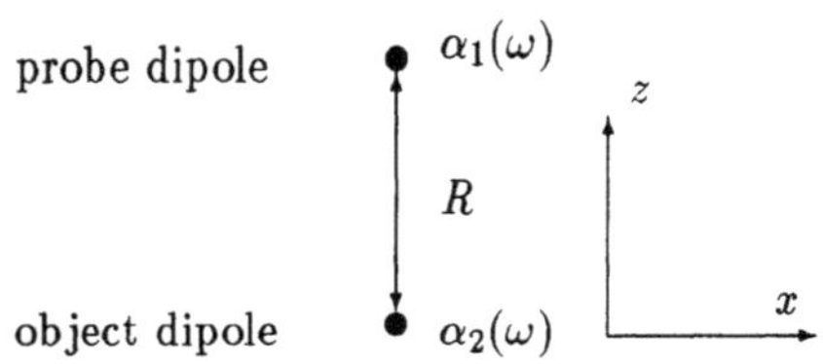


**Figure 1.** Schematic diagram showing the geometry of the system of two dipoles.

It can readily be seen from Eqs.(1) and (2) that for any positive real values of $\alpha_1(\omega)$ and $\alpha_2(\omega)$ there are two different resonance distances $R_0^z$ and $R_0^{x(y)}$ between the dipoles, at which the $z$- or $x(y)$-components of the local field are infinitely enhanced, i.e.

$$R_0^z = \left(\frac{\alpha_1(\omega)\alpha_2(\omega)}{(2\pi\epsilon_0)^2}\right)^{1/6} \quad , \tag{3}$$

$$R_0^{x(y)} = R_0^z/\sqrt[3]{2} \quad . \tag{4}$$

The resonances for the $x$- and $y$-components are coincident because of symmetry. Similar resonances have been found for more complicated systems of dipoles [8] and were referred to as configurational ones, since, for any system of point-like dipoles, there is at least one resonance configuration. Instead of searching for configurational resonances at a given frequency, one can look for resonance frequency at the given value of $R$. By inserting the given value of $R$ into Eqs.(3) and (4) the associated resonance frequencies can be obtained, and compared to the resonance frequencies in the limit ($R \to \infty$) when the dipoles are uncoupled, i.e. the intrinsic dipole resonances.

In a sense our system of two dipoles can be considered as a simple near-field resonator and Eqs.(3) and (4) express the condition that the field being scattered by one of the dipoles after *reflection* from the other dipole is equal to the driving field of the first dipole at resonance.

The presence of damping in the system changes the resonance positions in Eqs.(3) and (4) and leads to a finite height of the resonance peaks. In the following we are concerned with the resolution of the system and shall for simplification assume the polarizability of the object dipole is much smaller than that of the probe dipole $|\alpha_2(\omega)| \ll |\alpha_1(\omega)|$. In this case, the $z$-component of the local field takes the form $E_z = E_z^0/(1-\Lambda)$ (with $\Lambda \equiv \alpha_1(\omega)\alpha_2(\omega)/(2\pi\epsilon_0 R^3)^2$), near resonance ($\alpha_1\alpha_2 \simeq (2\pi\epsilon_0 R^3)^2$). The expression for $E_x$ near resonance is identical to that for $E_z$ apart from a factor of 4 in front of the second term in the denominator. The peak of the local field intensity as a function of the dipole separation is well pronounced if $ReA > |ImA|$, where $A = \alpha_1(\omega)\alpha_2(\omega)$. The normalized (i.e. $E_z E_z^* / (E_z^0 E_z^{0*})$) value of the peak intensity and the associated resonance distance between the dipoles are then given by

$$I_{max} \simeq \frac{|A|^2}{(ImA)^2} \quad and \quad R_0 = \left(\frac{|A|^2}{(2\pi\epsilon_0)^2 ReA}\right)^{1/6} . \tag{5}$$

The full width of the resonance peak at half maximum (FWHM) obtained when the probe dipole is approaching the object one ($\Delta R$) or scanning along $x$-axis in the plane $z = R_0$ ($\Delta x$) in the case of small damping in the system ($|ImA| \ll ReA$) becomes as follows:

$$\Delta R \simeq \frac{R_0}{3}\frac{|ImA|}{ReA} \quad and \quad \Delta x \simeq 2R_0\sqrt{\frac{|ImA|}{3ReA}} . \tag{6}$$

The values of $\Delta R$ and $\Delta x$ are related to the resolution of the SNOM near resonance. In general, the resolution in depth ($\Delta R$) is much higher than the lateral resolution ($\Delta x$) and both resolutions are limited only by the system damping. The wavelength dependence of the resolution appears only via the frequency dependence of $A = \alpha_1(\omega)\alpha_2(\omega)$. The resolution is also better for smaller values of polarizabilities, since $R_0 \sim (\alpha_1\alpha_2)^{1/6}$ (cf. Eq.(5)). As far as the polarizability of a small spherical particle is concerned there is no simple relation to its size [11].

## 3. Influence of the substrate

In this section we shall consider the situation where the object dipole is placed near the bulk at a distance $\delta$ (Fig.2). In this case, the self-consistent field at the sites of the dipoles can be expressed in an explicit form only when the projections of the dipoles on the surface are coincident. For example, the $z$-component of the local field at the site of the probe dipole is given by

$$\begin{aligned} E_z \; = \; & E_z^0 \left\{1 + \frac{\alpha_2}{2\pi\epsilon_0}\left[\frac{1}{R^3} + \frac{r^p}{(R+2\delta)^3} - \frac{r^p}{(2\delta)^3}\right]\right\} \times \\ & \left\{\left[1 - \frac{\alpha_2 r^p}{2\pi\epsilon_0(2\delta)^3}\right]\left[1 - \frac{\alpha_1 r^p}{2\pi\epsilon_0[2(R+\delta)]^3}\right] - \right. \\ & \left. \frac{\alpha_1\alpha_2}{(2\pi\epsilon_0)^2}\left[\frac{1}{R^3} + \frac{r^p}{(R+2\delta)^3}\right]^2\right\}^{-1} , \end{aligned} \tag{7}$$

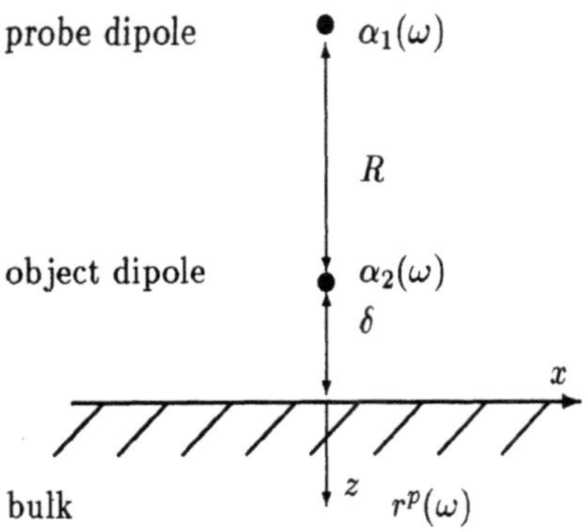


**Figure 2.** Schematic diagram showing the geometry of the system of two dipoles placed near a surface.

where $r^p = (\epsilon(\omega) - 1)/(\epsilon(\omega) + 1)$ is the amplitude reflection coefficient of the bulk for $p$-polarized light, and $\epsilon(\omega)$ is the local bulk dielectric constant. The resonances of this system can be obtained by setting the denominator in Eq.(7) equal to zero. Inside this denominator appear the resonances of the subsystems. These are single-particle ($\alpha_1, \alpha_2 \rightarrow \infty$) and surface-plasmon ($\epsilon \rightarrow -1$) resonances.

The general relation in Eq.(7) can be significantly simplified if we neglect the interaction between the probe dipole and the bulk, and the indirect coupling between the dipoles:

$$E_z = E_z^0 \left[1 - \frac{\alpha_2}{2\pi\epsilon_0}\left(\frac{r^p}{(2\delta)^3} - \frac{1}{R^3}\right)\right] / \left[1 - \frac{\alpha_2 r^P}{2\pi\epsilon_0(2\delta)^3} - \frac{\alpha_1\alpha_2}{(2\pi\epsilon_0 R^3)^2}\right] \quad . \tag{8}$$

The structure of Eq.(8) is similar to that of Eq.(2) near resonance and the relations in Eqs.(5) and (6) can be used with $A$ given by

$$A = \alpha_1(\omega)\alpha_2(\omega) / \left(1 - \frac{\alpha_2(\omega) r^p(\omega)}{2\pi\epsilon_0(2\delta)^3}\right) \quad . \tag{9}$$

In this case the influence of the bulk is described via a renormalization of the polarizability of the object dipole.

Specht et al. [10] recently have developed a scanning plasmon near-field microscope, where surface plasmons in silver film interacted with a sharp tungsten tip. The lateral resolution of $3nm$ has been achieved with a tip size of about $20nm$(!). This remarkable result can be understood by assuming the system to be near resonance and by using the relation in Eq.(6) for $\Delta x$ together with the long-wavelength expression for the polarizability of a sphere, i.e.

$$\alpha(\omega) = 4\pi\epsilon_0 a^3 \frac{\epsilon_s(\omega) - 1}{\epsilon_s(\omega) + 2} \quad , \tag{10}$$

where $\epsilon_s(\omega)$ is the relative dielectric constant of the sphere and $a$ its radius. Taking $R_0 = 13nm$ (the tip radius $10nm$ plus the tip-surface distance $3nm$) [10], the radius of the object dipole to be equal its distance from the bulk and the appropriate dielectric constants for silver and tungsten at $\lambda = 633nm$ [12] results with the help of Eqs.(6), (9) and (10) in a lateral resolution $\Delta x \simeq 6nm$. It has been noted [10], that the imaging is better after depositing silver on the tip by slightly touching the sample with it. Using Eqs.(6), (9) and (10) for two silver spheres and a silver substrate leads to the value $\Delta x \simeq 3nm$.

# 4 . Numerical results

The behaviour of the system of two dipoles placed near the bulk is very complicated, especially near resonance. Here, we shall consider various possibilities of resonance excitation of the system by employing the Drude formula to describe in turn one of the three dielectric constants of the bulk and the two dipoles:

$$\epsilon(\omega) = 1 - \frac{\omega_p^2}{\omega(\omega + i\gamma)} \quad , \tag{11}$$

where $\omega_p$ is the plasma frequency and $\gamma$ is a phenomenological damping rate. The polarizabilities of the dipoles were calculated by using Eq.(10) with the radius of the probe dipole $a_1 = 10nm$ and that of the object dipole $a_2 = 1nm$. If not specified the latter is separated from the bulk by the distance $\delta = 1nm$. The incident field was taken in the form $\vec{E}^0 = (0, 0, 1)$ and the intensity $I$ of the local field at the site of the probe dipole was calculated by solving a set of self-consistent algebraic equations established previously [8].

First, we shall consider the system response when the dielectric constants of two dipoles are equal to each other $\epsilon_{s1} = \epsilon_{s2} = 3$ and that of the bulk is given by Eq.(11). If the radiation frequency $\omega$ is close to the surface plasmon frequency $\omega_0 = \omega_p/\sqrt{2}$, then the reflectivity $r^p$ can be written in the form:

$$r^p(\omega) = \left(2\left(\frac{\delta\omega}{\omega_0}\right) - i\left(\frac{\gamma}{\omega_0}\right)\right)^{-1} \quad , \tag{12}$$

where $\omega = \omega_0 - \delta\omega$. For $\delta\omega = 0.02\omega_0$ the resonance distance $R_0$ between the dipoles has been found to be $R_0 \simeq 12.4nm$. Subsequently, the intensity $I$ was calculated as a function of the $x$-coordinate of the probe, scanning along the surface at various distances (near the resonance distance). It can be seen (Fig.3) that scanning at the resonance distance results in the best imaging and that the resolution increases with the decrease of damping. We would like to stress that the size of the probe dipole is $20nm$ but the lateral resolution of $3.5nm$ can be obtained when scanning at the resonance distance (Fig.3b). It should be mentioned that similar dependencies have been found for other distances $\delta$ between the object dipole and the bulk ($\delta > 1nm$).

Another possibility for resonance interactions in the system is connected with the resonance excitation of the probe dipole. To consider it we used Eq.(11) for the probe dipole dielectric constant with the frequency $\omega$ being close to the resonance frequency $\omega_0 = \omega_p/\sqrt{3}$ and the constants of the bulk and the object dipole have been chosen to be equal $\epsilon = \epsilon_{s2} = 3$. In this case the system response (Fig.4) appeared to be quite similar to that discussed previously (Fig.3). Again, the image contrast is the best when scanning the probe dipole along the surface at the resonance distance and the system resolution is better for smaller damping of the system. The qualitative features of the system response near resonance for other values of $\delta$ ($\delta > 1nm$) are the same as for $\delta = 1nm$ (Fig.4).

The behaviour of the system becomes completely different if the excitation frequency is close to the resonance frequency of the object dipole. This situation is studied by taking Eq.(11) for the object dipole dielectric constant, choosing the frequency $\omega \simeq \omega_0 = \omega_p/\sqrt{3}$ and letting the dielectric constants of the bulk and the probe dipole be equal, viz. $\epsilon = \epsilon_{s1} = 3$. For the object dipole placed on the surface ($\delta = 1nm$), the system response shows

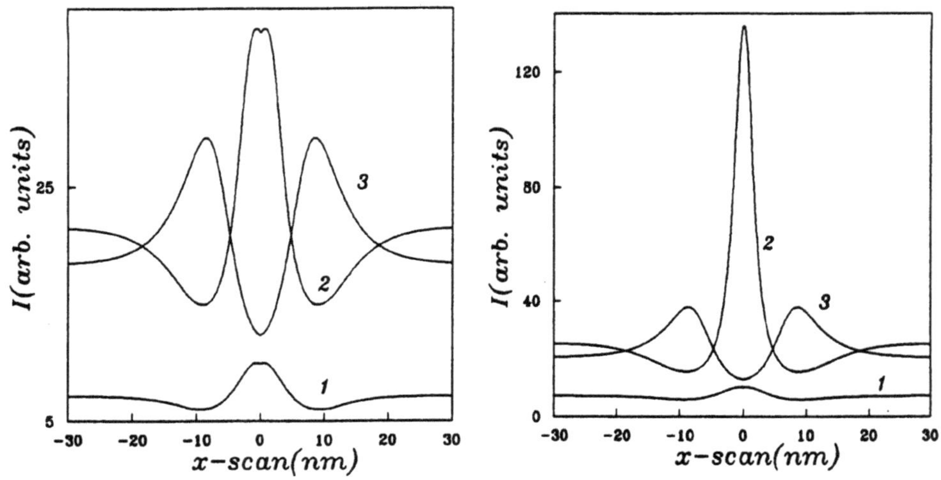


(a) (b)

**Figure 3.** Intensity of the local field at the site of the probe dipole as a function of its $x$-coordinate. The distances between the probe dipole and the surface are $R + \delta = 13nm$ (1), $13.4nm$ (2) and $13.8nm$ (3). The frequency $\omega$ is close to the surface plasmon frequency $\omega_0$: $\omega = 0.98\omega_0$ and the resonance distance is $R_0 + \delta \simeq 13.4nm$. The damping rate is $\gamma = 0.001\omega_0$ (a) and $\gamma = 0.0002\omega_0$ (b).

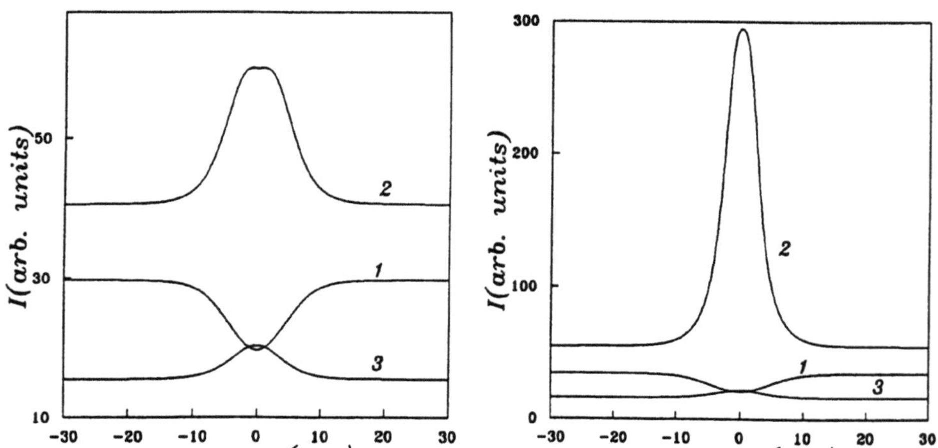


(a) (b)

**Figure 4.** Intensity of the local field at the site of the probe dipole as a function of its $x$-coordinate. The distances between the probe dipole and the surface are $R+\delta = 12.65nm$ (1), $12.85nm$ (2) and $13.05nm$ (3). The frequency $\omega$ is close to resonance frequency of the probe dipole $\omega_0$: $\omega = 0.97\omega_0$ and the resonance distance is $R_0+\delta \simeq 12.85nm$. The damping rate is $\gamma = 0.001\omega_0$ (a) and $\gamma = 0.0002\omega_0$ (b).

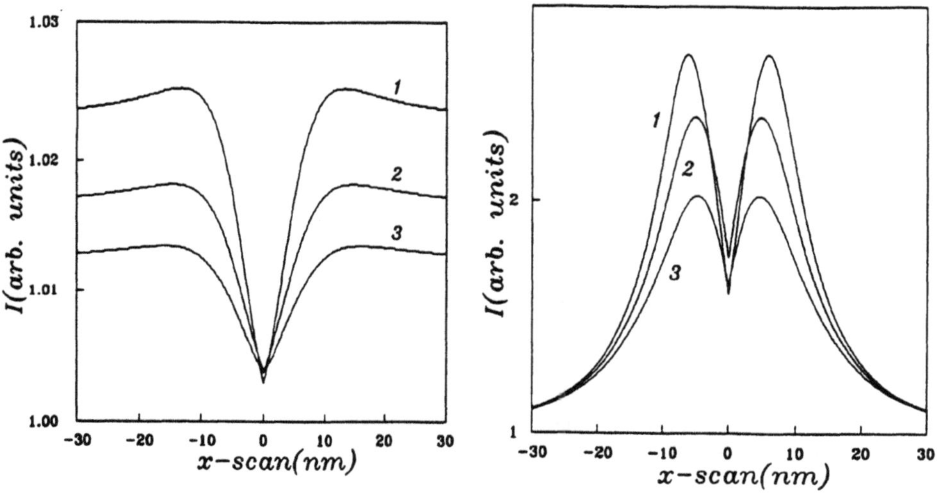


(a) (b)

**Figure 5.** Intensity of the local field at the site of the probe dipole as a function of its $x$-coordinate. The distances between the probe dipole and the surface are $R+\delta = 13nm$ (1), $14.5nm$ (2), $16nm$ (3) in Fig.5a ($\delta = 1nm$) and $R+\delta = 16.5nm$ (1), $18nm$ (2), $19.5nm$ (3) in Fig.5b ($\delta = 5nm$). The frequency $\omega$ is close to the resonance frequency $\omega_0$ of the object dipole: $\omega = 0.999\omega_0$ (a) and $0.9995\omega_0$ (b), and the resonance distances are $R_0+\delta \simeq 14.5nm$ (a) and $18nm$ (b). The damping rate is $\gamma = 0.0005\omega_0$.

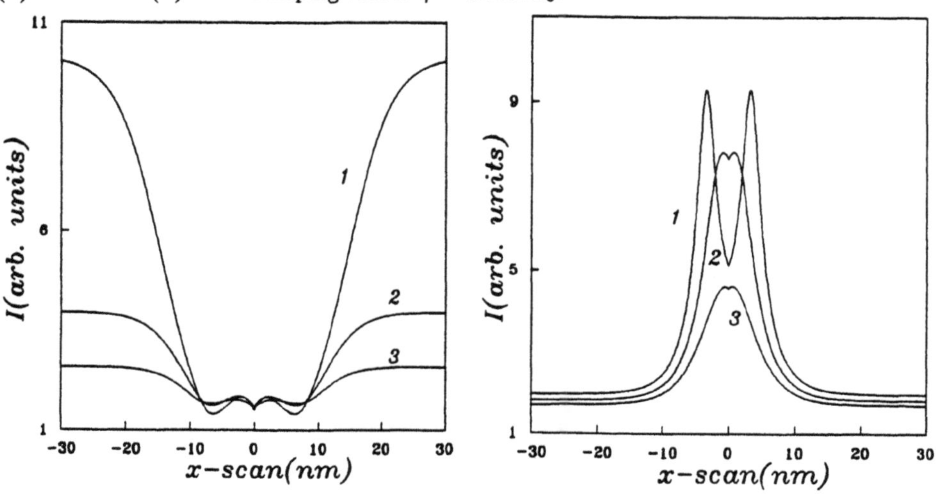


(a) (b)

**Figure 6.** Intensity of the local field at the site of the probe dipole as a function of its $x$-coordinate. The distances between the probe dipole and the surface are $R+\delta = 13nm$ (1), $14nm$ (2), $15nm$ (3) in Fig.6a ($\delta = 1nm$) and $R+\delta = 16.2nm$ (1), $16.7nm$ (2), $17.2nm$ (3) in Fig.6b ($\delta = 5nm$). The frequency $\omega$ is close to the resonance frequency $\omega_0$ of dipoles: $\omega = 0.97\omega_0$ and the resonance distances are $R_0+\delta \simeq 14nm$ (a) and $16.7nm$ (b). The damping rate is $\gamma = 0.005\omega_0$.

a contrast reversal and the resolution is better for smaller distances between the probe dipole and the surface (Fig.5a). If the object dipole is far from the surface ($\delta = 5nm$), it results in a double image with direct contrast (Fig.5b). In both cases there is no advantage in scanning the probe dipole at the resonance distance. This is connected with the very strong interaction between the object dipole and the bulk, an interaction which leads to a substantial increase of the $x$-component of the local field at the site of the probe dipole.

A similar behaviour of the system has been found when the dielectric constants of both dipoles were described by Eq.(11) with a frequency $\omega \simeq \omega_0 = \omega_p/\sqrt{3}$ and a bulk constant being equal to $\epsilon = 3$ (Fig.6). However, for the object dipole being far from the surface ($\delta = 5nm$), the double image is noticeable only for the scanning distance less than the resonance distance (Fig.6b). In this case the best image contrast and resolution have been found when scanning at the resonance distance between the probe dipole and the bulk (Fig.6b). It should be mentioned that the approximate relations in Eqs.(5), (6) and (9) can be used to estimate the characteristics of the resonance for the last configuration, since the influence of the substrate is relatively small.

We have used in our calculations an incident field of the form $\vec{E}^0 = (0, 0, 1)$ for several reasons. The main reason is that the first resonance occurring when approaching the probe dipole to the surface will be due to the coupling of $z$-component (cf. Eq.(4)). This also explains why $p$-polarized light is more efficient for imaging in SNOM [4]. The SNOM with the fiber tip of small aperture angle can be more sensitive to the field component parallel to the surface [4], but, near a well pronounced resonance, the resonant component of the field will dominate the interaction.

## 5 . Summary

Using the rigorous point-dipole approach developed previously [8] we considered with inclusion of damping the system of two dipoles (probe and object) placed near the bulk. Scanning the probe dipole along the surface at distances close to the resonance distance between the probe and the surface the system response was studied. Simple approximate relations for the characteristics of the resonances have been established for the system, in the case where the interaction between the object dipole and the surface and the indirect interaction between dipoles can be neglected. These relations account for the high laterial resolution recently demonstrated with the scanning plasmon near-field microscope [10]. Scanning the probe dipole along the surface the intensity of the local field at the site of the probe was calculated for various resonance configurations of the system. The best image contrasts and resolutions have been found when scanning the probe dipole at the resonance distances for systems with resonance excitation of the bulk or of the object dipole. In these cases the resolution of the system is limited by the system damping and not by the size of the probe nor by the probe-surface distance. Our results show how to choose the system parameters in order to obtain a high resolution of the SNOM. Using a radiation frequency close to the frequency of the resonance excitation of surface plasmons is one approach, which already has been successfully demonstrated experimentally by Specht et al. [10]. Another approach is to excite resonantly the probe by using, for example, the fiber tip with a single resonant atom or molecule at the very end of the tip or by placing there a small metallic sphere. In the latter case, a number of sufficiently strong subsidiary resonant peaks can be expected besides the main resonance at $\omega_p/\sqrt{3}$ [13]. This approach is still to be experimentally verified, but it seems more promising than the first one, since there is no limitation imposed on the sample to be investigated.

## References


[1] Ash, E. A. and Nicholls, G. (1972) "Super-resolution aperture scanning microscope", Nature 237, 510-512.

[2] Pohl, D. W., Denk, W. and Lanz, M. (1984) "Optical stethoscopy: image recording with resolution $\lambda/20$", Appl. Phys. Lett. 44, 651-653.

[3] Betzig, E., Trautman, J. K., Harris, T. D., Weiner, J. S. and Kostelak, R. L. (1991) "Breaking the diffraction barrier: optical microscopy on a nanometric scale", Science 251, 1468-1470.

[4] Girard, C. and Bouju, X. (1992) "Self-consistent study of dynamical and polarization effects in near-field optical microscopy", J. Opt. Soc. Am. B9, 298-305.

[5] Girard, C. (1992) "Plasmon resonances and near-field optical microscopy: a self-consistent theoretical model", Appl. Opt. 31, 5380-5387.

[6] Vigoureux, J. M., Depasse, F. and Girard, C. (1992) "Superresolution of near-field optical microscopy defined from properties of confined electromagnetic waves", Appl. Opt. 31, 3036-3045.

[7] Vigoureux, J. M. and Courjon, D. (1992) "Detection of nonradiative fields in light of the Heisenberg uncertainty principle and the Rayleigh criterion", Appl. Opt. 31, 3170-3177.

[8] Keller, O., Xiao, M. and Bozhevolnyi, S. (1992) "Configurational resonances in optical near-field microscopy: a rigorous point-dipole approach", Surf. Sci., in press.

[9] Markel', V. A. (1992) "Scattering of light from two interacting spherical particles", J. Modern Opt. 39, 853-861.

[10] Specht, M., Pedarnig, J. D., Heckl, W. M. and Hänsch, T. W. (1992) "Scanning plasmon near-field microscope", Phys. Rev. Lett. 68, 476-479.

[11] Zaremba, E. and Person, B. N. J. (1987) "Dynamic polarizability of small metal particles", Phys. Rev. B35, 596-606.

[12] Palik, E. D. (1985) "Handbook of optical constants of solids", Academic Press, London.

[13] Ruppin, R. (1992) "Optical absorption by a small sphere above a substrate with inclusion of nonlocal effects", Phys. Rev. B45, 11209-11215.

# EMISSION AND ABSORPTION OF LIGHT BY ELECTRONS OR ATOMS IN OPTICAL NEAR FIELDS.

J.M. VIGOUREUX
*Laboratoire de Physique Moléculaire, UA 772 CNRS*
*Université de Besançon*
*25030 BESANCON Cedex, France*

ABSTRACT. Although usually discussed in the field of Scanning Tunneling Optical Microscopy, Near Field Optics corresponds to fundamental processes which are not limited to microscopy and which can open the way to a number of different applications. I consider here interactions between electrons and optical near fields. Some results on the interaction between atoms and the optical near field are also presented.

## 1. Introduction

A free electron moving at uniform velocity in a dielectric medium can emit light only when its velocity is greater than the phase velocity $c/n(\omega)$ of the emitted light. This result leads to consider that an electron moving at uniform velocity in vacuum cannot emit any light. It also leads to think that emission of light by a free electron moving in a dielectric always needs relativistic electrons. That is not exact. An electron can be thought of as emitting evanescent waves whatever may be its velocity and even when moving in vacuum. It is right that a part of evanescent waves can be converted into propagating waves *by refraction processes* only when the Cerenkov condition ($v > c/n(\omega)$) is fulfilled. However, even when that condition is not fulfilled ($v < c/n(\omega)$) *a part* of the non-radiative field of the electron can also be converted into propagating waves by using optical processes other that refraction. I show that the Cerenkov condition can be overpassed by using subwavelength scattering centers, exactly as the Rayleigh criterion is overpassed in Near Field Microscopy. This can have interesting applications, for example in the problem of emission of light in Scanning Tunneling Microscopy. Some results on interactions between an atom an a non-radiative field are also presented.

## 2. Representation of electrons and photons

The simplest way to present all results with as few calculations as possible is to use figures. Our first step is then to represent the possible dynamic states of an electron and of a photon.



*D.W. Pohl and D. Courjon (eds.), Near Field Optics, 239–246.*

## 2.1 REPRESENTATION OF ELECTRONS

The possible dynamic states of a free electron moving at uniform velocity in vacuum are represented by the hyperbola

$$E = \sqrt{c^2 P^2 + m_o{}^2 c^4} > c\,P \qquad (1)$$

where E and P are the energy and the momentum of the electron respectively, and where $m_o$ is its rest mass. An electron moving at uniform velocity in vacuum can consequently be represented by a vector such **OA** (fig 1). To be clearer in following figures, instead of representing the dynamic state of the electron by the vector **OA**, I will represent it by the only point A on the hyperbola.

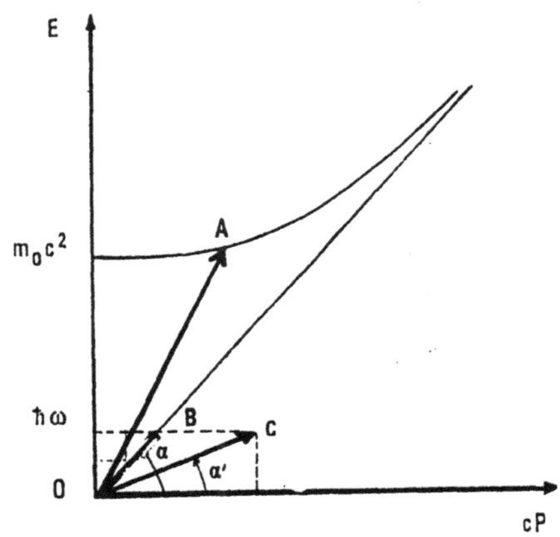


Fig. 1. Representation of the possible dynamic states of an electron (vector **OA**), of a free photon (vector **OB**) and of an evanescent mode.of the field (vector **OC**)

## 2.2 REPRESENTATION OF PHOTONS

Photons are also represented by vectors

Free photons correspond to vectors such **OB** ; (E, cP = E) lying on the light cone ($\omega$=ck $\Leftrightarrow$ E = cP). They can be characterized by their "characteristic angle $\alpha$" which obviously verifies tg $\alpha = E / c\,P = \omega / ck = 1$)

Evanescent modes are represented by vectors **OC** ; (E, cP > E) pointing out of the light cone ($\omega < ck \Leftrightarrow E < cP$). Their characteristic angle verifies tg $\alpha' < 1$.

## 2.3. REMARK

The electromagnetic field generated by a moving electron is described in [1,2]. Interaction between matter and light near a surface has been considered in [3]. However, representing evanescent waves by "photons" may be *a priori* surprising and has consequently to be explained. The problem of the quantization of evanescent waves which appear in the angular spectrum representation of the electromagnetic field in a half space was first discussed by

Carniglia and Mandel[4]. These authors have shown that, although evanescent waves are associated with material sources, scatterers, etc, it is possible to quantize the electromagnetic field, including evanescent waves, effectively as a free field. They showed that the field hamiltonian can be expressed as the sum of independent harmonic-oscillators hamiltonian. Using this result, the quantization of the electromagnetic field is then straightforward and proceeds in the usual manner. It has similarly been shown[5] that the field momentum parallel to the interface reduces to the sum of momentum of the same independent harmonic oscillators. These two results justify describing interactions between matter and an evanescent field as emission or absorption of elementary quanta, each one having the energy $h\,\omega$ and the momentum $hk_{||} > h\omega/c$ . It is however to be underlined that in this quantum description of the field, each mode of the field consists of *three components* (corresponding respectively to the incident, to the reflected and to the transmitted wave on the interface) which have to be considered *together*. Evanescent waves constitute only one component of a mode; they cannot be separated from the two others components. Because of this, we must not speak of *evanescent photons* but of *evanescent modes* (having one evanescent component).

## 3. Interaction between electrons and the optical near field

### 3.1. THE EMISSION (OR THE ABSORPTION) OF A PHOTON BY A FREE ELECTRON MOVING AT UNIFORM VELOCITY IS IMPOSSIBLE.

As is well known, a free electron moving at uniform velocity in vacuum cannot absorb or emit a photon because momentum and energy cannot possibly be both conserved in the process. Let us consider, for example, the emission of a photon ($h\omega$, $\mathbf{hk}$). Noting ($E_I$, $\mathbf{P}_I$) and ($E_F$, $\mathbf{P}_F$) the energy-momentum of the electron in its initial state and in its final state respectively, energy and momentum conservation require that

$$E_F = E_I - \hbar\,\omega \tag{2}$$

$$\mathbf{P}_F = \mathbf{P}_I - \hbar\,\mathbf{k} \tag{3}$$

Taking the square of these relations and using eq.(1) lead to

$$\hbar\,\omega\,E_I = \hbar\,c^2\,P_I\,k = \hbar\,c^2\,P\,k\,\cos(\mathbf{P}_I\,,\mathbf{k}) < \hbar\,c^2\,P_I\,k \tag{4}$$

so that, using (4) together with the fundamental equation for a free photon $\hbar\,\omega = \hbar\,c\,k$, we obtain

$$E_I < c\,P_I \tag{5}$$

This result contradicts eq(1) and shows that the emission of photon by a free electron moving at uniform velocity cannot both conserve energy and momentum (eqs. 2,3). The emission is consequently impossible.

This result can be easily understood by looking at the fig(2a). By emitting the photon **OB**, the dynamic state of the electron (initially in state A), would become represented by the point

B'. This point B' is not on the hyperbola of its *possible* dynamic states; consequently, the emission (or the absorption) of the photon **OB** by the free electron is **impossible.**

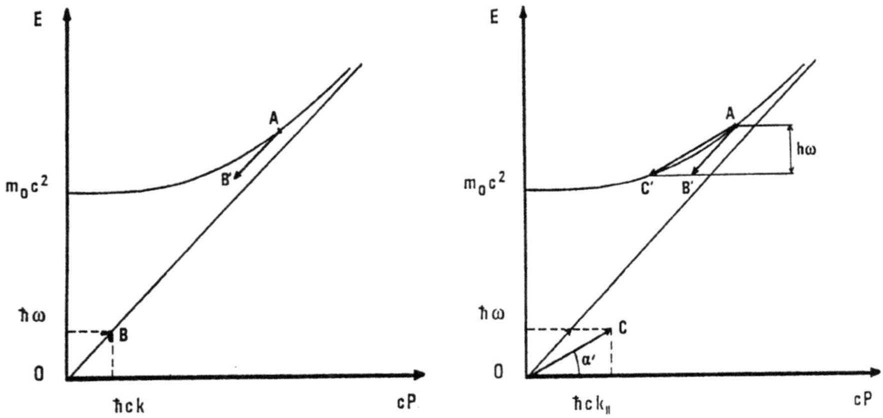


Fig. 2. a) After having emitted the free photon **OB** the dynamic state of the electron would be represented by the vector **AB'**; The point B' being not on the hyperbola, that emission is impossible.

b) After having emitted the evanescent mode **OC**, the dynamic state of the electron is represented by the vector **AC'**. The point C' is on the hyperbola of the possible dynamic states of the electron; the emission of the evanescent mode **OC** is consequently possible

## 3.2. THE EMISSION (OR THE ABSORPTION) OF AN EVANESCENT MODE BY A FREE ELECTRON MOVING AT UNIFORM VELOCITY IS POSSIBLE.

By emitting an evanescent mode represented by the vector **OC** (fig.(2b)) the final state C' of the electron lies on the hyperbola so that the emission is possible. It can be seen on the figure that an electron can emit evanescent modes *whatever may be its velocity.* Fig(2b) also shows that the smaller is the velocity of the electron, the smaller is the characteristic angle $\alpha'$ of the emitted evanescent mode. That means that slow electrons can only emit evanescent modes verifying $\hbar k_{||} > \hbar\omega / c$ . In usual words of optics, such modes correspond to very high spatial frequencies and consequently to evanescent waves which decay very quickly with increasing distances from the electron. (The larger is the value of $k_{||}$, the more rapidly do these waves decay).

## 3.3. REMARKS

Fig(2b) corresponds to the interaction of evanescent modes with free electrons. Similar results also hold in the case of interaction between "evanescent electrons" (tunneling electrons) and free photons. In both cases, the interaction which is impossible between a *free electron* and a *free photon* (because energy and momentum cannot possibly be both conserved in the process) is made possible by the presence of a "larger than usual" momentum.

This may provide a new approach of the emission of light in Scanning Tunneling Microscopy[6,7,8].

## 4. How to detect the near field emitted by the electron ?

Parts II and III show how an electron can interact with optical near fields. Our aim now is to explain how the light emitted by the electron can be detected and to show that the Cerenkov condition can be overpassed exactly as the Rayleigh criterion is overpassed in Near Field Microscopy.

### 4.1 REFRACTION

For a very high value of the electron velocity V ($A_1$, on fig. (3a), corresponds to a high value of the momentum P), the evanescent mode $OC_1$ emitted by the electron (for a given value of $\omega$) can be very near the light cone. It can verify $k_{||} < n\omega/c$ and consequently, it can be converted into a propagating wave *by refraction.*

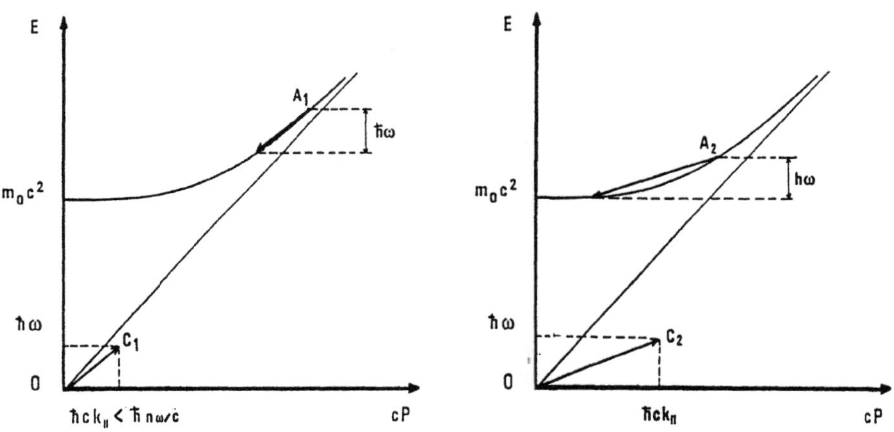


Fig. 3. a) When the velocity of the electron is very high (point $A_1$) the emitted evanescent mode can have a $k_{||}$ component less than $n\omega/c$. The evanescent wave can consequently be converted into propagating wave by refraction on the surface of a dielectric of index n.

b) When the velocity of the electron is small (point $A_2$), the emitted evanescent mode has very high value of $k_{||}$. The conversion of the evanescent wave into propagating wave is however possible by using diffraction by subwavelength objects.

The Snell-Descartes law of refraction in fact show that all modes of the field for which $k_{||} \leq n\omega/c$ (In other words, all modes satisfying $1/n < \text{tg}\ \alpha' < 1$) can be converted upon interaction with a dielectric into homogeneous propagating waves by refraction. By using refraction, the field emitted by the moving electron can consequently be converted into a propagating wave in the dielectric and detected in the far field by usual technics. This constitutes the basic effect of the Toraldo di Francia's model for the explanation of the Cerenkov radiation[9]. The electromagnetic field of the electron can be thought of as being made up of evanescent waves of all frequencies which decay away from the electron. To explain the Cerenkov effect, Toraldo di Francia assumed the electron to move in the narrow gap between two blocks of dielectric medium. Evanescent waves created by the uniformly moving electron are refracted into propagating waves at the boundary of the two dielectric blocks. If the gap becomes vanishingly small, we obtain the situation where an electron is moving uniformly in an infinite dielectric. The homogeneous waves emitted by the electron then correspond to the Cerenkov radiation and lead to the same quantitative results[10] as the ones of Frank and Tamm[11] who were (jointly with Cerenkov) awarded the Nobel prize for having given the first theoretical understanding of the phenomena.

Calculations show that the condition $k_{||} < n\omega/c$ can only be verified when $V > c/n$ that is to say when the velocity of the electron exceeds the phase velocity of light in the dielectric. This condition constitutes what is called the *Cerenkov condition* or the *Cerenkov limit.*

## 4.2. DIFFRACTION

When $V < c/n$ (point $A_2$ in fig(3b)) the evanescent mode $OC_2$ emitted by the electron verifies $k_{||} > n\omega/c$. Its emitted field only contain very high spatial components which cannot be converted into propagating modes *by refraction*. The Cerenkov condition can however be overpassed exactly as the Rayleigh criterion is overpassed in Scanning Tunneling Optical Microscopy[12].

A small object lit by a propagating wave always diffracts both propagating ($k_{||} < \omega/c$) and evanescent ($k_{||} > \omega/c$) waves. They correspond to radiative components of the field (which propagate and can consequently be captured at great distance) and to non radiative components, respectively. The smaller is the size of the object, the larger can be the spatial frequencies of the diffracted evanescent waves. Conversely, by applying the reciprocity theorem, a small object placed in an evanescent field can convert a part of an evanescent field into propagating waves. The smaller is the object, the larger are the spatial components of the field which can be converted by diffraction into propagating waves. The non radiative field of the electron can thus be converted into propagating waves *by using diffraction instead of refraction.* As in the STOM, the system to be used is a scattering center of subwavelength size (a small tip, a sub-wavelength hole, little spheres, colloïds ... or the surface roughness itself). With such systems, the light emitted by the electron can be converted into propagating waves (and be detected in the far field) *even when the electron goes at velocity less that $c/n$.* The Cerenkov condition $v > c/n$ is overpassed.

Because of the exponentially decreasing amplitude of evanescent waves, the diffraction of the near field of the electron into propagating waves will of course be all the more easy that the electron trajectory will pass nearer of the scattering center.

## 4.3. REMARKS

It can be shown that detecting a part of the non radiative field of an electron moving at the uniform velocity V = 10 000 km/s needs a diameter tip smaller or equal to $\lambda/30$. The extremity of the tip used in STM usually consists of only one atom. Its distance to the surface is less than one nanometer. This shows that evanescent modes emitted by an electron can easily be converted into propagating homogeneous waves by the tip of a STM even when the velocity of the electron is very small.

The possible role of the Cerenkov effect in the emission of light by STM has also been considered in [13].

**5. Interaction between atoms and the optical near field.**

Similar results also hold with atoms. Some of them have been presented in [12]. I only give here some examples

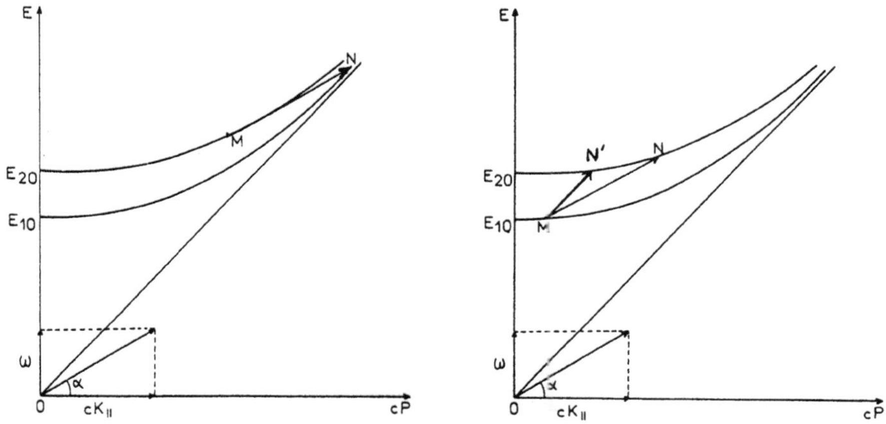


Fig. 4. a) An excited atom can "jump" into its ground states by *absorbing* (instead of emitting) an evanescent mode. It can similarly "fall" into an excited states by *emitting* (instead of absorbing) an evanescent mode.

b) Normal absorption of an evanescent mode of the field. The absorbed frequency is $(E_N-E_M)/\hbar$; in a free field it would be $(E_{N'}-E_M)/\hbar < (E_N-E_M)/\hbar$ leading to a doppler shift smaller than in a near field.

Some "exotic" effect can be considered. For example, as shown in fig(4a), an excited atom can "jump" into its ground state by *absorbing* (not emitting) an evanescent mode. Similarly, a new type of fluorescence can be considered; the atom can "fall" into an excited state by *emitting* an evanescent mode. It can then emit another photon (a free photon or an evanescent mode) to come back in its ground state. Of course, these two processes can only be realized by using atoms moving in a near field having very high spatial components; however it is to be noted that very high spatial components are always present in the near field of very small object...

More usual interactions can also be considered. For example usual absorption of light by an atom can be studied in the near field. It is to be noted that in that case, the "greater than usual" momentum of the evanescent mode will lead not only to unusual doppler shift (fig4b) but also to more efficient cooling of atoms.

**6. Conclusion.**

I have presented the case of the interaction of *a free electron* moving near a dielectric. It is to be recalled that a similar reasoning also holds for the interaction between *an evanescent electron* (a tunneling electron) *and a free photon*. In both cases, the interaction is made possible by the presence of a "larger than usual" momentum.

Another remark can be made. I have considered evanescent waves, that is to say, modes existing very close to a *dielectric*. Similar results can also be obtained near metallic surfaces; In that case, evanescent modes are to be replaced by plasmon modes which interacts similarly. Quantization of plasmon modes have for example been considered in[14]. As in the above discussion, a free electron moving near a metal can absorb or emit a plasmon mode of the field.

---

BIBLIOGRAPHY


[1] Asby, R. and Wolf, E. (1971) 'Evanescent waves and the electromagnetic field of a moving charged particle' J. opt. Soc. Amer. 61, 52-59.

[2] Lalor, E. and Wolf, E. (1971) 'New model for the interaction between a moving charged particle and a dielectric and Cerenkov effect' Phys Rev Lett 26(20) 1274-1277.

[3] Costa de Beauregard, O. ; Imbert, C. and Ricard, J. (1971) 'Energy momentum quanta in Fresnel's evanescent waves' Int. J. of Theor. Phys. 4, 125-135.

[4] Carniglia, C. K. and Mandel, L. (1971), 'Quantization of evanescent electromagnetic waves' Phys. Rev D 3, 280-296.

[5] Vigoureux, J. M. , D'Hooge, L. and Van Labeke, D. (1980) 'Quantization of evanescent electromagnetic waves. Momentum of the electromagnetic field very close to a dielectric medium' Phys. Rev. A 21, 347-355.

[6] Gimzewski, J.K., Sass, J. K., Schlitter, R. R.and Schott, J. (1989) Europhys. Lett. 8, 435.

[7] Berndt, R., Baratoff, A. and Gimzewski, J. K. (1990) in Proc. NATO ASI "Basic concepts and applications of STM and related techniques." Erice (Italy) April 17-19, 1989. NATO ASI Series E: Applied Sciences **184**, 269-280 (Kluwer, Dordrecht, 1990).

[8] Johansson, P., Monreal R. and Apell, P. (1990) 'Theory of light emission from a STM' Phys. Rev. B 42, 9210-9213.

[9] Toraldo di Francia, G. (1960) 'On the theory of some Cerenkovian effects' Nuov Cimento. 16, 61-77.

[10] Vigoureux, J. M. and Payen, R. (1975) 'Interaction matière onde évanescente. Absorption par un atome au voisinage d'un dioptre plan' J. de Phys. 36, 631

[11] Frank, I. E. and Tamm, I. M. (1937) Compt. Rend. Acad. Sci. URSS 14, 169-171.

[12] Vigoureux, J. M. and Courjon, D. (1992) 'Detection of non-radiative fields in light of the Heisenberg uncertainty principle and Rayleigh criterion' Appl. Optics. 31(16) 3170-3177.

[13] Smolyaninov I. I. (1992) in the present Proceeding

[14] Grossel Ph. (1984) Thèse, Université de Reims.

# A PERTURBATION-THEORY APPROACH TO SCANNING NEAR-FIELD OPTICAL MICROSCOPY (SNOM)

ULRICH SCHOLLWÖCK and HERBERT WAGNER
*Sektion Physik der Universität München,*
*Theresienstr. 37, W-8000 München 2, Germany*

ABSTRACT. We describe a multiple scattering approach to SNOM with the permittivity of the object and the scanning tip taken as perturbations. In the leading order showing sub-$\lambda$-resolution, the expression for the observed signal is found to be a convoluted product of an object form factor and a modulation function characterizing the imaging capabilities of the microscope. For a particular setup we discuss the dependence of the resolution and the faithfulness of the imaging on the probing device and on the scattering geometry. The appearance of secondary structures in SNOM-images is noted. The second-order predictions of the theory on sub-$\lambda$-resolution are tested numerically with objects of simple shape.

## 1. Introduction

Scanning microscopy overcomes the diffraction limit of resolution $\propto$ wavelength $\lambda/2$ in optics by exploiting the local interaction between object and scanning device. Several types of SNOM have evolved[1]: scanning by a sub-$\lambda$-aperture, by a fibre or by a scattering tip. The local interaction may be considered as photon tunneling or as multiple scattering, depending on the particular setup. We shall concentrate on a scattering scheme employed by Specht et al. [2] and sketched in Fig. 1, where a STM tip scans a surface + object at constant height, which is illuminated by evanescent transmission. The reflected far-field intensity is measured during scanning to yield the signal.

The crucial feature of this form of SNOM is the appearance of a second length scale $l$ (besides the incident wavelength $\lambda$), which is the size of the effective antenna made up by the object and the scanning tip. The scattered radiation in the far field is determined by the dipole moment of this antenna. Its effective dipole moment results from the induced multiple scattering between object, tip and substrate and is estimated to vary like $l^{-3}$ (in leading order: nonretarded dipolar interaction tip-object) with the tip position. Consequently, the correlation of the amplitude modulation in the observed signal with the scanning motion of the tip, being controllable on the scale $l \ll \lambda$, gives rise to a sub-$\lambda$-picture of the object. The phase factor of the signal will behave as $\exp i(l/\lambda)$, $l/\lambda \ll 1$, and thus carries no information by itself. In this sense, SNOM is *amplitude-modulated* microscopy.

Our aim is to focus attention primarily on those terms in the scattered intensity which vary during scanning and to report on the results of a simple approach yielding an expres-



*D.W. Pohl and D. Courjon (eds.), Near Field Optics*, 247–254.

sion for the observed signal which is easy to interpret.

## 2. Outline of Approach

An ab initio calculation of the signal, by modeling tip, object and substrate with a large number of molecular dipoles ([3], for other approaches see [4]), requires considerable numerical efforts and has to be repeatedly performed for each individual object. To gain physical insight, we pursued a complementary albeit approximate approach based on a truncated perturbation series for the multiple scattering interaction tip-object, with the local permittivity contrast between object or scanning tip and embodying layer as perturbations. The scattering amplitude is obtained as a convolution of an object form-factor and a modulation function which governs the object-independent imaging features of the microscope and provides criteria for sub-$\lambda$-resolution.

The obvious deficiency of a truncated perturbation series is that it neglects possible resonances. These may occur at sufficiently small tip-object distances for particular material parameters and are therefore not generic[5]. With a trivially soluble toy model where the antenna tip + object is replaced by 1 + 2 dipoles with frequency dependent polarizabilities we checked the validity of our second-order approximation and found that in praxis there is a wide range of scanning heights which yield sub-$\lambda$-resolution without significant interference of resonance effects.

## 3. Perturbation Series

We briefly describe the essential technical aspects without going into details[6]. Our model is a glass-silver-air multilayer system perturbed by the scanning STM tip and by objects at the surface (Fig. 1). Starting from the solution for the ideal multilayer ("zeroth-order"), the terms in a perturbation series are ordered by the number of scattering processes occuring before the wave leaves the system.

We choose the x-axis normal to the multilayer planes, with $\mathbf{r} = (y, z)$. The local permittivity is $\epsilon(x, \mathbf{r}) = \epsilon_n(x) + \eta(x, \mathbf{r})$, where $\eta$ denotes the permittivity variation in the layer $(n)$. A coherent plane wave (channel $e$) $\exp(ik_{nx}^e x + i\mathbf{k}^e \mathbf{r} - i\omega t)$, with $\epsilon_n(\omega/c)^2 = (k_{nx}^e)^2 + (\mathbf{k}^e)^2$, enters the multilayer at an angle $\theta^e$ with the surface (y-z plane). Maxwell's equations read, in units of $\omega^{-1}$ and $(\omega/c)^{-1}$,

$$\nabla \times \mathbf{E}(x, \mathbf{r}) = i\mathbf{H}(x, \mathbf{r}) \tag{1}$$

$$\nabla \times \mathbf{H}(x, \mathbf{r}) = -i[\epsilon_n(x)\mathbf{E}(x, \mathbf{r}) + \mathbf{P}(x, \mathbf{r})], \quad \mathbf{P}(x, \mathbf{r}) = \eta(x, \mathbf{r})\mathbf{E}(x, \mathbf{r}) \tag{2}$$

We use a formalism described by Bussemer et al. [7] and extend their work to higher order scattering. It amounts to a conventional Green function approach but conveniently adapted to the geometry of a multilayer device.

The fields **E**, **H** and **P** are Fourier-transformed in the y-z plane. Furthermore, we decompose into *s*- and *p*-polarised light; then each vector-wave can be described by a scalar. Our theory is symbolized by diagrams as in Fig. 2.

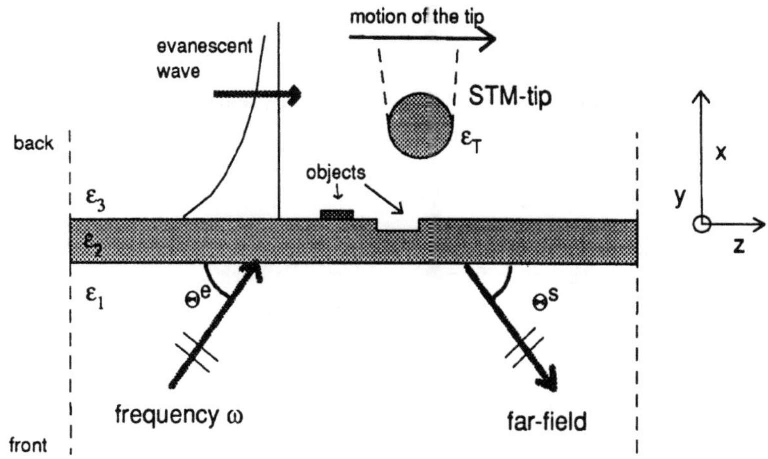


Figure 1: Model device. Three layers with permittivities $\epsilon_n(\omega)$. The incident plane wave $(\omega)$ excites an evanescent wave in layer (3) due to surface plasmons. In the numerical implementation the layers are glass (1), silver (2) and air (3) with a silver tip.

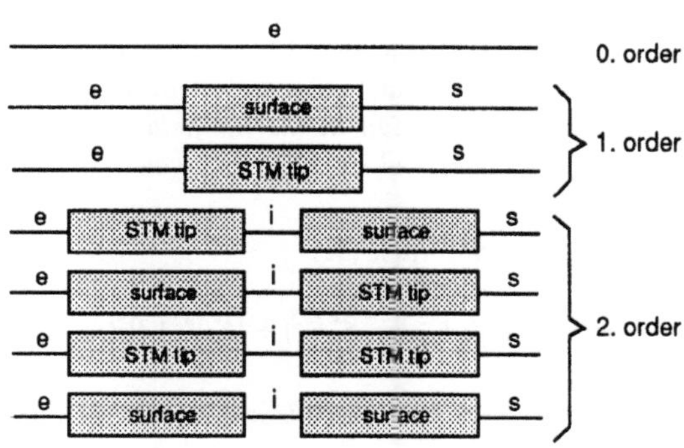


Figure 2: Zeroth- to second-order interaction diagrams. The lines represent plane waves. The boxes represent scatterers ("surface": object at the surface), with the left line denoting the incoming wave and the right line the outgoing (scattered) wave. $e, i, s$: exciting, intermediate, scattered wave.

*Zeroth-order: unperturbed multilayer:* We find the standard results including plasmon excitation. A remarkable result is that the reflected ray is shifted (like in the Goos-Hänchen effect[8]) and elliptically deformed.

*First-order:* We choose one wave vector $(k^s_{nx}, \mathbf{k}^s)$ as scattering channel $s$. After separating the electric field into incident and scattered field, we approximate

$$\mathbf{P} = \eta(\mathbf{E}^e + \mathbf{E}^s) \rightarrow \mathbf{P} = \eta\mathbf{E}^e \tag{3}$$

Thus, within regions where $\eta \neq 0$, the scattered field is neglected. The wave due to the tip will *interfere* with the wave from the object. As both wave vectors are radiative, the momentum transfer $\mathbf{k} = \mathbf{k}^s - \mathbf{k}^e$ gives $(2\pi|\mathbf{k}|)^{-1} \gtrapprox \lambda/2$ for the lateral resolution, which is the diffraction limit.

*Second-order: near-field interaction:* In this order, we have to deal with interactions between different perturbations as well as "self-interactions". Only the former depend on the tip position and will therefore vary in the scanning process; subsequently we only keep these terms.

To calculate a second order scattering amplitude, we have to sum the product of the scattering coefficient from the first perturbation and that from the second perturbation over all possible polarisations $p_i$ and we have to integrate over all possible intermediate momenta, with the incident and scattered momenta fixed.

The high resolution information arises from the fact that the intermediate momentum transfer may be arbitrarily large in principle. Higher order expressions may now simply be constructed by adding further integrations. The theory will certainly be most useful for multilayered surfaces, where the formalism is quite versatile. The higher-order contributions are expected to decrease quickly due to the small size of object and tip perturbation, provided the tip-layer and tip-object resonances do not invalidate the series truncation. By comparison with the previously mentioned and selfconsistently soluble toy model we argue that in the experimental situation considered here there are no tip-object resonances which couple to tip-layer resonances, so that higher orders may be safely dropped.

*Extraction of amplitude information in the far-field:* With $S_n$ denoting the scattering amplitude at $n$-th order, the observed intensity is given

$$\begin{aligned} I_{obs.} &= \int |S_0 + S_1 + S_2|^2 d\mu(\mathbf{k}^i;\Omega) \\ &= \int \{|S_0|^2 + |S_1|^2 + 2\mathrm{Re}(S_0^* S_1) + |S_2|^2 + 2\mathrm{Re}[(S_0 + S_1)^* S_2]\} d\mu(\mathbf{k}^i;\Omega), \end{aligned} \tag{4}$$

where the integral is taken over the intermediate momenta and the solid angle of observation. The first three terms $|S_0|^2, |S_1|^2, 2\mathrm{Re}(S_0^* S_1)$ do not vary on sub-$\lambda$ scale with the tip position; $|S_2|^2$ is an amplitude information; $2\mathrm{Re}[(S_0 + S_1)^* S_2] \equiv 2|\tilde{S}_2|(\mathbf{r}_{tip})\xi(\mathbf{r}_{tip})$ is the dominant sub-$\lambda$-varying term. The coefficient $\xi$ can be shown not to vary on a sub-$\lambda$-scale. Thus, the amplitude of second order scattering is faithfully observable on a sub-$\lambda$-scale. The scanning process is therefore essential to cancel out phase-dependent interference terms.

## 4. The Modulation Function

We will now extract a modulation function, which characterizes the experimental setup

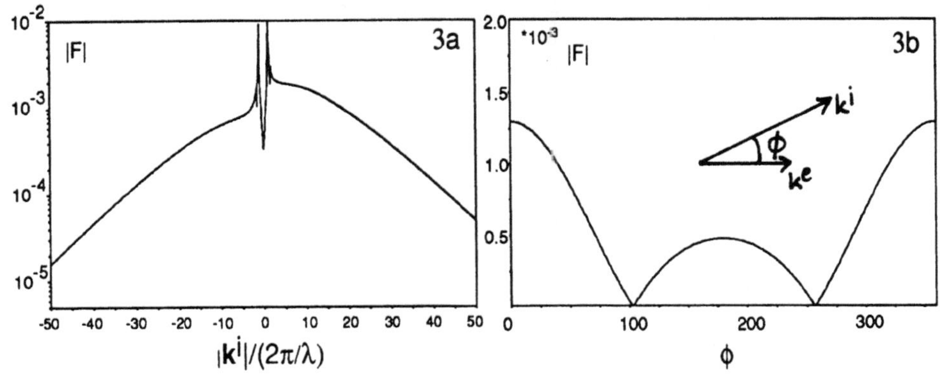


Figure 3: $|F|$ for tip radius 10 nm; scanning height 3 nm; silver tip ($\epsilon = -17.8 + 0.67i$). $\lambda$ = 632.8 nm; $p$-polarisation for exciting, intermediary and scattered ray; silver layer thickness 51.4 nm; incidence under plasmon excitation angle 46.0°; object in silver layer (e.g. dip). (a) radial dependence: $\mathbf{k}^i \parallel \mathbf{k}^e \parallel \mathbf{k}^s$. Momentum in units of $2\pi/\lambda$. (b) angular dependence: $|\mathbf{k}^i|/(2\pi/\lambda) = 15.0$; $\mathbf{k}^e \parallel \mathbf{k}^s$

independently of the object.

If we assume the surface perturbation to be *shallow* (height or depth $\lessapprox$ 2 nm), which excludes topographical effects on the resolution, we may write the two relevant second order interactions as

$$A^s_{out,1}(\mathbf{r}) = \sum_{p_i} e^{-i\mathbf{k}^s\mathbf{r}} \int d^2\mathbf{k}^i f(-\mathbf{k}^i) F_1(\mathbf{k}^i; \mathbf{k}^e, \mathbf{k}^s) e^{i\mathbf{k}^i\mathbf{r}} A^e_{in} \tag{5}$$

$$A^s_{out,2}(\mathbf{r}) = \sum_{p_i} e^{+i\mathbf{k}^e\mathbf{r}} \int d^2\mathbf{k}^i f(\mathbf{k}^i) F_2(\mathbf{k}^i; \mathbf{k}^e, \mathbf{k}^s) e^{-i\mathbf{k}^i\mathbf{r}} A^e_{in} \tag{6}$$

where $A^{e,s}_{in,out}$ is the scalar field amplitude of the ingoing and outgoing (backscattered) field for momentum channel $e, s$, $\mathbf{k}^{s,e} - \mathbf{k}^i \approx -\mathbf{k}^i$ and $\mathbf{r} = \mathbf{r}_{tip}$ is the position of the scanning tip in the y-z plane; the dependency on the polarisation states is suppressed.

The *ideal imaging function* $f$ is the form factor

$$f = \int d^2\mathbf{r} dx \eta(x, \mathbf{r}) e^{-i\mathbf{k}^i\mathbf{r} - ik^i_x x} \tag{7}$$

which depends on frequency and the geometric shape of the object.

The *modulation functions* $F_{1,2}(\mathbf{k}^i; \mathbf{k}^e, \mathbf{k}^s)$ introduced in (5) and (6) are products of multilayer propagators, surface transmission coefficients and of the form factor of the scanning tip (defined as in (7)).

In the experiment, the intensity in a small solid angle near normal reflection is measured for each position of tip and plotted vs. $\mathbf{r}$.

The Fourier integrals in (5) and (6) would yield the original perturbation in real space, if $F_{1,2}$ = const., and we would have perfect resolution. In general, *the modulation function $F_{1,2}$ controls the imaging and the lateral resolution.* An *absolute* modulation (i.e. global scaling factors in the modulation function) will influence resolution by determining the general signal/noise ratio, a *relative* modulation normalized to the maximum one, which attributes different weights to different Fourier components, will suppress certain Fourier components (Fig. 3a and 3b).

Let us make some general remarks which govern lateral resolution:

(i) A large cross section of the object and of the probing tip will favour a good signal/noise ratio, but too small a tip will give signals which are no longer detectable.

(ii) A small lateral extension of the tip and a small tip-surface distance minimize the suppression of high momentum coefficients by $F$ and thus improve lateral resolution. This suppression is due to the evanescence of waves with high lateral momentum transfer (since $\epsilon(\omega/c)^2 = (\mathbf{k}^i)^2 + (k_x^i)^2$, $k_x^i$ will have an increasing imaginary part for large $\mathbf{k}^i$) and due to the decrease of the cross-section for high momentum transfer scattering from large objects; for an example see Fig. 3a.

(iii) The layout of the multilayer is important to achieve a favourable signal/noise ratio. An important feature of the present device is that surface plasmon excitation leads to field enhancement at the metal/air interface, which induces a stronger scattering signal from objects in its vicinity.

For observation in the direction of the reflected ray ($\mathbf{k}^e = \mathbf{k}^s$), $F_1 = F_2 \equiv F$, so that we may concentrate on either scattering process. Furthermore, the main contribution then comes from the scattering of $p$-polarised light into $p$-polarised light, so that we do not discuss the other contributions here. They introduce nothing new. As expected, an increase of the tip radius and the scanning height further reduces high momentum Fourier coefficients, since the evanescence of high momentum transfer waves becomes more important and the form factor decreases faster for high momentum transfer, if the tip is larger. For shallow dips in the surface, the strong angular dependence (Fig. 3b) and phase jumps in $F$, due to the angular dependence of scattering, lead to an angular dependence of the imaging quality and of the resolution. $F$ acts as an angularly distorted weight function for the Fourier components. The distortion and also the weight is least in the direction normal to that of incidence (Fig. 3b): we therefore expect most faithful imaging (i.e. least distortions) normal to the direction of incidence, and highest resolution (sharpest contrast) in direction of incidence. It can be shown that protusions of a surface are better imaged than dips because of a much reduced angular variation of the modulation function $F$.

## 5. Illustrative Numerical Results

We calculate pictures as predicted by our theory to check results obtained from the analysis of the modulation function (Fig. 4; $\epsilon_1 = 2.2$; $\epsilon_2 = -17.8 + 0.67i$; $\epsilon_3 = 1$). Our prediction that imaging is better for protrusions ((a)–(c)) than for dips ((d),(e)) and that in the latter case resolution is best in direction of incidence ((d) vs. (e)) is confirmed, even if we calculate for not really shallow dips. We observe sub-$\lambda$ secondary structures (f) depending on the actual surface structure due to the asymmetry of the scattering processes ($\mathbf{k}^e \neq -\mathbf{k}^s$ in (5) and (6)). We also found that out-of-reflection observation ($\mathbf{k}^e \neq \mathbf{k}^s$) introduces distortions, because then $F_1 \neq F_2$. Finer surface structures vanish critically with probing device size.

## 6. Conclusion

For the SNOM device considered here, we conclude that the main resolution limit is the probing tip size. Resolution and imaging quality depend on the angle of the object with direction of incidence. As a rule of thumb, resolution is best in direction of incidence, imaging faithfulness is greatest normal to the direction of incidence. It would therefore be advisable to rotate the object under the incident ray, if possible. Observation out of reflection may introduce severe imaging errors.

Our approach is convenient to generate numerically scanning pictures for a large class of objects. The introduced modulation function is a characteristic of the experimental setup; it gives access to an approximative treatment of the inverse scattering problem, which may be precise enough for applying image enhancement procedures.

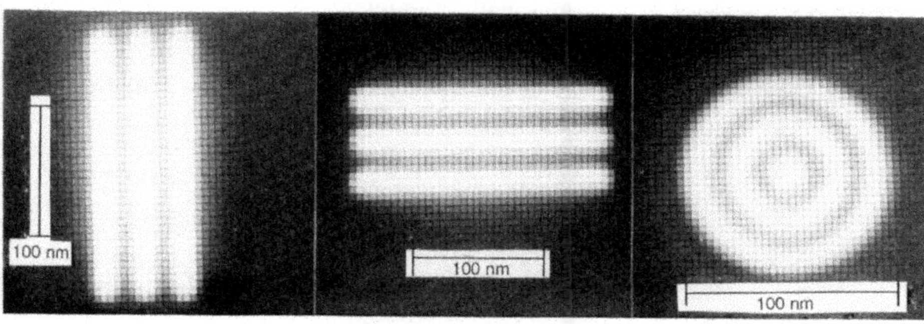


(a),(b): 3 tracks (200 x 15 nm), height 2 nm, distance 15 nm; (c) 3 concentric circles, width & distance 10 nm, height 2 nm, (a)-(c) protrusions.

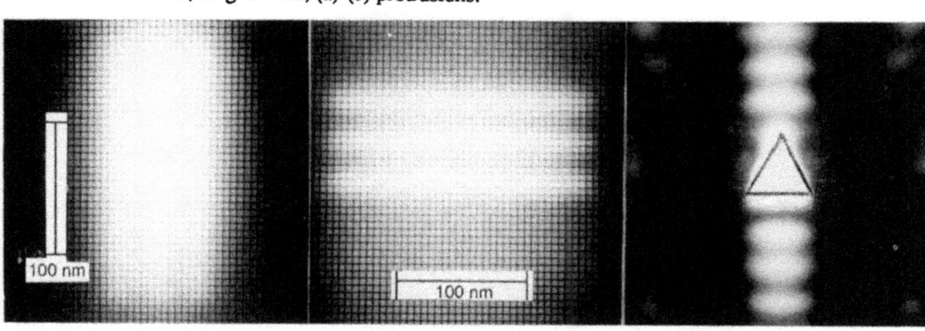


(d),(e): 3 tracks (200 x 15 nm), depth 5 nm, distance 15 nm dips; (f) (overexposed) triangle (width and heigth 600 nm)

Figure 4: Images of surface structures as predicted by our theory. Direction of incidence of light in y-z-plane (0 – 180°-axis in Fig. 3b) vertical from the top. Other parameters as in Fig. 3.

# References

[1] For an overview see the historical introductions by D. Courjon, H. Rohrer, D. Pohl and D. van Labeke, these proceedings.

[2] M. Specht, J.D. Pedarnig, W.M. Heckl and T.W. Hänsch, Phys. Rev. Lett. **68**, 476 (1992)

[3] J.-M. Vigoureux, C. Girard and D. Courjon, Opt. Lett. **14**, 1039 (1989), C. Girard and M. Spajer, Appl. Opt. **29**, 3726 (1990), C. Girard, D. Courjon, Phys. Rev. **B 42**, 9340 (1990), C. Girard, X. Bouju, J. Opt. Soc. Am. **B 9**, 298 (1992).

[4] A. Dereux and D. Pohl, D. van Labeke and D. Barchiesi, F. Pincemin and J.J. Greffet, O. Keller, S. Bozhevolnyi and M. Xiao, these proceedings.

[5] Resonance effects have been observed by U.Ch. Fischer and D. Pohl, Phys. Rev. Lett. **62**, 458 (1989)

[6] U. Schollwöck, Diploma thesis (Ludwig-Maximilians-Universität München), 1992

[7] P. Bussemer, K. Hehl and S. Kassam, Waves in Random Media, **1**, 207 (1991)

[8] F. Goos and H. Hänchen, Ann. Phys. (6) **1**, 333 (1947), J.J. Greffet and C. Baylard, these proceedings.

# THE TETRAHEDRAL TIP AS A PROBE FOR SCANNING NEAR-FIELD OPTICAL MICROSCOPY

U. C. FISCHER
*Physikalisch- Technische Bundesanstalt*
*D 3300 Braunschweig*
*FRG*

ABSTRACT. In scanning near field optical microscopy (SNOM), a tip whose dimensions are small compared with the wavelength serves as a light-emitting antenna. Light is transmitted from a source via a waveguide structure to this antenna. An important part of an SNOM probe - apart from the detailed structure of the tip - is the link between the macroscopic waveguide and the antenna of subwavelength dimensions. A partially metal-coated tetrahedral tip is considered to be a structure fulfilling these functions. Motivations for the choice of this structure are models of metal waveguide structures without cutoff and the fact that surface plasmons can be excited on planar metal films and along edges as well as on tips. Aspects of probe fabrication and first results of tests of their optical properties are reported.

## Introduction

In a transmission-type scanning near-field optical microscope (SNOM) as shown schematically in fig. 1, a tip, whose dimensions are small compared with the wavelength serves as a light-emitting antenna. It is excited by light from a source via a macroscopic waveguide such as a glass fibre and a link which serves to transmit light between the macroscopic waveguide and the tip. Light emitted from the tip and transmitted through the object is converted by a detector into an electrical signal which is used for imaging. The probe is mounted on a piezoelectric scanner which scans the probe relative to the object. The tip and the link have specific near-field optical functions and therefore are the characteristic parts of a near-field optical microscope. Tapered, metal-clad monomode fibres ending in a tip with an aperture of a diameter of 10 - 100 nm at its apex are very successfully used as probes for SNOM (1). The metal-clad conical taper constitutes in this case the link between the fibre and the aperture. This link is expected to be a rather inefficient one due to severe cutoff in regions where the cone diameter is significantly smaller than the wavelength (2). The concept of the coaxial tip (3,4,5) with an additional conical metal core tries to overcome this limitation based on the model of such a tip - assuming an ideally conducting metal - , which supports a TEM (transverse electric and magnetic) mode without cutoff, and which in principle allows electromagnetic energy to be focussed to dimensions which are small compared with the wavelength. An attempt to realise this concept for optical frequencies has been started (5), but a complete fabrication scheme has not yet been found.



*D.W. Pohl and D. Courjon (eds.), Near Field Optics,* 255–262.

Similar considerations lead us to the concept of the tetrahedral tip as a probe for the SNOM.

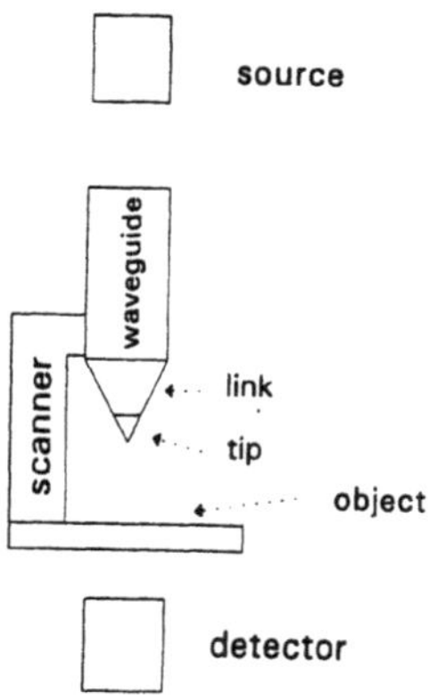


Fig.1
Scheme of a transmission-type of SNOM

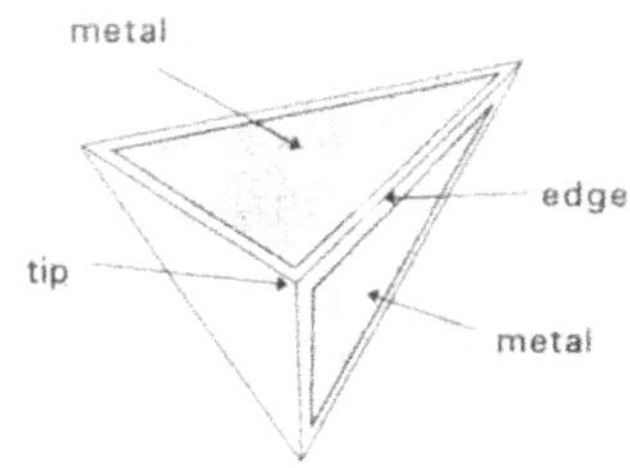


Fig.2
Schematic view of the tetrahedral tip

## The Concept of the Tetrahedral Tip

A tetrahedral tip is shown schematically in fig. 2. The body consists of a transparent material such as glass. It has three faces and edges which converge to a common tip. Two adjacent faces are coated with a thin film of metal, leaving the edge between the metal films and the tip itself free. For the model case of ideally conducting metal films this structure supports a TEM mode travelling parallel to the edge as shown in fig.3. Such a mode has a diverging intensity at the edge (rho=0) which is transmitted to the tip. It would be strongly damped in real metals at optical frequencies, but may still serve to transmit energy efficiently over a short distance in the order of one $\mu$m. For metals such as silver or gold, it would be expected that resonant surface plasmons

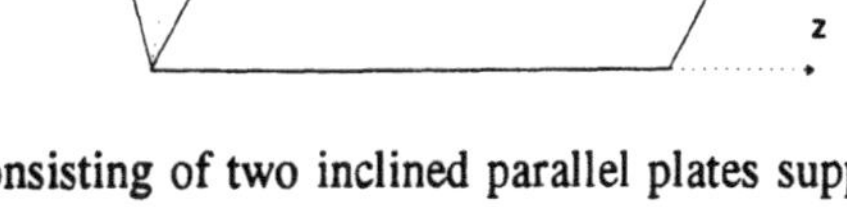


Fig. 3
A waveguide consisting of two inclined parallel plates supports a TEM mode with the electric field E:

$$E^{-/+}_{\phi}\,(\text{rho},z) \sim (1/\text{rho})\exp(-/+ikz)$$

In this expression rho,z and $\phi$ are the cylindrical coordinates. The indices "-" and "+" signify waves travelling inwards or outwards in relation to the tip. k = $2\pi$ lambda, where lambda is the wavelength in the dielectric body.

could be excited in the visible spectral range either on the planar metal films, or along the edge in analogy to surface plasmon modes on cylindrical wires (6). The excitation of surface plasmons is possible - in analogy to the Kretschmann configuration for exciting surface plasmons on thin metal films by attenuated total internal reflection (7) - by irradiating the metal films from within with a monochromatic beam of light at a selected angle to the metal coatings or possibly to the edge. These surface waves may serve to transmit electromagnetic energy efficiently to the tip, from where they are emitted as radiation into non-specular directions. A resonant field enhancement due to coupled surface plasmon modes of the metal films and of the edge, and of local surface plasmon modes of the tip, may additionally lead to a strong field enhancement at the tip as in the case of protrusions from a gold film (8). The emitting tip is meant to serve as an SNOM probe.

## Fabrication of Tips

A simple, well established laboratory scale technique is adopted for fabricating tetrahedral tips. Ultramicrotome glass blades for cutting thin sections of embedded biological material of a thickness down to 30 nm are routinely made by biologists. The method of breaking such a blade is shown schematically in fig.4. A very well-defined glass blade with a radius of curvature in the nm range is formed in this way. The method can easily be modified to make the edge end in an equally well-defined corner, as shown in fig. 5. This corner serves as the body of the tetrahedral tip. The two freshly formed fracture faces are coated with a metal such as aluminium, silver or gold at an oblique angle relative to the edge, leaving the third face of the tetrahedral tip, the edge and the corner, uncoated. The high resistance between the metal films indicates their true separation (10). The ultramicrotome blade is attached to a glass prism by an epoxy resin, as shown schematically in fig. 6. In this way it is possible to irradiate the tetrahedral tip from within by a slightly focussed beam of light which is inclined at an angle of approximately $45^0$ to the edge. With its tetrahedral tip this device is planned to serve as a probe for SNOM in a transmission mode.

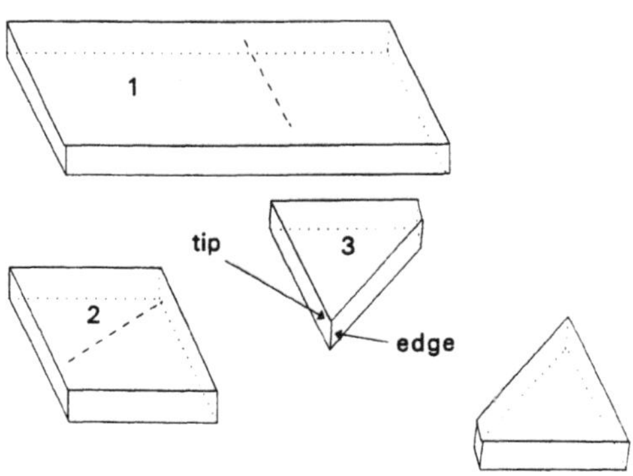


Fig 4
Fabrication of an ultramicrotome blade. From a glass bar /1/ of rectangular cross section of a thickness of 6mm and a width of 20 mm, a piece /2/ is broken off and this piece is then broken along a line slightly deviating from the diagonal into two triangular pieces /3/. Two freshly formed fracture planes intersect in a very well-defined edge with a radius of curvature in the nm range, forming the ultramicrotome blade(9).

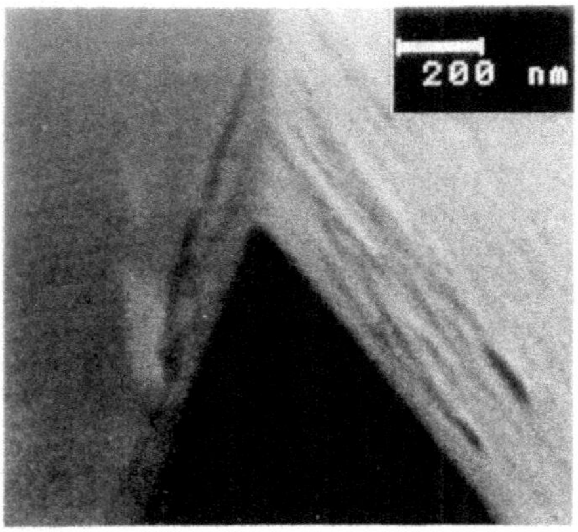


Fig. 5
Scanning electron-micrograph of the corner of an ultramicrotome blade, forming the body of a tetrahedral tip.

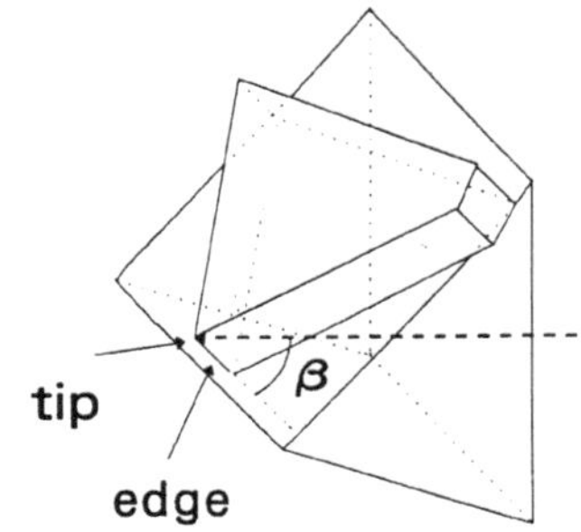


Fig.6
Schematic view of the ultramicrotome knife mounted on a prism in such a way that it can be irradiated from within the body of the tetrahedral tip at an angle of approximately 45° relatively to the edge

## Optical Properties

In order to investigate the optical properties of the tetrahedral probe it is mounted in a SNOM set-up schematically in fig.7.

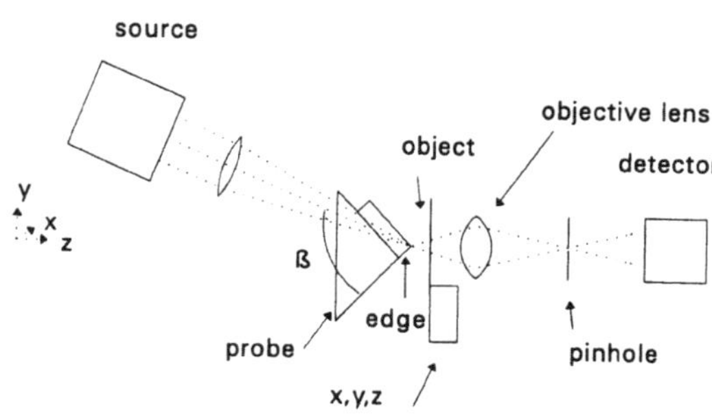


Fig. 7
Scheme of the SNOM with tetrahedral tip.

Light from a source, which may be a laser or a high-pressure arc lamp, is focussed at a numerical aperture in the order of 0.1 and at an angle ß to the edge. This angle may be varied over a range of 25° to 65° by the optical illumination path pivoting around the tip. A microscope cover glass supports the object. On this support it is intended to prepare objects such as thin evaporated films, thin organic layers, thin sections, biological membranes, biopolymers and other filamentous polymer structures. The cover glass is mounted on a combined mechanical translation stage and piezoelectric scanner by which coarse adjustments of the x and y positions of the object and the x-y scan, and the z adjustment of the object during the scan, are actuated. The coarse adjustment of distance between probe and object is made by means of a linear stage which moves the probe. The tip is adjusted to the focus of a light microscope using an objective lens with a numerical aperture of 0.6. A light micrograph of the light emission from a freshly prepared tip being irradiated at an angle ß of ca $60^0$ with light of 480 nm from a high pressure xenon arc is shown in fig. 8 .

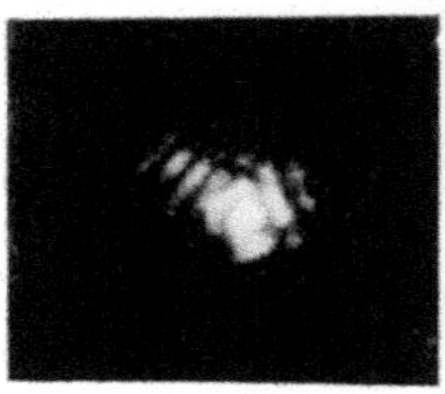

Fig. 8
Micrograph of the diffraction limited spot of light emission from the tetrahedral tip.

The coating of the tetrahedral tip consisted of a 50 nm silver film. A bright diffraction-limited spot of light is observed at the tip. This spot is surrounded by a diffraction pattern of smaller intensity that seems to reflect the tetrahedral geometry of the three edges converging to the tip. Similar images are obtained for wavelengths between 650 and 460 nm. The central spot is masked in the image plane by a pinhole of a diameter of 75 $\mu$m corresponding to a spotsize of 2 $\mu$m in the object plane, and its intensity is measured using a photomultiplier or a photodiode as a detector. When the probe was irradiated with the beam of a He Ne laser of an intensity of 250 $\mu$W at a wavelength of 633nm, the intensity collected from the central spot was 1.5 nW, corresponding to a transmission factor of at least $0.6 \times 10^{-5}$. In several experiments the intensity of the central spot was recorded as a function of distance between the tip and a cover glass. These experiments were performed in order to test whether a characteristic near-field behaviour of the transmitted signal can be obtained which can be used as a signal for

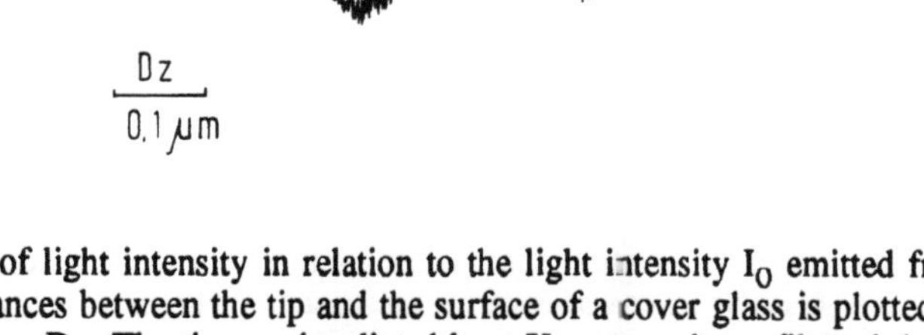


Fig. 9
The change DI of light intensity in relation to the light intensity $I_0$ emitted from a tetrahedral tip at large distances between the tip and the surface of a cover glass is plotted as a function of change in distance Dz. The tip was irradiated by a Xenon arc lamp filtered through a bandpass filter of 10 nm bandwidth at 520 nm. The assumed mechanical contact is indicated by the broken line.

the SNOM. In most the cases a strong interference pattern in the signal of the transmitted intensity was observed, as shown in fig.9. No pronounced near-field phenomenon was observed. The amplitude of the oscillation for a new tip is only a few per cent and it decays rapidly over two periods. Repeated mechanical contact of the tip

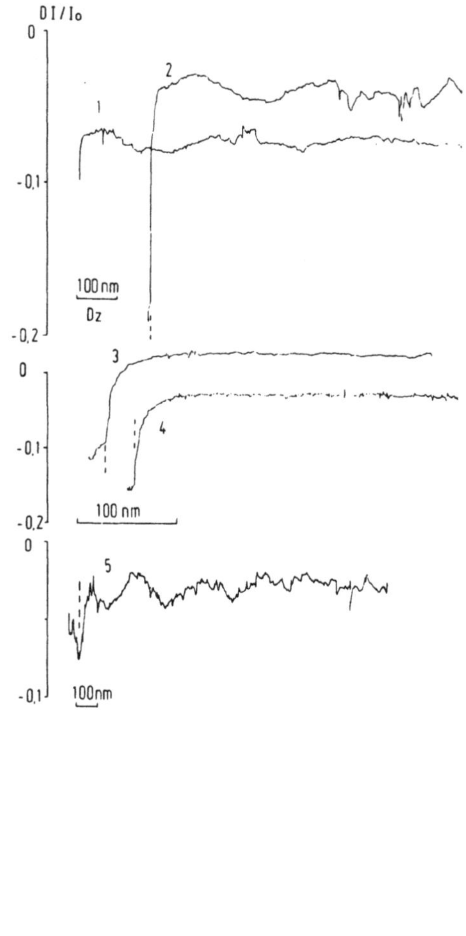


Fig.10
The same kind of plots as in Fig.9 for 5 subsequent approaches of a tetrahedral tip versus a glass reference surface. A HeNe laser was used for the irradiation at an intensity in the order of 100 $\mu$W at a wavelength of 633nm. The polarisation was parallel to the edge (for a further explanation of irradiation conditions see text) in scans 1-4 and perpendicular in scan 5. The approach was stopped before mechanical contact in scan 1. The assumed mechanical contact is indicated for the other scans by the broken line.

and the cover glass leads to an increase of intensity by up to two orders of magnitude and the interference patterns extend to large distances of up to ten microns. This kind of behaviour as observed for tips coated with aluminium, silver and gold is strongly dominated by a far-field interference effect and is reminiscent of the behaviour which was reported for SNOM tips consisting of a metal-clad tapered glass fibre with an aperture at the apex (11, 12). It was explained as being due to interfering waves between the flat end of the tip with a width in the order of a few 100 nanometers and the reference surface (12). In the case of of the tetrahedral tip with a silver coating, a condition was, however, found where a pronounced near-field signal is obtained as shown in fig. 10. An angle ß of about $60^0$ was chosen and the incident beam was polarised within the plane, which is determined by the direction of incidence and the edge of the tetrahedral tip ( see fig 6 for a visualisation). A small interference effect of an amplitude of approximately 2% is seen, which falls off after about two periods. Over a distance in the order of only 15 nm from contact a steep signal decrease of about 10% of the signal was repeatedly observed in four subsequent approach experiments. No comparable change was observed for the perpendicular polarisation. In the course of trying to repeat this approach experiment the tip was damaged, leading to a concomitant rise of the signal. Approach characteristics with strong interference oscillations and no pronounced near-field behaviour were then obtained.

## Conclusions

From these first investigations into the optical properties of the tetrahedral tip, it is concluded, that conditions can be found, under which a pronounced near-field effect is observed, as in the case of SNOM antennas in the form of protrusions from a metal film (8). Such an effect may well be used to operate the SNOM in a constant intensity mode, as also has been previously demonstrated (8). However, more experiments must be performed in order to define reproducible conditions for obtaining this near-field effect and in order to determine whether it is associated with local surface plasmon resonances of the tetrahedral tip. The rather high transmission factor of $0.6 \times 10^{-5}$ may indicate quite a high efficiency of the probe, as anticipated from qualitative considerations of the model. This value can however, only be evaluated in a meaningful way if one succeeds in relating it to an effective size of the emitting tip as estimated from the lateral resolution limit in a SNOM image, which has yet to be obtained. The concept of the tetrahedral probe may well be extended to other polyhedral geometries and to other materials such as piezoelectric, electro-optic or semiconductor materials. An electric field may be applied between the metal coatings and then be applied to modulate the light emission or to integrate optoelectronic functions such as light emission or detection into the probe (10).

## Acknowledgements

The author wishes to thank M. Rohde from the Gesellschaft für Biotechnologische Forschung, Braunschweig, for teaching him the art of breaking glass knives, to U. Dierksen and R. Lauer for their help in metal film evaporation, and to W. Mirandé, E. Buhr and R. Krüger for stimulating discussions.

## References

1)Betzig E., J.K. Trautman. (1992). 'Near-Field Optics: Microscopy Beyond the Diffraction Limit.'Science 257, 189-195. See also these proceedings.

2)Roberts A. (1991) 'Field Detection by Subwavelength Aperture probes. SPIE 1556 Scanning Microscopy Instrumentation, 11-18. See also these proceedings.

3)Keilmann F. (1988) German Patent DE 3837389 C1. See also these proceedings.

4)Fee M., S. Chu, T.W. Hänsch (1989). Optics Communications 69, 219 - 224.

5)Fischer U.Ch., M. Zapletal (1991). 'The Concept of the Coaxial Tip as a Probe for Scanning Near Field Optical Microscopy and Steps towards a Realisation.' Ultramicroscopy 42-44, 393-398)

6)Ruppin R. (1982). 'Spherical and Cylindrical Surface Polaritons in Solids'. In A. D. Boardman Ed., Electromagnetic Surface Modes', John Wiley & Sons Ltd, 345-397.

7)Raether H. (1988). 'Surface Plasmons on Smooth and Rough Surfaces and on Gratings'. in G. Höhler Ed. : Springer Tracts in Modern Physics 111. Springer, Berlin.

8)Fischer U.Ch., D.W. Pohl (1989). 'Observation of Single Particle Plasmons by Near-Field Optical Microscopy'. Phys. Rev. Lett. 62, 458-461.

9)Robinson,D.G., U.F.Ehlers, R.Herken, B.Herrmann, F.Mayer, F.W.Schürmann, (1985). 'Präparationsme-thoden der Elektronenmikroskopie'. Springer Verlag, Berlin, 54 - 59.

10)Danzebrink H.U., U.C. Fischer, these proceedings.

11)Pohl D.W., U.Ch. Fischer, U. Dürig. (1988). 'Scanning Near-Field Optical Microscopy (SNOM): Basic Principles and some Recent Developments', Proc. SPIE 565,56 - 61.

12)Cline J.A., H.Barshatzky, M.Isaacson (1991). 'Scanned-Tip Reflection-mode Near-field Scanning Optical Microscopy.'Ultramicroscopy 38, 299-304.

# DEVELOPMENT OF A SCANNED NEAR-FIELD OPTICAL MICROSCOPE FOR MAGNETO-OPTIC KERR IMAGING OF MAGNETIC DOMAINS WITH 10 NM RESOLUTION

T. J. Silva and S. Schultz
*Center for Magnetic Recording Research and Department of Physics*
*University of California, San Diego*
*9500 Gilman Drive*
*La Jolla, CA 92093-0319*

ABSTRACT. Work is described on the development of a scanning near-field optical microscope (SNOM) for the primary purpose of imaging magnetic systems with resolution on the order of 10 nm. Since many magnetic materials are optically opaque, it is desired to have a probe which is appropriate for the reflection mode. The near-field probe must be linearly polarizable, since the magneto-optic kerr effect (MOKE) is utilized for magnetic imaging. Data will be presented on the characterization and use of ~30 nm diameter Ag particles as probes for MOKE-sensitive SNOM. Such small metal particles exhibit a localized plasmon resonance in the visible, which greatly enhances their optical scattering cross-section. We have made a systematic study of the light scattering properties of single Ag particles from 20 to 30 nm in size with a high-sensitivity micro spectrophotometer system, for several reasons: (1) To compare the light scattering properties of individual particles with the predictions of Mie theory; (2) to understand the polarization sensitivity of the optical scattering properties of real aspherical particles; and (3) to optimize the probe-sample separation distance for various magnetic materials of interest. Ag particles prepared by colloidal chemistry techniques have been deposited on Bi-doped YIG and silica-coated permalloy to measure the magnitude of near-field MOKE. Near-field MOKE polarization rotations of 1.5 and 6.5 milliradians were measured for permalloy and YIG, respectively.

## 1. Near-field Microscopy for Micro-magnetic Investigations

### 1.1. POLARIZATION-SENSITIVE REFLECTION-MODE SNOM

There is a need in the magnetic recording industry for a magnetic imaging technique which combines the ease and accessability of Kerr optical microscopy, but with a spatial resolution of 1/50 of the wavelength of light. Polarization-sensitive SNOM would seem to satisfy that need. Indeed, the work of Betzig, et al, have used SNOM employing tapered optical fibers to detect and even write magnetic domains in the 10 nm size range [1,2]. Unfortunately, the present implementation of polarization-sensitive SNOM as demonstrated by Betzig, et al, can only be applied to optically transparent magnetic materials which are magnetized perpendicular to the film's surface. The vast majority of magnetic materials used in the conventional magnetic recording industry are opaque and have longitudinally-oriented magnetization patterns. Thus we have endeavored to develop a polarization-sensitive version of SNOM which operates in a reflection mode [3].

Present implementations of polarization-sensitive SNOM rely on passing light through an aperture of sub-wavelength dimension in a conducting film in order to obtain high resolution images. An alternative form of SNOM uses an optical scatterer of sub-wavelength dimension as the near-field probe [4]. In the Rayleigh limit, a scattering probe will behave as a single dipole transducer with a well characterized polarization state. On the other hand, an apertured probe may be composed of multiple dipole sources. These manifold sources may transmit linearly polarized light into the far-field when properly compensated [2], but could still represent a complicated polarization pattern in the near-field when used in a reflection mode. Thus it would seem that an implementation of SNOM which relies on a sub-wavelength scattering probe rather than an apertured probe may be more suitable for the goal of high-resolution magnetic imaging of opaque surfaces.



*D.W. Pohl and D. Courjon (eds.), Near Field Optics*, 263–272.

1.2. PLASMON SNOM

The principle advantage of an aperture-source SNOM is the achievement of highly efficient optical coupling out of a sub-wavelength metal orifice [5]. It is believed that this is due to an adiabatic transition for guided waves in the optical fiber from a propagating to an evanescent regime. Fundamentally, the apertured source still suffers from the attenuation effects associated with Rayleigh scattering, i.e. the optical cross-section of a sub-wavelength scatterer is proportional to $a^6/\lambda^4$, where a is the scatterer dimension and $\lambda$ is the wavelength of the illuminating light [6]. There is a class of optical scatterers, however, which do not exhibit Rayleigh scattering behavior. These are particles composed of the noble metals; principally silver. Such particles experience a surface plasmon resonance (SPR) mode in the visible which greatly enhances their optical scattering cross-sections [7]. The resonance is essentially a collective conduction band electron resonance, with the internal depolarizing fields providing the restoring force necessary for an oscillation. The effect is largest for silver, due to its unusually low dissipation at the resonance frequency, though it also occurs to a lesser degree in gold and copper. The SPR has been exploited in previous implementations of SNOM [4,8], but they have relied on the interaction of surface plasmons in films and particles ("localized surface plasmons") to generate a contrast mechanism. In our instrument we eliminate the need for planar surface plasmons, and use only the localized surface plasmon resonance associated with 20-30 nm diameter Ag particles. The SPR has also been considered in the theoretical modeling of SNOM [9,10].

The program for our study begins with a comparison of the real optical behavior of nanoscopic Ag particles with the predictions of Mie theory. In particular, we are interested in measuring the "Q" and frequency of the resonance under different environmental conditions. This has to be performed on a per-particle basis since it is the behavior of a single particle, which is to be used as the SNOM probe, in which we are ultimately interested. We then need to know the magnitude of magneto-optic Kerr rotation induced through the near-field coupling of the Ag particle and a magnetic sample. Conventional MOKE is a function of wavelength, polarization and angle of the incident light which can be readily determined. For a near-field geometry, there exists no simple or intuitive method of calculating the polarization rotation induced in the light scattered from a single particle in virtue of near-field coupling to a magnetic substrate. It simply must be measured. Having addressed these issues, it may then be possible to develop an appropriate mounting for the Ag particle in a manner suitable for raster-scanned image formation.

1.3. NEAR-FIELD MOKE

Simply stated, the MOKE represents a rotation of the polarization of linearly polarized light when reflected off of a ferromagnetic surface. In general, the effect is small, ranging from 1 to 10 milliradians (3 to 30 arc seconds) for most materials of interest. At optical frequencies, the phenomena finds its origins in spin-orbit coupling within the magnetic materials [11], but may be phenomenologically described in terms of off-diagonal elements in the macroscopic permittivity tensor describing the optical reflectivity[12]. Such a tensor couples the electric field vector of linearly polarized incident light which is orthogonal to the magnetization vector into the other remaining direction orthogonal to the magnetization. It is not known whether a macroscopic permittivity tensor is completely valid in a geometry involving highly non-uniform applied electric fields and/or magnetization distributions, as is often the case for SNOM. Recent work by Schaefer and Hubert shows evidence for a new magneto-optic effect which is sensitive to components in the magnetization gradient tensor [13]. In the magneto-optic recording work of Betzig, et al, sub-wavelength magnetic domains written with a SNOM probe show unusual contrast qualities that could be the result of both the large magnetization and electric field gradients involved in the experiment [2].

In our experiments to be described, we believe the geometry to be sufficiently simple to show the fundamental behavior of near-field MOKE. The Ag particle should behave as a single dipole scatterer due to its extremely small size in comparison to the wavelength used. Our illuminating optical beam is relatively uniform over the Ag particle, avoiding any possible secondary effects due to gradients in the applied electromagnetic field. The surfaces to be used will be thin films, and the magnetization patterns studied are assumed to be uniform, at least in their saturated states. However, it is not possible to verify that the magnetization is completely uniform throughout the thickness of the magnetic films employed. The extreme surface sensitivity of near-field techniques makes comparison with other more conventional MOKE measurement techniques somewhat troublesome.

## 2. Single Ag Particle Photo-Spectroscopy

### 2.1. QUANTITATIVE OPTICAL ULTRA-MICROSCOPE

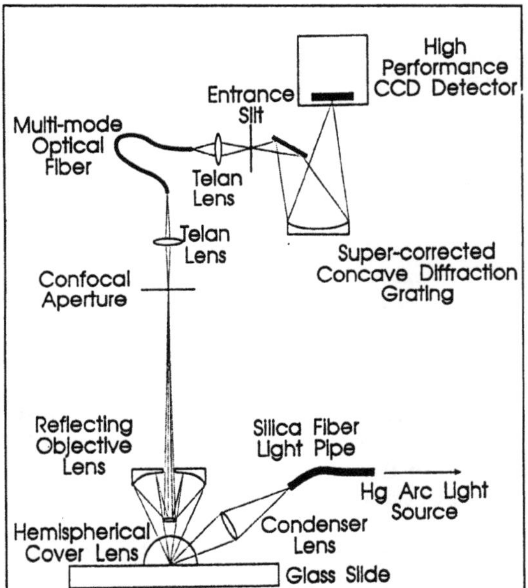


Figure 1. Quantitative Optical Ultra-Microscope.

In Fig. 1 a diagram is shown of the optical microscope system used to perform spectroscopic analysis on the light scattered from individual Ag particles of 20-30 nm size [7]. The apparatus relies heavily on the fact that individual noble metal particles scatter light strongly enough to be visible under dark-field illumination even under modest magnification, as was discovered by Richard Zsigmondy at the turn of the century[14]. This allows one to align the image of a single particle with a multi-mode "light-pipe" which in turn feeds the light to a multi-channel analyzer spectrometer system. A Kodak KAF4200 high-performance CCD was used with a Photometrics CCD9000 camera system for the optical measurements. The CCD device was chilled to a nominal temperature of 237° K (-36° C) to reduce dark-noise. Exposure times of 15 minutes were typical to measure the spectrum of light scattered from individual Ag particles with maximum signal-to-noise.

### 2.2. SAMPLE PREPARATION

Ag particles were produced by colloidal chemistry techniques using commercially available reagents [15]. The recipe for producing colloidal Ag in the 20-40 nm range is as follows:

1. Mix 2 mg. gelatin in 20 ml distilled water.
2. Bring to boil while stirring constantly. Boil for 3 minutes to sterilize, then cool to room temperature.
3. Add 1 ml BioCell 2 nm colloidal gold.
4. Add 0.1 ml "Initiator" solution from the BioCell Silver Enhancement Kit.
5. Stirring rapidly but avoiding aeration, add 10-100 ml "Enhancer" solution from the BioCell Silver Enhancement Kit. The more solution that is added, the larger the resulting Ag particles, and the more optically opaque the solution will become. Within one minute, the solution will begin turning a noticeable yellow color. The color change should come to a halt in 5-10 minutes.
6. The colloid should remain stable for weeks to months if kept refrigerated and exposure to air is kept to a minimum.

The particles are in fact composed of ~2 nm Au nuclei surrounded by Ag shells. This has been verified by observing the resultant particles with a TEM. The gold particles act as nucleation sites for the neutral silver atoms produced when the "Initiator" and "Enhancer" solutions are mixed, resulting in a colloid with a uniform size distribution[16]. The ratio of gold to silver is so small by mass, that the gold should have negligible optical effect. This is verified by Mie theory calculations which we performed that agreed with the experimental observations. A TEM micro graph of two representative particles is shown in Fig. 2. Notice that the particle labeled #1 is quite spherical in shape and appears to be composed of a single crystal. Particle #2, on the other hand, is neither spherical nor single-crystalline. The ability to discriminate between the variations in Ag particle morphology based on their optical scattering characteristics will be important in order to fabricate the ideal probe for reflection-mode MOKE-sensitive SNOM.

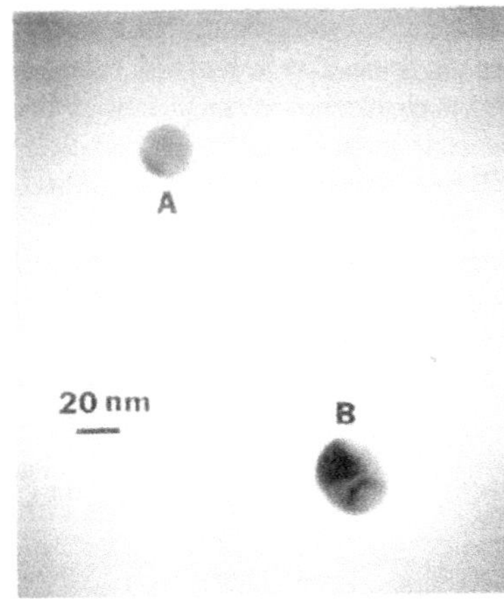


Figure 2. TEM micrograph of Ag particles made by colloidal chemistry technique.

The particles contained within the resulting colloids appear to be highly attracted to glass surfaces, making sample preparation for TEM and optical microscopy easy. Drops containing a suitably dilute concentration of particles were placed on either a clean glass surface or an $SiO_2$-coated surface, then rinsed off with distilled water after a few seconds time. The individual Ag particles remain firmly attached to the surface at a density proportional to the concentration of the colloid used. It is necessary for the glass surface to be extremely clean for adhesion to occur. Surfaces were rinsed with 1,2-dichloroethane, acetone, methanol and distilled water, in that order. This removed most of the contaminants as was confirmed by observation with a dark-field optical microscope.

When prepared for optical microscopy, we find that it is necessary to provide a uniform dielectric environment for the Ag particle. A drop of immersion oil (Nikon, n=1.515) placed on the glass surface supporting the Ag particles works very well. A hemispherical cover-lens is then placed on top of the sample to reduce the fresnel reflection of obliquly incident light and to reduce chromatic aberation when collecting the scattered light for spectroscopy.

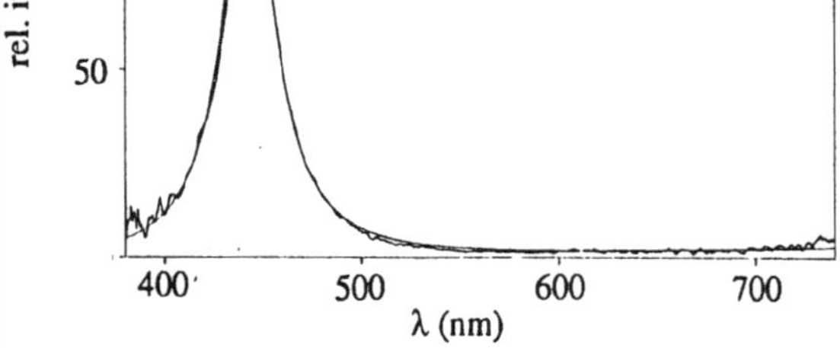


Figure 3. Corrected spectrum of light scattered from single Ag particle. Lorentzian curve-fit is superimposed on data.

Spectrum data is corrected for the spectral inhomogeneity of the light source (100 Watt Hg arc lamp) and the spectral response of the detector by dividing the particle spectrum by the spectrum obtained when scattering light from a sample of pressed PTTF powder, an achromatic diffuse scatterer [17].

### 2.3. MEASUREMENT RESULTS

A representative spectrum for the light scattered from a single particle is shown in Fig. 3. Note the solid Lorentzian curve-fit superimposed on the actual data. The peak scattering wavelength of ~420-440 nm agrees well with the theoretical predictions of Mie theory for a 30 nm, spherical particle. We have observed peak wavelengths over the range from 400 to 700 nm, though the majority tend to be in the range of 400 to 450 nm. Some of the measured spectra revealed double peaks. We believe that both the wide range of peak wavelengths and multiple peaks are due primarily to shape effects.

Figure 4 shows an example of what we believe to be a shape effect; in particular, the splitting of the plasmon resonance into two orthogonally polarized modes. In Fig. 4a we present the wavelength dependence of the light scattered from a single Ag particle without polarization analysis when the light incident on the particle was polarized. Notice the two peaks at 430 nm and 600 nm. The particle appeared violet to the eye since it was a mixture of purple and orange scattered light. The particle was oriented in the incident light to maximize its violet appearance. In Fig. 4b, the spectra is from the same particle, but an analyzer of dichroic sheet polarizers is inserted into the path of the scattered light. The analyzer was adjusted to maximize the intensity of the orange peak @ 600 nm. Rotation of the analyzer by 90° then blocks the light from the 600 nm peak but maximized the light in the 430 nm peak, as seen in Fig. 4c. We believe these data confirm that the particle is elliptical in shape, with its long axis parallel to the analyzer axis when set to maximize the red peak. The red-shift of the surface plasmon mode in this direction finds its origin in the reduced depolarizing fields within the particle in the long-axis direction of a prolate spheroid [11].

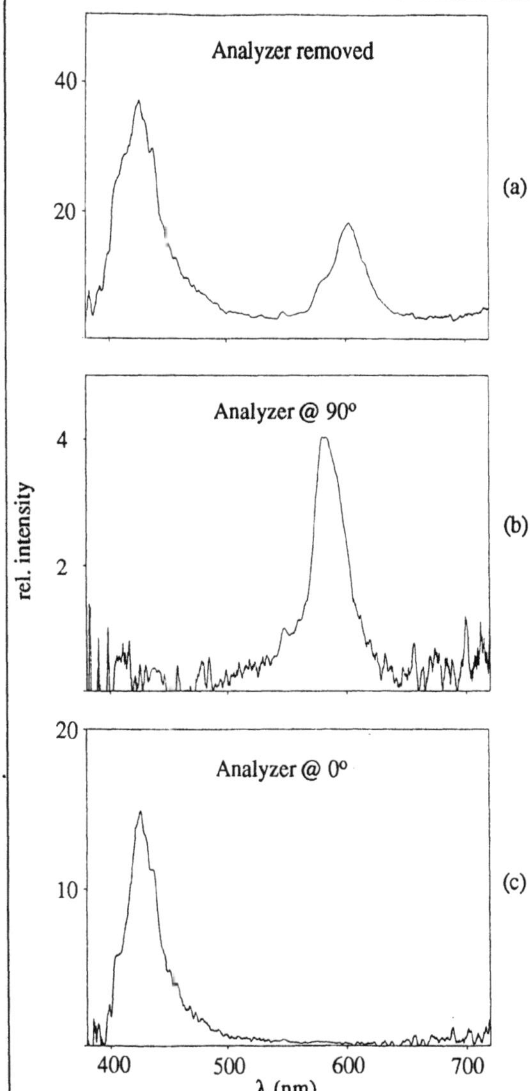


Figure 4. Corrected spectrum of light scattered from single Ag particle. a) Analyzer removed from path of scattered light. b) Analyzer in path of scattered light and oriented at 90 degrees. c) Analyzer in path of scattered light and oriented at 0 degrees.

With the ability to perform quantitative photo-spectroscopy on individual Ag particles, it is now possible to perform numerous experiments measuring the shift in the plasmon resonance induced by changing the local environment of the particles, either by changing the dielectric constant of the immersion oil or by

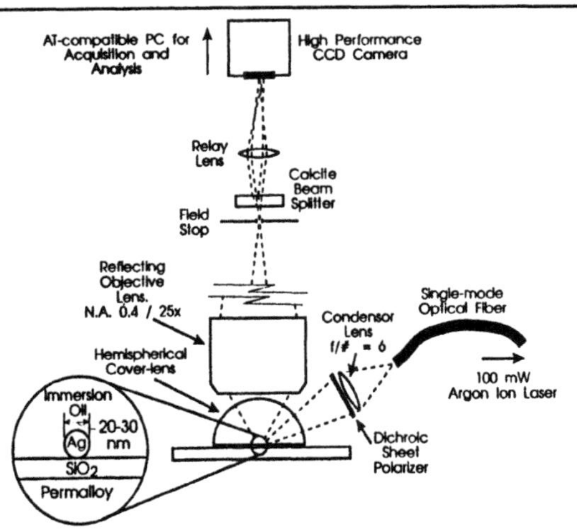


Figure 5. Near-field MOKE measurement system.

moving the particles close to various dielectric substrates using piezo-electric positioning devices. Such measurements will be reported elsewhere.

## 3. Near-field MOKE

### 3.1. NEAR-FIELD MOKE POLARIZATION ROTATION MEASUREMENT APPARATUS

In Fig. 5 is shown the optical system used to measure the linear polarization angle of light scattered from an individual Ag particle. The illumination source used for the polarization studies was a 100 mW air-cooled argon-ion laser, operating in single-line mode. The 514.5 nm line was used for all the data to be presented. A low N.A. condenser lens was used to focus the laser beam to a ~7 micron spot on the sample surface. The low f/# optics provides a relatively uniform linear polarization across the beam width at the beam waist. A field stop is used in the microscope to isolate the light from a single Ag particle. It is followed immediately by a calcite beam displacer nominally oriented at 45° to the polarization of the scattered light. 5x relay optics form a split image of the diffraction-limited spot from the particle on the same high-performance CCD camera used in the spectroscopy studies. The data is then captured and stored on an AT-compatible personal computer. Saturation of the CCD (80,000 e-/pixel) was achieved with exposure times on the order of 1-10 seconds. This corresponded to a total detected intensity of $10^7$-$10^8$ photons per second. Since the data transfer of a single image from the CCD camera requires ~1 sec, a brighter illumination source would not significantly improve the data rate. Also, since the incident flux on the Ag particle is believed to be < 1000 Watts/cm$^2$, higher incident intensities could lead to optically induced damage in either the sample and/or the particle.

The intensity of each spot imaged by the CCD is integrated and the polarization computed by using differential analysis similar to that used in commercial magneto optic storage devices. Kasiraj, et al, have found the technique to exhibit excellent signal-to-noise when employed in a conventional scanning Kerr effect microscope [18]. The technique makes maximum use of available light, which is important for shot-noise limited conditions. It is also immune to intensity noise commonly associated with laser sources. However, it was found that the location of the spots on the CCD would drift with time. These drifts are most likely due to thermal expansion of the microscope's mechanical parts. It is desirable to measure the intensity at only these pixels illuminated by light scattered from the Ag particle of interest in order to minimize dark-noise, read-noise and background scatter contributions to the measured signal. Thus, it is necessary to determine the exact center of the image spot for each exposure to compensate for the drift. This is accomplished by finding the position of maximum correlation between the spot data and an "ideal" spot function. The "ideal" spot function is an easily calculated two-dimensional function which approximates the intensity profile of the measured spot. The correlation calculations are performed on an AT-compatible personal computer employing an Intel 33 MHz 80386 CPU with an Intel 80387 floating point coprocessor. The calculations required 30 to 60 seconds per data set.

3.2. POLAR MAGNETIZATION GEOMETRY: BISMUTH-DOPED YIG

The first attempts at measuring a near-field MOKE in reflection mode were with a Bi-doped YIG sample, owing to its extremely large Kerr rotation. The sample used had a Kerr rotation angle of 16 milliradians (0.9°). The measurement was made with 436 nm light (Hg source) in a conventional polar MOKE arrangement For the near-field Kerr measurement, Ag particles were deposited directly onto the YIG surface, much in the same manner as is done on glass surfaces. The scattering efficiency of the Ag particles was greatly attenuated, probably due to the strong absorption of YIG in the blue. The particles appeared green to the eye and scattered most strongly when illuminated with the 514.5 nm line of the argon laser.

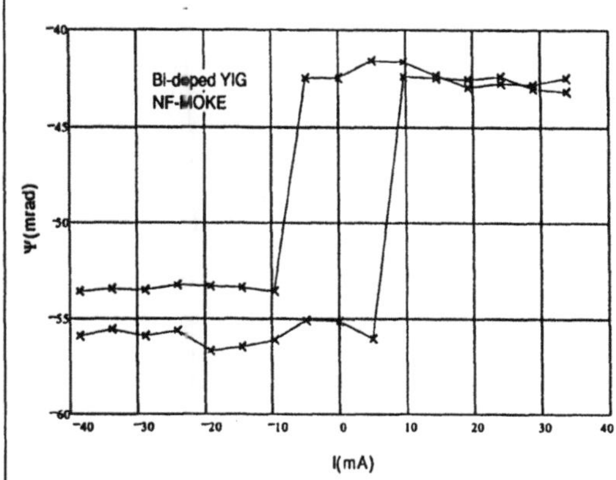


Figure 6. Near-field MOKE hysteresis loop from Bi-doped YIG sample.

A near-field MOKE hysteresis loop measured from a single particle on the YIG film is shown in Fig. 6. The magnet used to apply the reversing field produced ~10 Oe / mA. The coercivity of the film as measured in a Vibrating Sample Magnetometer (VSM) was found to be 119 Oe. This agrees fairly well with the zero-crossing of the data in Fig. 6. Exact agreement with the VSM data was not expected since the near-field hysteresis loop measured the magnetic behavior over a very small region of the sample surface, whereas the VSM data is an average result for the entire sample. Note the magnitude of the polarization rotation was found to be only 13 mrad, peak-to-peak. This is roughly half that measured in a conventional polar MOKE geometry for this sample using the same polarization measurement system. This is not surprising, given the strong dependence of the MOKE in YIG on wavelength [19]. A spectroscopic study of the Kerr rotation vs. wavelength has not been performed for our sample, but similar materials exhibit a maximum rotation between 400 and 450 nm light, with a rapidly decreasing effect for longer wavelengths [20].

3.3. LONGITUDINAL MAGNETIZATION GEOMETRY: $NI_{80}FE_{20}$ (PERMALLOY)

Permalloy is a magnetic metallic alloy commonly used in the magnetic recording industry for recording/read back heads. It is magnetically "soft", with a low coercivity, and can be made with negligible magnetic anisotropy in the film plane. The samples used in this study were 50 nm thick sputtered films on Si substrates and had a coercivity of 5 Oe.

It was found that Ag particles did not adhere to the surface of metal films in the same way as for glass surfaces. Thus it was necessary to coat the Permalloy with an $SiO_2$ spacer layer to provide an adhesion surface (see inset in Fig. 5). The $SiO_2$ coating was also needed because the plasmon effects were sufficiently damped by the proximity of the Permalloy surface to make visibility of the particles under conventional dark-field illumination with a Hg source impossible. For that reason, a 15 nm thickness was chosen for the measurements presented here. Once the Ag particles were deposited, a protective layer of 100 nm of $SiO_2$ was sputtered over the sample.

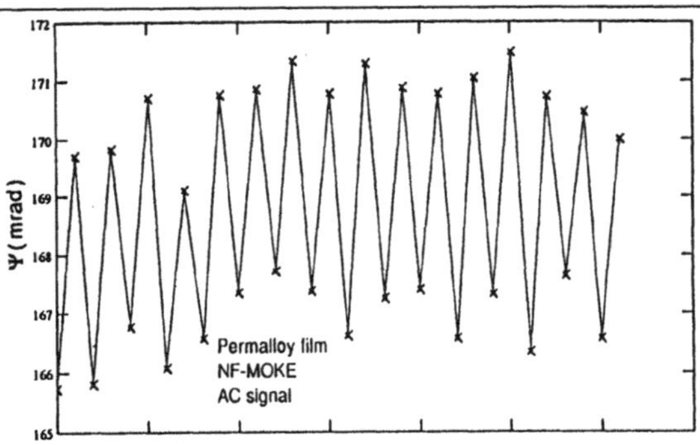


Figure 7. Near-field MOKE AC signal from permalloy. Magnetization of permalloy sample is reversed between successive data points.

The damping of the plasmon resonance by absorptive samples is an inherent problem with plasmon SNOM. In the case of Permalloy, with an ε" of 5.15 at 546 nm [12], the problem is not extremely severe if sufficient spacing between the sample and particle is provided. The 15 nm spacing employed is not incompatible with the needs of high resolution SNOM. Imaging of more absorptive magnetic materials, such as magnetic ferrites, may turn out to be extremely difficult.

The magnetization of the Permalloy film was reversed with a small in-situ electro-magnet placed between the glass hemisphere and the Permalloy surface. The magnet could not be calibrated since the extremely small gap (~1 mm) precluded the use of a conventional Hall-probe Gaussmeter. Measurement of the polarization state of the scattered light from a particle was measured after each magnetization reversal, to form the ac signal shown in Fig. 7. Fourier analysis of the signal showed the component at the reversal frequency to be 1.6 milliradians, 0-to-peak. This is 4x smaller than the signal measured from the YIG sample, but 2x larger that the signal expected from Permalloy in an optimized geometry for conventional longitudinal MOKE [21].

To test whether the ac signal detected was due to changes in the magnetization of the Permalloy or due to mechanical motion coupled to the applied field, the magnitude of the ac response was measured as a function of the maximum current applied to the electro-magnet on the sample. If the signal was due to magneto-mechanical effects, the signal should be a monotonically increasing function of current. If the signal was magneto-optic in origin, then it should be constant when the current achieves a level sufficient to saturate the Permalloy film. Fig. 8 shows the results of such a test. Notice the curve is flat, except for a discontinuous jump between 160 and 180 mA. The flat shape is taken as evidence for a magneto-optic origin to the measured rotation. The discontinuity may be due to a change in the magnetic domain pattern in the Permalloy sample. The discontinuous behavior was repeatable.

## 4. Future Direction

More tests shall be performed to verify the authenticity of the results obtained with the Permalloy film. In particular, the magnet domain patterns in the Permalloy sample shall be imaged by conventional MOKE microscopy methods [22] to verify the discontinuous magnetic behavior mentioned above.

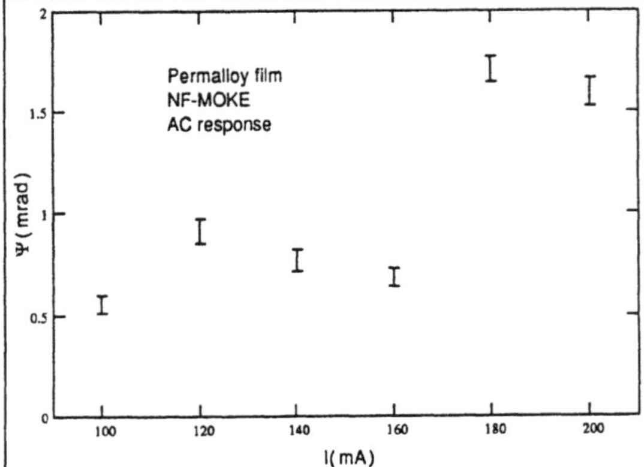


Figure 8. Magnitude of near-field MOKE AC signal from permalloy as a function of maximum current applied to electromagnet.

Various optical components can be brought to bear to optimize the signal-to-noise present for polarization measurements. The ellipticity of the reflected light will be compensated for using a quartz wedge to maximize the measured rotation. An extra-long-working-distance refractive objective lens will produce a more concentrated diffraction pattern, further reducing the necessary data acquisition time.

Now that it is known that it is possible to measure the reflective MOKE signal with adaquate signal-to-noise in a near-field geometry, work has begun to measure the effect as a function of spacing between the Ag particle and magnetic substrate. It is planned to mount the particles to a hemispherical transparent substrate which may be brought arbitrarily close to a magnetic sample via a piezo-electric translator. If optimum separation is found to be in the 10's of nanometers, then the hemispherical substrate may be the best mechanical mount for an Ag particle probe when it is attempted to raster-scan the particle across a magnetic sample. Servoing may be provided by monitoring the interference pattern established between the hemispherical surface and the reflective flat surface of the sample, i.e., "Newton rings." Closer separations would require AFM or shear-force servoing techniques similar to that used by other workers in SNOM [23,24].

A study of the effect of particle ellipticity on near-field MOKE is also planned.

## 5. Conclusion

20-30 nm diameter Ag particles have been shown to be potential candidates as probes for reflective-mode polarization-sensitive SNOM. They have been shown to exhibit excellent light scattering properties. They can be optically characterized on an individual basis by microspectrophotometry. The existence of a MOKE in a near-field geometry has been verified for both a perpendicular and longitudinally magnetized medium. Sufficient signal-to-noise exists to use a Ag particle to form images via conventional scanning techniques.

## 6. Acknowledgments

Bi-doped YIG samples were provided by Bernie Argyle, IBM. Permalloy samples were provided by Jay Freeman, Eastman Kodak Company. $SiO_2$ sputtering was provided by Matt Carey, CMRR. This research has been supported by the National Science Foundation, grants number DMR-90-10908 and EID-87-11688 and The Center for Magnetic Recording Research (CMRR).

## 7. References

[1] Trautman, J. K., Betzig, E., Weiner, J. S., DiGiovanni, D. J., Harris and T. D., Hellman, F. (1992) 'Image contrast in near-field optics', J. Appl. Phys. 71, 4659-4663.
[2] Betzig, E., Trautman, J. K., Wolfe, R., Gyorgy, E. M. and Finn, P. L. (1992) 'Near-field magneto-optics and high density data storage', Appl. Phys. Lett. 61, 142-144.
[3] Schultz, S. and Silva, T. J. (1992) 'Development toward magneto-optic Kerr scanned near-field optical microscope with 10 nm resolution', Srinivas Manne (ed.), Scanning Probe Microscopies, SPIE Proceedings Vol. 1639, 31-35.
[4] Fischer, U. Ch. and Pohl, D. W. (1989) 'Observation of Single-Particle Plasmons by Near-Field Optical Microscopy', Phys. Rev. Let. 62, 458-461.
[5] Betzig, E., Trautman, J. K., Harris, T. D., Weiner, J. S. and Kostelak, R. L. (1991) 'Breaking the Diffraction Barrier: Optical Microscopy on a Nanometric Scale', Science 251, 1468-1470.
[6] Jackson, J. D. (1975) Classical Electrodynamics, John Wiley and Sons, New York, pp. 405-427.
[7] Bohren, Craig F and Huffman, Donald R. (1983) Absorption and Scattering of Light by Small Particles, John Wiley and Sons, New York.
[8] Specht, M., Pedarnig, J. D., Heckl, W. M. and Haensch, T. W. (1992) 'Scanning Plasmon Near-Field Microscope', Phys. Rev. Let. 68, 476-479.

[9] Girard, C. (1992) 'Plasmon resonances and near-field optical microscopy: a self-consistent theoretical model', Applied Optics 31, 5380-5387.

[10] Denk, Winfried and Pohl, Dieter W. (1991) 'Near-field optics: Microscopy with nanometer-size fields', J. Vac. Sci. Technol. B 9, 510-513.

[11] Argyres, Petros N., (1955) 'Theory of the Faraday and Kerr Effects in Ferromagnetics', Phys. Rev. 97, 334-345.

[12] Robinson, C. C., (1963) 'Longitudinal Kerr Magneto-Optic Effect in Thin Films of Iron, Nickel, and Permalloy', J. Opt. Soc. Amer. 53, 681-689.

[13] Schaefer, R. and Hubert, A., (1990) 'A New Magnetooptic Efffect Related to Non-Uniform Magnetization on the Surface of a Ferromagnet', phys. stat. sol. (a) 118, 271-288.

[14] Zsigmondy, R. (1914) Colloids and the Ultramicroscope, John Wiley & Sons, New York, Chapters 4 & 5.

[15] BioCell Research Laboratories, Cardiff Business Technology Centre, Senghenydd Rd, Cardiff CF2 4AY, Tel: 0222 238750.

[16] Jirgensons, B. and Straumanis, M. E., (1962) A Short Textbook of Colloid Chemistry, The Macmillan Company, New York.

[17] Weidner, V. R., Hsia, J. J., and Adams, B. (1985) 'Laboratory intercomparison study of pressed polytetrafluoroethylene powder reflectance standards', Applied Optics 24, 2225-2230.

[18] Kasiraj, Prakash, Shelby, Robert M., Best, John S., and Horne, Donald E. (1986) 'Magnetic Domain Imaging with a Scanning Kerr Effect Microscope', IEEE Trans. Mag. MAG-22, 837-839.

[19] Winkler, Gerhard (1981) Magnetic Garnets, Vieweg, Braunschweig, Wiesbaden, pp. 253-259.

[20] Wittekoek, S. and Popma, T. J. A., (1973) 'Magneto-optic Kerr rotation of bismuth-substituted iron garnets in the 2-5.2-eV spectral range', J. Appl. Phys. 44, 5560-5566.

[21] Tanaka, Shun-ichi (1963) 'Longitudinal Kerr Magneto-Optic Effect in Permalloy Film', Jap. J. of Appl. Phys. 2, 548-555.

[22] Argyle, B. E., Petek, B. and Herman, D. A., Jr. (1987) 'Optical imaging of magnet domains in motion", J. Appl. Phys. 61, 4303-4306.

[23] Betzig, E., Finn, P. L. and Weiner, J. S. (1992) 'Combined shear force and near-field scanning optical microscopy', Appl. Phys. Lett. 60, 2484-2486.

[24] Toledo-Crow, R., Yang, P. C., Chen, Y. and Vaez-Iravani, M. (1992) 'Near-field differential scanning optical microscope with atomic force regulation', Appl. Phys. Lett. 60, 2957-2959.

# Scanning Plasmon Near-Field Microscope

*J. D. Pedarnig, M. Specht, W. M. Heckl and T. W. Hänsch*
*Sektion Physik der Universität München, Schellingstr. 4/III,*
*D-8000 München 40, W. Germany*

ABSTRACT. Recently we have introduced the scanning plasmon near-field microscope (SPNM) which achieves a resolution of a few nanometers. Resonantly excited extended surface plasmons generate an optical near-field which is locally probed by a sharp tip. The technique is applied to imaging of dye clusters and photooxidation of dyes. Here we present another type of optical near-field microscope, the scanning near-field ellipsometric microscope (SNEM).

## I Introduction

Optical near field microscopy [1,2] has become an intriguing scanning probe method due to its high resolution, low probing forces and spectroscopic capabilities. There are two main implementations of an optical near-field microscope. Historically the first was to scan a subwavelength aperture very close to an object. Superresolution can be achieved by measuring the intensity of the light that passes through the hole [3-11]. In an alternative setup a dipole is brought in close proximity to the object and the effective scattering cross section of the dipole as a function of its position above the sample surface is measured [12,13].

Recently we have introduced the scanning plasmon near-field microscope (SPNM) [13] which is of the second type and has achieved a resolution of a few nanometers thus targeting molecular resolution. The SPNM is based on the interaction of extended surface plasmons with a sharp tip that is scanned across the surface of the object.

To us the most exciting application of the technique would be to spectroscopically identify molecules adsorbed on a substrate surface. Therefore our experiments aim into that direction. We review the state of the art of SPNM when applied to metal surfaces and to dye molecules adsorbed to such surfaces. We will show that particularly photooxidation, a chemically interesting and technically important process, can be studied with SPNM.

Additionally we will present a new optical near-field microscope, the scanning near-field ellipsometric microscope (SNEM) which is also of the second type of optical near-field microscopes. The main advantage of this microscope compared to SPNM is that any type of transparent material can be used as a substrate.



*D.W. Pohl and D. Courjon (eds.), Near Field Optics*, 273–280.

## II Scanning Plasmon Near-Field Microscope (SPNM)

The SPNM is based on the interaction of extended surface plasmons (SP) with a sharp metal tip which is placed close to the surface of the object [14,15]. From a theoretical point of view this interaction can be understood mainly in terms of elastic plasmon scattering [16] as well as radiationless energy transfer from the tip to the sample. These processes depend very strongly on the distance between tip and sample and are the key for a high lateral resolution capability. Therefore by raster scanning the tip over the sample surface, nanometer scale maps of interaction strength related to the topography of the sample can be recorded.

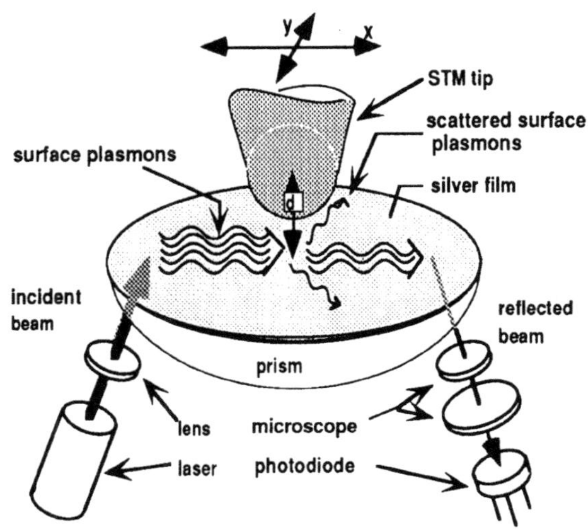


Fig.1. Schematic of the experiment: Surface plasmons, excited by an incident laser beam, interact with a thin tungsten tip positioned very close to a silver surface.

The experimental setup (fig. 1) contains three microscopes on different scales: an STM using the stylus as a tunneling tip, the SPNM using the stylus as an optical near-field sensor and an optical microscope [17]. Consequently the setup allows for overlapping and complementary in situ characterisation of the sample over many orders of magnitude (0.1 nm to 1mm).

The incident laser light is coupled to the SP-mode at the silver-air interface via a prism in the usual Kretschmann configuration [18,19]. The excitation of SP is recognized as a minimum in the reflected light intensity (Attenuated Total Reflection (ATR) - minimum) [19]. When the tungsten tip approaches the silver surface, a bright spot and a less intensive tail like structure appear in the image plane of the optical microscope. An aperture preselects the reflected light from this area on the surface. Changes in the reflected light intensity are measured with a photodiode while the tip is scanning across the surface. Although the light intensity is measured in the far-field the variations originate from a localized near-field interaction between a nanometer sized tip and the sample allowing for a resolution beyond the diffraction limit.

The superresolution capability of the near-field microscope is demonstrated in figure 2. These images are all taken at the same place on the sample surface and show the topography of a silver surface on a scale of 600 nm x 600 nm. The hole pierced into the surface measures in the STM as well as the SPNM image about 20 nm. Such a hole can be considered a suitable test sample for an estimate of the lateral resolution of an optical near-field microscope because it is one of the simplest geometrical objects to allow resolution measurements in two dimensions. From a comparison of the edge sharpness between the STM and the SPNM image the resolution is estimated to be 3 nm.

The contrast in these images can be understood as refractive index contrast, because in the SPNM a corrugated surface is equivalent to a flat sample with refractive index changes. Therefore a hole filled with e. g. air or gold can be viewed as a feature with different refractive index in an homogeneous index matrix and is expected to provide image contrast in SPNM.

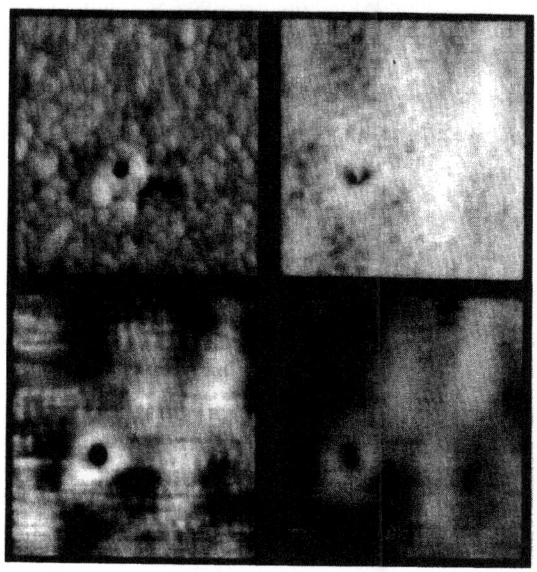

Fig. 2. 600 nm x 600 nm scan of a silver surface (Specht et al. [13]); (upper left) STM image; (lower left and lower right) SPNM images taken at the same area with a mean tip sample spacing of 3 nm and 10 nm, respectively; (upper right) image recorded in the "STM-SPNM hybrid mode".

## III Experiments on dye films

In a first experiment we demonstrate the capability of SPNM to image clusters of dye molecules adsorbed to a silver surface by switching from the SPNM to the STM mode during recording an image of this surface.

The sample was prepared by evaporating a 50 nm thick silver film on a glass slide and subsequently dipped into a dye solution of nile blue ($10^{-4}$ M in ethanol). Nile blue has its absorption maximum near the 633 nm wavelength of the He-Ne laser.

Figure 3 shows an SPNM image of the sample surface recorded in the constant height mode where the tip was scanned at a fixed distance of 20 nm above the silver surface. The optical near-field image has detailed structure showing large areas of different contrast on a scale of approximately 100 $nm^2$ and additional thin channel-like features a few 100 nm long. We interpret this image as aggregated dye molecules forming inhomogeneous patches. The lower part of the image is recorded in the conventional STM mode. No indication of dye molecules is discernible at all. The nonconductive molecule layer has been pierced by the tip revealing the underlying granular silver structure of typically 20 nm grain size. This clearly demonstrates the capability of SPNM to image organic coatings which can not be imaged in the STM mode.

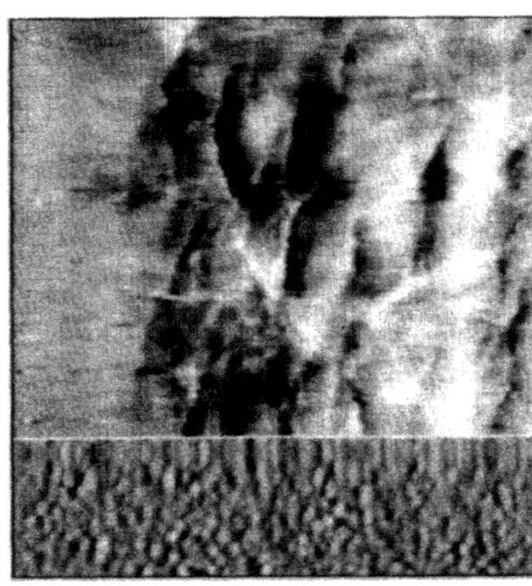

Fig. 3. 1 µm x 1 µm scan of a dye coated silver surface (Pedarnig et al. [20]); (upper part) image taken in the SPNM mode showing clustered dye molecules; (lower part) image taken in the conventional STM mode showing the underlying granular structure of the silver substrate.

In a further experiment the image contrast in SPNM is shown to depend on the oxidation state of the adsorbed chemical species under investigation. A dye film of "stains all" (from Sigma) has been prepared on a silver substrate by first coating the silver film on a glass slide with a bilayer film of Cd-Arachidate by the Langmuir-Blodgett technique. Subsequently two mixed films of arachidic acid and stains all (1:1 in molarity) are transferred onto the Cd- Arachidate films. The first bilayer acts as a spacer in order to avoid quenching of the dye molecules due to the otherwise close proximity to the silver surface.

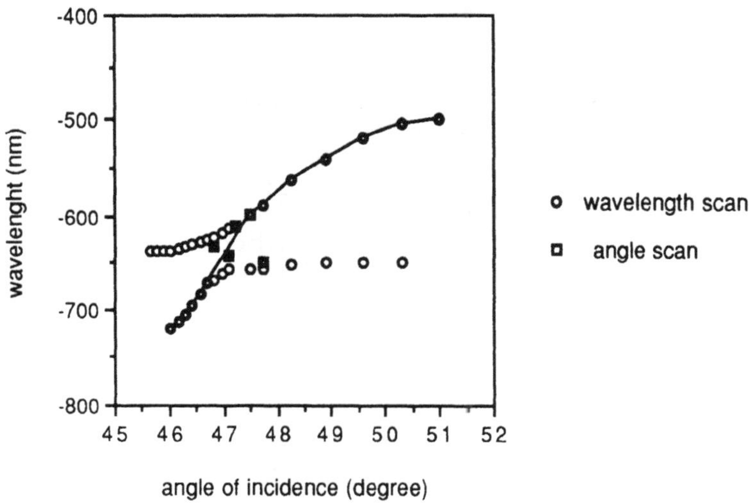


Fig. 4. Dispersion relation: ATR - minimum for fixed angle and variable wavelength (wavelength scan) and fixed wavelength and variable angle (angle scan), respectively.

Figure 4 shows the dispersion relation for this system. The minima of the reflected light intensity have been measured by scanning the angle of incidence at constant wavelength or the wavelength at constant angle. Additionally to the normal plasmon behaviour (solid curve) the horizontal branch corresponds to the absorption of the coupled dye - surface plasmon excitation. The measurements in figure 5 are made in the overlap region where the repulsion of the two curves and the typical backbending in the angle scan is observed [19,21,22].

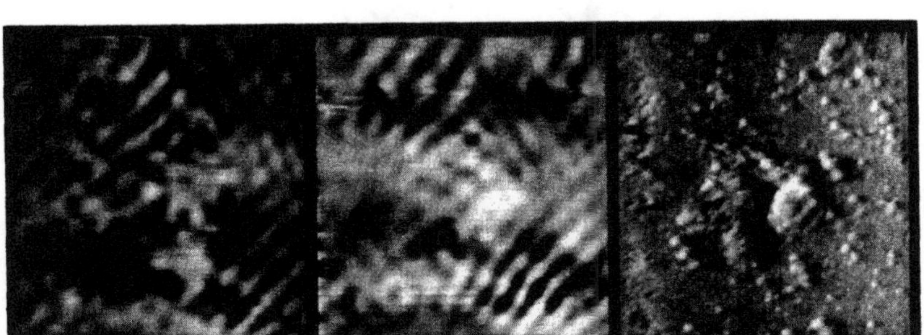

Fig. 5. Photooxidation: 3 μm x 3 μm scan of "stains all" dye in nitrogen atmosphere (left), after purging with oxygen (middle) and STM image for comparison (right).

Figure 5 shows three images of this sample at an identical location. The left image is recorded in the SPNM mode under nitrogen atmosphere thus avoiding oxidation. In the middle one the sample was purged with oxygen. We interpret the obvious change in the image contrast as being due to photooxidation of the dye molecules. This oxidation occured within one minute and was irreversibel. The right image shows the STM picture for comparison, again revealing the structrure of the underlying silver coated substrate. Prominent features like the big silver cluster in the center are visible in all three images. Interference strucutres in the two SPNM images originate in the superposition of the incident surface plasmons with the ones scattered by surface defects. After photooxidation coupled dye-surface plasmon excitations dissapear. The change in image contrast therefore demonstrates the sensitivity of SPNM to spectroscopic characteristics of the sample.

## IV Scanning Near-Field Ellipsometric Microscope SNEM

We introduce a new optical near-field microscope, the scanning ellipsometric near-field microscope (SNEM) which is based on the scattering of a small perturbation in the evanescent field of a sample surface.

A laser beam is totally reflected on a sample surface generating an evanescent field which is locally perturbed by a sharp tip (fig. 6). As in SPNM an optical microscope and an aperture are used to magnify and preselect the relevant areas of the surface. An ellipsometer is used to cancel out the unperturbed background intensity. The perturbation leads to an local increase in the reflected intensity which is recorded with a photodiode.

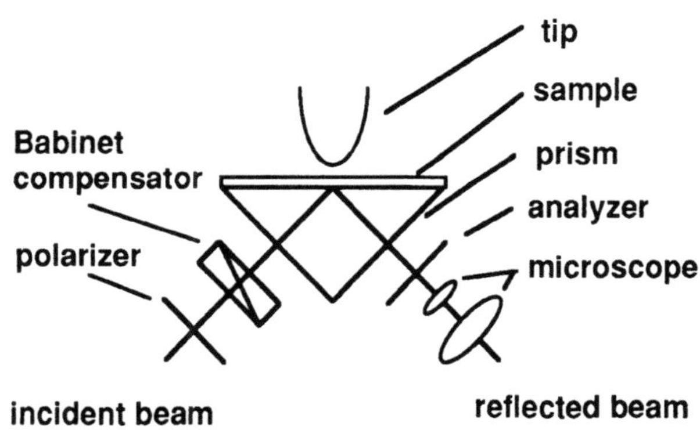


Fig. 6. Schematic of the experiment : in contrast to the setup of figure 1 an ellipsometer is used to cancel out the unperturbed specular reflected light and no surface plasmons are excited.

The upper part of figure 7 shows an Indium Tin Oxide (ITO) surface (500 nm x 500 nm) which has been imaged with the SNEM. The optical near-field image shows the corrugated surface due to the ITO microcrystallites. The lower part of figure 7 is a conventional STM image.The superresolution capability of SNEM can be understood in terms of near-field interactions between the scattering tip and the sample leading to a locally varying scattering cross section.

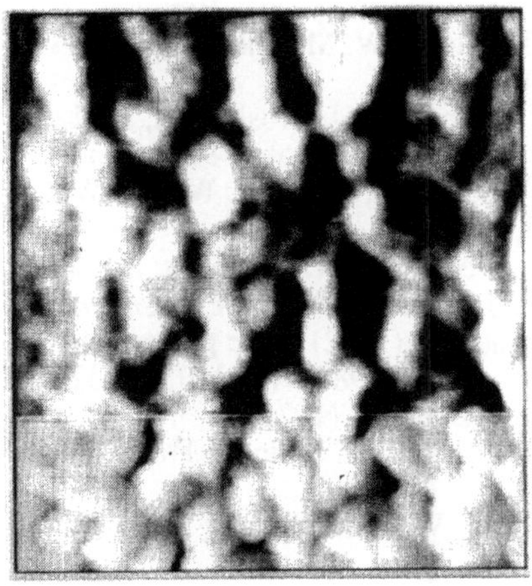

Fig. 7. Image of Indium Tin Oxide (500 nm x 500 nm): (upper part) SNEM image taken in the STM-SNEM hybrid mode; (lower part) STM image.

Compared to SPNM where metal films are used to couple surface plasmons into interesting samples like molecular dye layers SNEM may be applied to any transparent substrate because it does not use surface plasmons.

## Summary

We have undertaken first steps to demonstrate the use of SPNM and SNEM as a near-field optical microscope revealing topographic as well as spectroscopic information on samples coated with organic molecules. We hope that improvements in tip geometry as well as sample preparation together with a better theoretical understanding will lead to the ultimate target of molecular spectroscopic resolution.

## References


[1] Pohl, D. W, in Advances in Optical and Electron Microscopy (1990), Sheppard, C. J. R. and Mulvey, T. , Eds. (Academic Press, London) pp. 243-312.

[2] Betzig E. and Trautmann, J. K. (1992), Science, 257, 189

[3] Synge E. H.(1928), Philos. Mag. 6, 356.

[4] Ash, E. A. and Nicholls, G. (1972), Nature 237, 510

[5] Lewis, A. et al. (1984), Ultramicroscopy 13, 227.

[6] Pohl D. W., Denk W., and Lanz M. (1984), Appl. Phys. Lett. 44, 651.

[7] Reddick R. C. ,Warmack R. J. and Ferrel T. L.(1989), Phys. Rev. B 39, 767.

[8] Fee, M., Chu, S.and Hänsch, T. W. (1989), Opt. Comm. 69, 219.

[9] Courjon D., Vigoureux, J.-M., Spajer, M., Sarayeddine, K. and Leblanc, S. (1990), Appl Opt. 29, 3734

[10] Liebermann, K., Harush, S., Lewis, A., R. Kopelman (1990), Science 247, 59.

[11] van Hulst, N. F., Segerink, F. B.(1992), European Microscopy and Analysis 13.

[12] Fischer U. Ch. and Pohl D. W. (1989), Phys. Rev. Lett. 62, 458
[13] Specht M., Pedarnig J. D., Heckl W. M., and Hänsch T. W. (1992), Phys. Rev. Lett. 68, 476.
[14] Möller, R., Albrecht, U., Boneberg, J., Koslowski, B., Leiderer, P. and Dransfeld, K. (1991), J. Vac. Sci. Technol. B 9, 506 and Kroo, N., Thost, J.-P., Völcker, M., Krieger, W. and Walther, H. (1991), Europhys. Lett. 15, 289.
[15] Marti, O., Bielefeldt, H., Hecht, B., Herminghaus, S., Leiderer, P. and Mlynek, J. : *Near Field Optical Measurement of the Surface Plasmon Field* (in press)
[16] Schollwöck, U. and Wagner, H. (1992) these proceedings.
[17] Rothenhäusler, B. and W. Knoll (1988), W., Nature 332,615.
[18] Kretschmann, E., Z. Phys. 241, 313 (1971).
[19] Raether H. (1988), Surface Plasmons, Springer Tracts in Modern Physics Vol. 111, Springer, Berlin
[20] Pedarnig, J. D., Specht, M., W. M. Heckl, and T. W. Hänsch (1992), Applied Physics A, 55, 476.
[21] Wähling, G., Möbius, D. and Räther, H. (1978), Z. Naturforsch. Teil A, 33, 907
[22] Pockrand, I. and Swalen, J. D. (1978), J. Opt. Soc. Amer. 68, 1147.

# DIRECT MEASUREMENT OF THE FIELD ENHANCEMENT CAUSED BY SURFACE PLASMONS WITH THE SCANNING TUNNELING OPTICAL MICROSCOPE

H.Bielefeldt, B.Hecht, S.Herminghaus, J.Mlynek, and O.Marti
Faculty of Physics
University of Konstanz
P.O.Box 5560
D-W-7750 Konstanz
Germany

ABSTRACT. A scanning tunneling optical microscope (STOM) is used to directly determine the electromagnetic field amplitude of optically excited surface plasma waves on a silver film. The field amplitudes are measured as a function of the angle of incidence of the exciting light beam. The excitation of surface plasmons on the silver film causes a characteristic field enhancement compared to evanescent waves on uncoated surfaces. The measured values for the resonance angle, resonance width and the field enhancement ratio at various wavelengths are compared with theory and with simultaneous measurements of attenuated total internal reflection (ATR). The sensitivity of near-field optical methods to the film thickness is comparable to that of the ATR method. However, while ATR measurements are limited by diffraction, the lateral resolution of STOM can be of the order of 50 nm.

## 1. Introduction

The excitation of surface plasmons is a powerful tool for the study of thin metal films [1,2]. This method exhibits an extremely high sensitivity to surface roughness or adsorbate coverage on the metal surface [3]. Usually surface plasmons are generated by coupling to the evanescent wave that arises from total internal reflection of a light beam at a glass surface. By changing the angle of incidence, the wave vector of the evanescent wave can be varied. So the resonance condition for the excitation can be fulfilled: a part of the energy of the illuminating beam is transferred into the surface plasmon excitation. Therefore, in the totally reflected beam an attenuation can be observed.

So far, the plasmons usually are observed indirectly by measuring this attenuation of total internal reflection (ATR). In this contribution we report on a different approach using a Scanning Tunneling Optical Microscope (STOM) [4,5] and measure their evanescent amplitude directly.



*D.W. Pohl and D. Courjon (eds.), Near Field Optics*, 281–286.

## 2. Experiment

The experimental arrangement is shown in Fig. 1: Surface plasmons are excited optically on a silver film surface in a Kretschmann geometry. The exponentially decaying light field in the air halfspace is probed by a fiber tip. Simultaneously, the intensity of the internally reflected light can be measured separately in order to obtain an independent determination of the surface plasmon resonance angle. The upper part of the apparatus in fig. 1 is essentially a STOM.

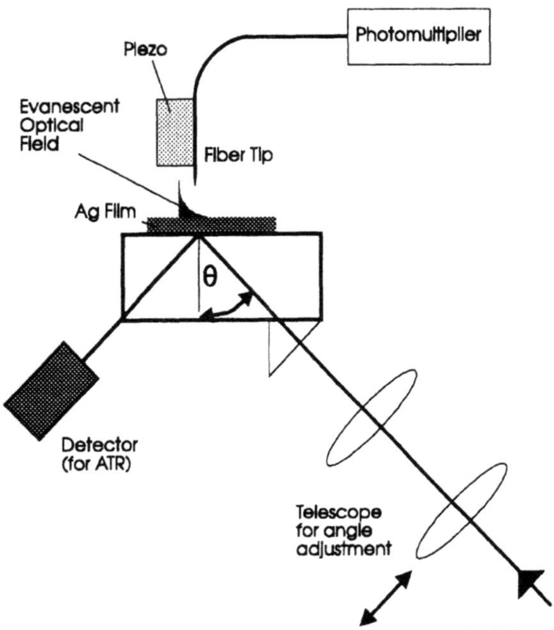


Figure 1. Experimental arrangement to measure the field enhancement. Surface plasmons are excited by total internal reflection in the Kretschmann geometry and detected from above by a STOM.

The fiber tips are prepared by heating and pulling a monomode fiber designed for a wavelength of 633 nm. The typical radius of curvature of the tips measured by scanning electron microscopy is about 1 $\mu$m.

The angle of incidence of the laser beam can be adjusted without changing the position of the focal spot with respect to the tip by means of a telescope arrangement. The angle scale is calibrated with reference to the onset of total internal reflection. The diameter of the laser focal spot is ≈ 0.3 mm. The light intensity extracted by the fiber and detected by a photomultiplier tube was in the range of several nW, while the sample was illuminated with 3 to 17 mW.

The magnitude of the evanescent field was measured as follows: as the fiber tip is approached to the sample, the intensity shows an exponential increase with decreasing separation as can be seen in Fig. 2. On a bare glass plate the exponential decay length is given by the index of refraction, the wavelength and the angle of incidence $\Theta$ [4,5,6]. When the tip reaches the extreme near-field of the sample surface the signal saturates and exhibits a complicated and not very reproducible behavior. The intensity at the onset of saturation ($I_0$) is a direct measure of the magnitude of the evanescent light field.

Because the fiber tip fabrication and therefore the tip shapes and characteristics are not very reproducible, the ratio of the evanescent field coupled into the tip is not known. Hence, the absolute value of $I_0$ has no direct meaning. A quantitative measure can be gained by comparing the intensity of the plasmon field with the intensity coupled from the evanescent field on a bare glass substrate under the same illumination conditions, i.e. identical tip and beam geometry. Practically, only a part of the microscope slide is covered by the silver film and the rest remains uncovered. By a lateral displacement of the slide which is in optical contact to the prism by an index matching fluid, an area with or without silver film can be chosen without changing the position of the laser spot under the tip.

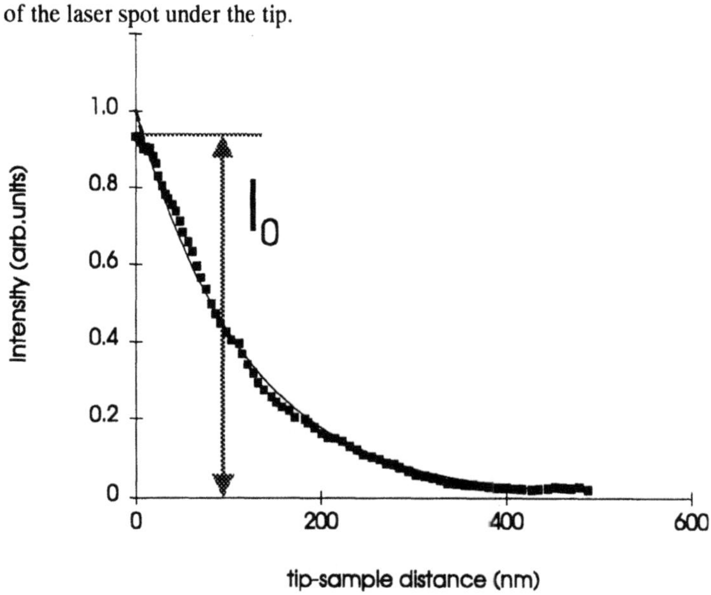


Figure 2. Typical behavior of the evanescent field above a glass plate, measured by varying the tip-sample distance z and monitoring the fiber output. The decay length of the exponential fit (solid line) is 116 nm, while the theoretically expected value is 138 nm for the given geometry.

## Results and Discussion

Three different laser sources are used to measure plasmon resonance curves: a 514 nm $Ar^+$ laser, a 633 nm HeNe laser, and a 670 nm laser diode. The power is kept below 17 mW to avoid damage to the silver film. The resonance angles were in agreement with simultaneous ATR measurements.

Fig. 3 shows a typical result: The intensity was calculated numerically using the transfer matrix representation of Fresnel's formulae [6,7] (see full lines). The data for the optical properties of silver were taken from Ref. [8]. The experimental data points are scaled with respect to the calculations so that the measured evanescent field on the glass at the resonance angle has the calculated intensity. The angle scale was adjusted with the onset of total internal reflection;

therefore the height, width and position of the resonance are no adjustable parameters. The noise level of the photomultiplier signal in Fig. 3 is comparable to the intensity of the evanescent field ("triangles").

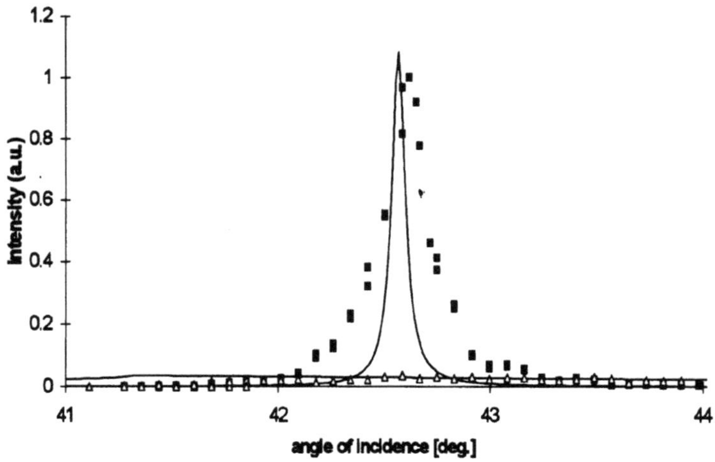


Figure 3. Detected intensity of the evanescent field at the wavelength 670 nm. The "squares" correspond to the field on the silver film (plasmon field), while the "triangles" are the measurements on the bare glass substrate. The solid lines represent the numerically calculated intensities. Theory and experimental data are scaled such that the evanescent waves on bare glass have the same intensity.

We define the field enhancement ratio F as the ratio between the maximum intensity of the evanescent field with plasmon excitation and the field intensity of the evanescent wave on the bare glass surface:

$$F = \frac{|\textit{field with metal film at resonance}|^2}{|\textit{field without metal film}|^2}$$

Table 1 shows the measured and calculated field enhancements F as well as the resonance angles θ and the widths Δθ of the resonances.

| wavelength (nm) | 514 | 633 | 670 |
|---|---|---|---|
| $F_m$ (measured) | 17.0±2.3 | 12.8±0.9 | 34.2±9.3 |
| $F_c$ (calculated) | 22.1 | 17.0 | 38.6 |
| $F_m/F_c$ | 0.77 | 0.75 | 0.88 |
| $\theta_m$ (measured) | 43.90° | 42.89° | 42.67° |
| $\theta_c$ (calculated) | 43.89° | 42.83° | 42.57° |
| $\Delta\theta$ (measured) | 0.45° | 0.21° | 0.25° |
| $\Delta\theta$ (calculated) | 0,23° | 0,15° | 0,075° |
| film thickness used in calculation (nm) | 70 | 70 | 70 |

**TABLE 1. Comparison of measured and calculated values of the plasmon field enhancement factor, resonance width and angle for different laser wavelengths. For details see the text.**

The angle measurement is accurate to about 0.15°. The good agreement of the measured resonance angles is obvious. The experimental values of the resonance width are 2 to 3 times larger than the calculated values, while the enhancement factors are only 70 to 80 % of the expected values.

Several explanations can be found for the broadening of the resonances and the smaller enhancement values: It is well known that resonance width and field enhancement are sensitive to the film thickness and quality. Besides thickness inhomogenities, aging processes in the silver film (structural changes and chemical reactions with the air) seem to play an important role. The films seem to undergo a rapid change within the first hours after preparation. Within some days, the field enhancement factors decrease to about 20 % of the theoretically predicted values.

The beam divergence that varies for the different laser types used was not taken into account in the calculations and therefore is another reason for the broader resonances. The calculations were performed assuming an ideal 2-dimensional system without regarding the influence of the fiber tip. The tip causes a local damping of the plasmon, as it couples out a non-negligible amount of light from the evanescent field: For the incident intensity of $\approx$100 nW/$\mu m^2$ and an assumed probe area of 1 $\mu m^2$, the detected intensity of 1 nW at the end of the fiber corresponds to 1 % of the intensity in the probe volume. This local damping is not seen in the ATR measurement, as the latter integrates over the total spot size.

## Conclusions

The measurements performed with the STOM are in reasonable agreement with the theoretical calculations as well as with the measurements of ATR. While experiments based on the interaction of an STM tip with surface plasmons [9,10] make use of the scattering of the plasmons caused by the tunneling tip or the rectification of the tunneling current, the STOM directly measures the local optical plasmon field on the surface with a detector placed in the near field region.

The spatial resolution of STOM is roughly given by the fiber tip diameter [4,6]. The technique presented here could be used to extract information about the propagation of the surface plasmons on the film surface with high lateral resolution, especially in combination with topographic information. The instrument can be expanded to a combination of STOM and reflection SNOM [11] to get an independent measure of topography: While the reflection SNOM is used for topography mapping, the plasmon excitation can be investigated simultaneously using a second laser source with different wavelength. Such local measurements might be useful as a test for theories of plasmon propagation on rough surfaces [12]. The high adsorbate sensitivity of the plasmon excitation could be used for locally resolved studies of e.g. dielectric materials brought onto the silver film. Moreover, the light extracted from the plasmon field could be analyzed spectroscopically.

The combined technique has a number of possible applications including, for example, the characterization of mirrors for neutral atoms based on evanescent fields [13,14]. Atoms interacting with the strong field gradient of the plasmon field are deflected by the induced dipole force. Since the atoms will approach the plasmon carrying surface to a distance less than an optical wavelength, the near field structure of the plasmon field will become important for the performance of the mirror. For instance a metal surface with nanometer-sized corrugation has a considerable variation in the plasmon near field and might produce some unwanted scattering of neutral atoms.

## Acknowledgements

The authors gratefully acknowledge discussions with V.Balykin and J.Mertz. S.Hahn and S.Eggert helped building the apparatus. The fiber tips were imaged in the SEM by J.Künzel. This work was supported by the Land Baden-Württemberg and by the Deutsche Forschungsgemeinschaft (SFB 306).

## References

[1] C.F.Eagen, W.H.Weber, Phys.Rev. B19 (1979) 5068
[2] S.Herminghaus, P.Leiderer, Appl.Phys.Lett. 58 (1991) 352
[3] H.Raether, "Surface Plasmons", Springer Tracts in Modern Physics 111, Springer Verlag Berlin,Heidelberg 1988
[4] D.Courjon, K.Sarayeddine, M.Spajer, Opt.Comm. 71 (1989) 23
[5] R.C.Reddick, R.J.Warmack, T.L.Ferrell, Phys.Rev.B 39 (1989) 767
[6] M.Born, E.Wolf, "Principles of Optics", Pergamon, Oxford 1970
[7] Z.Knittl, "Optics of Thin Films", Wiley, New York 1976
[8] P.B.Johnson, R.W.Christy, Phys.Rev.B 6 (1972) 4370
[9] R.Möller, U.Albrecht, J.Boneberg, B.Koslowski, P.Leiderer, and K.Dransfeld, J.Vac.Sci.Technol. B9 (1991) 506
[10] M.Specht, J.D.Pedarnig, W.M.Heckl, T.W.Hänsch, Phys.Rev.Lett. 68 (1992) 476
[11] M.Spajer, D.Courjon, K.Sarayeddine, A.Jalocha, J.-M.Vigoureux, J.Phys.III 1 (1991) 1
[12] A.R.McGurn, A.A.Maradudin, V.Celli, Phys.Rev. B31 (1985) 4866
[13] V.I.Balykin, V.S.Letokhov, Yu.B.Ovchinnikov, A.I.Sidorov, Phys.Rev.Lett. 21 (1988) 2137
[14] M.A.Kasevich, D.S.Weiss, S.Chu, Opt.Lett. 15 (1990) 607

# SENSING TIP MORPHOLOGY IN SCANNING OPTICAL MICROSCOPES

‡*M.A. PAESLER, †*E.L. BUCKLAND, ‡*P.J. MOYER, and
*B. I. YAKOBSON
**Department of Physics and ‡Precision Engineering Center,*
*North Carolina State University,*
*Raleigh, NC 27695-8202*
*USA*
*†Sumitomo Electric Fiber Optics Corporation,*
*Research Triangle Park, NC 27709*
*USA*

ABSTRACT. The lateral spatial resolution of a collection-mode scanning optical microscope is discussed in terms of the electromagnetic mode solutions of the probe tip. Analysis of glass rheology during tip fabrication suggests a region in experimental parameter space which optimizes tip structure.

## 1. Introduction - the Importance of Tip Morphology

Tip morphology in scanning probe microscopes determines the nature of signal transduction and instrument resolution. An understanding of this mechanism is crucial to tip-based advances in design and in image interpretation. Tip morphology and how to control it provide the foci of this paper. Though distinct, they are related in that optimization of instrument design can only proceed with knowledge of how the tip transduces the optical signal and how one can control the structure of the tip to optimize this process.

## 2. Analysis of Signal Transduction

### 2.1. MODE STRUCTURE OF THE DIELECTRIC PROBE

Currently, it is generally assumed that the probe does not substantially perturb the field around the sample. Given this, the mode structure forms a complete orthogonal set and the total field is the sum of fields over a finite set of bound or propagating radiation modes, plus an integral over an infinite set of radiation modes[1]. For a step index fiber with fixed core radius, $\rho$, and infinite cladding, the normalized frequency parameter V at a wavelength $\lambda$ is given by: $V = (2\pi\rho/\lambda)A_n$ where $A_n$ is the numerical aperture of the fiber[1]. The waveguide will support multiple propagating modes for $V > 2.405$. For all imaginable circumstances for NSOM (near-field scanning optical microscopy) collection mode or PSTM (photon scanning tunneling microscopy) configurations, $V < 2.405$. Thus waveguide analysis must be concerned with the fundamental $HE_{11}$ mode only, and for a fiber with non-absorbing cladding, the $HE_{11}$ mode has no cutoff no matter how small the tip radius. But as the tip size is decreased, the fraction of guided power which is contained in the core or within the bounds of the probe tip, decreases monotonically with normalized tip radius, $\rho/\lambda$.



*D.W. Pohl and D. Courjon (eds.), Near Field Optics, 287–294.*

## 2.2 SIGNAL TRANSDUCTION THROUGH FIELD COUPLING

Given the singular mode structure of the tip, one must next address signal transduction through field coupling. The simplest signal that can be sensed would be a delta function. If the source is such a delta function defined at the plane of the probe tip, z = 0, with polarization in the (x-y) plane of the sample (with unit vectors **i** and **j**), then the electric field associated with this source is

$$\mathbf{E}_s = E_{xy}\delta(x-x_o, y - y_o)(\mathbf{i} + \mathbf{j}).$$

This expression which represents the total field at the probe tip, may be expanded in terms of the orthogonal fiber waveguide modes. In such an expansion, the first term is a sum over bound modes with amplitudes $a_i$, and these are the only modes which can propagate to the detector. As the tip is sufficiently small to support only the $HE_{11}$ bound mode, for the delta function source, the only non-zero amplitude is

$$a_i = \left[\int_\infty \delta(x-x_o, y-y_o)(\mathbf{i} + \mathbf{j}) \cdot \mathbf{E}_F\, dA\right]^2 / N_F$$

$$a_i = \frac{E_F^2(x_o,y_o)}{N_F}$$

where

$$N_F = \int_\infty (\mathbf{E}_F \times \mathbf{H}_F^*) \cdot \mathbf{k}\, dA ,$$

and the subscript F on the electric and magnetic fields refers to the fields associated with the fiber.

As the delta function is scanned by the probe, an image of the field intensity distribution is obtained. The maximum amplitude results when the source $\mathbf{E}_S$ is on axis with the tip. This amplitude, $a_i$, varies quite strongly with index difference $\Delta = (n_{core}^2 - n_{cladding}^2)/2n_{core}^2$ and with normalized core radius $\rho/\lambda$ as shown in figure 1. In this figure it is clear that for very small tips and all $\Delta$, the peak intensity of the mode decreases sharply with decreasing normalized core radius, $\rho/\lambda$, as the fraction of the power in the tip decreases and the field width increases. For unclad glass fibers $\Delta = 0.3$.

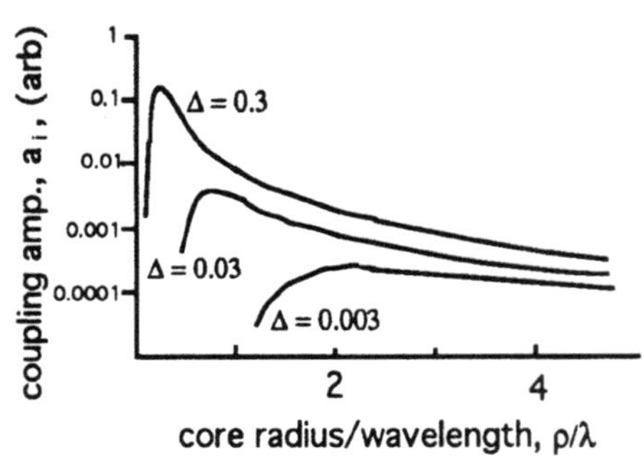


Figure 1. Amplitude for coupling, $a_i$, between a delta-function source and the $HE_{11}$ mode of the probe tip.

## 2.3 RESOLUTION

We consider resolution in terms of the signal-to-noise ratio between maxima and minima of the intensity distribution in the scan of a closely spaced regular two-dimensional square delta function array. A peak should be detected as the tip passes over an array point, and a minimum should result between points. As the nearest neighbor spacing is reduced, the signal intensity at the midpoint increases, and the image becomes blurred along the nearest neighbor directions. The image of the two-dimensional array takes on the form of parallel lines. (This definition of resolution is a two-dimensional analog of the Sparrow criterion.) With this definition, we show in figure 2 the resolution for a silica fiber with index differences $\Delta$. We presume in this figure - as is usually the case - that these unclad fibers are sufficiently small that only the lowest mode, $HE_{11}$, exists.

Figure 2. The minimal achievable resolution from a presumed regular 2-dimensional square delta-function array into the $HE_{11}$ mode as a function of normalized tip radius for three different index differences, $\Delta$.

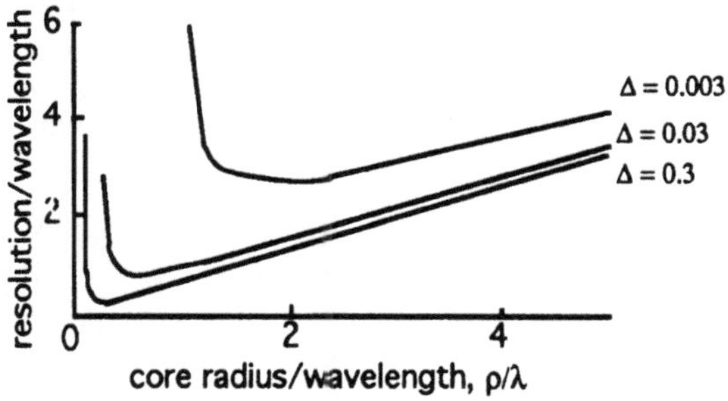


From figure 2, if the signal transduction mechanism involves modal coupling into a small ($\rho/\lambda < 1$) tip, there is a fundamental limit on instrument resolution that is - even in the best of cases - only slightly improved over the Abbe barrier. Without commenting on the resolution claims for any particular microscope or configuration, insofar as it can be demonstrated that a microscope provides spatial resolution finer than the Abbe barrier, modal coupling does not provide the transduction mechanism for such a microscope.

## 2.4 METAL CLAD FIBER

A metal cladding applied to the fiber alters the waveguide characteristics and the field coupling mechanism. The lowest bound mode is $TE_{11}$, and for $\rho/\lambda \leq 1.841/(2\pi n)$ the mode is cutoff. This limit always obtains for tips of interest, thus in terms of direct modal coupling, metal cladding is of little use. (This ignores the effect of the finite conductivity of the cladding, which would allow for modal coupling into somewhat smaller apertures.) The distinct cutoff of bound modes is advantageous, however, in reducing competition between these mechanisms and secondary coupling through scattering, discussed in § 2.5.

## 2.5 SCATTERING MECHANISMS

Coupling into local bound modes represents coherent oscillations of the molecular structure of the probe tip. In addition to these excitations, there are independent excitations or scattering events which can lead to secondary coupling. Rayleigh scattering in silica fibers has a characteristic $\cos^2\vartheta$ angular dependence and $\lambda^{-4}$ wavelength dependence. To determine the resolution and power coupled through scattering, one must first determine the fraction of power scattered and then the fraction of this power which subsequently couples into waveguide bound modes, i.e. the capture fraction.

2.5.1 *Power Coupled through Scattering.* If one neglects ambient scattering and edge effects, one must nevertheless make assumptions concerning the scattering volume involved in signal transduction. Any of a number of reasonable choices for the scattering volume result in the same conclusion[2]: with scattering as the principal signal transduction mechanism, resolution improves with decreasing tip size, surpassing the limits demonstrated above for $HE_{11}$ modal coupling.

2.5.2 *The Capture Fraction.* The capture fraction is a function of the internal critical angle. The limiting cases to consider are i) a propagating incident wave in the near field (e.g. normal transmission NSOM), and ii) a purely evanescent field over the sample surface (PSTM). Perhaps more simply put, the scattering wave vector may be resolved into components either parallel or perpendicular to the plane of the sample. A simple calculation shows that the capture fraction for these two extreme cases never differs by more than a factor of two[2]. The taper of the tip also affects the capture fraction since coupling out of tightly bound lower order modes is minimized in a slowly varying taper. Of course the mechanical disadvantage of a taper that is too gentle must be considered.

Using reasonable assumptions about fiber index and tip radius and taper the relative role of scattering and modal coupling for tips typically in use[2] is shown in figure 3. As the tip continues to decrease in size, scattering eventually dominates the detected signal, though the intensity may be as many as nine orders of magnitude below the peak obtainable through bound-mode coupling. (The bound mode intensities here are those calculated using the coupling amplitude calculations outline in section 2.2.) If dielectric scattering were to play a role in signal transduction, resolution would once again be a decreasing function of tip size below the temperature where scattering dominates. In figure 3 for $\Delta = 0.3$, this occurs for tip radii below 30 nm at wavelength $\lambda = 600$ nm. The likelihood of detecting a scattering signal at an intensity this low is, however, quite small. Thus, one may conclude that for metal clad fibers, dielectric scattering in the tip does not provide the signal transduction mechanism.

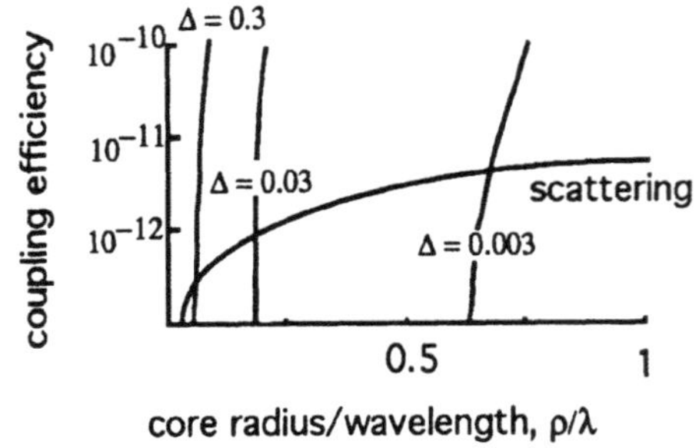


Figure 3. Comparison of transduction efficiencies for $HE_{11}$ bound mode coupling and scattering mode coupling. The $TE_{11}$ mode of a metal clad fiber cuts off at a normalized radius of 0.2

These analyses lead us to the conclusions that for metal clad tips pulled to a tip radius on the order of a few tens of nanometers or less, neither modal coupling nor dielectric scattering provides the means to detect the high-resolution signal observed. If modal coupling were operative, the *resolution* would not be nearly as good as it is. If dielectric scattering were the signal transduction mechanism, the *intensity* would be drastically less than what is observed.

Our discussion to this point has ignored the effects of the aperture in the aluminum coating that exists at the tip. Indeed, we neglected previously (§2.5.1) edge scattering from this structure. Although we make no quantitative analysis here of the magnitude of this effect, others [3] have studied the resolution which might be expected from an aperture. A treatment of the magnitude of the signal obtained from an aperture awaits further study.

## 3. Tip Fabrication

### 3.1 THE TWO STEP TIP FORMATION PROCESS

Given the importance of tip size and morphology, it is reasonable to closely examine current sensor tip fabrication techniques. These typically involve an adaptation of a heating-and-pulling method developed in the pipette industry, which has accumulated considerable experience[4] and obtained results satisfactory for quite different needs. The formation process which determines the final tip shape and sharpness includes two stages: i) the pulling and rupture of a fiber heated to some elevated temperature $T_h$ and ii) subsequent cooling and relaxation until viscous flow is negligible.

3.1.1. *Pulling and Rupture.* The initial radius at the moment of rupture, $r_i$, can be estimated by considering it as a result of two competing phenomena: a) a uniform thinning caused by the externally applied tension p; and b) an intrinsic capillary instability of a cylinder due to the surface tension $\gamma$ which tends to neck off the fiber into separate parts with smaller total surface area. Competition between these effects can be described in terms of their characteristic times.

The elongation of a fiber segment under external stretching force results in its thinning. The instantaneous rate of the radius decrease should be proportional to the pulling pressure p and inversely proportional to the glass viscosity $\upsilon$. Thus,

$$(1/r)\cdot dr/dt = - p/\upsilon ,$$

and a characteristic time of thinning $\tau_t = \upsilon/p$ can be introduced. In the absence of other phenomena leading to rupture and tip formation one could expect unlimited thinning up to a cylinder of atomic diameter. The cylinder, however, is not a minimal free energy shape for a given quantity of material. In the case of interest, when surface tension and viscosity are paramount, any surface perturbation with a wavelength greater than $2\pi r$ will grow. There is no finite mode of maximum instability, which means that a fiber (even homogeneously heated) would give way at a few distant places rather than break up into a sequence of drops. The characteristic time associated with this capillary instability [5] is given by $\tau_c = (\upsilon/\gamma)\,r$.

Fiber thinning continues homogeneously until the interference of the instability is negligible, that is, until the thinning characteristic time, $\tau_t$, is shorter than that of capillary necking, $\tau_c$. The path representing the actual pulling might follow the curved line on figure 4, which depicts characteristic thinning time versus the instantaneous fiber radius, provided that tensile force p is defined for every r for some regime of pulling. The initial tip will be formed when this path intersects the dashed capillary "pinch-off" line ($\tau_c$ versus r). An important restriction related to glass rheology should be kept in mind: the tension (or, equivalently, the strain rate) must be lower than certain threshold value $p_o$ to avoid brittle fracture of the fiber. This condition holds above the horizontal dashed-dotted line representing $\tau_t = \upsilon/p_o$ on figure 4. An optimal (i.e. smallest $r_i$) initial tip is formed when the curved line meets $\tau_c$ at the intersection of the line segments representing $\tau_c$ and $\tau_t$. The initial tip radius, which is given by $r_i \approx \gamma / p$, is generally well above its lower bound $\gamma / p_o$. For a glass surface energy of 1 J/m$^2$ and strength of 1GPa one obtains a 1 nm radius limit. Thus, at this first stage an ultimately atomically sharp tip may be possible provided a low surface tension and maximum fiber strength during the pulling are attained.

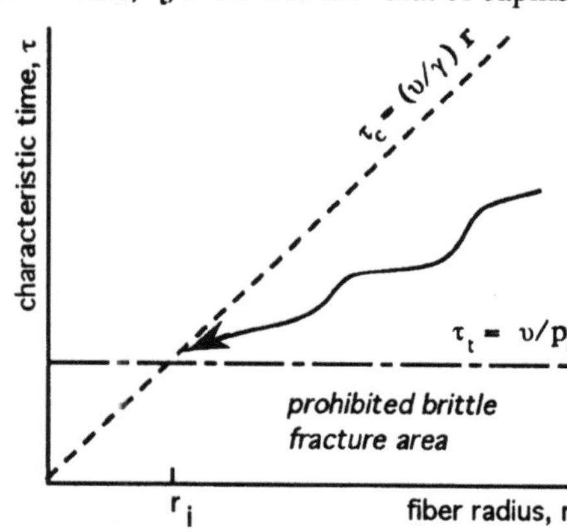


Figure 4. Characteristic times vs. radius for two competing effects in fiber pulling. Pulling path is the curved line.

3.1.2. *Cooling and Relaxation.* The substantial Laplace pressure $\gamma/r$ combined with the low viscosity of the still-hot glass, will result in tip blunting after the formation of the initial radius $r_i$. Very roughly, the geometry of the tip can be characterized by one variable: tip curvature radius $r \geq r_i$ at every moment. The rate of the relative increase of this radius must be proportional to the "driving force" of the blunting (the Laplace pressure) and inversely proportional to the viscosity, $(1/r) \cdot dr/dt = (\gamma/r)/\upsilon$. So, the simple relation follows,

$$dr/dt = \gamma / \upsilon \qquad (1)$$

This process practically stops as soon as the tip cools down to the glass transition temperature $T_g$ and viscosity increases tremendously. Elementary estimates show that among the three channels of the heat dissipation thermal radiation is greater than convective heat transfer to the air and much larger than heat conduction through the glass itself. This means that, with the heater (typically a laser beam) off, the temperature of the tip follows the equation,

$$(dT/dt)\, c\, r^3 = \sigma T^4 r^2 \qquad (2)$$

where c is the specific heat (per unit volume), $\sigma$ is the Stefan-Boltzmann constant, and the backward radiation flux from the environment is neglected. (Here $T_o$ is the ambient temperature and $T_h$ is the temperature to which the tip is heated during pulling.) Assuming that the heat source is switched off at the very moment of fiber rupture, one can eliminate time from the eqs. (1) and (2) and obtain after integration,

$$\ln r_f = \ln r_i + \int_{T_o}^{T_h} \frac{c\,\gamma}{\sigma T^4 \upsilon(T)}\, dT .$$

In this integrand, $\upsilon$ has by far the strongest temperature dependence, so one can pull all other terms out of the integral and presume an empirical extrapolation for the viscosity[6] $\upsilon = \upsilon(T) \approx \upsilon_o \exp(A/T)$. Here $\upsilon_o$ and A are material-dependent parameters which describe the $\upsilon(T)$ behavior in the region of interest - i.e. near the softening point of the glass. This approximation allows one to proceed with integration [7] in the second term, representing tip relaxation, to obtain:

$$\ln r_f = \ln r_i + (\gamma c / \sigma) \frac{1}{\upsilon_o A^3} \left\{ [(A/T_h)^2 + 2(A/T_h) + 2] \exp(-A/T_h) \right\} . \qquad (3)$$

The expansion of the tip radius due to post-rupture cooling is represented by the second term on the right hand side of equation 3. It is instructive to express the above relation in the form:

$$\frac{r_f}{r_i} = \exp \left\{ [(\gamma c / \sigma) \frac{1}{\upsilon_o A^3}]\, \Theta(T_h) \right\} . \qquad (4)$$

From equation 4 it is clear that the amount of post-rupture relaxation is represented by the product of two terms in the argument of the exponential, one involving material parameters, and the other, $\Theta(T_h)$, incorporating the temperature dependence of thermal relaxation.

To evaluate the contribution of each of these terms to the final tip radius, we first assume that $\gamma \approx 1$ J/m$^2$, $c \approx 2 \cdot 10^6$ J/m$^3$/K (typical values for glasses) and $\upsilon_o = 0.63 \cdot 10^{-7}$ Poise, and $A = 0.63 \cdot 10^5$ K for $SiO_2$. With these values, the temperature independent factor is quite large. Indeed, we find

$$(\gamma c / \sigma) \frac{1}{\upsilon_0 A^3} = 10^7.$$

The temperature dependent term, $\Theta(T_h)$ is, however, a strongly varying function of the heating temperature which is in most applications very small. The temperature dependence of $\Theta(T_h)$ normalized to a temperature A is given in figure 5. It is clear from the figure that $\Theta$ is a particularly strong function of temperature near the glass transition temperature, $T_g$, of $SiO_2$. For $SiO_2$ at and below the "working point" the temperature relaxation factor is on the order of $10^{-10}$ to $10^{-12}$ so that the relaxation contribution to the final tip radius is on the order of a few percent at most. Only if the heating temperature, $T_h$, were extremely high - on the order of 3000K - would relaxation play a role in tip radius determination. In most conceivable laboratory situations in tip-pulling settings, such a high temperature would never be achieved. Furthermore, to keep the prefactor of this term small, minimization of the surface tension, $\gamma$, is desirable. The latter condition might be realized through introduction of an ambient other than air during the pulling process.

Our analysis suggests that post-rupture relaxation hardly ever comes into play in normal tip manufacture. It does not address, however, the observed shape of drawn tips. Normally the approximation that the shape is spherical at the very tip breaks down under close scrutiny. A flattening of the tip is often observed in the sharpest tips when examined in a high resolution electron microscope. The rather naive model presented here does not, however, suggest how such a shape comes about. Full explanation awaits a detailed examination of rupturing. Also ignored is the shape of the sensor past its tip. Mechanical and optical analysis of this shape leads to suggestions with regards to such items as ideal taper angle. These issues are not addressed here.

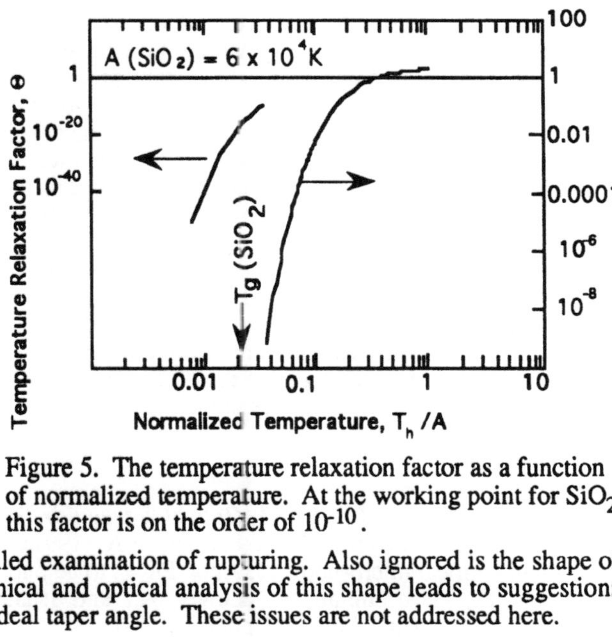


Figure 5. The temperature relaxation factor as a function of normalized temperature. At the working point for $SiO_2$, this factor is on the order of $10^{-10}$.

## 3.2. RECOMMENDATIONS FOR TIP MANUFACTURE

This analysis appeals to several approximations and simplifications but does identify the mechanisms involved in tip fabrication. It provides recommendations for optimal tip fabrication. In particular, the initial radius of curvature $r_i$ may be minimized by: i) employing maximum tension below the glass rupture threshold; and ii) decreasing the surface tension $\gamma$. Once rupture is complete, very little relaxation occurs during most pulling operations. If relaxation does come into play, the final radius $r_f$ may be minimized if the surface tension is as low as possible. Since both stages of fabrication may be optimized by decreased $\gamma$., introduction of an ambient may decrease $r_f$. For systems currently in use, the main conclusion of this study is the recommendation that a large pulling force should be employed. These results are consistent with what is observed in the laboratory[8], i.e. large pulling forces generally result in fine points, and the flattened ends often observed in pulled tips suggest minimal post-rupture relaxation.

## 4. Discussion and Conclusions

With respect to the *role* of tip morphology on signal transduction, we have demonstrated that the dielectric nature of the tip is merely incidental to signal transduction. If this transduction were related to the dielectric nature of the tip explicitly, then either modal coupling or scattering must provide the means for transferring image information from the sample to the detection apparatus. But if modal coupling were this means then the present analysis suggests that the lateral spatial resolution would not be nearly as fine as has been experimentally demonstrated. And if scattering were the means, the amplitude of the signal would be many of orders of magnitude smaller than what is observed. Studies of imaging should therefore center on the aperture that is held by the sharpened dielectric tip rather than the tip itself. With respect to imaging, the dielectric tip merely provides a convenient way to hold a small aperture near the sample.

Other theoretical works [9-11], which have different approaches to the problem of imaging with scanning microscopes, hold some promise for deepening our understanding of signal transduction. Unlike these studies, the present study has decidedly negative results. That is, we suggest what the transduction mechanism is not. We have, however, taken care to make contact with experiment. Thus - for example - the curves of figure 2 represent obtainable resolution through modal coupling for fibers generally in use. And the ordinate in figure 3 represents absolute coupling efficiency for actual microscope tips.

With respect to the *control* of tip morphology, we have considered only a simple analysis that treats the radius of a tip that is created when a fiber is locally heated and subjected to tensile stress. This analysis suggests that a larger pulling tension results in a smaller initial radius, and that during cooling the tip radius does not increase appreciably.

Research performed through grants from the Army Research Office (DAAL03-91-G-0053) and the industrial affiliates of the NCSU Precision Engineering Center. One of us (EB) expresses thanks to colleagues in the Lightwave Lab at Sumitomo.

## 5. References

1. Snyder, A., & Love, J.D. (1983) *Optical Waveguide Theory* (Chapman & Hall, NY)
2. Buckland, E.L., Moyer, P.J. and Paesler, M.A. (1992) "Resolution in collection-mode scanning optical microscopy", J. Appl. Phys., to be published, February 1993.
3. Betzig, E., Harootunian, A., Lewis, A., and Isaacson, M. (1986) "Near-field diffraction by a slit", Applied Optics 25, 1890.
4. Many NSOM laboratories adapt commercially available pipette-pullers (such as those available from Sutter Instruments) to make NSOM sensing elements.
5. Chandrasekhar, S., (1961) *Hydrodynamic Stability* (Oxford Press, NY), Chap. 12.
6. Doremus, R.H.(1973) *Glass Science* (Wiley, New York), Chaps. 6, 15.
7. Yakobson, B.I., Moyer, P.J. and Paesler, M.A., (1992) "Kinetic limits for sensing tip morphology in near-field scanning optical microscopes", J. Appl. Phys., to be published, March 1993.
8. Betzig, E. and Trautman, J.K., private communication.
9. Vigneron, J.P., Lambin, Ph., Lucas, A.A., Derycke, I., and Dereux,A., this volume.
10. Dereux, A., and Pohl, D.W., this volume.
11. Girard, C., and Bouju, X., this volume.

# FIBER TIPS FOR REFLECTION SCANNING NEAR-FIELD OPTICAL MICROSCOPY

P. TOMANEK
*Department of Physics*
*Faculty of Electroengineering*
*Technical University*
*Areal VUT-Kravi hora, 602 00 Brno*
*Czech Republic*

ABSTRACT. Nanometer sized tips of fiber probes for reflection scanning near-field optical microscopy were fabricated by selective chemical etching of telecommunications optical fibers. The manufacturing of reproducible probes was controlled by varying different etching parameters.

## 1. Introduction

A great number of research projects have been done in the domain of scanning near -field optical microscopy (SNOM) having submicron or even nanometer resolution [1-8].

In order to achieve a higher resolution the principal problem is to obtain a submicron size of aperture in the form of fiber stylus to detect evanescent waves modified by sample surfaces.

In the present paper the characteristics of 571 fiber probes are reported.

## 2. Principle

It is well known that for higher coupling efficiency between a luminous source and an optical fiber the chemical process to obtain a microlens on the extremity of the fiber could be used [9].

Similary the appropriate chemical action on the doping concentration is an excellent way to the manufacture of the fiber probe tips [10-13]. There are two techniques to obtain an appropriate probe according to the object illumination. If the light diffused or reflected from the object surface is collected by the fiber, the cone must be sharp [13]. If the light is transmitted forward to the object and backward to the detector, the probe angle must be 90°[14].

The radius of curvature as well as a cone angle represent main parameters of the fiber probe tips. They cause a major problem in the lateral resolution of the microscope



*D.W. Pohl and D. Courjon (eds.), Near Field Optics*, 295–302.

because the resolution is proportional to the dimension of the tip. In the ideal case, the tip could be monoatomic.

Moreover, because optical fiber probes are fragile in practical applications, it is necessary to supervise and control the reproducible manufacturing of the probes to ensure equal measurement results on the sample surfaces.

## 3. Experimental

Different methods exist to obtain the etching of the optical fiber probes with a submicron or even with a nanometer sized aperture.

The first involves heating, pulling and chemical etching of the fiber: a taper of 300 μm long with a tip typically 70 nm large was obtained with quite good reproductibility [10]. Simpler methods use the properties of fiber doping [11-14].

Single-mode and multimode fibers for telecommunications can be processed by selected etching technologies using solutions of different ratios of ammonium fluoride to hydrofluorid acid. Because of the difference in the doping concentration, the etching speed of the cladding is higher than that of the core and, consequently, the fiber stylus extremity becomes conical. The diameter of the cone base is equal to the core diameter, and its height and tip curvature can be adjusted.

In the present study we have tried to respond to the following questions:

- What is the influence of the composition and concentration of the chemical solution on the processing quality and speed?
- What is the influence of the temperature of the solution on the processing?
- What processing time should be chosen to obtain the best probes?
- How can the manufactured probes be protected against mechanical strains and forces?
- Are the probes manufactured in this manner reproducible and, so, with what precision?

## 4. Results and discusion

The probes from nine types of fibers have been manufactured and tested:

A - single-mode fiber (Société Lyonnaise de Télécommunications), 125/8 μm of diameter, wavelenght 1,3 μm, attenuation 0.31 dB/km,

B - single-mode fiber (Laboratoire de Marcousis), type M 8323 N,

C - single-mode fiber (Laboratoire de Marcousis), type M 8231,

D - single-mode fiber (CNET Lannion), type FPG 2015,

E - single-mode fiber (from Japan) without specifications,

F - multimode fiber (from France), type EE 5314, without specifications,

G - multimode GI fiber, without specifications,

H - multimode PCS fiber, Czech production, without specifications,

I - single-mode fiber (from Russia), without specifications

The multimode fibers were eliminated in the early stages of this study because of their large modal noise.

## 4.1. INFLUENCE OF THE CONCENTRATION OF THE SOLUTION

Fig.1 shows a theoretical profile of the refractive index of the fiber and the evolution of the fiber shape after different moments in the chemical etching process.

Fig.2 presents a real profile of the refractive index of fiber (A) which has been the most important one for our purposes.

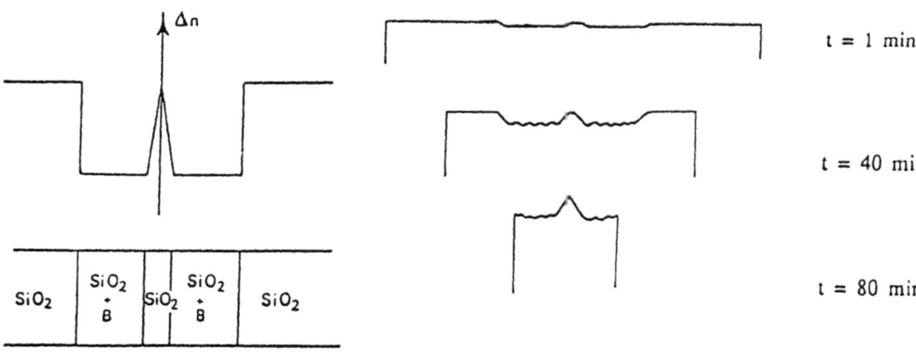


Fig.1.: Boron doped fiber: ideal profile of refractive index and the doping distribution (left), evolution of fiber cross section vs. processing time (right)

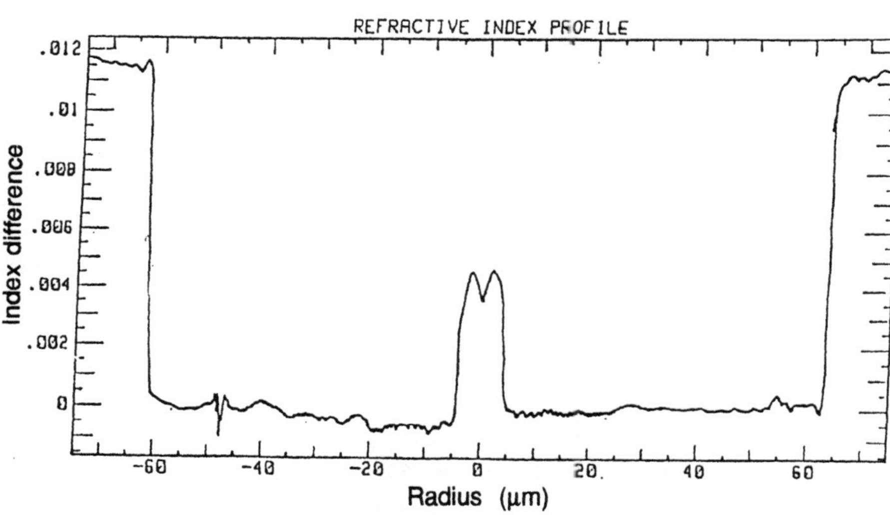


Fig.2.: Real profile of refractive index of fiber (A)

During the chemical processing part of the doped zone of the fiber cladding must be removed to achieve a significant probe tip with a cone angle about 90°.

No variation in the shape of the tip was observed in function of the solution concentrations using (at all times) the ambient temperature. The etching speed only varies according to the concentration of solution.

The Fig.3 presents some curves corresponding to the various compositions of etching solution at ambient temperature $T_1 = 18°$ C. The curves 1 and 4 with a negative cone angle on the extremity of the fiber are not interesting for the further use in SNOM.

The best proportion of the etching solution for all of tested fibers is made up of one part 40% hydrofluorid acid (HF) buffered by one part of 40% amonium fluoride ($NH_4F$) (see curve 2 of Fig.3).

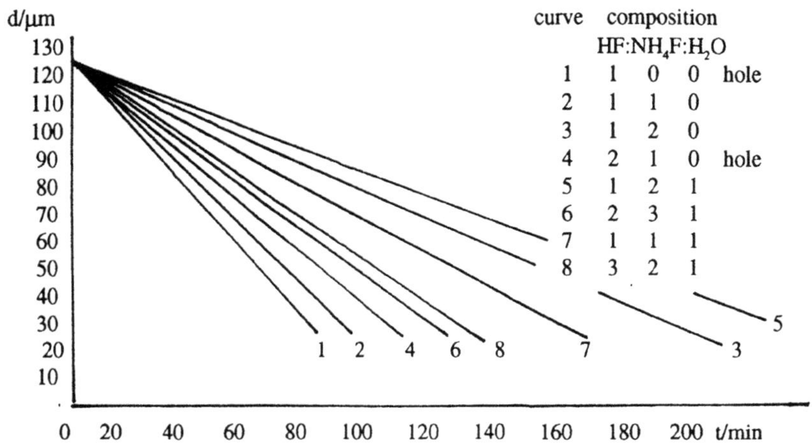


Fig.3.: Tapering of outer diameter vs. processing time for different compositions of solution

The results of the comparison of the processing of fibers (A), (B), and (C) are given in Fig.4.

Fig.5 shows an electron microscope image of the tip of fiber (A) and Fig.6 presents the statistical processing of some 112 samples of the same fiber (A) at $T_1 = 18°C$ and $T_2 = 50°C$ respectively. The $\pm 3\sigma$ bars of normal distribution are indicated for a certain number of processing time. In this case 5% of uncertainty in the tip shape was reached.

## 4.2. INFLUENCE OF THE PROCESSING TEMPERATURE

Contrary to the first case, strong variations in fiber profile were observed in function of the temperature of the solution while using a stable concentration.

The curves of temperature level vs. processing time for some outer diameters of the fiber are given in Fig.7. Here, the best relation between tip shape and its quality vs. processing speed is on the bend of the curves.

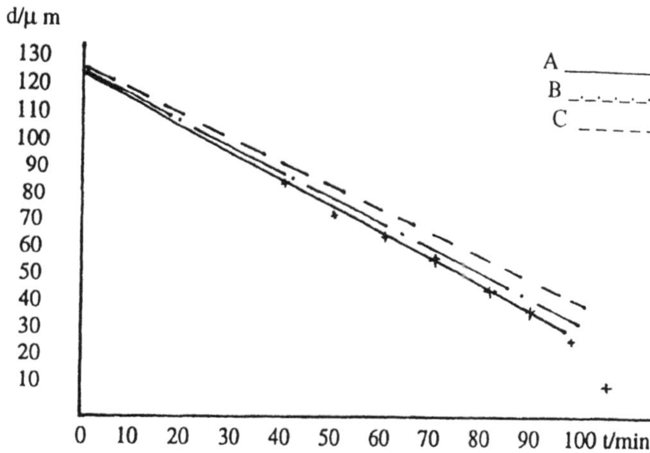


Fig.4.: Reduction of outer diameter vs. processing time for fibers A,B,C at ambient temperature T =18 C

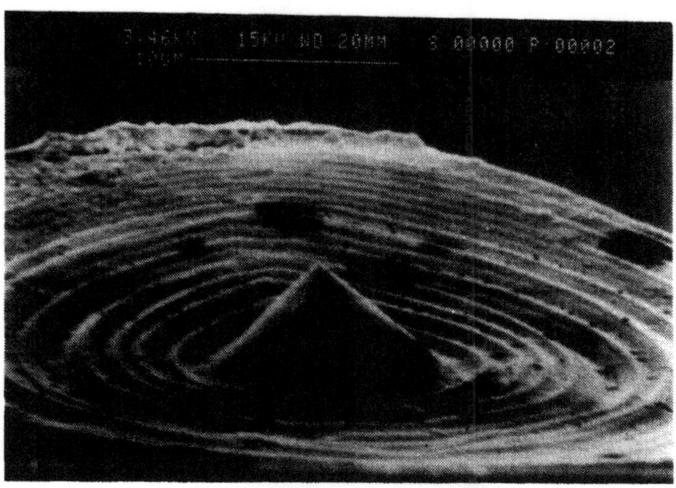

Fig.5.: Dimension of fiber tip (M 3460x), processing time 70 min.

## 4.3. INFLUENCE OF THE PLASTIC CLADDING PROTECTION OF THE FIBER

Because the external diameter of these fiber probes is about 40 µm, the fiber stylus is fragile rather against mechanical strains and forces. Therefore, it is necessary to protect the fiber stylus (Fig.8) in the following manner:

- cut the fiber some 10-25 µm from the end of the plastic cladding,
- put the fiber extremity (1-2 mm) in the chemical solution.

Since the plastic cladding is made from two layers of different Young' s modules, the inner layer is removed more quickly than the outer one. This outside cladding is flexible and resistant and can protect the fiber against mechanical strains and serve as a connector to the transmission fiber.

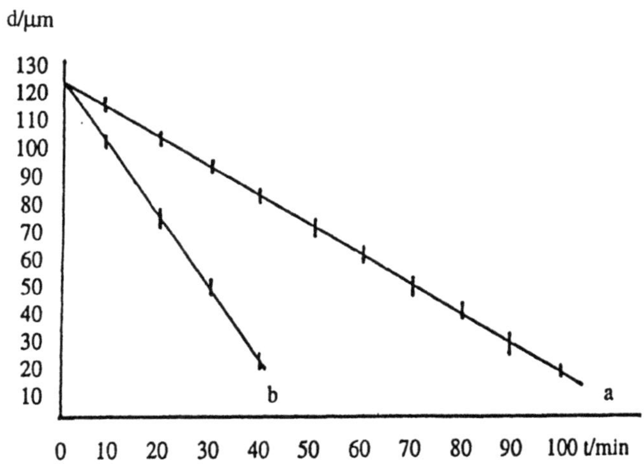


Fig.6.: Statistical processing of fiber A at a) T = 18°C, b) T = 50°C

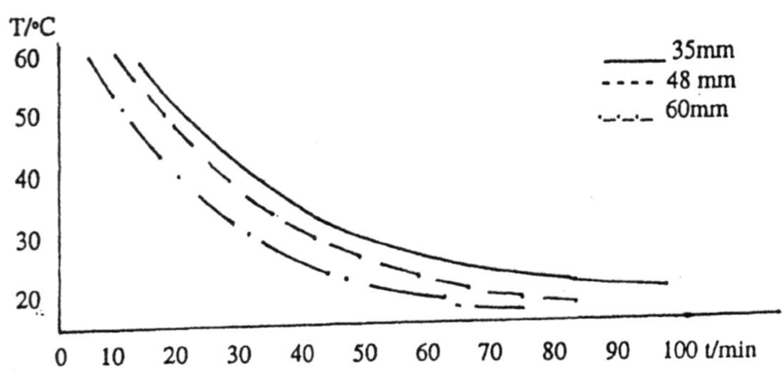


Fig.7.: Influency of temperature vs. processing time on the outer diameters

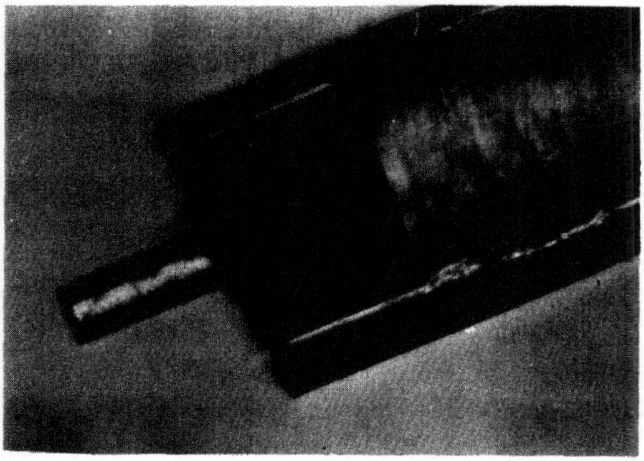

Fig.8.: Shape of the extremity of fiber with flexible plastic protection

## 5. Conclusion

Results obtained with a heated, pulled and chemically etched sharp taper as a probe exhibit a resolution of about 70 nm [10]. This method is suitable for scanning tunneling optical microscopy (STOM).

A second method of probe manufacturing under the influence of different parameters (processing time vs. temperature, composition and concentration of chemical solution) on the characteristics of the probe forms was systematically tested on 571 probes manufactured from 9 types of fibers.

With regard to the forward and backward transmission of the light through the fiber, the tip peak angle could reach 90° to eliminate the noise inside the fiber. Then this probe can be used for the external reflection SNOM. The lateral resolution of these fiber sondes is 50-60 nm for the fiber (A), (B) and (C) respectively.

The best results have been obtained with the single-mode fiber (A) of Société Lyonnaise de Télécommunications processed during 40 minutes at the constant temperature of 35° C with the etching solution having the parameters of the curve 2, Fig.3.

## 6. Acknowledgments

This study sponsored by French Ministery for Research and Space grant was started in Laboratoire d' Optique P.M.Duffieux in Besancon. I wish to thank Prof.J.P.Goedgebuer, director of Laboratory, Dr.D.Courjon and Dr.M.Spajer for their constant help and fruitful discussions. Thanks also to Dr.M.Blondy of Limoges University for the measurement of real fiber profiles (Fig.2).

## 7. Bibliography

1. Pohl,D.W., Denk,W.,Lanz,M. (1984), ' Optical stethoscopy:Image recording with resolution λ /20' , Appl.Phys.Lett., 44, 651-653
2. Reddick,R.C., Warmack,R.J., Ferrell,T.L. (1989) ' New form of scanning optical microscopy ' , Phys.Rev.B., 39, 767-770
3. Courjon,D., Sarayedinne,K. Spajer,M. (1989) ' Scanning tunneling optical microscopy' , Opt.Comm.,71, 23-28
4. Fornel,F. de. et al. (1989) ' An evanescent field optical microscope' , ECO2 Opt.Storage Scanning Technology, 77-84
5. Hulst,N.F. van et al. (1992) ' Operation of a scanning near-field optical microscope in reflection in combination with scanning force microscope' , in: Scanning probe microscopies, Proc.SPIE, vol. 1639
6. Courjon,D. et al.(1992) ' Near-field optical microscopy and optical tunneling detection' , in: Scanning probe microscopies, Proc.AIP Conference,vol.241
7. Betzig,E. (1992) ' Recent develoments in near-field scanning optical microscopy' , in: Pohl,D.W. and Courjon, D. (eds) Near-field optics - SNOM, Kluwer Academic Publ. (in press)
8. Spajer,M., Jalocha,A. (1992) ' The reflection optical near-field microscope: an alternative to SNOM' , in: Pohl,D.W. and Courjon, D. (eds) Near-field optics - SNOM, Kluwer Academic Publ. (in press)
9. Jeunhomme, L.B. (1983) ' Single mode fiber optics. Principles and application' , Marcel Dekker, New York, Basel
10. Jalocha, A. et al (1990) ' Etude des capteurs utilises en microscopie tunnel opti que' , OPTO 1990, ESI Publ., Paris, 525-529
11. Spajer,M. et al. (1991) ' Microscopie en champ proche par reflexion' , J.Phys.III, 1-12
12. Jiang,S. et al. (1991) ' A photon scanning tunneling microscope using an AlGaAs laser' , Jpn. J. Appl. Phys., 30, 2107-2111
13. Pangaribuan, T. et al. (1992) ' Reproducible fabrication technique of nanometric tip diameter fiber probe for photon scanning tunneling microscope' , Jpn. J. Appl. Phys. , 31, 1302-1304
14. Tománek,P. and Spajer,M. (1992) ' All-fiber compact near-field scanning optical microscope' , in: Specification and measurement of optical systems, Proc. SPIE, vol. 1781 (submitted)

# THE CONCEPT OF AN OPTOELECTRONIC PROBE FOR NEAR FIELD MICROSCOPY

H.-U. DANZEBRINK, U. C. FISCHER
*Physikalisch-Technische Bundesanstalt*
*Bundesallee 100*
*3300 Braunschweig*
*Germany*

ABSTRACT. A new geometry and functional principle for submicroscopic sensors and light sources as potential probes for near field microscopy is presented. The basic idea is to use cleaved semiconductor crystals as probe material. Two advantages are envisaged : first, the fact that by cleaving crystalline material very sharp and well defined edges and perhaps corners can be fabricated, and secondly, the optoelectronic conversion can be integrated into the probe itself. Micrographs of a light-emitting structure and the related current-voltage characteristics are shown. The structure is formed by a Schottky contact consisting of GaAs as the semiconductor and aluminium as the metallic contact material on two adjacent cleavage faces. The emission from the edge can be tentatively explained as the result of an avalanche breakdown effect in the semiconductor due to the high electric fields between the narrow contacts with distances of about 100 nm.

## 1. Introduction

One of the main problems in near field optical microscopy is the reproducible fabrication of efficient probes. There are several concepts which are described elsewhere [1]. These probes normally function as passive waveguides which transmit electromagnetic waves from macroscopic to submicroscopic dimensions or vice versa. In order to avoid the high losses which are for example expected when tapered cylindrical waveguides are used, we chose different geometries such as a tetrahedral tip with two adjacent metallized faces because propagation modes may exist with their maximum of intensity in the very edge or corner of such a structure [2]. Our experience in producing very sharp glass blades for ultramicrotomy which are formed by two fracture planes led us to the idea of cleaving semiconductor crystals to fabricate sharp edges and corners and also make use of the optoelectronic characteristics of the semiconductor.



*D.W. Pohl and D. Courjon (eds.), Near Field Optics*, 303–308.

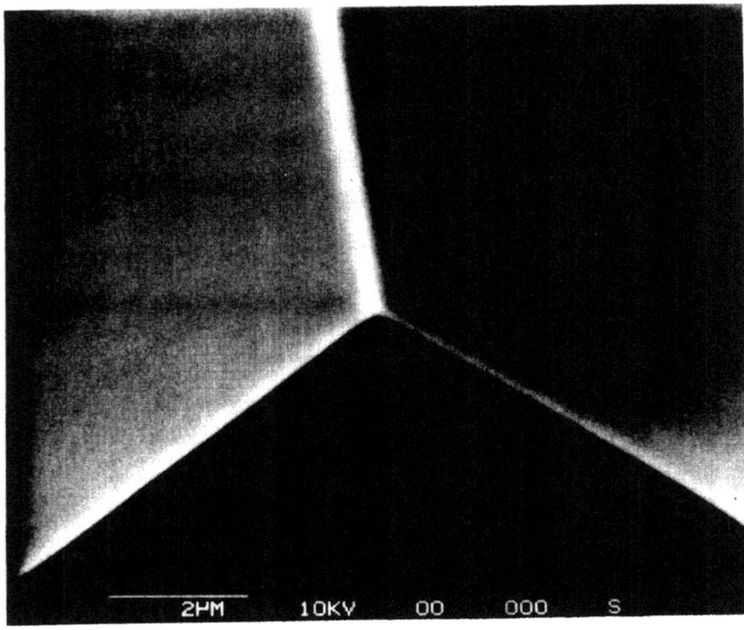


Figure 1: Scanning electron micrograph of a cleaved GaAs corner

## 2. Concept of a submicroscopic light source

Using the results of investigations into the light emission behaviour of planar GaAs Schottky contacts [3-5], we intend to develop a tetrahedral probe as a submicroscopic light source by extending the concept of light emission to the three-dimensional structure of our probe. The tetrahedral tip is fabricated by cleaving a (100) orientated GaAs wafer twice along a cleavage direction. In this way we obtain two (110) cleavage planes and the polished wafer surface joining at three edges to form a common corner, which constitutes our tetrahedral tip. The angles between the edges are defined by the specific cleavage properties of the material. It should be mentioned that corners of cleaved Si wafers were previously used [6] as tips for electron tunneling microscopy.

Fig. 1 shows a scanning electron micrograph of such a tetrahedral GaAs corner. The two upper faces are cleaved planes, the third plane in the lower part of the image belongs to the wafer surface. In order to define a submicroscopic localisation of light emission, we evaporated the aluminium electrodes onto the two cleaved planes in a way shown schematically in Fig. 2 (left side). Adjusting the angle "$\alpha$" of evaporation of about $15^o$ to the edge, we succeeded in keeping the edge itself free of aluminium.

Using a tetrahedral structure made of glass [2] instead of GaAs we actually showed a separation distance of about 150 nm between such electrodes (see Fig. 3). The transition resistance was still higher than 50 M$\Omega$.

The example of the metal-on-glass evaporation shows the capability of this simple and reproducible fabrication process. We tried to measure the width of the gap in the

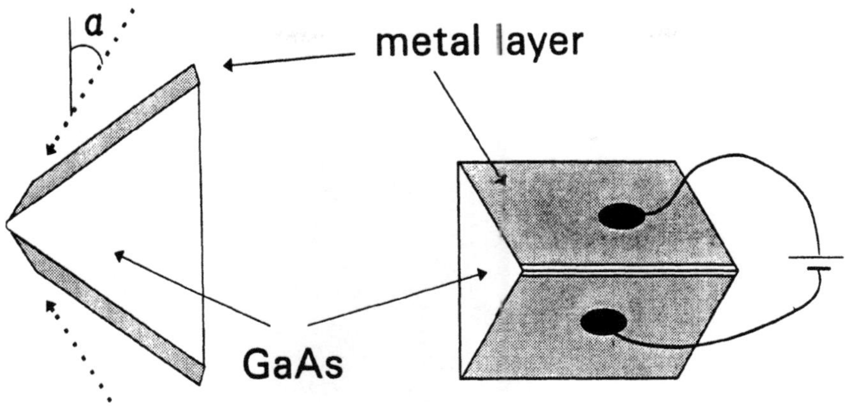


Figure 2: On the left side the principle of evaporation of the metallic contacts is shown, the dotted arrows represent the metal rays; on the right side a sketch of the 3-dimensional probe is shown

case of metal on GaAs using a SEM, but the contrast between the metal electrodes and the GaAs was very bad, so we could not resolve the electrode separation. While the geometry and the evaporation angles of the two electrode fabrications (on GaAs & on glass) are very similiar, we assume that we will obtain comparable electrode separations.

A metal-semiconductor combination of Al and GaAs was chosen which in our geometry is expected to result in a series connection of two opposing Schottky contacts. If a DC voltage is applied between the two Al electrodes (see Fig. 2 right side), then one of the Schottky contacts is driven in the high-resistance, the other one in the conducting direction. The narrow gap of a width in the order of 150 nm leads to a field strength of about 6 V/150 nm = $4 \cdot 10^5$ V/cm at the gap. An avalanche-breakdown is expected to set in above a certain bias voltage at the location of the highest field. References [3-5] deal with detailed experimental investigations and a theoretical interpretation of light emission accompanying this breakdown for the case of planar electrode configurations.

## 3. Realization and first results

After fabricating a probe as described above, thin copper wires are fixed on the metallized faces by conductive adhesive. The current-voltage characteristics are measured by means of a stabilized voltage power supply and a digital ammeter. The light emission is simultaneously observed in a photomicroscope.

One of the resulting I/V characteristics of a GaAs substrate with an n-doped

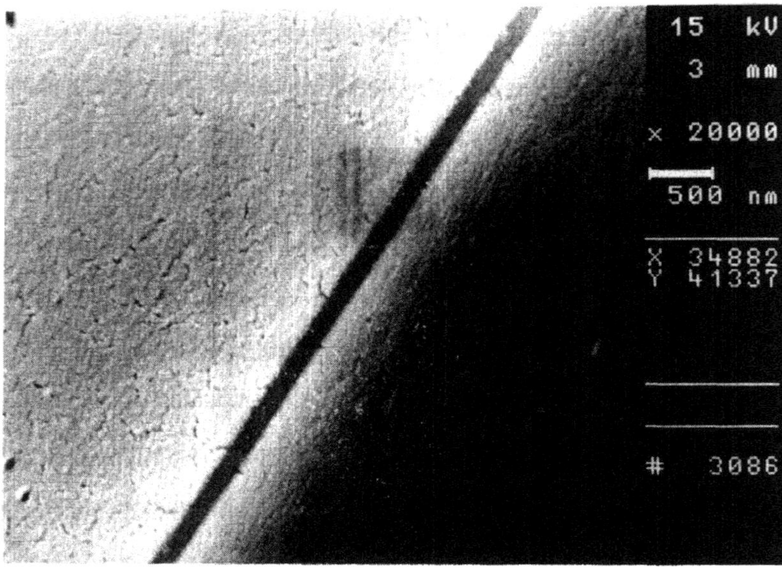


Figure 3: Slit between two evaporated gold electrodes on a tetrahedral glass blade "viewed" from above

($N_D = 2{\cdot}10^{17}$ 1/cm$^3$) epitaxial layer 7 $\mu$m thick is shown in Fig. 4. For voltages above 6 V a breakdown behaviour sets in. We think that the steep ascent of current is a strong indication of an avalanche breakdown through the reverse biased Schottky barrier. A polarity reversal leads to the same qualitative behaviour of the diode characteristic as is to be expected for a symmetrical structure.

There is a sudden jump of the current from about 500 $\mu$A to about 1.6 mA at a constant voltage. This process is accompanied by the onset of light emission which continues for up to several hours. At the same time a gradual decay of the current in the order of 150 $\mu$A/hour is measured. When the experiment is repeated the breakdown voltage increases slightly (see Fig. 4) and the same qualitative decay of current is observed. We assume this irreversible process to be due to influences of local heating or electro-migration on the semiconductor material, the metal electrodes or the metal semiconductor interface.

Fig. 5 shows a micrograph of light emission of the epitaxial probe (the same as in Fig. 4). We see one of the cleaved planes in the upper part of the image with the edge adjusted horizontally (compare Fig. 2 right side). At the edge near the corner we observe a thin line with a sharp bright structure in the middle and a blurred surrounding. The emission is not monochromatic and has a white appearance of a not yet known spectral distribution.

The intensity of the captured emission is weak. It was not yet measured and can only be estimated roughly by the sensitivity of photographic exposure to be in the order of $10^4$-$10^5$ photons/second. Focusing the image was problematic but we have the

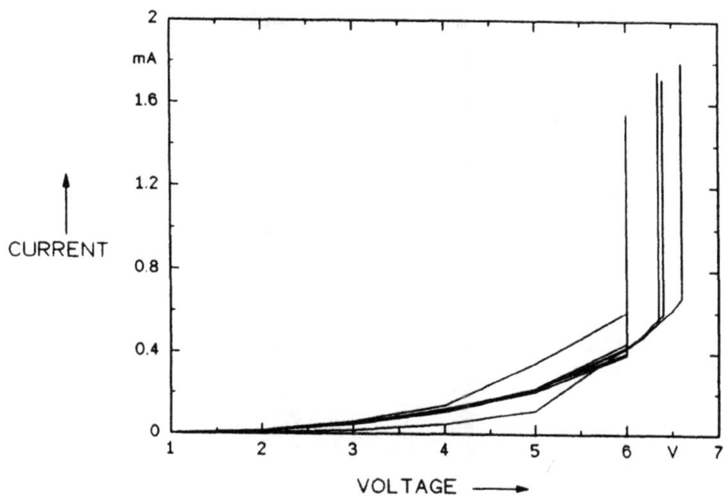


Figure 4: I/V characteristics of the Al-GaAs-Al structure

impression that the inner bright line is essentially diffraction limited at a numerical aperture of 0,54 of the objective lens.

Additionally we made I/V measurements for different evaporation angles "$\alpha$". For "$\alpha$" below $15^o$ we got an ohmic characteristic due to an overlap of the metal electrodes.

One may assume that similiar I/V characteristics and emission behaviour arises due to a metal insulator metal (MIM) contact at the narrow gap between the electrodes. A MIM contact of similiar geometry is formed in the case of the glass blade mentioned above. There, only occasionally did we observe very short-lived light emission behaviour at voltages of about 10 V. No breakdown of the current versus voltage was measured. We therefore believe that it is very unlikely that an unspecified MIM contact over the gap could be responsible for the light emission.

## 4. Conclusions and outlook

Our investigations show that the principle of an optoelectronic probe as a light emitter works. Until now the emission intensity is very weak. A higher voltage is expected to increase the current-dependent emission [4]. In order to reduce destructions caused by the high electric field it may be advantageous to use short voltage pulses with low repetition rates as excitation. A very sensitive lock-in-detection system can be built up based on this idea.

One of the main interests in the future will be to develop and investigate submicroscopic sensors instead of sources using the same geometry as the tetrahedral structure described above. In submicroscopic sensors the sensing principle is based on the photoconductive effect, and the measurement of the spatial intensity distribution of the near field of a radiating object with subwavelength resolution is achieved

Figure 5: Micrograph of the light emitting GaAs edge

by means of the aperture, which is formed by the submicroscopic slit between the electrodes.

## 5. Acknowledgement

The authors would like to thank Dr. M. Bischoff and Dr. N. Sotnik for helpful discussions concerning the phenomena of hot electrons and microplasmas, and Dr. R. Lauer and U. Dierksen for their helps in the evaporation of metal films.

## 6. References

[1] Pohl, D.W. (1990), Scanning near field optical microscopy (SNOM), in C.J.R.Sheppard and T.Mulvey (eds.), Advances in optical and electron microscopy, Academic Press, London.

[2] Fischer,U.C. (1992), these proceedings

[3] Adachi,H. and Hartnagel,H.L. (1982), Experimental data on light emission behaviour from GaAs Schottky contacts, Int.J.Electronics 52, 89-93.

[4] Adachi,H. and Hartnagel,H.L. (1981), GaAs Schottky light emitters for the study of surface avalanching and electroluminescence, J.Vac.Sci.Technol.19(3), 427-430.

[5] Huber,E.J. (1983), Untersuchung der Lichtemission beim Lawinendurchbruch bei GaAs-Schottkydioden, Dissertation TH Darmstadt.

[6] Anders,M., Thaer,M., Mück,M. and Heiden,C. (1988), Scanning tunneling microscopy imaging of microbridges under scanning electron microscopy control, J.Vac.Sci.Technol.A6 (2), 436-439.

# FIRST SPECIFICATIONS OF A PSTM WORKING IN THE INFRARED

A. PIEDNOIR and F. CREUZET,
*Laboratoire CNRS/Saint-Gobain, "Surface du Verre et Interfaces"*
*B.P. 135, 39 Quai Lucien Lefranc, F-93303 Aubervilliers, France*

C. LICOPPE,
*Laboratoire de Bagneux, CNET*
*196 Av. H. Ravera, B.P. 107, F-92225 Bagneux, France*

F. DE FORNEL,
*Groupe de Microscopies Tunnel de l'Université de Bourgogne,*
*B.P. 138, Faculté des Sciences, F-21004 Dijon, France*

ABSTRACT. One of the most promising extension of the photon scanning tunneling microscope (PSTM) is its use as a locally resolved spectroscopic tool. The infrared (IR) spectral window offers a wide variety of interesting applications since many materials have a well-defined signature in this range. We have undertaken the experimental modifications that are necessary to operate the existing PSTM's in the IR domain. In this work, we present preliminary data obtained with two IR light sources : an extended source (Globar type) and the free electron IR laser source (CLIO) available at LURE (Orsay). In both cases, the evanescent field has been detected with a fluoride glass optical fiber and the characteristic decay is reproducibly observed. On the basis of these results, we discuss the current limits of spatial and spectral resolutions. Mainly, we conclude that a spectral resolution of about 10 $cm^{-1}$ over an area of 1µm x 1µm is possible to achieve.

## 1. Introduction

During the last ten years, scanning tunneling microscopy (STM) and related techniques (Proc. of the 6th ICSTM, (1992)) have undoubtedly brought new insights to the field of surface science. For conducting materials, STM permits to image surfaces at the atomic scale ; furthermore, a local spectroscopy based on the properties of the Fermi surface is achievable. For insulators, atomic force microscopy (AFM) has more or less the same potential, but the spectroscopy is generally not straightforward because the total interaction between the tip and the surface is rather complex to analyze. In many cases, these two near-field microscopies have strong limitations, due to either mechanical deformations of the substrate or the absence of a well-defined signature for the spectroscopy.

The photon scanning tunneling microscope (PSTM) has the potential to circumvent these problems. Although the resolution does not reach the atomic level, it lies well below the Rayleigh criterion which characterizes the far field optical microscopies (Courjon et al (1989), Reddick et al (1989)). The first advantage of the PSTM comes from the rather large distance



*D.W. Pohl and D. Courjon (eds.), Near Field Optics*, 309–316.

between the optical fiber tip and the surface which prevents from damaging the sample. More importantly, a local spectroscopy may be performed and easily compared with well-established techniques. Because the infrared (IR) spectral window is known to be very rich for most of the materials, the extension of the PSTM to the IR range appears to be one of the interesting modifications of this microscope and represents some kind of a compromise between resolution and spectroscopy.

We can illustrate this introduction with the example of adhesion of polyurethane (PU) on glass. An AFM study (Creuzet et al (1992)) of peeling residues has evidenced islands which grow after ageing of the PU-glass assembly. These heterogeneities have a typical size of 1 µm and are identified as residues of PU by other methods (Berquier et al (1992)). However, wettability data and AFM suggest that the bare substrate is not cleared between the islands and that an organic matter still covers the substrate, even without ageing of the assembly. Obviously, a PSTM working in the IR domain will allow the complete description of the residue, which is necessary to understand the role of the glass composition in this particular case.

Based on the above example, we may give the specifications that are desirable for this locally resolved spectroscopy. The spatial lateral resolution should be in the 1 µm range . The spectral resolution would ideally be 10 $cm^{-1}$ over the spectral window [1000-10000 $cm^{-1}$]. Finally, in order to detect the IR adsorption signal of a 10 nm layer, the signal to noise ratio must be 1000 or so.

## 2. Experimental

The description of the different set-ups for PSTM in the visible range can be found elsewhere (De Fornel et al (1990), Spajer et al (1991)). We have chosen the total internal reflection (TIR) arrangement which gives the best resolution and is easier to interpret. In the following, we will present the modifications that are required to work in the IR domain, namely the source, the substrate, the optical fiber and the detector.

We have used two different IR sources. One is a conventional Globar source, which might be approximated to a black body at 1300 K. The total power over the full spectral window (roughly from 0.5 to 100 µm) is about 150 mW. The light is unpolarized. The angular dispersion is large and may be estimated to be 1 steradian. The size of the source point is 1 $cm^2$ or so. The other source is similar to what is commonly used with a PSTM. The free electron laser operated at LURE (the so-called CLIO source) behaves as a tunable IR laser (Ortéga et al (1986)). At the time being, the specifications are as follows. The IR window where the laser can be tuned is 3-10 µm, with a narrow distribution ($\Delta\lambda/\lambda \approx 10^{-3}$). The light is s-polarized before reaching the substrates. The beam is collimated and its section is of the order of 10 $mm^2$. The time average power vary from 10 to 40 mW ; however, the time dependence of the intensity is somewhat complex and very narrow (≈ 10 ps) and intense (≈ 50 MW) peaks are generated by the lasing procedure.

Mainly, we have used a ZnSe hemispherical crystal as a substrate for total internal reflection on the flat surface. This first choice is especially convenient when the Globar source is used, since the angular dispersion is large after the focusing mirror. With the CLIO source, we have also used a silicon wafer polished at an angle of about 21°. This angle is critical to satisfy the two constraints : i) total reflexion within this high index material and ii) frustration of the evanescent field by the third medium (the fiber) which has a low optical index in comparison to the substrate.

Like in most of the PSTM, the probe is a glass optical fiber. Because we work in the IR range, the silica glass fiber has been replaced by a fluoride glass fiber (from "Le Verre Fluoré", Rennes) which has a cutoff close to 8 µm. As explained in the following sections, the radius of curvature is not a concern and a typical size of 10 µm or less can be achieved.

The signal has been detected with a HgCdTe sensor commercially designed to be coupled to an optical fiber (Nicolet, USA) in order to reduce the signal to noise ratio. The final stage was a lock-in amplifier when using the Globar source with a chopper, whereas we simply used an analog to digital converter to measure the amplitude of the CLIO pulses. Each data point reported in this paper has been obtained without any long time averaging : typical acquisition time is 100 ms (few time constants of the lock-in amplifier) for the Globar source and 5 s for the CLIO source.

Figure 1 sketches the experimental set-up used to detect the evanescent field when the source is the Globar. In the case of the CLIO laser source, the arrangement is very similar to the one which is commonly utilized in the visible range.

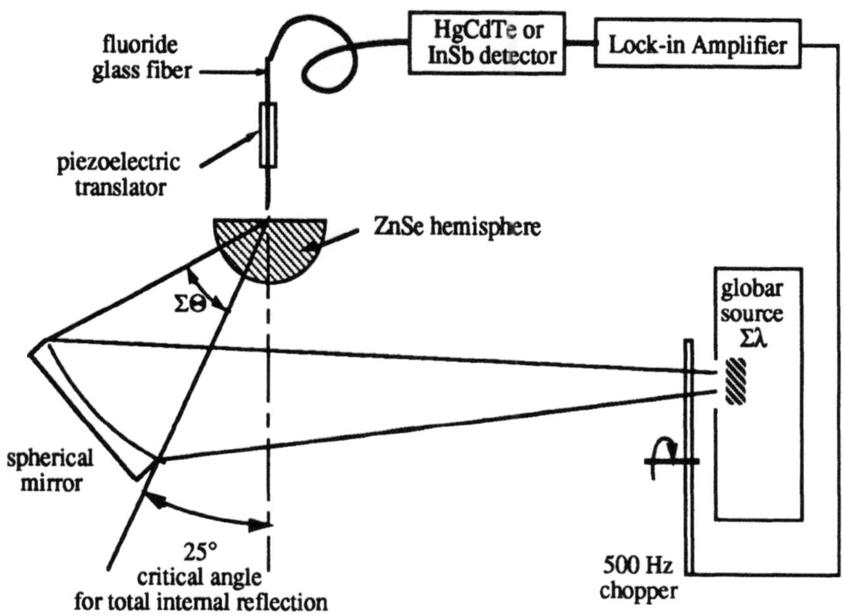


Figure 1. Scheme of the experimental set-up used with the Globar source.

## 3. Results

We have observed the evanescent field in different ways which will be described in the following sections. When possible, the data are successfully compared to the theoretical profile.

## 3.1. FIRST APPROACH TO THE PSTM IN THE IR RANGE

The first experiment has been designed to test the feasibility of this project. We have used the Globar source and the ZnSe crystal to generate the evanescent field. The optical fiber was simply moved with a precision screw giving a resolution of about 0.5 μm. The signal to noise ratio is of the order of 30 and typical data are displayed in Figure 2.
At long distance, the background is due to the diffusion of light and we have checked that this level is unchanged when the fiber is far away from the surface. The main feature is the large increase observed when the fiber is in the vicinity of the surface. In order to evaluate whether the length scale is in the right range or not, we have undertaken a simulation of the data based on the calculations described by Salomon (1991). Basically, the theoretical framework is to evaluate the intensity of the evanescent field when frustrated by a third medium. The full study shows that the exact shape of this third medium is not relevant and does not influence the profile of the transmittance. Note that, in this particular case, the electromagnetic field has been integrated over both the full range of experimental angles and the whole spectral window where the instrument operates. The very good agreement between theory and experiment has been reproducibly found with this set-up.

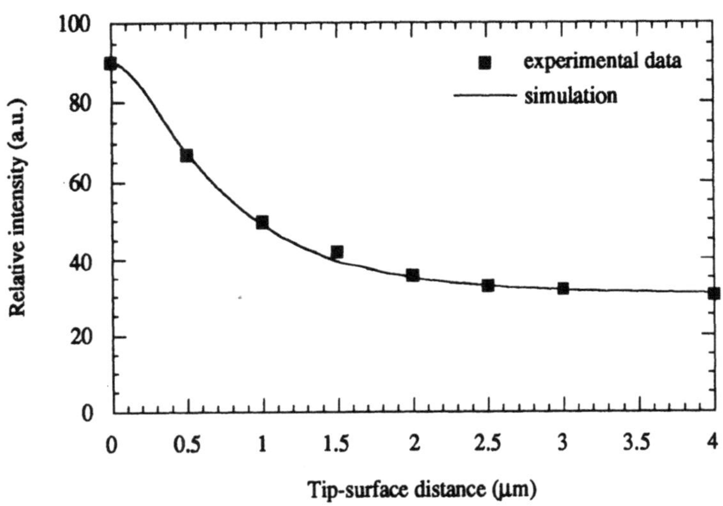


Figure 2. Decay of the evanescent field when frustrated by the optical fiber. The incident beam on the TIR surface has a large angular dispersion (0.6 st) around the mean value of 50° for the angle of incidence. The considered wavelength domain is [1.35 μm, 8 μm].

A quantitative evaluation of the tip radius can be derived from the magnitude of the transmitted field, when the fiber is closed to the surface. The focusing mirror collects 10 mW which converge onto the TIR surface with a spot size of 50 $mm^2$. Therefore, the light power per area unit of the ZnSe surface is 2 $10^{-10}$ W/$\mu m^2$ over the all spectrum. The magnitude of the transmitted field detected by the fiber is 3 $10^{-8}$ W. We can infer that the fiber tip radius is of the order of 6 μm. We want to strongly emphasize that this value is not a limitation of the fiber tip manufacturing. Due to the rather poor intensity per area unit that is available with this set-up, we had to significantly damage the shape factor of the fiber. This aspect will be briefly discussed at the end of the paper.

## 3.2. CHARACTERISTICS OF THE EVANESCENT FIELD AS MEASURED WITH THE GLOBAR SOURCE

In an attempt to accurately characterize the evanescent field generated by the same set-up (Globar + ZnSe crystal), we have mounted the fiber on a piezoelectric actuator, as it is usually done in near field microscopy. The results are shown in Figure 3 where forward (motion towards the surface) and backward (motion away from the surface) data points have been plotted after removal of the background. The decay is unambiguously observed but a rather large hysteresis appears between the two traces. This effect is more likely due to the properties of the piezoelectric ceramics ; for instance, the magnitude of the hysteresis measured on the same kind of material by Kaizuka et al (1988) accounts for the discrepancy between the two traces. This phenomenon is important when working in the IR domain since the piezo has to be extended on a large scale in order to acquire the full profile. Therefore, a comparison with the theory is difficult to perform unless a calibration of the piezo is done. In the next paragraph, we will present a way to correctly record the tip-surface distance.

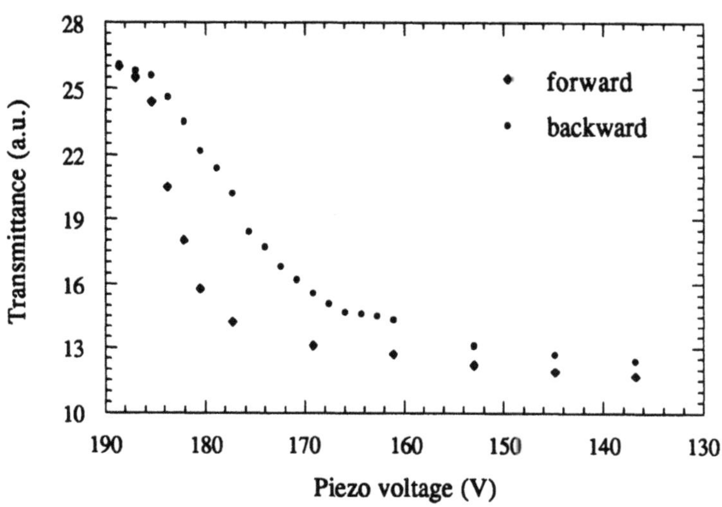


Figure 3. Observed hysteresis between forward and backward data points. The mean distance/voltage ratio is 35 nm/V.

As it is well known for PSTM, the characteristic length of the decay is very sensitive to the angle of incidence. This is particularly true here because this angle is not unique and is widely distributed. We have examined this effect by slightly changing the position of the focusing mirror. As expected, the decay is modified in agreement with the prediction of the theoretical calculation.

### 3.3. CHARACTERIZATION OF THE EVANESCENT FIELD AS MEASURED WITH THE CLIO SOURCE

As pointed out in the experimental section, the CLIO source has the net advantage to be intense and tunable. As far as spectroscopy is concerned, this will impose to image the surface at various wavelengths which are differently absorbed at the interface. Unfortunately, we could not completely exploit the full power of the source because of arcing problems in the ZnSe crystal. The evanescent field is observed with a signal to noise ratio of 50 (at $\lambda = 5$ µm) without any kind of filtering, a value which is much better than the one found with the Globar source if we take into account the narrow spectral window that is involved here. Again, the radius of curvature of the fiber tip had to be modified in order to increase the magnitude of the transmitted field with respect to the background. We expect a considerable improvement of the signal to noise ratio after removal of the background since it will be possible to efficiently average the signal. Moreover, the running projects at CLIO will permit to increase the available power without increasing the peak power. As an example, Figure 4 displays the observed decay for two different wavelengths, together with the simulations.

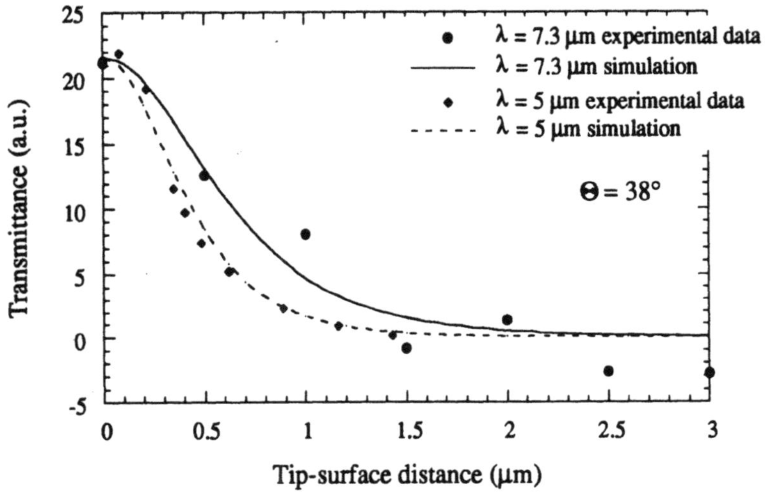


Figure 4. Decay of the transmitted field for two typical wavelengths used at CLIO. The indices used for the simulation are : substrate 2.4, air 1, fiber 1.5.

In order to get rid of the possible fluctuations of the CLIO source, the transmittance is defined as the ratio of the fiber signal to the incoming power. The tip surface distance has been estimated from the piezo hysteresis curve and systematic errors occur around a distance of 0.25 µm. At 7.3 µm, the water vapor absorbs a large amount of power, a phenomenon which does explain the poor signal to noise ratio. If necessary, this problem will be suppressed by working in completely dry air. Nevertheless, the shift between the two traces can only be attributed to the wavelength difference, since everything else remained unchanged between the two measurements. Future work will focus on the short wavelength range (down to 3 μm) in order to improve the comparison between theory and experiment.

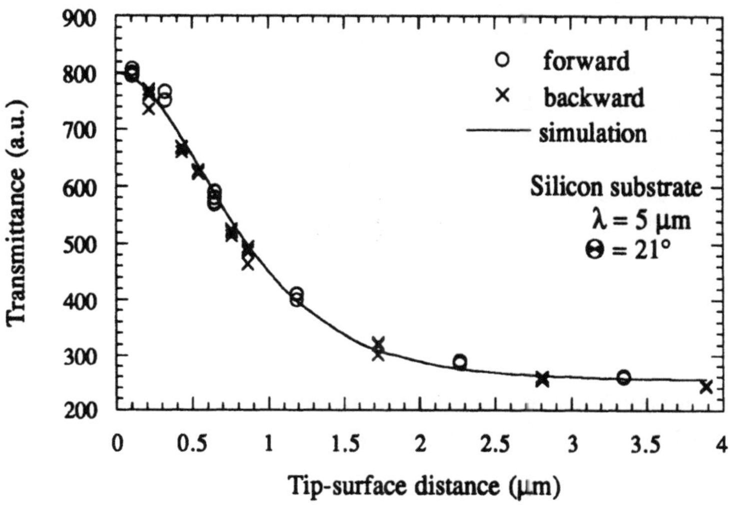


Figure 5. Decay of the evanescent field recorded with a capacitive sensor. The indices used for the simulation are : substrate 3.4, air 1, fiber 1.5.

Finally, we present a modification of the experiment which will significantly improve the results. In order to circumvent the hysteresis problems that have been shown to be important in this IR range, we have implemented a capacitive sensor which continuously records the fiber position. We have tested this technique with the silicon wafer as a substrate. Results for a wavelength of 5 μm are displayed in Figure 5. Note that the decay cannot be directly compared to the one of Figure 4 because the optical indices are different. The hysteresis is correctly suppressed and both forward and backward data can be reconciled with the theoretical simulation (Zhu et al (1986)). This last calculation is appropriate when the three media have different optical indices.

## 4. Discussion and conclusion

In this paper, we have shown that it is possible to reproducibly record and analyze the frustrated evanescent field in the IR domain. By many aspects, the behavior is in good agreement with the theory developed for the PSTM in the visible range. At the time being, we are facing two different problems that will have to be solved in the next future.

For all experiments, a rather large background has prevented the observation of the decay with a small fiber. We believe that the diffused light is collected over a large uncoated area near the tip. We are currently testing different procedures to significantly reduce this effect without damaging the tip.

Taking into account the Noise Equivalent Power of the HgCdTe detector, the total power per area unit must be close to $10^{-8}W/\mu m^2$ if a complete spectrum is to be acquired with a resolution of 10 $cm^{-1}$ after one hour. This implies that the intensity of the Globar source is too low by a factor 100. A possible way to gain this factor is to use the cascade arc, a IR source which has been recently developed at the Technical University of Eindhoven. This source may be approximated to a blackbody at 12000 K, the source point is as small as 1 $mm^2$ and the angular dispersion is less than 50 mrad. As already mentioned, another way to perform a locally resolved spectroscopy is to use the CLIO source at different wavelengths. In this case, the signal to noise ratio should be increased by one order of magnitude. Averaging techniques associated with the power increase that is scheduled at CLIO will permit to achieve this goal.

Finally, we want to emphasize that, in many cases, the requirements given as a starting point of our work might be less constraining. In particular, it will be possible to work on a reduced spectral window which contains the signature of the materials to be observed.

## 5. Acknowledgements

We wish to thank Nicolet (USA) for providing us with part of the equipment involved in these experiments. The valuable assistance of the CLIO research group is warmly acknowledged.

## 6. References

Berquier J.M., Buffeteau T., Chartier P., Desbat B. and Arribart H. "Détermination par spectroscopie infrarouge de la localisation de la rupture d'adhésion verre-élastique" accepted for publication in Compte-rendu de l'Académie des Sciences (1992).

Courjon D., Sarayeddine K. and Spajer M. " Scanning tunneling optical microscopy" Opt. Com 71 (1989), 23.

Creuzet F., Ryschenkow Y. and Arribart H. "A new tool for adhesion science : the atomic force microscope" accepted for publication in J. of Adhesion (1992).

De Fornel F. and Goudonnet J.P. European Patent submitted on Oct 31, 1990 "Procedé d'analyse spectroscopique ponctuelle de la lumière diffractée ou absorbée par une substance placée dans un champ proche".

Kaizuka H. and Siu B. " A simple way to reduce hysteresis and creep when using piezoelectric actuator" Jap. J. Appl. Phys. 27 (1988), 773.

Ortéga J.M. and Billardon M."Le laser à électrons libres" Pour la science (1986), 91.

Proceedings of the Sixth International Conference on Scanning Tunneling Microscopy, Edited by Descouts P. and Siegenthaler H. Ultramicroscopy 42-44 (1992)

Reddick R.C., Warmack R.J. and Ferrell T.L. "New form of scanning optical microscopy" Phys. Rev. B 39 (1989), 767.

Salomon L. " Théorie et mise au point d'un microscope à effet tunnel optique" Thesis, Université de Bourgogne, February 1991.

Spajer M, Courjon D., Sarayeddine K., Jalocha A. and Vigoureux J.-M. "Microscopie en champ proche par réflexion" J. Phys. *III* 1 (1991) 1-12

Zhu S., Yu A.W., Hawley D. and Roy R. " Frustrated total internal reflection : a demonstration and review" Am. J. Phys. 54 (1986), 601.

# FAR-INFRARED NEAR-FIELD SPECTROSCOPY OF TWO-DIMENSIONAL ELECTRON SYSTEMS

F. KEILMANN and R. MERZ
*Max-Planck-Institut für Festkörperforschung*
*D-7000 Stuttgart 80*
*F. R. Germany*

ABSTRACT. We have investigated the use of near-field optics to spectroscopy in the far infrared. At these long wavelengths of 0.1 to 1 mm a focusing waveguide tip with an aperture of 1 to 10 μm represents an enormeous gain in spatial resolution. Such a tip can be formed with simple mechanical means. As a technical advance, our experiments demonstrate that a tapered coaxial waveguide strongly enhances the coupling of a free-space beam to a sub-wavelength aperture.

An interesting application of near-field optics lies in the spectroscopy of two-dimensional electron systems. Far-infrared radiation can strongly interact with these electrons and thus give information about the system's structure and dynamics. A first experimental study will be discussed, where we can directly demonstrate the nonthermally populated edge states of a GaAs/AlGaAs heterostructure in the quantum-Hall regime.

## 1. Introduction

Optical methods provide a wealth of information on all materials, but the characteristic absorption is often outside the visible region. Semiconductor physics, for example, requires a broad spectral coverage to characterize the fundamental gap excitation in the near-visible, or the lattice vibrations in the mid-infrared regions. Still lower frequencies in the far-infrared and microwave regions give selective information on the mobile charge carriers. Clearly these long-wave spectroscopies are key methods which have successfully supplemented d.c. and a.c. electrical transport methods in characterizing quantum effects[1] in modern structured semiconductor systems.[2] We point out that in these applications a sub-wavelength spatial resolution can lead to valuable new insight. Our contribution discusses our experimental approach and describes a first result.[3]



*D.W. Pohl and D. Courjon (eds.), Near Field Optics*, 317–324.

## 2. Limit of metal waveguide focusing

Microwave technology uses low-loss transmission lines in the form of hollow metal tubes. These waveguides have an inner dimension of about 0.8 λ which has the double advantage of compact size and of well-defined single mode propagation. Such guides are commercially available up to 300 GHz (equiv. to λ=1 mm) where the attenuation length becomes significantly short (15 cm). Due to scattering and metal absorption the attenuation increases at still higher frequencies in the infrared where single-mode metal guides have only been used in special applications which require extremely short lengths.[4]

A conically expanding form ("taper") can be used to keep the propagation in the fundamental mode of an increasingly "oversized" metal waveguide and from there into free space ("horn antenna"), with virtually negligible reflection. Used in reverse, such a conic taper system can focus a suitable free-space beam into a fundamental metal waveguide of about 0.8 λ without loss. Let us discuss the use of this system as a near-field scanning tip in the infrared.

The propagation of the fundamental mode in a conically converging circular guide can be modelled by differential equations for forward and backward travelling fields $E_f$ and $E_b$ [5]

$$dE_f / dz = - ikE_f - Q(E_f - E_b),$$
$$dE_b / dz = ikE_b - Q(E_f - E_b),$$

with $Q = (1/2n)\, dn/dz$, $k = n\omega/c$, and a refractive index depending on the diameter d

$$n(z) = \mathrm{sqrt}\,(1 - (d_c/d)^2),$$

where $d_c = 3.68\ c/\omega$ is a critical or "cutoff" diameter (Fig 1).

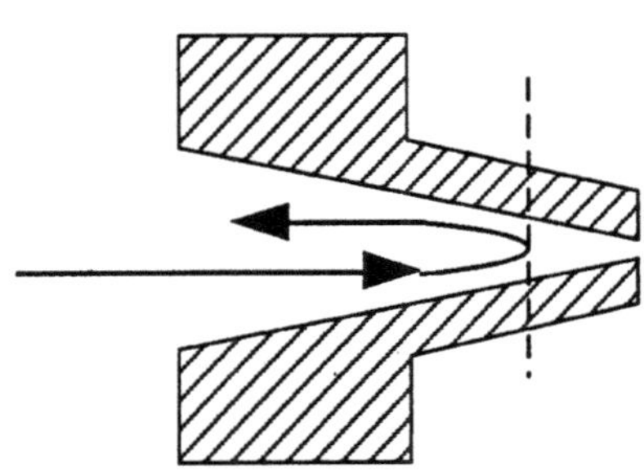

Figure 1. Conical hollow metal waveguide exhibiting total reflection at the "cutoff" diameter $d_c = 0.6\ \lambda_{vac}$.

A numerical solution[5] shows that, for $\lambda_{vac}$ = 392 μm, a cone (full angle 28°) ending with an aperture of 250 μm transmits 85% of the input power. The transmittance decreases drastically to only 12% when the end aperture is slightly decreased to 230 μm, the cutoff point, and further with even smaller diameter.

Conical waveguide focusing therefore has a limiting aperture, unto which practically all input radiation is transmitted, of about 0.65 λ. Drastic losses can however occur when the input wave is not matched to the fundamental guide mode, as is the case with metal coated glass tips employed in early optical near-field microscopy.[6] The high sensitivity of optical detection still allows considerable sub-wavelength resolution. A great signal increase was noticed when single mode optical guides with careful tapering were introduced.[7] Any further aperture reduction below 0.6 λ introduces severe reflection losses since the wave propagation becomes evanescent.

## 3. Application to a quantum-Hall system

The current distribution in a two-dimensional electron gas is far from homogeneous under quantum-Hall conditions, i.e. at low temperature of a few K and strong out-of-plane magnetic field of several T. Since the electrons exhibit cyclotron rotation at several THz they can be resonantly heated by far-infrared radiation. Localized heating through a small, scanned near-field aperture then can serve as the basis to probe the inhomogenous current distribution.

The geometry used in quantum-Hall studies is outlined in Fig. 2. The two-dimensional electron gas is buried about 100 nm beneath the surface of a GaAs crystal, and is laterally confined by etching that forms a rectangular bar. Electrical contacts allow to impress a current flow and to measure the longitudinal as well as the transverse voltage drops, which

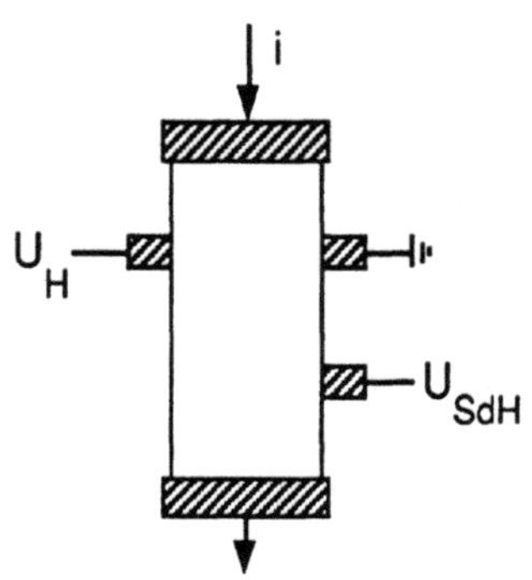


Figure 2. Typical layout of two-dimensional electron gas sample ("Hall bar") with contacts to inject d.c. current, and to measure longitudinal (SdH - Shubnikov-de Haas) and transversal (Hall) voltages.

both exhibit characteristic signatures when the magnetic field is varied in the range of several T (Fig. 3). Especially the Hall resistance (ratio of transverse voltage to longitudinal current) exhibits plateaus given by fundamental constants, $R_H = h/e^2n$, where n is an integer. When such a plateau occurs in a magnetic field scan, the longitudinal voltage vanishes (Shubnikov-de Haas oscillation). According to the"edge state" model[8] the current does not flow over the breadth of the Hall bar but rather in narrow one-dimensional channels which are formed very close to the bar's edges, where the electronic potential rises due to confinement and crosses the Fermi energy (Fig. 4, note that several electronic (Landau) states exist, separated by the cyclotron energy). The current direction in an edge state depends on the magnetic field direction and is opposite for the two sample

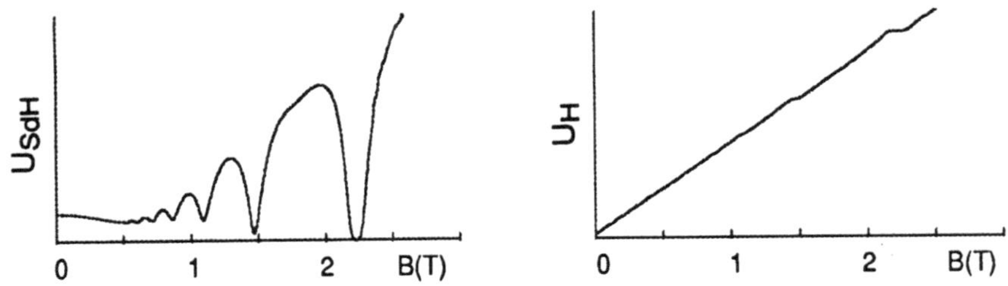


Figure 3. Typical Shubnikov-de Haas (left) and Hall voltage (right) *vs*. magnetic field.

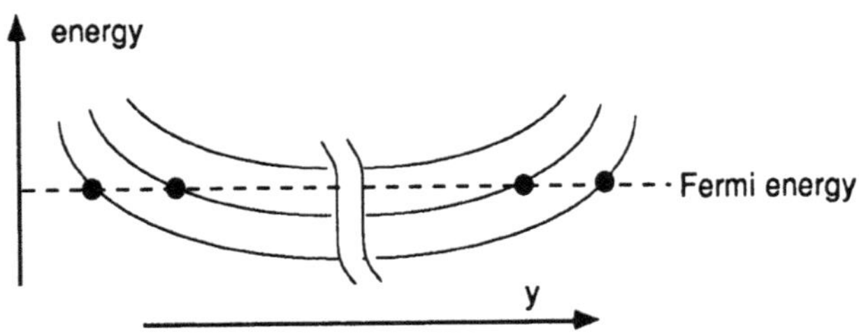


Figure 4. Sketch of left and right Hall bar edge regions showing two edge states.

edges. The model explains the occurence of Hall plateaus with a suppression of scattering between channels at opposite edges when the inner states become localized. While the edge state model is supported by many transport measurements no experiment has directly demonstrated the predicted, extremely inhomogeneous current distribution.

Our experiment is based on the recent observation that the cyclotron-resonance far-infrared photoconductivity becomes enhanced when the edge-state population is out of equilibrium.[9] Such a non-equilibrium is thought to be achieved *at one of the sample edges only* by decoupling the innermost channel from the other ones, either by using a quantum-point-contact[10] or a scattering phenomenon[11]. The decoupling then is suppressed by resonant far-infrared pumping and leads to enhanced conductivity, through a mechanism not known in detail. Our experiment is designed to confine the infrared irradiation and to test whether an increased response is observable only at the very edge predicted to carry non-equilibration population.

The beam of a far-infrared laser is focused by an f/6 lens through two Mylar foil windows into the cryostat. A conical waveguide with end aperture of about 210 μm is positioned in close distance (50 μm) from the Hall bar sample which can be mechanically moved in three dimensions, from outside the cryostat (Fig. 5). Since the width of the Hall bar is 700 μm we can expect to resolve four significant spots when scanning across the bar. Metallic top layers outside the channel serve to sense—by monitoring electrical contact—the approach of the waveguide tip during the mechanical scan.

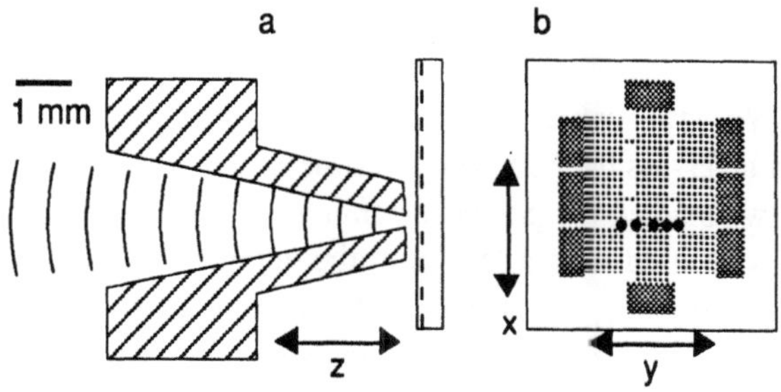


Figure 5. Scale drawing of taper guide focusing λ=392 μm radiation onto a GaAs crystal (a); lateral motion of the sample scans the focus area (black dots) across the Hall bar (b).

Indeed we observe an enhanced photosignal in the SdH voltage when one of the sample edges is irradiated.[3] When we invert the magnetic field direction the enhancement occurs at the opposite side, as predicted. Therefore the existence of a non-homogenous current distribution under the condition of non-equilibrated population of the edge states is clearly demonstrated.

## 4. Development of coaxial line focusing

The cutoff limit of conical hollow metal waveguides at 0.6 λ can be overcome by using other types of waveguide for near-field microscopy. To this end a coaxial metallic transmission line—which has no cutoff—was independently suggested by two groups.[12,13] Fee[12] demonstrated the use of commercial coaxial cable which is limited however to microwave frequencies below about 20 GHz; for higher frequencies they proposed an ultraminiature microstip patch antenna which has the potential of sub-μm resolution throughout the infrared region.

Our own development[13] uses conical coaxial waveguides. These are made by first machining a tapered hollow metal guide (full angle 30°) and then indroducing a central wire as illustrated in Fig. 6.[14] In practice we drill lead down to an inner diameter near

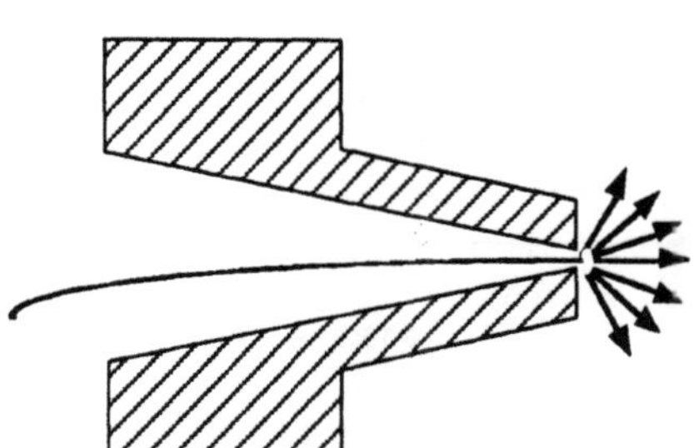

Figure 6. Conical coaxial waveguide

100 μm. The fine end of the taper is then made by pushing a sharp steel needle while observing the end surface in a microsope. In this way end apertures down to 3 μm diameter are achieved. A 25 μm Wo wire—which is sharpened to about 1 μm radius and coated with a thin SiO isolation layer—is then introduced as the center conductor.

In a first experiment we studied the power transmittance of a conical coaxial tip. A far-infrared laser beam is focused with a lens onto the entrance of the cone. A Golay detector is placed in 5 mm distance, to subtend, with its 5 mm aperture, a sizable solid angle to detect part of the transmitted power. The tungsten wire is axially moved on a micrometer stage. The result (Fig. 7) shows that sizable power is transmitted and varies with a $\lambda/2$ period, because the wire protruding from the aperture acts as an antenna.

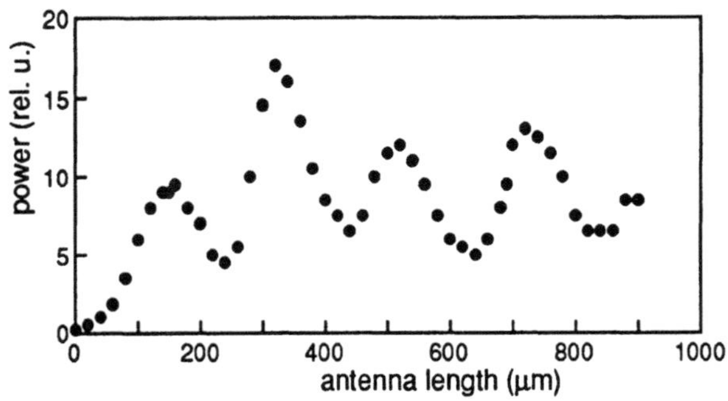


Figure 7. Transmittance of conical coaxial guide at $\lambda = 392$ μm *vs.* "antenna" length of center conductor which protrudes beyond the end aperture of 100 μm diameter.

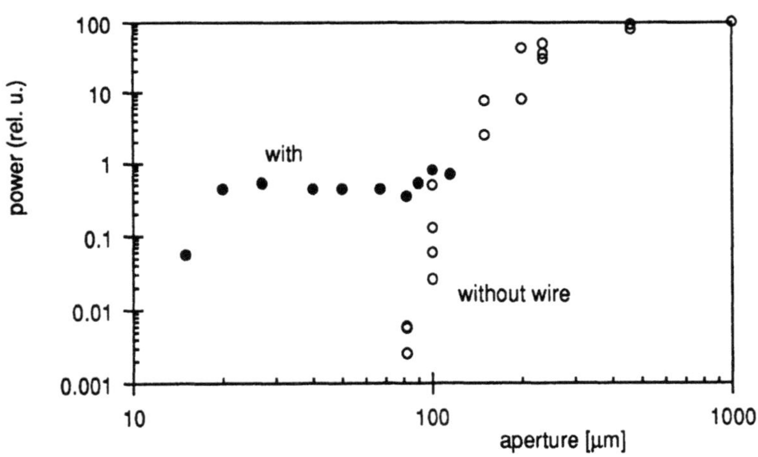


Figure 8. Transmittance at $\lambda = 320$ μm *vs.* diameter of end aperture.

In a series of further measurements we systematically varied the end diameter. The wire was positioned near λ/2 protrusion to give maximum signal. The result (Fig. 8) clearly demonstrates that the transmittance is independent of the end aperture diameter in the range from λ/2 down to λ/10. The measurement at λ/20 gives a reduced signal possibly because of an unintended shorting. In contrast the cones without wire exhibit unmeasurably small transmittance at small aperture, as expected from the cutoff effect. At higher end diameters the presence of the wire does not influence the power transmittance. Part of the increase for end diameters between λ/2 and λ comes from the narrowing of the end aperture's radiation pattern, which increases the coupling to the detector.

Stable coaxial focusing tips are produced by fixing the inner conductor—with a typical protrusion length near 10 to 50 μm—with PMMA. The spatial resolution was measured by scanning metallic apertures (metal mesh) and found to be of the order of the protrusion length or end aperture, whichever was largest. The coaxial tips were tested in low-temperature cryostats in several experiments which did not yet involve scanning. Both transmittance measurements and photoconductivity measurements with unstructured two-dimensional electron systems are being performed. The frequencies used so far range from 50 MHz to 3 THz. Cylotron resonance has been seen both in photoconductance as in chapter 3 and also in transmission. Shubnikov-de Haas oscillations have been seen in transmission, i.e. without the necessity of any electrical contacts to the sample. Interesting further applications are foreseen when the nearfield method will be applied to observe laterally structured systems such as a single quantum dot.

## 5. Acknowledgements

We are grateful to K.W. Kußmaul, W. Böhme and D. Stöckle for their expert technical assistance, and thank A. Wieck, D. Heitmann, R.J. Haug, K. Ploog and D. Weiß for discussions.

## 6. References

[1] Klitzing, K. von, Dorda, G., Pepper, M. (1980) "New method for high-accuracy determination of the fine-structure constant based on quantized Hall resistance", Phys. Rev. Lett. 45, 494 - 497;

Wees, B.J. van, Houten, H. van, Beenakker C.W.J., Williamson, J.G., Kouwenhoven, L.P., Marel van der, D. and Foxon, C.T. (1988) "Quantized conductance of point contacts in two-dimensional electron gas", Phys. Rev. Lett. 60, 848 - 850.

[2] Ando, T., Fowler, A.B. and Stern, F. (1982) "Electronic properties of two-dimensional systems", Rev. Mod. Phys. 54, 437 - 672.

Heitmann, D. (1987) "Far-infrared spectroscopy in two-dimensional electronic systems" in NATO ASI Series B, Physics 170, 317-346, ed. by E.E. Mendez and K. v. Klitzing;

Kuchar, F. (1988), "Microwave investigations of the quantum Hall effect in GaAs/AlGaAs heterostructures", Festkörperprobleme 28, 45-62.

[3] Merz, R. , Keilmann, F., Haug, R.J. and Ploog, K. (1992) "Non-equilibrium edge-state-transport resolved by far-infrared microscopy", Phys. Rev. Letters 1993.

[4] Keilmann, F. (1981) "Infrared high-pass filter with high contrast", Int. J. Infrared and Submillimeter Waves 2, 259-272.

[5] Tutter, M., Max-Planck-Institut für Plasmaphysik, D-8046 Garching, private communication (1992).

[6] Pohl, D.W., Denk W. and Lanz, M. (1984) "Optical stethoscopy: image recording with resolution $\lambda/20$", Appl. Phys. Letters 44, 651-653;

Dürig, U., Pohl, D.W. and Rohner, F. (1986), "Near-field optical-scanning microscopy", J. Appl. Phys. 59, 3318-3327;

Betzig, E., Lewis, A., Harootunian, A., Isaacson, M. and Kratschmer, E. (1986) "Near-field scanning optical microscopy (NSOM)", Biophys. J. 49, 269-279.

[7] Betzig, E., Trautmann, J.K., Harris, T.D., Weiner, J.S. and Kostelak, R.L. (1991) "Breaking the diffraction barrier: optical microscopy on a nanometric scale", Science 251, 1468-1470.

[8] Büttiker, M. (1990), "Edges, Contacs and the Quantized Hall Effect", in Festkörperprobleme 30, ed. by U. Rössler, Vieweg, Braunschweig, pp. 41-52;

Haug, R.J. (1992) "Edge-State transport and its experimental consequences in high magnetic fields", to be publ. in Semiconductor Sci. Tech.

[9] Dießel, E., Müller, G., Weiss, D., Klitzing, K. von, Ploog, K., Nickel, H., Schlapp, W. and Lösch, R. (1991), "Novel far-infrared-photoconductor based on photon-induced interedge channel scattering", Appl. Phys. Letters 58, 2231-2233.

[10] Wees, B.J. van, Willems, E.M.M., Harmans, C.J.P.M., Beenaker, C.W.J., Houten, H. van, Williamson, J.G., Foxon, C.T. and Harris, J.J. (1989), Phys. Rev. Letters 62, 1181-1184.

[11] Komiyama, S., Hirai, H., Sasa, S., Hiyamizu, S. (1989), "Violation of the integral quantum Hall effect: influence of back-scattering and the role of voltage contacts", Phys. Rev. B40, 12566-12569.

Alphenaar, B.W., McEuen, P.L., Wheeler, R.G., Sacks, R.N. (1990) , "Selective equilibration among the current-carrying states in the quantum Hall regime", Phys. Rev. Letters 64, 677-680.

[12] Fee, M., Chu, S. and Hänsch, T. (1989) "Scanning electromagnetic transmission line microscope with sub-wavelength resolution", Opt. Commun. 69, 219-224.

[13] Keilmann, F. (1991, filed 1988) "Scanning tip for optical radiation", U.S. patent no. 4,994,818.

[14] Merz, R. (1992) "Nahfeldmikroskopie an einem zweidimensionalen Elektronensystem", doctoral thesis, Universität Stuttgart.

# TUNNELING WITH COUPLING TO SURFACE PLASMONS

C. BAUR, B. KOSLOWSKI, R. MÖLLER AND K. DRANSFELD
*Universität Konstanz, Fakultät für Physik; Postfach 5560, D-7750 Konstanz, Germany*

ABSTRACT. To investigate the influence of a high frequency field on the tunneling current surface plasmons have been excited in a silver sample. An additional, plasmon induced current has been observed. The plasmon induced tunneling signal has been measured as a function of several parameters (lateral displacement, bias, tunneling current, gap-width). Several effects which might induce such a signal will be discussed. A comparison to the rectification of a low frequency bias suggests that the plasmon induced current is due to the rectification of the optical field by the tunneling junction.

## Introduction

A few years after the invention of scanning tunneling microscopy [1], the experimental techniques have been extended to investigate the interaction of the tunneling junction with photons. Gimzewski et.al. [2] have observed the emission of light from a tunneling junction. They have explained the effect by the excitation of localized plasmons in the silver sample and the tip. The emitted light is generated by the decay of the localized plasma oscillations. On the other hand side several experiments have been performed where laser radiation is coupled into the tunneling junction. Amer et. al. [3] reported the thermal expansion of tip and sample caused by the illumination with visible laser light. Mixing of two lasers by the tunneling junction has been demonstrated by Arnold and Krieger [4]. They also observed a dc current at zero bias. These effects have been explained by mixing/rectification of the lights field by the nonlinearity of the junction. Möller et. al. [5] and Kroo et. al. [6] have coupled the light into the tunneling junction by means of surface plasmons. An additional tunneling current induced by the plasmons has been found. It is assumed that the plasmon induced signal (PS) is due to rectification of the ac field related to the surface plasmon. The aim of this work is to verify this by comparing the plasmon induced tunneling current to the rectification of a low frequency bias.



*D.W. Pohl and D. Courjon (eds.), Near Field Optics*, 325–331.

**Experiments**

To excite the surface plasmons the method of Kretschmann [7] is used. π -polarised light is reflected internally at the hypotenuse face of a glass prism coated with a thin silver film (fig. 1). At a critical angle of incidence the parallel momentum of the light matches the momentum of the surface plasmons. Most of the light is converted to surface plasmons at the silver-vacuum interface. This leads to an attenuation of the total reflection. There is no excitation of surface plasmons for σ-polarisation of the incident light. In our experiment an electro optical modulator (EOM) switches from π to σ-polarisation to control the excitation of the surface plasmons. A photodiode detects the reflected light and monitors the excitation of the plasmons.

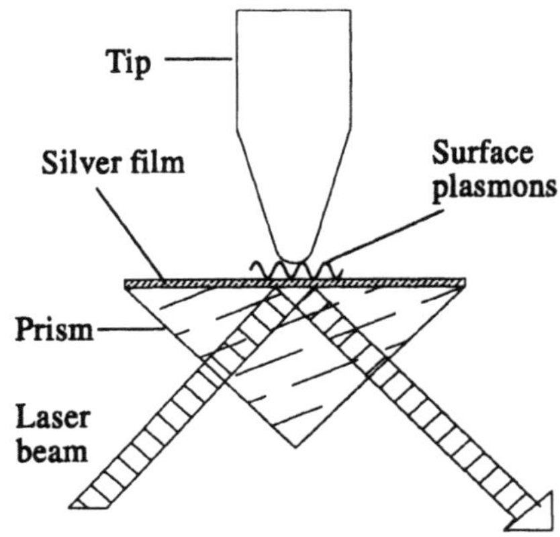


Fig. 1 Kretschmann setup for the excitation of surface plasmons

The experiments have been performed by a pocket size STM described in detail elsewhere [8]. The coarse approach is achieved by a spring reduction system yielding a very good mechanical stability of the instrument. Vibrational damping is provided by a stack of metal plates spaced by pieces of viton. The microscope is operated in ultra high vacuum at a pressure of $2*10^{-10}$ mbar. The tunneling tips have been prepared by electrochemical etching of a tungsten wire of 0.5 mm diameter. Prior the measurement the tips have been cleaned by fieldemission against a tantalum sample as counter electrode.

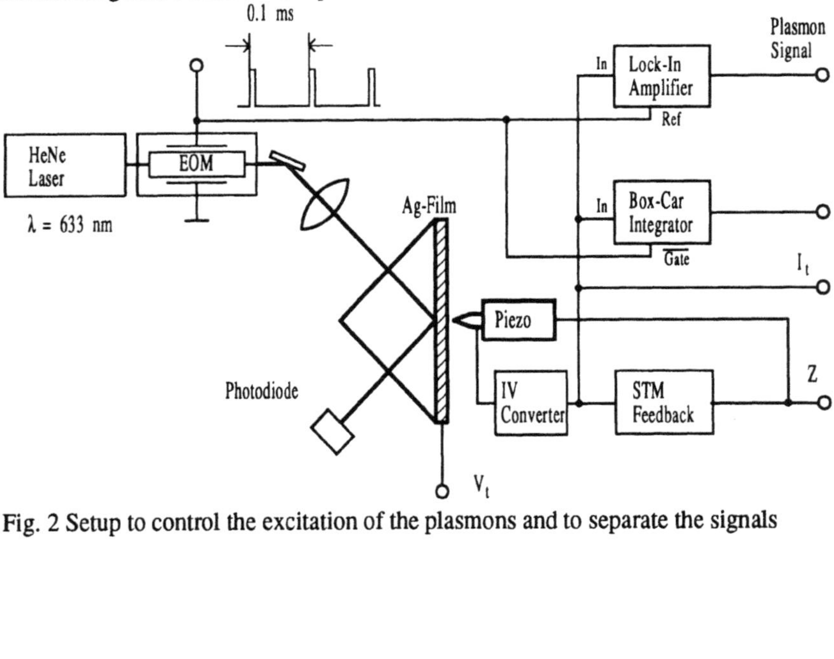


Fig. 2 Setup to control the excitation of the plasmons and to separate the signals

The experimental setup is sketched in fig. 2. The STM electronics and the computer system allow to acquire images and I/V-spectra. The plasmons are excited periodically for 10 μs at a rate of 10 kHz. A lock-in amplifier detects the PS. A boxcar integrator masks out the PS to obtain he undistorted tunneling current.

**Results**

Fig. 3 shows a time record of the tunneling current, the signal at the photo diode and the boxcar gate. In the off-phase of the gate the EOM switches the laser beam to π-polarisation. The attenuated reflection leads to the dips in the photo diode signal. Simultaneously sharp peaks appear in the tunneling current. For the chosen tunneling voltage the current reverses the sign excluding that the plasmon induced current results from thermal expansion.

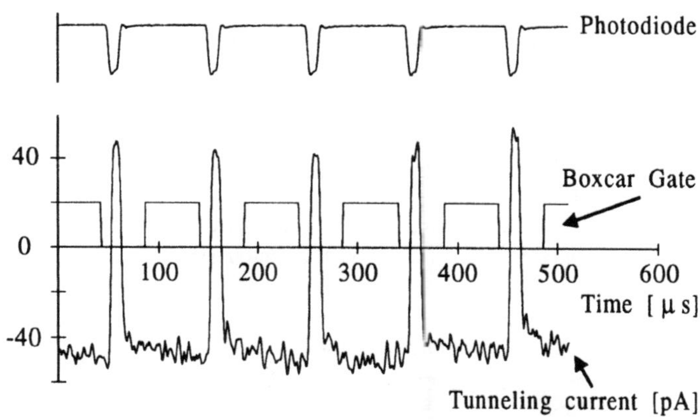


Fig. 3 Time record of the tunneling current and related signals ($V_t$ = 10 mV).

To find the dependence of the PS on the tunneling voltage the PS has been recorded simultaneously to an I/V-characteristic (fig. 4). The latter fits well to a polynomial of third order (shaded line). The plasmon signal (fig. 4b), however, exhibits a linear dependence proportional to the second derivative of the I/V-curve (shaded line).

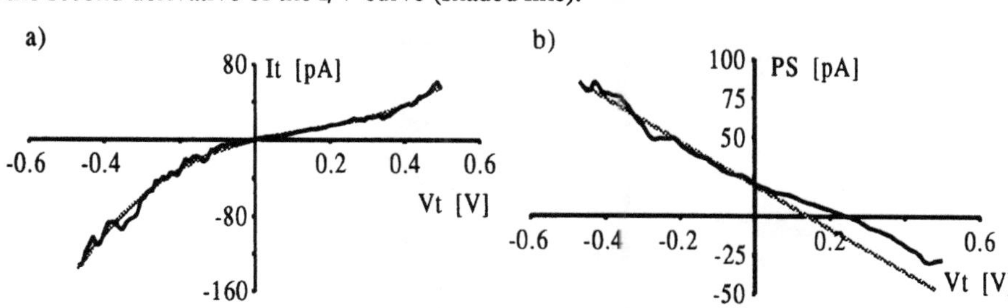


Fig. 4 Tunneling current and plasmon signal as a function of tunneling voltage (Tip)

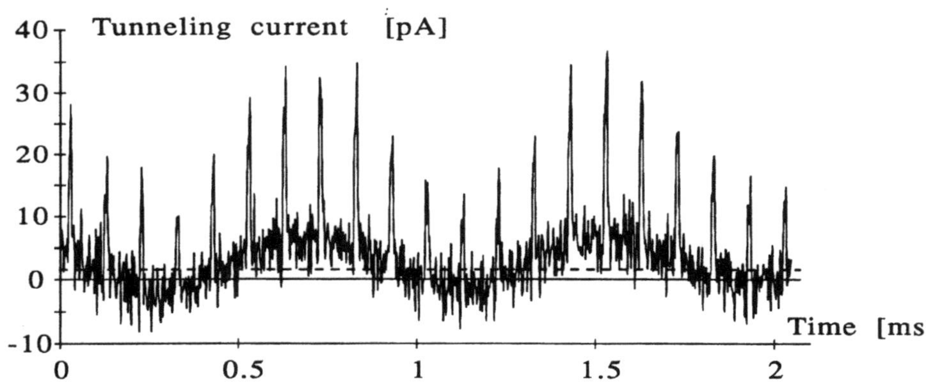


Fig 5. A time record of the tunneling current when tunneling with an ac bias. The dashed line marks the offset current.

Fig. 5 displays a time record of the tunneling current, if the junction is biased by an ac-voltage of 40 mVpp (f=1kHz). The plasmons are excited periodically giving rise to the positive and narrow peaks. The second contribution is proportional to the ac tunneling voltage corresponding to a tunneling resistance of 3 GΩ. In addition, there is an offset current due to rectification of about 2 pA. The boxcar averager evaluates this offset, giving a measure of the rectification of the low frequency bias (RF).

The lateral distribution of the signals and the topography of the sample has been measured at the same time. The topography shows the typical granular structure of a polycrystalline silver film. In the PS a strong variation is found. Some of the features can be related to the topography, e.g. the PS is smaller at grain boundaries. The RF has a quite similar behavior, i.e.the ratio between PS and RF is mainly constant.

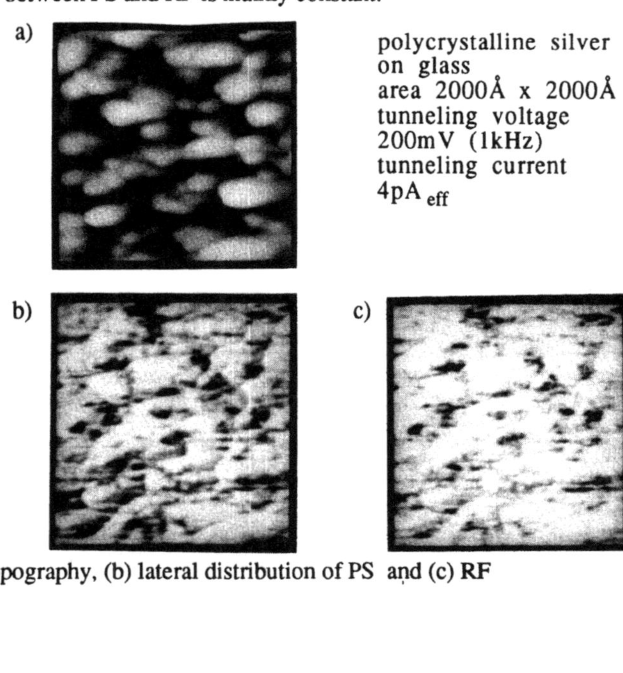


Fig. 6 (a) topography, (b) lateral distribution of PS and (c) RF

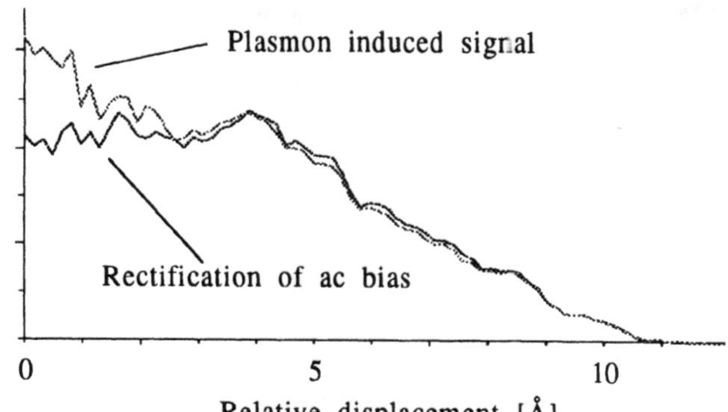


Fig. 7 PS and RF as function of relative Z-displacement

The dependence of PS and RF on gap-width is plotted in fig 7. Both signals have a complex structure. They do not decay exponentially as expected for the tunneling current. At small tip-sample distances the two signals differ from each other. The difference vanishes for displacements larger then 0.8 nm.

**Discussion**

There are several effects which may contribute to the PS: thermal expansion of sample and tip, thermovoltage, photoemission and rectification of the plasmon field.
Since the surface plasmons decay mainly by ohmic losses there is a heating of the silver film. Heat transfer from the sample to the tip and the illumination by the evanescent field of the surface plasmon cause heating of the tip. Only a small part of the tip is exposed to the light, compared to experiments, where the light is shined from the tip side into the tunneling gap. Consequently the thermal expansion is only a fraction of that for these experiments. To minimize the expansion the modulation frequency has been chosen as high as possible. Fig. 3 shows that the main part of the PS does not arise from thermal expansion of tip and sample. The tunneling current changes its sign, however thermal expansion of the tip or the sample alters the tunneling resistance, prohibiting a reversal of the current.

A thermovoltage can be caused by the different materials of tip and sample. Since the thermo-power of silver and tungsten differ just a little, a voltage not larger then 50 μV is expected for a temperature increase of the junction below 100 K. For a tunneling resistance of 1 GΩ the resulting current is smaller than $5*10^{-17}$ A. Støvneng and Lipavský [9] predicted a thermopower of the tunneling gap itself. This effect has the same order of magnitude as the thermovoltage mentioned above and can be neglected, too. Furthermore, a thermovoltage offsets the tunneling voltage leading to a signal proportional to $\partial I/\partial V$. This is in contradiction to the data of fig. 4.

Photoelectron emission has been observed by Gimzewski et. al. [10]. The tip acts as a collector of the emitted electrons and a current has been detected at a tip sample separation of more then 100 nm. This is in contradiction to the measurement of fig. 7. At a distance two orders of magnitude smaller the PS vanishes. H. Chen et. al. [11] observed photoemission caused by surface plasmons. To achieve a electron yield of $10^{-9}$ they had to use a light intensity of 6 MW/cm$^2$ which is much higher than in our experiment. Finally, photoemission cannot explain the dependence of the PS on the tunneling voltage.

If the PS arises from the rectification of the high frequency field by the static I/V-curve, it should be proportional to its second derivative:

$$I_{PS} = \frac{V_{PS}^2}{4} \cdot \frac{\partial^2 I}{\partial V_t^2},$$

where $I_{PS}$ is the plasmon signal, $V_{PS}$ is the plasmon induced voltage and $V_t$ is the tunneling voltage.

This is supported by the data plotted in fig. 4: the I/V-curve has been differentiated two times, the resulting rectification is in good agreement with the PS when $V_{PS}$ is set to 500 mV (dashed line in fig. 4b). Such a large amplitude can be understood, if there is a field enhancement of several orders of magnitude. Krieger et. al. [12] calculated an appropriate field enhancement using long wire antenna theory applied for the tip. This method is not valid for light at a wavelength of 633 nm. However, a field enhancement is inherent to the surface plasmons. If one assumes that the plasmon does not penetrate the tunneling tip the electric field amplitude increases in the order of 100, since the potential drops over the tunneling gap instead of the decay length in free space. An additional enhancement may arise from localized plasma oscillations of tip and sample. Unfortunately no calculations or measurements have been performed for the system silver film and tungsten tip in the vicinity of running surface plasmons.

The lateral distribution and the dependence on tip-sample separation of the PS and the rectification at low frequency corroborate the rectification model since, except for small gap-width, the behavior is essentially the same. The measurements have been performed at constant amplitude of the low frequency bias. As a consequence the induced high frequency voltage $V_{PS}$ or the field enhancement should be independent on the lateral position of the tip. Furthermore it follows that $V_{PS}$ is independent of the tip-sample separation at large gap width. It is somewhat surprising that neither the PS nor the RF have an exponential distance dependence as one would expect.

The distance dependence of PS and RF is also interesting from another point of view. If there would be a time scale defined by the tunneling process the rectification is expected to depend on the frequency of the ac voltage [13]. Provided the traversal time of electrons is a function of the tunneling barrier length [13,14] the ratio between PS and RF becomes a function of the tip-sample separation. We have not found a deviation of the rectification at 1 kHz and $5*10^{14}$ Hz and conclude that for this case the "tunneling time" is smaller than 2 fs.

## Conclusion

We coupled light into the tunneling junction of an STM via surface plasmons. Thermal effects are small compared to experiments exposing the tip directly to the laser beam. The plasmons generate an additional current which is superimposed to the tunneling current. The comparison to a low frequency bias supports the model of rectification of the optical field by the non-linearity of the junction. The estimation of the induced voltage leads to a field enhancement which has not been explained so far. An effect of a "traversal time" for tunneling electrons has not been observed.

**Acknowledgements**

Part of the work was supported by the Deutsche Forschungsgemeinschaft (SFB 306) and the Land Baden-Württemberg. We thank S. Herminghaus, J. Boneberg, U. Albrecht and P. Leiderer for stimulating discussions. O. Marti provided the software for image processing. One of us (C.B.) thanks the Schering AG for financial support.

**References**

[1] Binnig, G. and Rohrer, H. (1982) 'Scanning tunneling microscopy', Hel. Phys. Acta **55**, 726

[2] Gimzewski, J. K., Sass, J., Schlittler, R. R. and Schott, J. (1989) 'Enhanced Photon Emission in Scanning Tunneling Microscopy', Europhys. Lett. **8**, 435

[3] Amer, N. M., Skumanich, A. and Ripple, D. (1986) 'Photothermal modulation of the gap distance in scanning tunneling microscopy', Appl. Phys. Lett. **49**, 137

[4] Arnold, L., Krieger, W., Walter, H. (1987) 'Laser-frequency mixing in the junction of a scanning tunneling microscope', Appl. Phys. Lett. **51**, 786

[5] Möller, R., Albrecht, U., Boneberg, J., Koslowski, B., Leiderer, P. and Dransfeld, K. (1991) 'Detection of surface plasmons by scanning tunneling microscopy', J. Vac. Sci. Technol. B **9**, 506

[6] Kroo, N., Thost, J.P., Völcker, M., Krieger, W. and Walther, H. (1991) 'Decay length of surface plasmons determined with a tunnelling microscope.', Europhys. Lett. **15**, 289

[7] Kretschmann, E. (1971) 'Die Bestimmung optischer Konstanten von Metallen durch Anregung von Oberflächenplasmonen', Z. Physik **241**, 313

[8] Möller, R., Esslinger, A. and Koslowski, B. (1990) 'Thermal noise in vacuum scanning tunneling microscopy at zero bias voltage', J. Vac. Sci. Technol. A **8**, 590

[9] Støvneng, J. A. and Lipavský, P. (1990) 'Thermopower in scanning-tunneling-microscope experiments', Phys. Rev. B **42**, 9214

[10] Gimzewski, J. K., Berndt, R. and Schlittler, R. R. (1992) 'Observation of local photoemission using a scanning tunneling microscopy',Ultramicroscopy **42-44**, 366

[11] Chen, H., Boneberg, J. and Leiderer, P., submitted to Phys. Rev. B

[12] Krieger, W. (1992), talk 'Nonlinear Interactions of Laser Radiation in the STM' given at the workshop 'photons and scanning probe microscopies', Konstanz

[13] see for example:
Leavens, C. R. and Aers, G. C.,'The Time-Modulated Barrier Approach to Traversal Times from the Bohm Trajectory Point of View', Solid State Comm. **78**, 1015 (1991)
Büttiger, M. and Landauer, R. (1982) 'Traversal Time for Tunneling', Phys. Rev. Lett. **49**, 1739
Huang, Z. H., Cutler, P. H., Feuchtwang, T. E., Kazes, E., Nguyen, H. Q. and Sullivan, T. E. (1990) 'Model studies of tunneling time', J. Vac. Sci. Technol. A **8**, 186
Raedt, H. de, Garcia, N. , Huyghebaert J. (1990) 'Tunneling through time modulated barriers: is there a crossover frequency ?', Solid State Comm. **76**, 8969

[14] see for example
Hartmann, T. E. (1962) 'Tunneling of a Wave Packet', J. Appl. Phys **33**, 3427
Büttiger, M. (1983) 'Larmor precission and the traversal time for tunneling', Phys. Rev. B **27**, 6178

# OPTICAL SPECTROSCOPY AND MICROSCOPY USING SCANNING TUNNELING MICROSCOPY

J.K. GIMZEWSKI, R. BERNDT* and R.R. SCHLITTLER
*IBM Research Division*
*Zurich Research Laboratory*
*8803 Rüschlikon, Switzerland*

A.W. McKINNON, M.E. WELLAND and T.M.H. WONG
*Department of Engineering*
*University of Cambridge*
*Trumpington St.*
*Cambridge, UK*

Ph. DUMAS, M. GU, C. SYRYKH and F. SALVAN
*GPEC, URA/CRNS 783,*
*Fac. de Luminy,*
*Marseille, France*

A. HALLIMAOUI
*CNET-Meylan*
*Meylan, France*

ABSTRACT: Light emission arising from an electron making a transition between two energy levels provides one of the most important sources of information on the electronic structure of matter. In optical microscopy and spectroscopy, the spatial resolution achievable is generally limited by the wavelength of the photons involved. Two approaches developed to overcome this limitation are near-field optic probes to spatially confine the electromagnetic fields used and the use of sub-nm sized electron excitation probes. The latter approach, developed by us using the STM, has similarities and differences with the former approach which forms the main part of the present workshop. We discuss the technique with examples of optical spectroscopy and, comparisons of topography and photon maps for metal, semiconductor and adsorbate systems.

## Introduction

In this paper we highlight a variety of optical spectroscopic and optical microscopic measurements obtained using what can be conceptually viewed as a near-field optic probe that can be readily achieved as an extension of scanning tunneling microscopy. In the current experimental realizations of near field optics (Fig. 1) a conventional photon source ($S$) is focused by conventional optics ($O$) onto a geometric construction

*Present address: Université de Lausanne, Institut de Physique Expérimentale, 1015 Lausanne, Switzerland



*D.W. Pohl and D. Courjon (eds.), Near Field Optics,* 333–340.

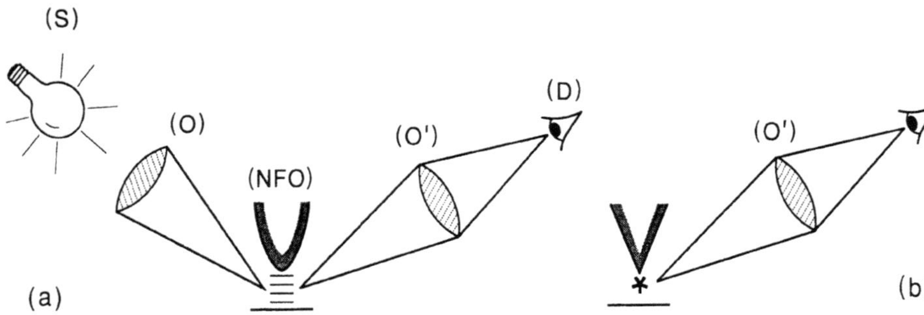


Figure 1. (a) Schematic of NFO. (b) Schematic of photon emission from STM.

(NFO) which spatially confines or 'squeezes' the photon to sub-wavelength dimensions. The resulting electromagnetically confined system interacts with its surroundings and then radiates into the far-field where conventional optics ($O'$) and photon detection schemes are employed ($D$) [1].

Our arrangement, pioneered in 1988 [2-4] is shown schematically in Fig. 1(b). For metals, tunneling electrons, which are spatially confined to dimensions of several angstroms in width with tip-surface gap distances of ~1 nm, define the excitation volume of what we observe as photons in the far field. The excitation in the near field results in enhanced spontaneous radiation due to the presence of tip-induced plasmon (TIP) modes [5] shown schematically in Fig. 2. It is a matter of personal taste as to whether one considers the process as two steps i.e. excitation of TIP modes and then emission or just as the emission of a photon by the tunneling electron [5-7].

The spectrum of light emitted from the tunnel junction is determined not only by the optical properties (dielectric functions) of the tip $\varepsilon_L$ and surface $\varepsilon_s$ and medium $\varepsilon_V$ but also by the geometric properties of the junction [6-8]. The advantages of the experiment are that optical spectroscopy and microscopy can be achieved with high lateral resolution readily using the well developed methods of STM where atomic resolution is the norm.

For metallic systems, photons in the visible are created by a process known as inelastic tunneling (IET). In STM, elastic tunneling dominates the measured current and occurs as an electron transfer process from one atom $A$ to another $B$ via a potential barrier of width, $S$, and height $\bar{\phi}$ under the influence of an applied voltage ($V$) [see Fig. 2(a)]. Inelastic tunneling processes dominate in photon emission [5,7]. The inelastic transmission probability is governed by similar parameters $\phi'$, $S'$ and $V$ as well as the local density of states in the sample situated at an energy $h\nu$ below that of $E_F$ of the tip. In addition, the coupling efficiency of the tunnel electron to TIP modes enters into the photon emission probability. This latter aspect closely links our approach to those of photon induced NFOS. In fact clusters $A'$ and $B'$ must be considered as a collective system defined by their dielectric properties, and the local geometry of the junction.

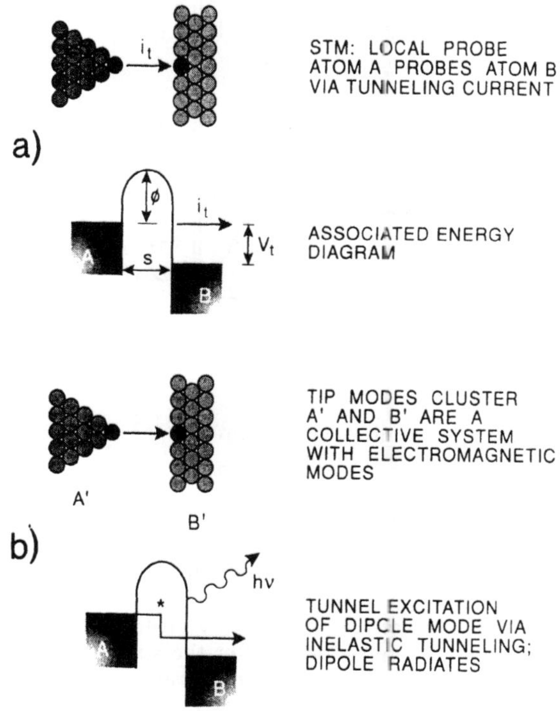


Figure 2. (a) STM and (b) TIP modes. See text for explanation.

## Fluorescence Spectra and Isochromat Spectra

A variety of spectroscopic modes are available to us. The most useful, and time consuming in terms of data collection, is fluorescence spectroscopy. Cooled CCD or diode arrays for detection and low *f* number grating spectrographs are employed typically covering $\lambda \approx 180$-900 nm. Figure 3 provides an overview of fluorescence spectra obtained from (a) Ag(111), (b) Au(110) and (c) Cu(111) surfaces using W tips. All spectra show a single maximum of detected emission which exhibits a sharp cutoff at short wavelength and a broad tail in the red region of the optical spectrum. On Cu, a shoulder is observed on the short wavelength side of the emission maximum. Emission was not detected on Ag at $\lambda \sim 330$ nm corresponding approximately to the bulk and surface plasmon of Ag. The large dynamic intensity range of the spectra shown in Fig. 3(a) places an upper limit on the detected emission from propagating surface plasmons which is at least $10^2$ times smaller than the observed emission [9].

Good agreement between experiment and theory [5] in the positions of the maxima and in signal intensity provides persuasive evidence that tip-induced modes are predominant in the emission process.

Most of the structure in calculated spectra results from a frequency-dependent field enhancement factor which describes the response of the tip-sample region. The

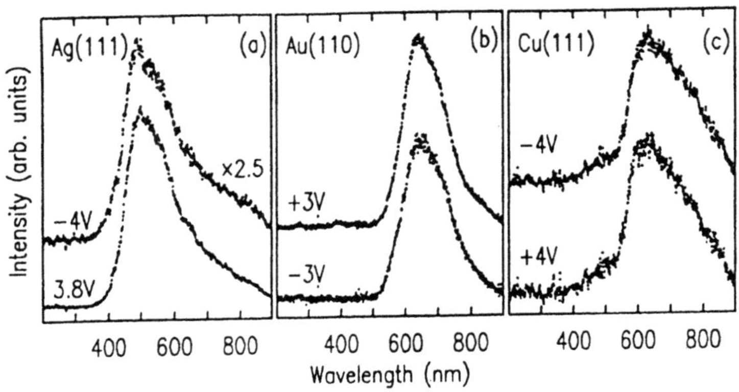


Figure 3. Fluorescence spectra acquired with W tips on (a) Ag(111), (b) Au(110) and (c) Cu(111) for positive and negative tip voltages $V_t$.

structure of this enhancement is dominated by the dielectric properties of the material with the lower damping, i.e. the noble metal when a tungsten tip is used. The role of the tungsten tip is to break lateral symmetry, thus giving rise to the TIP modes, in addition to broadening the resonance peak.

Figure 3 also displays the effect of tip bias polarity on fluorescence spectra. Spectra at positive $V_t$ have been obtained immediately after the spectra at negative $V_t$ on Au and Cu. No significant dependence of the emission characteristics on the polarity of $V_t$ is observed. The variation in intensity observed on Ag for slightly different $|V_t|$ is due to the steep intensity-voltage relation in this voltage range [10]. The lack of a polarity dependence is interpreted as a supplementary indication that the TIP modes are excited by tunneling electrons *inside the gap*. For a coupling mechanism *inside the electrodes* as assumed in the hot-electron model in which excitation occurs within a mean free path of the material, one expects the different dielectric properties of the materials of tip and sample to have an important influence.

For most metals – except those exhibiting free electron-like behavior in the optical range where TIP resonances tend to occur – the dielectric functions of the tip $\varepsilon_t$, sample $\varepsilon_g$ and interviewing medium $\varepsilon_v$ determine the spectrum. Hence spectra will generally yield information on local optical properties of the cavity formed by tip and sample. Further details of the theory are addressed in the contribution of Johansson in these proceedings [11].

Whereas a single maximum is observed in fluorescence spectra when using a tip material with a large imaginary part of its dielectric function (eg. W, Ir), for non-lossy materials i.e. an Ag coated W tip and Ag sample sharp photon emission peaks, often multiple peaks due to different modes are observed. The energetic position and higher intensities observed experimentally are extremely geometry dependent. This should eventually permit the study of electromagnetic modes in different sized and shaped clusters of atoms. At present our degree of control over the cluster preparation and stabilization, which may involve employing low temperatures for the latter aspect [12] are experimentally in the development stage.

Isochromat spectroscopy is widely used for inverse photoemission (IPE) studies in UHV by surface physicists [13] to obtain information on the occupied density of states of the sample. They tend to work in the uv-region and our first photon emission results

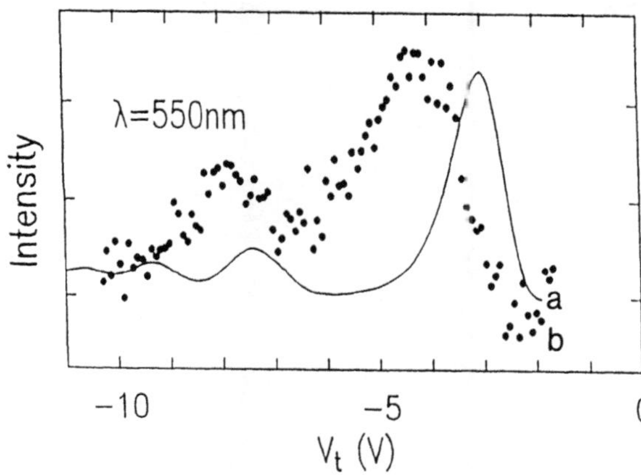


Figure 4. Isochromat spectra from clean (solid curve) and oxidized (dotted curve) Ti at $\lambda$ = 550 nm. Intensity scale is adjusted to make both curves visible on the same plot.

with STM [2] used such a detector. For STM studies a photo multiplier equipped with a band pass filter in the visible provides a more suitable energy range for tunneling electrons. Figure 4 shows isochromat spectra recorded from clean and oxidized Ti (0001) facets. The onset of photon emission at $V = 2.2$ V for the clean metal reflects an unoccupied density of states down to $E_F$ as expected whereas oxidation shifts this onset to below $E_F$ consistent with charge transfer from the Ti to O in forming the oxide. Similar data using IPE shows an encouraging similarity [14]. In STM however the area probed is of sub-nm dimensions. Additional features in the spectra in particular at $V_t > 5$ V are discussed in detail elsewhere [10].

**Photon Mapping**

Photon mapping is one of the exciting aspects of photon emission from STM. In addition to conventional topographic imaging, photon maps provide another channel of information related to the chemical properties of surfaces. Returning to the oxidation of Ti, discussed with respect to isochromat spectroscopy, photon mapping of the metallic areas of partially reacted Ti films have been obtained [15]. Figure 5 shows both the topography (a) and photon map (b) of a partially oxidized Ti surface. The topographic image contains terraces separated by single and multiple atomic steps. Superimposed on the terraces one observes an array of protrusions with apparent heights of 0.5-2 Å with diameters varying from ~10-20 Å. The density of these structures was found to increase with oxygen exposure and their height depends on tip polarity indicating that they are oxygen induced. In Fig. 5(b) a direct correlation is found between the protrusions and photon emission intensity comparable with the isochromat spectral characteristics discussed earlier. The photon maps themselves clearly show the distribution of metallic and oxidized areas with subnanometer spatial resolution and indicate also increased reactivity of the step structures.

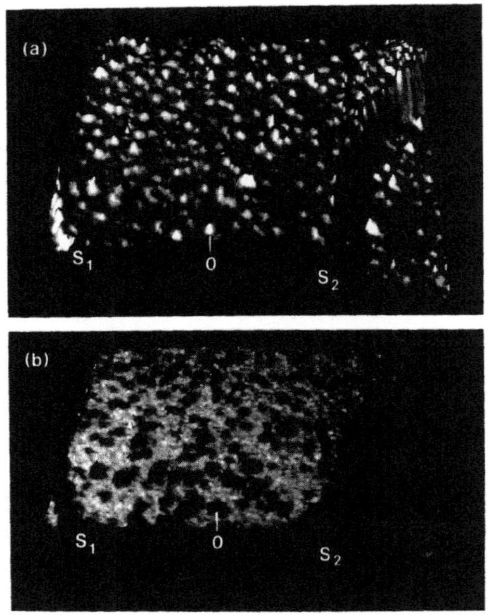


Figure 5. (a) STM topograph of a 1000 Å × 1000 Å area of a Ti film, that has been exposed to 20 L of $O_2$ at room temperature, rendered as a pseudo-illuminated image. The vertical extension is 20 Å. The STM was operated in constant-current mode with $i_t = 50$ nA and $V_t = -3.7$ V. (b) Integral photon intensity (photon map) recorded simultaneously with the topography shown in (a). The intensity values are represented by a grey-scale used to shade the topographic image. The intensity scale ranges linearly from 0 to 1000 counts per second. From Ref. 15.

Luminescence from semiconductors [16] can also be investigated using the tip as a source of electrons or holes. The mechanism for light emission differs in the sense that radiative recombination in the bulk occurs. A recent study of optical spectroscopy and photon mapping from CdS $(11\bar{2}0)$ surfaces is described elsewhere [17]. There is a growing interest in quantum size effects in semiconductors occurring on the nanometer scale.

Porous silicon (PS) is an interesting example of such a material since its luminescence properties in the visible range are believed to be determined by size effects [18]. It is postulated that columns of nanometer sized dimensions show a direct gap at $k = 0$ [19]. Recently we have investigated both the topography [20] and local photon emission characteristics of PS using STM. Figure 6 shows an example of topography (a) and simultaneous photon maps from an 85% PS layer of a photoluminescent active layer. The topography of the PS layer consists of small protrusions of variable size. In the photon map areas of the PS layer are observed to emit light more strongly than others. These emitting areas have characteristic dimensions of ~3-6 nm, consistent with TEM images of the column dimensions. Although these results are preliminary, they give an indication of the possibilities to study the local emission properties of quantum-confined systems and map a property (cathodoluminescence) on a length scale relevant to its nanostructural origin.

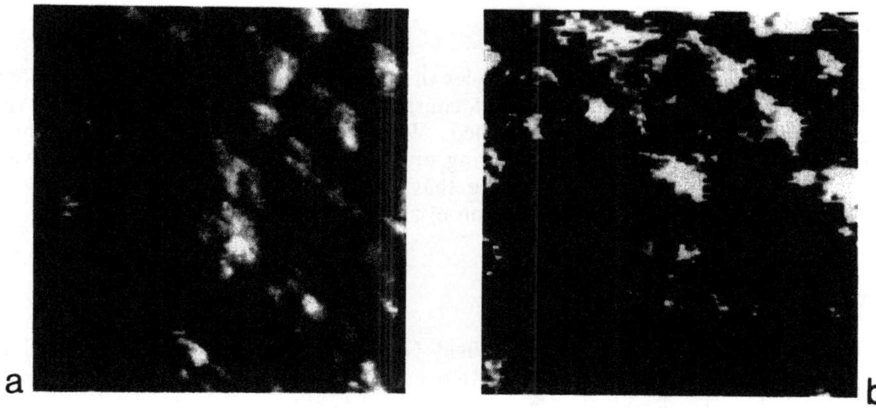


Figure 6. (a) STM topograph of a 500 × 500 Å area of 85% porous silicon. (b) Corresponding photon map recorded in the constant tunnel current mode.

The final example to conclude this paper concerns photon mapping of thin layers of Ag on Si(111), where a fascinating contrast mechanism in photon emission mapping, apparently unrelated to topography, has been observed. Figure 7(a) shows a topographic image of an area of the surface prepared by deposition of ~10 Å of Ag on Si(111) which exhibited the (7×7) reconstruction. The topography contains a fine granular structure.

The corresponding photon map shows no correlation with surface topographic features rather a series of bright bands is observed that run across the image. These results indicate an interference effect. The bands themselves can have widths of several angstroms only. Further work is required to establish their exact nature but their fine structure suggests that interesting and as yet unknown physics lies in front of us.

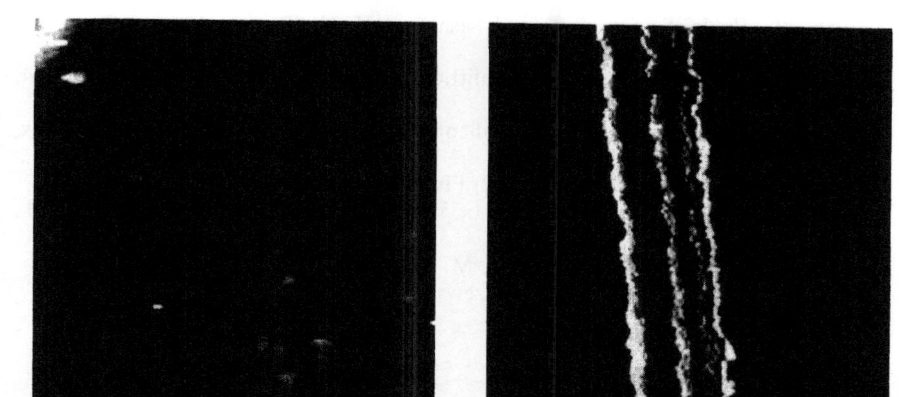


Figure 7. (a) Topograph and (b) photon map of a Si(111) (7×7) surface after deposition of ~10 Å of Ag. The granular structure in (a) has a mean roughness of several angstroms only.

## Concluding Remarks

We have demonstrated via selected examples that optical spectroscopy and microscopy can be performed through studying light emission from STM and that much of the physics can be understood and modelled. We also show a few examples where interesting and unexpected effects occurring on the nanometer scale can be observed. Both aspects provide persuasive evidence that further developments in the study of photons emitted from the tip-surface region of an STM are foreseeable.

## References

[1] For a review of Scanning Near-Field Optical Microscopy see D.W. Pohl in Advances in Optical and Electron Microscopy, Vol. 12 (1991) p. 243.
[2] Gimzewski, J.K., Reihl, B., Coombs, J.H. and Schlittler, R.R. (1988) Z. Phys. B 72, 497.
[3] Coombs, J.H., Gimzewski, J.K., Reihl, B., Sass, J.K. and Schlittler, R.R. (1988) J. Microscopy 152, 325.
[4] Gimzewski, J.K., Sass, J.K., Schlittler, R.R. and Schott, J. (1988) Europhys. Lett. 8, 435.
[5] Berndt, R., Gimzewski, J.K. and Johansson, P. (1991) Phys. Rev. Lett. 67, 3796.
[6] Johansson, P., Monreal, R. and Apell P. (1990) Phys. Rev. B 42, 9210.
[7] Persson, B. and Baratoff, A. (1990) Bull. Am. Phys. Soc. 35, 634; (1990) Phys. Rev. Lett. 68, 3227.
[8] Berndt, R., Baratoff, A. and Gimzewski, J.K. (1990) in "STM and Related Methods," NATO ASI Series E, Vol. 184, p. 269.
[9] Berndt, R. and Gimzewski, J.K. (1992) Surf. Sci. 269/270, 556.
[10] Berndt, R. and Gimzewski, J.K. (1992) submitted to Annalen der Physik.
[11] Johansson, P. (1992) these proceedings.
[12] Gaisch, R., Gimzewski, J.K., Reihl, B., Schlittler, R.R., Tschudy, M. and Schneider, W. (1992) Ultramicroscopy 42-44, 1621.
[13] Reihl, B. (1985) Surf. Sci. 162, 1.
[14] Konishi, R., Ikeda, S., Osaki, T. and Sasakura, H. (1990) Jpn. J. Appl. Phys. 29, 1805.
[15] Berndt, R., Gimzewski, J.K. and Schlittler, R.R. (1992) Ultramicroscopy 42-44, 355.
[16] Holt, D.B. (1981) "Microscopy of Semiconducting Materials," Oxford, Inst. Phys. Conf. Ser. 76, 316.
[17] Berndt, R. and Gimzewski, J.K. (1992) Phys. Rev B 45, 14095.
[18] Canham, L.T. (1990) Appl. Phys. Lett., 57, 1046; Cullis, A.G. and Canham, L.T. (1991) Nature (London), 353, 335.
[19] Buda, F., Kohanoff, J. and Parrinello, M. (1992) Phys. Rev. Lett., 69, 1272.
[20] Dumas, Ph., Gu, M., Syrykh, C. Gimzewski, J.K., Hallimaoui, A. and Salvan, F., in preparation.

# CALCULATION OF RESONANTLY ENHANCED LIGHT EMISSION FROM A SCANNING TUNNELING MICROSCOPE

PETER JOHANSSON[(a)]
*Institute for Studies in Interface Sciences,*
*Facultés Universitaires N.-D. de la Paix,*
*61 rue de Bruxelles, B-5000 Namur, Belgium*

R. MONREAL AND PETER APELL[(b)]
*Departamento de Fisica de la Materia Condensada CXII,*
*Universidad Autonoma de Madrid, E-28049 Madrid, Spain*

ABSTRACT. We have calculated the intensity of emitted light from a scanning tunneling microscope (STM) equipped with an Ir or W tip, probing the surface of a noble metal film. In the calculation we model the tip by a sphere. To describe the electromagnetic properties of the materials we have used experimentally measured dielectric functions. In agreement with recent experimental results we find that the light intensity is considerably enhanced compared with the yield in ordinary inverse photoemission experiments. Our theory also, in reasonable agreement with the experiments, predicts characteristic peaks in the light intensity at certain photon energies (≈2.5 eV for Ag, ≈2.1 eV for Au and Cu). The enhanced spontaneous emission is the result of the coupling between the tunneling electrons and an interface plasmon mode localized to the region between the tip and the metal film. This plasmon mode is radiative due to the tip-induced loss of translational invariance along the sample surface.

## 1. Introduction

The scanning tunneling microscope (STM) has undergone an enormous development during the ten years that have passed since its invention. The STM is today a standard tool when studying the topography of a variety of surfaces. The development of the STM has also lead to new applications. One of these is the study of light emission from an STM[1-4]. Gimzewski, Berndt, and co-workers have in a series of experiments studied the light emission from an STM with a tip of W (or Ir) probing the surface of a noble metal sample. They found that the intensity of the emitted light is resonantly enhanced in certain frequency ranges. For Ag-samples the light emission had a peak at ≈2.5 eV and for Au- and Cu-samples the light emission showed a maximum at ≈2.1 eV[1-3].

Earlier, one has studied light emission from solid-state tunnel junctions. The typical chain of events there, is that a tunneling electron excites an interface plasmon which then decays by sending out light. The second step is the weak link in this process. Photons in

[(a)]Present address: NORDITA, Blegdamsvej 17, DK-2100 Copenhagen, Denmark.
[(b)]Permanent address: Institute of Theoretical Physics, Chalmers University of Technology, S-412 96 Göteborg, Sweden.



*D.W. Pohl and D. Courjon (eds.), Near Field Optics, 341–152.*

the visible range are long-wavelength excitations compared with interface plasmons, which typically have a wavelength of a few hundred Å. Therefore interface plasmons on a perfectly smooth surface cannot radiate. If, on the other hand, the translational invariance along the surface is lost, some of the interface plasmons will decay into photons. In the experiments this was achieved in different ways. One used rough top electrodes[5] or top electrodes that consisted of small metal particles[6,7], or the whole tunnel juntion was fabricated on top of a grating[8,9].

Meanwhile theories that addressed the different experimental situations were developed. Davis[10] calculated the excitation rate of interface plasmons. Laks and Mills[11] studied the radiative decay of interface plasmons due to interface roughness. This work was later extended by Arya and Zeyher[12] to take into account multiple scattering of the interface plasmons by the surface roughness. Rendell and Scalapino[13] presented a theory that dealt with light emission from tunnel junctions with a top electrode consisting of small metal particles.

In the following we will present an explanation of the observed light emission from an STM[14,15]. The interface plasmons play an important role also in this case. The presence of the STM tip close to the sample surface leads to the formation of a localized plasmon in the cavity between the tip and the sample. The frequency of this localized mode, and consequently of the emitted light, depends on the geometry and the materials in the tip and sample. Since the translational invariance along the surface is lost due to the tip the localized plasmon can radiate. This leads to an enhanced spontaneous emission of light. We find that typically 1 out of $10^4$ tunneling electrons emits a photon in the visible range.

Persson and Baratoff[16,17] have also studied light emission from an STM theoretically. They calculated the intensity of the emitted light due to different processes when an electron is injected into a metal particle. The results show that inelastic tunneling is $\approx 10^3$ times more effective in causing photon emission than is hot-electron decay.

## 2. Theory

The light emission from an STM is caused by transitions, from states in one electrode, to states in the other electrode. The energy difference between these states equals the photon energy. The effects of a transition can for our purposes be described by a classical current, $zj(\mathbf{r}',w)$. We use the reciprocity theorem[18] to determine the radiated electric field $\theta E(\mathbf{r},\omega)$ far away from the tip. Therefore we actually calculate the z-component of the electric field, $\mathbf{z}\cdot\mathbf{E}_{ind}(\mathbf{r},\mathbf{r}',\omega)$, in the vicinity of the tip caused by a delta-function current source with strength $j_0$ at $\mathbf{r}$ directed along the local $\theta$-axis. The reciprocity theorem then states,

$$E(\mathbf{r},\omega) = (j_0)^{-1} \int d^3r' \, \hat{\mathbf{z}}\cdot\mathbf{E}_{ind}(\mathbf{r},\mathbf{r}',\omega)\, j(\mathbf{r}',\omega). \tag{1}$$

The problem is now split into two parts, the determination of $\mathbf{E}_{ind}(\mathbf{r},\mathbf{r}',\omega)$ and of $j(\mathbf{r}',\omega)$. Most of the interesting features of the results originate from $\mathbf{E}_{ind}(\mathbf{r},\mathbf{r}',\omega)$. For later convenience we write,

$$\hat{\mathbf{z}}\cdot\mathbf{E}_{ind}(\mathbf{r},\mathbf{r}',\omega) = \frac{i\omega j_0}{c^2}\frac{e^{iqr}}{r}\, G(\theta,\mathbf{r}',\omega). \tag{2}$$

The first factor is the amplitude, at the origin (i.e. below the tip), of the almost plane wave that the delta-function source sends out[19]. The second factor, $G(q,\mathbf{r}',\omega)$, describes the electromagnetic response, including plasmon resonances, of the tip-sample system. If the tip was not present G would take values ~1.

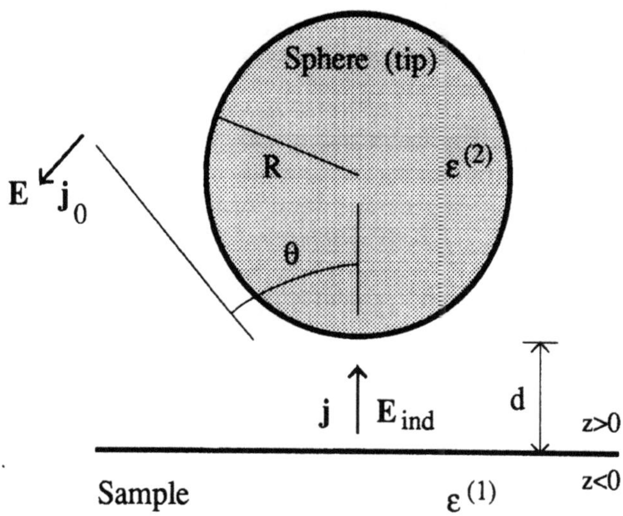


Figure 1. Schematic illustration of the model geometry considered in our calculation.

The exact geometry of the STM tip and the sample is not known. To proceed with the calculation we use the model geometry illustrated in Fig. 1. The sample is replaced by a semi-infinite medium and the tip is modeled by a sphere of radius R separated from the sample by a distance d. This distance is chosen such as to give a certain tunnel current. The dielectric properties of the tip and sample materials are taken into account by using experimentally measured dielectric functions in the calculations[20]. While the model geometry is a simplification, it has the virtue of giving an accurate description of the geometry in the region where the tip and the sample face each other. This is where the important interaction between the electromagnetic field and the tunneling electrons takes place. The calculation of $\mathbf{E}_{ind}(\mathbf{r},\mathbf{r}',\omega)$ is best understood if we split the field into a transverse part, $\mathbf{E}_T$, that varies slowly in space, and a rapidly varying longitudinal part $\mathbf{E}_L$, thus $\mathbf{E}_{ind}=\mathbf{E}_T+\mathbf{E}_L$ and $G=G_T+G_L$. When calculating $\mathbf{E}_T$ we ignore the presence of the sphere. The quantity we need in the following is the z-component of the transverse field just above the sample surface and just below the surface. The Fresnel formulae yield,

$$G_T(z=0^+) = \sin(\theta)\, 2\varepsilon^{(1)} q_0 / (\varepsilon^{(1)} q_0 + q_1), \quad (3a)$$

$$G_T(z=0^-) = \sin(\theta)\, 2\, q_0 / (\varepsilon^{(1)} q_0 + q_1). \quad (3b)$$

Here $\varepsilon^{(1)}$ is the dielectric function of the sample material and $q_0$ and $q_1$ are the z-components of the electromagnetic wave vector in the vacuum and sample respectively. Next we calculate the polarization effects due to the sphere. They are contained in $\mathbf{E}_L$. In doing this we neglect retardation effects and assume the z-component of $\mathbf{E}_T$ to be constant over the volume of interest, i.e. essentially the sphere. This is the major approximation of the calculation. It is reasonable since the largest sphere radius we use, 300 Å, is considerably smaller than the wavelength of visible light. We point out that retarded calculations on related problems give qualitatively similar results[21]. The electric field can now be determined from the boundary conditions at the sphere-vacuum interface and the sample-vacuum interface; the **E**-field parallel to the interfaces and the **D**-field normal to the interfaces should be continuous. The longitudinal electric field can of course be written as the gradient of a scalar potential, $\phi_L$, and moreover, since we now assume the z-component of the transverse field to be piecewise constant it can be written as the gradient of another scalar potential, $\phi_T$. These potentials are solutions to Laplace's equation in the three different regions of space. The above mentioned boundary conditions are thus,

$$\phi_L + \phi_T \text{ continuous across the interfaces,} \tag{4a}$$

$$\varepsilon\left(\frac{\partial \phi_L}{\partial n} + \frac{\partial \phi_T}{\partial n}\right) \text{ continuous across the interfaces.} \tag{4b}$$

Equations (4a) and (4b) can be solved in a bispherical coordinate system[22], in which both the sample surface and the surface of the sphere are coordinate surfaces. Laplace's equation is almost separable in these coordinates. To determine $\phi_L$, from which $\mathbf{E}_L$ and subsequently $G(\theta,\mathbf{r}',\omega)$ can be calculated, one has to solve a tridiagonal equation system. For more details see Refs. [13] and [15].

To calculate the tunnel current we have used an essentially one-dimensional model of the tunnel junction. We have, however, taken into account corrections due to the curvature of the tip. We do this by introducing an effective area, A, through which the tunnel current passes. The wave functions we use can be written,

$$\Psi(\mathbf{r}) = \frac{1}{\sqrt{A}}\, e^{i\mathbf{k}_{\parallel}\cdot\mathbf{r}_{\parallel}}\, \varphi(z). \tag{5}$$

A is determined by considering the elastic current,

$$I_{el} = A j_{el}(d) = \int d^2 r_{\parallel}\, j(\Delta(\mathbf{r}_{\parallel}, d)). \tag{6}$$

Here $j_{el}(d)$ is the current density on the symmetry axis and $j_{el}(\Delta)$ the current density through a barrier with thickness $\Delta$. When calculating these current densities both the tip and the sample are modeled as free electron metals. The barrier is assumed to be triangular in shape, thus the potential drop between the tip and the sample is linear. An electron tunneling inelastically from an initial state $|\Psi_i\rangle$ to a final state $|\Psi_f\rangle$ gives a contribution to the (classical) current density driving the light emission[19],

$$j_{fi}(\mathbf{r}',\omega) = \langle\Psi_f|2\,\hat{j}_z(\mathbf{r}')|\Psi_i\rangle =$$

$$= -\frac{ie\hbar}{m}\left\{\frac{\partial\Psi_f^*}{\partial z'}\Psi_i - \Psi_f^*\frac{\partial\Psi_i}{\partial z'}\right\}(\mathbf{r}'). \tag{7}$$

The energy difference between the two states should of course equal the photon energy, $\varepsilon_i - \varepsilon_f = h\omega$. The radiated electric field caused by the transition is,

$$E_{fi}(\mathbf{r}) = \frac{e\hbar\,\omega}{mc^2}\frac{e^{iqr}}{r}\int d^3r'G(\theta,\mathbf{r}',\omega)\left\{\frac{\partial\Psi_f^*}{\partial z'}\Psi_i - \Psi_f^*\frac{\partial\Psi_i}{\partial z'}\right\}(\mathbf{r}')$$

$$\approx \frac{e\hbar\,\omega}{mc^2}\frac{e^{iqr}}{r}G(\theta,\mathbf{r}'=0,\omega)\,\delta_{\mathbf{k}_{\|i},\mathbf{k}_{\|f}}\int_0^d dz'\left\{\frac{\partial\varphi_f^*}{\partial z'}\varphi_i - \varphi_f^*\frac{\partial\varphi_i}{\partial z'}\right\}. \tag{8}$$

To get the second line we have made two approximations. The small variations of G in the region between the tip and the sample have been ignored. Secondly we neglect the contributions to the light emission originating from within the tip and/or sample (hot-electron decay). In the present calculation this effect would only make a difference of a few percent. This is because the electric field of the interface plasmon is mostly located in the vacuum, since the magnitude of the dielectric constant is much larger in the metals. When considering tunneling through an oxide layer between two metals this may no longer be true; then the interface plasmon can also penetrate into the metals[10]. To calculate the total emitted power per unit photon energy and unit solid angle we add the contributions from different transitions incoherently. From electrodynamics we get[19],

$$\frac{d^2P}{d(\hbar\,\omega)d\Omega} = \frac{c}{8\pi}\sum_{i,f}|rE(\mathbf{r},\omega)|^2\,\delta(\varepsilon_i - \varepsilon_f - \hbar\,\omega). \tag{9}$$

Transforming the sum over states to k-space integrals yields (the first factor of 2 comes from the spin summation),

$$\frac{d^2P}{d(\hbar\,\omega)d\Omega} =$$

$$= 2\frac{c}{8\pi}\,\frac{e^2\hbar^2\,\omega^2}{m^2c^4}\,A\int\frac{dk_{zi}}{2\pi}\int\frac{dk_{zf}}{2\pi}\int\frac{d^2k_{||}}{(2\pi)^2}\,|M_{fi}|^2\,\delta(\varepsilon_i - \varepsilon_f - \hbar\,\omega), \qquad (10a)$$

$$M_{fi} = G(\theta,\mathbf{r}'{=}0,\omega)\int_0^d dz'\left\{\frac{\partial\varphi_f^*}{\partial z'}\,\varphi_i - \varphi_f^*\frac{\partial\varphi_i}{\partial z'}\right\}. \qquad (10b)$$

Three of the four k-space integrals can be done analytically, while the last one and the integral in the expression for $M_{fi}$ must be done numerically.

It is interesting to note that the largest contributions to the $M_{fi}$-integral come from points that lie close to the collector electrode. That is, an electron tunneling inelastically traverses most of the barrier at its initial energy before emitting a photon. A similar effect leads to the observation of a bias asymmetry in inelastic tunnel spectroscopy of dopant molecules in metal-oxide-metal tunnel junctions. There inelastic tunneling occurs more easily if the molecule is situated at the interface between the oxide and the collector electrode than in the opposite case[23].

We can get a considerably simpler expression for the differential power by making some more approximations. The differential power depends on $\omega$ through the prefactor $\omega^2$, through $G(\theta,\omega)$, and through the available phase space in the k-integrals. The phase-space dependence approximately yields a factor $(1\text{-}h\omega/eU)$, where U is the bias voltage. It is therefore reasonable to write,

$$\frac{d^2P}{d(\hbar\,\omega)d\Omega} \approx 2\frac{c}{8\pi}\,\frac{e^2\hbar^2\,\omega^2}{m^2c^4}\,A\,(1 - \frac{\hbar\,\omega}{eU})$$

$$\int\frac{dk_{zi}}{2\pi}\int\frac{dk_{zf}}{2\pi}\int\frac{d^2k_{||}}{(2\pi)^2}\,|M_{fi}|^2\,\delta(\varepsilon_i - \varepsilon_f). \qquad (11)$$

A transfer Hamiltonian calculation of the elastic current yields (provided the barrier is opaque),

$$I_{el} = 2\,\frac{2\pi e}{\hbar}\,A\int\frac{dk_{zi}}{2\pi}\int\frac{dk_{zf}}{2\pi}\int\frac{d^2k_{||}}{(2\pi)^2}$$

$$\left|\frac{\hbar^2}{2m}\left\{\frac{\partial\varphi_f^*}{\partial z}\,\varphi_i - \varphi_f^*\frac{\partial\varphi_i}{\partial z}\right\}(z_1)\right|^2\,\delta(\varepsilon_i - \varepsilon_f). \qquad (12)$$

The point $z_1$ should be situated in the barrier. A comparison of Eqs. (11) and (12) shows that we can write the differential power,

$$\frac{d^2P}{d(\hbar\,\omega)d\Omega} \approx \frac{I_{el}}{4\pi^2}\frac{e\omega^2}{\hbar\,c^3}\, d^2 \left| G(\theta,\mathbf{r}'=0,\omega) \right|^2 \left(1 - \frac{\hbar\,\omega}{eU}\right). \tag{13}$$

The numerical results obtained from Eq. (13) are larger than those we get from Eq. (10a), because the transition matrix elements are overestimated in Eq. (11). However, if we multiply the right hand side of Eq. (13) by a numerical factor (typically ~0.7) the agreement between the results is very good.

## 3. Results

In this section we discuss some of the theoretical results, emphasizing the qualitative aspects. For a more complete quantitative comparison between the theoretical and experimental results we refer to the paper by Berndt and Gimzewski in this volume.

As we mentioned above, almost all the essential features of the spectral distribution of the emitted light are contained in the electromagnetic response function, G, of the tip-sample system. In Fig. 2(a)-(c) the dependence of G on tip and sample material and geometry is illustrated. Figure 3 shows the spectral distribution of the emitted light in a typical case. Figure 2(a) has been calculated with a Ag-sample and an Ir-tip. The dielectric response has a broad peak at a certain frequency (photon energy). The resonance frequency varies with varying tip radius (radius of the sphere in our model). In Fig. 2(b) the tip material is W, and the sample consists of Au. The dielectric response has also in this case a resonant peak, but the resonance frequency is basically independent of the geometry (tip radius). In Fig. 2(c), finally, the tip consists of Ag and the sample of Au. Compared with the earlier cases the new feature there is that the resonance peak is much more narrow.

The dependence of the resonance frequency on geometry and material properties can be understood from a fairly simple model. Let us replace the sphere of tip material by a semi-infinite medium filling the part of space where z>d, as illustrated in Fig. 4. Let us also simplify the situation a bit more by assuming that the two semi-infinite media have identical dielectric properties and that they can be described by the Drude model,

$$\varepsilon(\omega) = 1 - \frac{\omega_p^2}{\omega^2}. \tag{14}$$

The short-wavelength interface plasmons existing in the vicinity of the two vacuum-metal interfaces can be described by a scalar potential, $\phi$, in a non-retarded model. We write $\phi$, satisfying Laplace's equation in the three regions of space as,

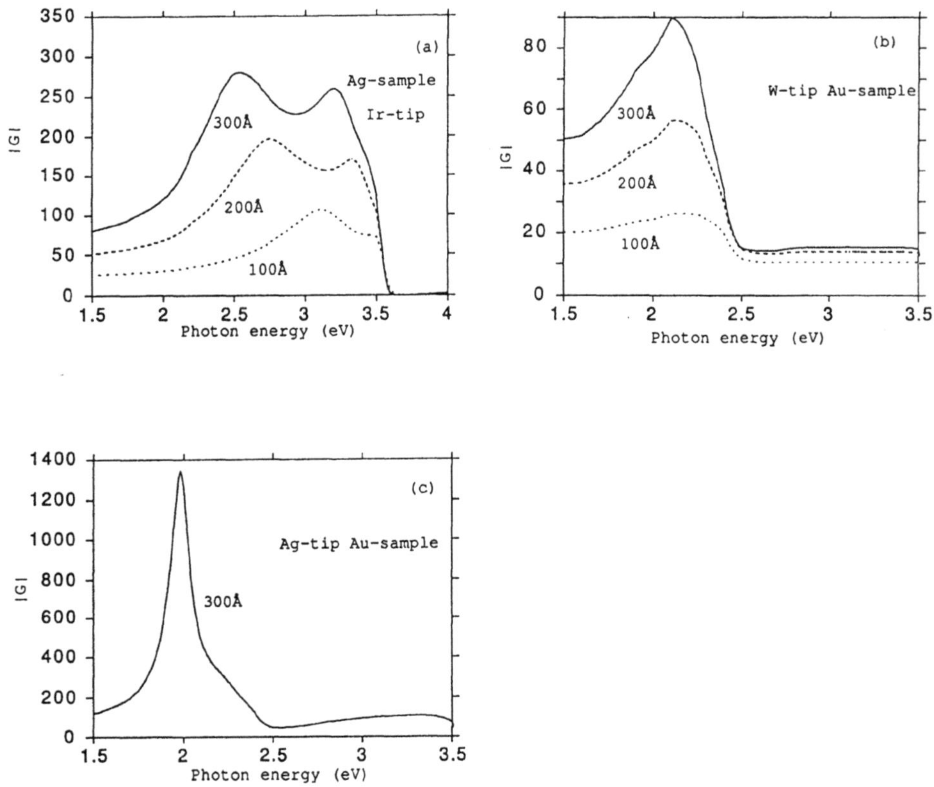


Figure 2. Calculated electromagnetic response function $G(\theta, \omega)$ for the tip-sample system for different tip radii and different materials. The distance between the tip and the sample is 5 Å and the observation angle is 45°. (a) Silver sample and iridium tip. (b) Gold sample and tungsten tip. (c) Gold sample and silver tip.

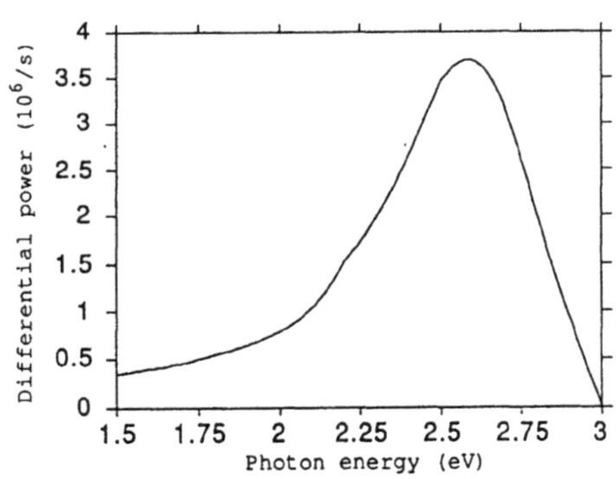


Figure 3. Calculated differential power, $d^2 P/d(\hbar\omega)d\Omega$, for the case of a silver sample and an iridium tip. The total current is 10 nA.

$$\phi = e^{i\mathbf{q}\cdot\mathbf{r}_{\|}}\,\varphi(z),$$

$$\begin{aligned} &\varphi(z) = A\, e^{qz} && \text{if } z<0, \\ &\varphi(z) = B\, e^{qz} + C\, e^{-qz} && \text{if } 0<z<d, \\ &\varphi(z) = D\, e^{-qz} && \text{if } z>d, \end{aligned} \tag{15}$$

where q is the absolute magnitude of **q**. In the same way as in Section 2 the **E**-field parallel to the interfaces and the **D**-field normal to the interfaces must be continuous and this gives us four equations from which we can eliminate the unknowns A, B, C, and D and find a condition which $\varepsilon(\omega)$ must satisfy in order to have a non-trivial solution (interface plasmon). We obtain

$$\varepsilon(\omega) = -\coth(qd) \pm \sqrt{\coth^2(qd) - 1}\ . \tag{16}$$

The solution with a plus-sign gives a plasmon that does not couple to the tunnel current. The other, low-frequency, interface plasmon is associated with oscillating surface charges that at each instant have opposite signs on the two surfaces. Thus, there is a strong electric field directed perpendicularly to the interfaces, and this leads to a strong coupling with the tunneling electrons. The resonances in the model with the spherical tip occur typically when half a wavelength of the low-frequency plasmon can be fit into the cavity between the sphere and the sample, i.e. $qL \approx \pi$. The linear size, L, of the cavity is a somewhat arbitrary quantity but a resonable definition is to say that the cavity ends where the distance between the sphere and the sample is 2d. This means that $L \approx (8dR)^{1/2}$, so the resonance condition becomes,

$$\varepsilon(\omega) = -\coth\left(\frac{\pi d}{\sqrt{8dR}}\right) \pm \sqrt{\coth^2\left(\frac{\pi d}{\sqrt{8dR}}\right) - 1}\ . \tag{17}$$

For a tip radius, R=300 Å, and with d=5 Å, $qd<1$, and we can use the approximation $\coth(qd) \approx 1/qd$. This gives us the following final expression for the resonance frequency,

$$\omega = \sqrt{\pi/2}\, (d/8R)^{1/4}\, \omega_p. \tag{18}$$

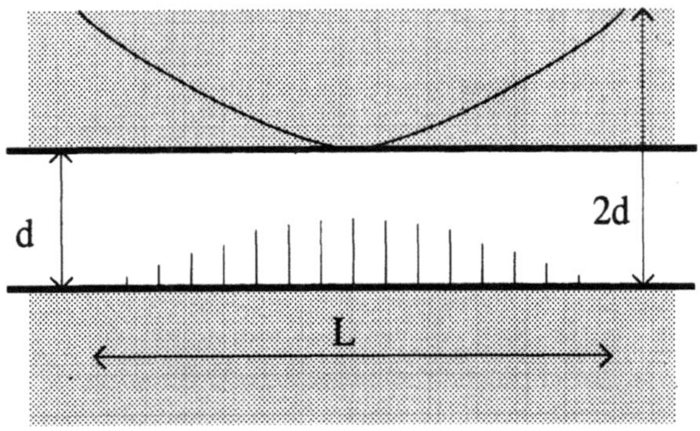


Figure 4. Schematic illustration of how the plasmon mode localized between the tip and the sample can be derived from the interface plasmon modes between two semi-infinite metal samples. The bars indicate the magnitude of the surface charge associated with an interface plasmon that fits into the cavity between the tip and the sample in the case of an STM.

From the above we can draw several conclusions. First of all, Eqs. (16) and (17) show that resonances occur for certain values of the dielectric function. The optical properties (dielectric function) of gold in the frequency range of interest are to a large extent goverened by the onset of interband transitions, from the narrow 5d-band into the empty states in the conduction band. This onset occurs at a photon energy ≈2.4 eV, but the real part of the dielectric function of gold is, in accordance with the Kramers-Kronig relations, also affected by the interband transitions for lower photon energies. This leads to a rapid increase of the real part of the dielectric function, from ≈-11 at 2.0 eV to ≈-4 at 2.4 eV. In all the cases we have considered, the localized plasmon resonance resulting in a peak in the response function, G, occur at values of the sample dielectric function in that interval. Thus, even if a change in the tip-sample geometry leads to a change in the resonance condition in terms of the dielectric function, the resonance frequency does not change much. This is what is seen in Fig. 2(b) when varying the tip radius. Silver, on the other hand, has a dielectric function that is well approximated by Eq. (14) over the frequency range 2-3 eV if one uses a plasma frequency $\omega_p \approx 9$ eV. One finds that the variation of the resonance frequency as a function of sphere radius in the model geometry in Fig. 1 is surprisingly well described by Eq. (18).

The broadening of the resonance peaks in Figs. 2(a) and 2(b) originate mainly from the damping of the localized plasmon that takes place in the sphere. The tip materials, Ir and W, are not very good conductors. When both the tip and the sample are made of noble metals this situation changes. In Fig. 2(c) the resonance peak is fairly narrow since both materials are good conductors.

Knowing the spectral distribution, shown in Fig. 3, and angular distribution of the radiation one can estimate the probability of photon emission per tunneling electron. The angular distribution has typically a broad maximum centered at $\theta=55°$. The radiation is distributed over a solid angle of ~3-4. We thus find that for the case considered in Fig. 3 the probability of photon emission is $\sim 0.5 \cdot 10^{-4}$. When both the tip and sample materials are noble metals the photon emission probability can increase by another order of magnitude.

## 4. Summary

In this paper we have presented a theoretical investigation of light emission from a scanning tunneling microscope probing the surface of a noble metal. In our calculation we have modeled the STM tip by a sphere, and therefore been able to calculate the local electric field in the vicinity of the tip within a non-retarded formalism. The presence of the tip close to the sample surface leads to the formation of a localized interface plasmon. Moreover, due to the tip the translational invariance along the sample surface is lost, so the plasmon mode is radiative. This leads to a strong resonant enhancement of the light emission for photon energies near the plasmon energy. Our calculation shows that the efficiency of the light generation is on the order of one emitted photon in the visible range per $10^4$ tunneling electrons.

## Acknowledgments

We wish to thank Göran Wendin, Bo Persson, Fernando Flores, and in particular Richard Berndt and James Gimzewski for many valuable discussions. The work in Göteborg was partly supported by the Swedish Natural Science Research Council. The research in Madrid was supported by the Comisón Asesora de Investigación Científica y Técnica. The work in Namur was supported by the Belgian Program on Interuniversity Attraction Poles initiated by the Belgian State-Prime Minister's Office-Science Policy Programming.

## References

1. Gimzewski, J. K., Sass, J. K., Schlittler, R. R. and Schott, J. (1989) 'Enhanced photon emission in scanning tunneling microscopy', Europhys. Lett. **8**, 435.
2. Coombs, J. H., Gimzewski, J. K., Reihl, B., Sass, J. K., Schlittler, R. R. and Schott, J. (1988) J. Microsc. **152**, 325.
3. Berndt, R., Gimzewski, J. K. and Johansson, P. (1991) 'Inelastic tunneling excitation of tip-induced plasmon modes on noble-metal surfaces', Phys. Rev. Lett. **67**, 3796.
4. Sivel, V., Coratger, R., Ajustron, F. and Beauvillain, J. (1992) 'Photon emission stimulated by scanning tunneling microscopy in air', Phys. Rev. B **45**, 8634.
5. Lambe, J. and McCarthy, S. L. (1976) 'Light emission from inelastic electron tunneling', Phys. Rev. Lett. **37**, 923.
6. Adams, A. and Hansma, P. K. (1981) 'Light emission from small metal particles and thin metal films excited by tunneling electrons', Phys. Rev. B **23**, 3597.
7. Bloemer, M. J., Mantovani, J. G., Goudonnet, J. P., James, D. R., Warmack, R. J. and Ferell, T. L. (1987) 'Observation of driven surface-plasmon modes in metal particulates above tunnel junctions', Phys. Rev. B **35**, 5947.
8. Kirtley, J. R., Theis, T. N. and Tsang, J. C. (1981) 'Light emission from tunnel junctions on gratings', Phys. Rev. B **24**, 5650 ; Kirtley, J. R., Theis, T. N., Tsang, J. C. and DiMaria, D. J. (1983) 'Hot-electron picture of light emission from tunnel junctions', *ibid.* **27**, 4601.
9. Sparks, P. D., Sjodin, T., Reed, B. W. and Stege, J. (1992) 'Light emission from the slow mode of tunnel junctions on short period diffraction gratings', Phys. Rev. Lett. **68**, 2668.
10. Davis, L. C. (1977) 'Theory of surface-plasmon excitation in metal-insulator-metal tunnel junctions', Phys. Rev. B **16**, 2482.

11. Laks, B. and Mills, D. L. (1979) 'Photon emission from slightly roughened tunnel junctions', Phys. Rev. B **20**, 4962; (1980) 'Roughness and the mean-free path of surface polaritons in tunnel-junction structures', *ibid.* **21**, 5175.
12. Arya, K. and Zeyher, R. (1983) 'Light emission from tunnel junctions: The role of multiple scattering of surface polaritons', Phys. Rev. B **28**, 4080.
13. Rendell, R. W., Scalapino, D. J. and Mühlschlegel, B. (1978) 'Role of local plasmon modes in light emission from small-particle tunnel junctions', Phys. Rev. Lett. **41**, 1746; Rendell, R. W. and Scalapino, D. J. (1981) 'Surface plasmons confined by microstructures on tunnel junctions', Phys. Rev. B **24**, 3276.
14. Johansson, P., Monreal, R. and Apell, P. (1990) 'Theory for light emission from a scanning tunneling microscope', Phys. Rev. B **42**, 9210.
15. Johansson, P. and Monreal, R. (1991) 'Theory for photon emission from a scanning tunneling microscope', Z. Phys. B **84**, 269.
16. Persson, B. N. J. and Baratoff, A. (1992) 'Theory of photon emission in electron tunneling to metallic particles', Phys. Rev. Lett. **68**, 3224; (1990) Bull. Am. Phys. Soc. **35**, 634.
17. Berndt, R., Baratoff, A. and Gimzewski, J. K., (1990) 'Scanning tunneling optical microscopy of silver nanostructures', in Behm, R. J., Garcia, N. and Rohrer, H. (eds.), Scanning Tunneling Microscopy and Related Methods, Kluwer Academic Publishers, Dordrecht, pp. 269-280.
18. Landau, L. D. and Lifshitz, E. M. (1960) Electrodynamics of Continuous Media, Pergamon, Oxford, pp. 288-290.
19. Jackson, J. D. (1975) Classical Electrodynamics, Wiley, New York.
20. Johnson, P. B. and Christy, R. W. (1972) 'Optical constants of the noble metals', Phys. Rev. B **6**, 4370; Weaver, J. H., Krafka, C., Lynch, D. W. and Koch, E. E. (1981) Physics Data No. 18-1, Fachinformationszentrum, Karlsruhe.
21. Takemori, T., Inoue, M. and Ohtaka, K. (1987) 'Optical response of a sphere coupled to a metal substrate', J. Phys. Soc. Jpn. **56**, 1587.
22. Morse, P. M. and Feschbach, H. (1953) Methods of Theoretical Physics, McGraw-Hill, New York, Vol. II, pp. 1298-1301.
23. Kirtley, J., Scalapino, D. J. and Hansma, P. K. (1976) 'Theory of vibrational mode intensities in inelastic electron tunneling spectroscopy', Phys. Rev. B **14**, 3177.

# LIGHT EMISSION FROM THE TUNNEL JUNCTION OF THE STM. POSSIBLE ROLE OF TCHERENKOV EFFECT.

I.I. SMOLYANINOV
*Institute of spectroscopy, Russian Academy of Sciences*
*142092, Troitsk, Moscow region*
*Russia.*
*(present address: CEMES-LOE/CNRS*
*B.P.4347, 31055 Toulouse Cedex,France.)*

ABSTRACT. Investigations of the light emission from the tunnel junctions between different noble metals in air and in liquid ambient are reviewed. A new mechanism of light emission by the tunneling electrons similar to the Tcherenkov effect is suggested. Experiments with copper phthalocyanine molecules in the Au-Au tunnel junction are discussed.

## 1. Light Emission From The Tunnel Junction Between Metals. Experimental Results.

The first observation of the light emitted from the tunnel junction of an STM was performed by Coombs et al.(ref.1). They used an iridium tip and a silver sample in an UHV chamber and a photomultiplier placed over a window of the chamber. They suggested that the mechanism causing this light emission should involve local surface plasmon generation.

Another method to detect the light from the tunnel junction of the STM was developed in works of Prof. M.S.Khaikin's group (ref.2-3). The end face of a 200 μm diameter quartz fiber coated on the outside with a metal film of 5 to 10 μm thickness, was used as the tip of the STM, and it operated as a light acceptor simultaneously. The photo of such a tip-light guide taken with an electron microscope is shown in fig.1. The geometry was chosen to provide the maximum solid angle wherein the photons emitted by the tunnel junction were detected.

Gold, silver and platinum tips and samples were used in the experiment. With constant voltage $1.5V< V_t <4V$ across the tunnel contact between the tip and the sample light radiation in the photon energy range 1.5-4.1eV was observed. The quantum efficiency of this process can be estimated as $10^{-5}$ photons per tunneling electron. The STM feedback circuit kept the tunnel current constant in the interval 10 - 100 nA. The sample topography and the intensity of light emission from the junction were measured simultaneously. An example of such measurements is shown in fig.2. These maps were observed with a silver covered tip and a gold sample at $V_t$=2.4V. The light emission features correlate well with grains in the gold film (One can see perfect images of gold films taken with the gold tips in ref.5 as well).

The light intensity depends strongly on the voltage $V_t$ applied to the tunnel contact. Data of measurements of the light intensity as a function of $V_t$ for different pairs of metals used for



*D.W. Pohl and D. Courjon (eds.), Near Field Optics, 353–360.*

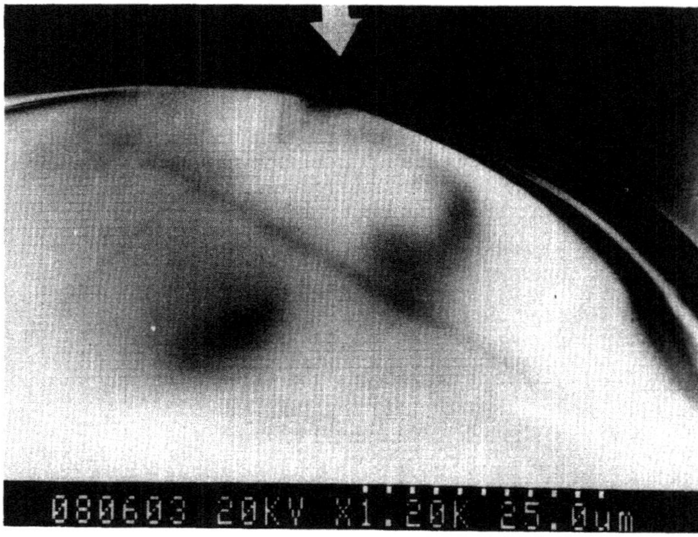


Figure 1. Electron microscope photo of the experimental tip-light guide. The protrusion at the edge of the fiber which is believed to be the tip of the STM is indicated by the arrow.

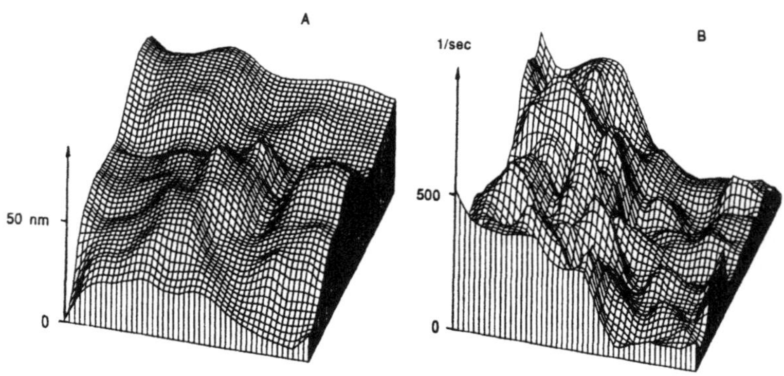


Figure 2. The 70*70 $nm^2$ topography map (a) and the intensity map of the light emission (b) obtained simultaneously with $V_{ts}$=2.4 V, $I_t$=30 nA for the gold surface with a silver covered tip.

the tunnel junction are shown in fig.3. In each case, the results of about 50 measurements, carried out with various tips, on different parts of different samples, were averaged. The positions of the intensity maxima in fig.3 agree well with the proposition that the mechanism behind the light emission involves surface plasmon generation. With materials which have larger plasma frequency, the maxima of the light emission are detected at larger voltages applied.

The spectra of the light emitted from the tunnel junction were investigated as well (ref.3). The investigations show that there are also some peaks in the spectrum corresponding to the light emission maxima. Fig.4 shows the position of peaks in the $I(V_t)$ curves as a function of the averaged plasma frequency $\omega_{12}$ (ref.3) of the pair of metals used for the tunnel contact

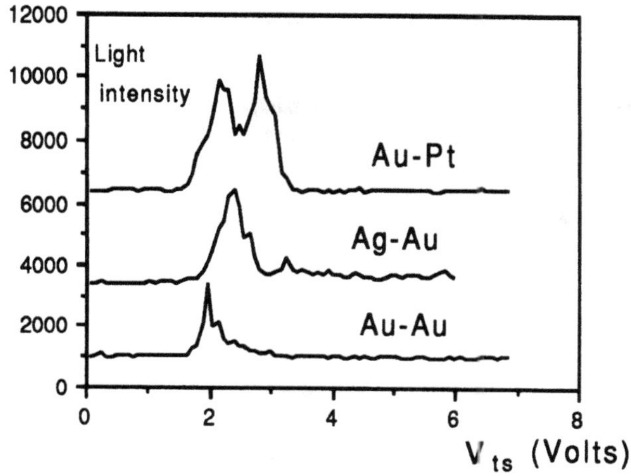


Figure 3. Light intensity versus the voltage between the tip and the sample for Au-Au, Ag-Au and Au-Pt tunnel junctions. The first metal of a pair is the tip material, and the second is the sample material.

$$\omega_{12} = \left( \frac{2\omega_1^2 \omega_2^2}{\omega_1^2 + \omega_2^2} \right)^{\frac{1}{2}} \tag{1}$$

where $\omega_1$ and $\omega_2$ are the plasma frequencies of the metals. Also the spectral maxima positions are shown. In both cases we can draw straight lines through the experimental points. These maxima could be explained as an excitation of different modes in the spectrum $\omega=\omega_{12}(n+0.5)^{1/2}(d/2R)^{1/4}$ (ref.6) of plasmons in the tunnel gap, where d is the gap and R is the tip radius. The localization region of these modes was estimated in ref.3 as 1.5 - 5nm.

The similar estimation (ref.4) follows from data of ref.7 for lifetimes of surface plasmons in small metal particles. The full width at the half maximum of the light emission features observed (about 0.2 eV) corresponds to the plasmons excited in the 3 nm diameter silver particles.

It should be also mentioned that one of the characteristic features of the light emission under consideration is the following. It occurs only at small voltages (less then 10 V) applied to the contact, and consequently at small distances between the tip and the sample. There is no detectable light emission in the 10-200V range (see ref.8) when the energy of electrons and the electric field intensity in the gap become even larger than that in the tunnel regime. There is no clear explanation of this fact elsewhere.

In what follows a possible mechanism of light emission from the tunnel junction between two metals is suggested which could explain the main features of the phenomenon.

## 2. Tcherenkov Effect In The Tunnel Gap.

It is well known (see for example ref.9) that an electron emits light when moving through the substance with the velocity V which is larger than the velocity of light U in the substance. This effect was discovered by Tcherenkov and Vavilov and explained by Tamm and Frank in 1937.

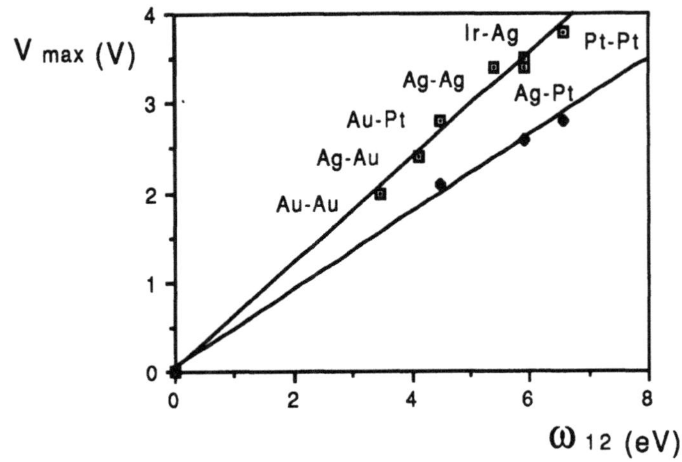


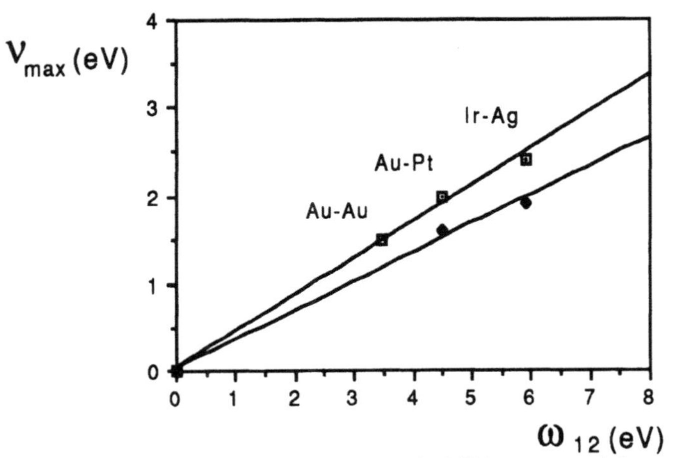


Figure 4. The positions of the maxima on the $I(V_t)$ curves, and the spectral maxima versus the average plasma frequency of the pair of metals used for the tunnel junction. The data for an Ir-Ag junction are taken from ref.1. Different symbols correspond to the different plasmon modes. $\omega_p$ equals 3.5, 5.4, 6.6 and 6.6 eV for Au, Ag, Pt and Ir. These values correspond to Drude's free electron theory.

The light is emitted in the direction in which the projection of the velocity of the electron equals the velocity of light. So the angle between the direction of motion of the electron and the direction in which the light is emitted equals $\cos\theta = U/V$. Traversing the dielectric, the electron loses the energy:

$$dE = \frac{\alpha h\omega}{2\pi C}\left(1 - \frac{U^2}{V^2}\right) S\, d\omega \quad \text{or} \quad dE = \frac{\alpha h\omega}{2\pi C} S\, d\omega \quad \text{when } U \ll V \;, \quad (2)$$

(ref.9) where e is the electron's electric charge, C is the velocity of light in vacuum, $\alpha=2\pi e^2/hC$, $d\omega$ is the spectral linewidth of light , and S is the projection of the path of the electron in the dielectric on the direction of the light emission.

We have a similar situation in the case of plasmons generated by electrons in the tunnel gap. The velocity of plasmons equals $U=CL/\lambda$, where L is the wavelength of plasmons and $\lambda$ is the wavelength of light with the same frequency. When L=1nm we have $U=10^{-3}C=3*10^5$m/s. Such a small wavelength of a plasmon may be possible because the light emission features with near atomic dimensions were observed in the light intensity maps of metal surfaces (see for example fig.8 in the ref. 1).

On the other hand it can be shown directly that the velocity of plasmons is very small in the narrow gap between two metals. The energy spectrum and the velocity of plasmons in the gap d between two plain metal surfaces with the plasma frequency $\omega_p$ are given by the expressions :

$$\omega = \omega_p \left(\frac{kd}{2+kd}\right)^{\frac{1}{2}}, \quad \text{and} \quad U = \frac{\omega}{k} = \omega_p \left(\frac{d}{k\,(2+kd)}\right)^{\frac{1}{2}} \tag{3}$$

(ref.6), where k is the wave vector of plasmons. When $d \rightarrow 0$ then also $U \rightarrow 0$. Thus the tunnel gap can be treated as a substance with a very large index of refraction.

The velocity of electrons equals the Fermi velocity ($V=1.4*10^6$ m/s in the case of gold) and it is larger than the velocity of plasmons. So, the emission of plasmons by the electrons traversing the tunnel gap similar to the Tcherenkov effect should occur. Electrons move almost perpendicular to the surface, but they can have the parallel projection of the velocity which is larger than the velocity of plasmons: $\arcsin(U/V) = \arcsin(3*10^5/1.4*10^6)=12°$.

One can use equation (2) to estimate the order of magnitude of the quantum efficiency of such a process. The path of the electron equals $S\approx 1$nm. The spectral linewidth of light equals $d\omega\approx 0.2$eV (ref.3). So we have $dE=10^{-5}h\omega/2\pi$. Thus the treatment of plasmon emission from the tunnel gap in terms of the Tcherenkov effect can give the efficiency about $10^{-5}$. But in order to

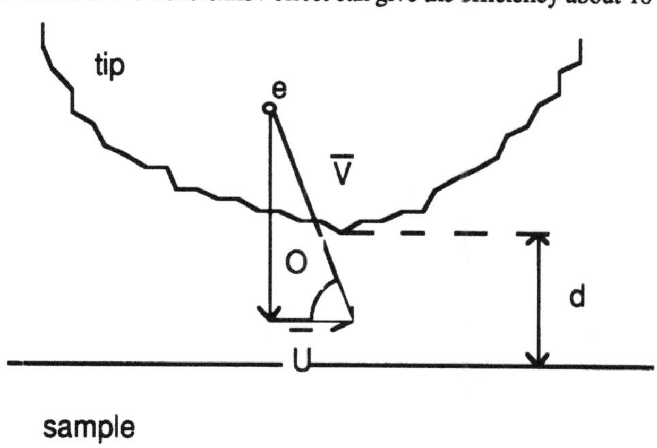


Figure 5. An electron traversing the tunnel contact. Projections of the electron's velocity are shown.

obtain the final quantum efficiency of the photon emission it is necessary to know the quantum efficiency of the plasmons into the photons transitions. In order to explain the experimental results this value should be of the order of 0.1-1.

The main problem of the previous works was the following. It cannot be explained why the light emission occurs only in the tunnel mode of operation of the STM. There is no light emission (ref.8) at higher voltages (from 10V to 200V) when the electric field in the gap and the energy of electrons become larger. This problem could be easily solved now.

The Tcherenkov effect takes place when $U<V\cos\theta$. So we have conditions (see equation 3):

$$\omega_p\left(\frac{Ld}{4\pi\left(1+\pi\frac{d}{L}\right)}\right)^{\frac{1}{2}} < V\cos\theta \quad \text{or} \quad \frac{Ld}{\left(1+\pi\frac{d}{L}\right)} < \frac{\xi^2}{\pi}\cos^2\theta \text{, where} \quad \xi = \frac{V}{\nu_p} \approx 1.6\ \text{nm} \tag{4}$$

and $\nu_p=\omega_p/2\pi$ . One can see that the Tcherenkov effect is possible when d and L are on the order of about 1 nm. At larger voltages the distance between metals increases and the light emission due to the Tcherenkov effect switches off.

### 3. New Possibilities.

The special interest of studying the light emission from the tunnel contact is the possibility of obtaining chemical surface data in addition to the usual information given by STM. This possibility was discussed and investigated in a preliminary way in ref.2-4. For example in ref.4 the possibility of nanometer scale gold particles identification by the STM assisted with the light detection system was demonstrated. The light emission was observed from the small particles of gold (the characteristic size of the particles was 10-100nm) which were situated on the graphite substrate. And no light was detected from the graphite. This result opens the possibility of determining the size and the chemical nature of metal microinclusions in conducting matrices.

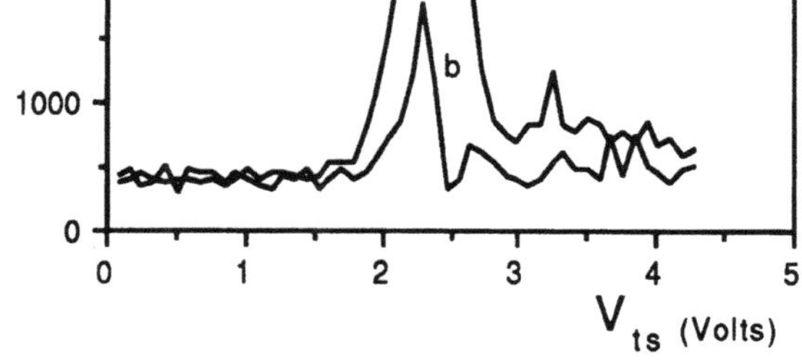


Figure 6. Light intensity versus the voltage between the tip and the sample for an Ag-Au tunnel junction in air (a) and in liquid decan (b).

Another application of the instrument developed is its use in different ambients. The light emission from the tunnel contact between two noble metals which were situated in liquid decan could be detected as well. One can see the results of such experiment in fig.6. No shift (within the experimental accuracy) was detected in the position of the light emission maximum. This fact was not evident apriori since the frequency of ligh. should have been smaller in the liquid due to its refraction index.

Another challenging possibility exists in the usage of the STM assisted with the light detection system. The small localization region of the plasmons excited should cause strong sensitivity of the light emission to the individual molecules adsorbed on the surface of the STM sample. Some experiments in this field were performed (ref.2).

Copper phthalocyanine (CuPc) molecules have a well known excitation spectrum which consists of two electronic transitions at 1.8 and 3.6eV. Approximately one monolayer of CuPc molecules was thermally evaporated onto a gold film. The light intensity versus tunnel voltage dependence was studied with the aid of a gold tip. The results of such measurements are shown in fig.6. The complicated curves include a plasmon feature at $V_t$=2V and an additional emission peak at

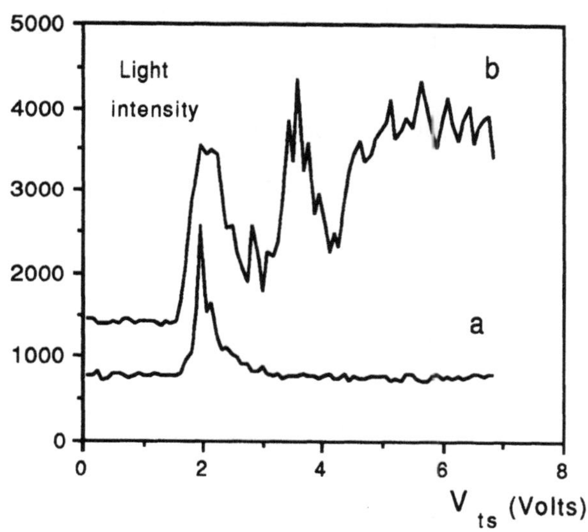


Figure 7. Light intensity versus tunnel voltage for a pure Au-Au junction (a) and for an Au-Au junction with approximately one monolayer of copper phthalocyanine molecules evaporated onto the Au sample (b).

$V_t$=3.6V. This new peak corresponds to the energy of the electronic transition in the CuPc molecule. The height of the new peak depends on the position of the tip. Two light emission maps taken one after another at $V_t$=3.6V are shown in fig.7. One can see some features with the size of approximately 1nm in these maps. But unfortunately we cannot obtain simultaneously topography maps with a good quality at such a large tunnel voltage in air. In order to clear the situation it is necessary to perform similar measurements under UHV conditions.

In conclusion, the results of investigations of the light emission from the STM tunnel junction between noble metals obtained in Prof. M.S.Khaikin's group have been reviewed together

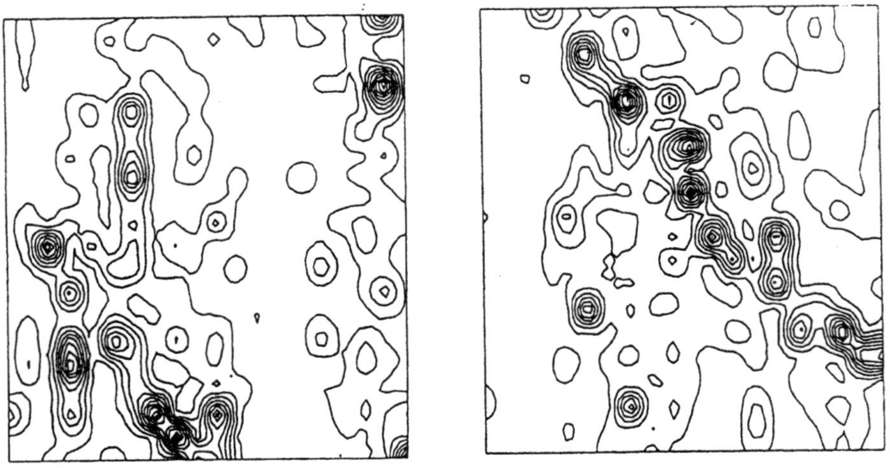

Figure 8. Two light emission contour 15*15 $nm^2$ maps taken with the golden tip one after another at $V_t$=3.6 V, $I_t$=300 nA for the Au sample surface with approximately one monolayer of copper phthalocyanine molecules. The time interval between (a) and (b) measurements was about 6 minutes. The difference between the neighbouring contours is 100 counts.

with some new results concerning the light emission in the liquid ambient. A new point of view on the mechanism of the light generation in the tunnel junction based on the Tcherenkov effect is proposed, which is in the good agreement with the experimental data.

## References


(1) Coombs, J.H., Gimzewski, J.K., Reihl, B., Sass, J.K., and Schlittler, R.R. (1988) J.Microsc.152, 325.
(2) Smolyaninov,I.I., Khaikin,M.S., and Edelman,V.S. (1990) Phys.Lett.A 149, 410.
(3) Smolyaninov,I.I., Edelman, V.S., and Zavyalov, V.V. (1991) Phys.Lett.A 158, 337.
(4) Smolyaninov, I.I., and Moskovets, E.V. (1992) Phys.Lett.A 165, 252.
(5) Sivel, V., Coratger, R., Ajustron, F., and Beauvillain, J. (1992) Phys.Rev.B 45 , 8634.
(6) Malshukov, A.G. (1990) Phys.Reports 194, 343.
(7) Tran Thoai, D.B., and Ekardt, W. (1982) Solid State Communications 41 , 687.
(8) Berndt, R., Gimzewski, J.K., and Johansson, P. (1991) Phys.Rev.Lett. 67 , 3796.
(9) Landau, L.D., Lifshitz, E.M. (1969) Theoretical Physics, Vol.8, Mir, Moscow
(10) Johansson, P., Monreal, R., and Appel, P. (1990) Phys.Rev.B 42 , 9210.
(11) Persson, B.N.J., and Baratoff, A. (1990) Bull.Am.Phys.Soc. 35 , 634.

# A TECHNIQUE FOR THE OBSERVATION OF MAGNETO-OPTICAL EFFECTS WITH NANOMETER RESOLUTION

S.F. Alvarado
*IBM Research Division*
*Zurich Research Laboratory*
*CH-8803 Rüschlikon*
*Switzerland*

ABSTRACT. A new technique is proposed to image magnetic domains on ferromagnetic thin films with nanometer lateral resolution. It is based on the observation of magneto-optical effects on metallic magnetic media by light excitation related to tunneling with the tip of an STM. Preliminary experimental results show that the tip of an STM can excite light emission with very high circular polarization in a ferromagnet. The radiation also exhibits a linear polarization component which is oriented along the tip axis. The polarization shows a dependence on the tunneling voltage as well as the properties of the tip and sample. Though the physical mechanism is not yet fully understood one can expect that the circular polarization can, at least in part, be accounted for by magneto-optical effects.

## 1. Introduction

Recently, the observation of resonant light emission from the tunneling region formed by a metal tip and a metal surface in STM experiments was reported [1]. Thereafter it was found that the emitted light in such experiments can exhibit a sizable degree of circular polarization [2]. In this report we now propose that the resonant excitation of polarized light emission be used to image magnetic domains of magnetic metals and alloys. The present approach is based on the observation of magneto-optical effects with nanometer resolution by local excitation of optical resonant surface modes at the tunneling junction formed between the tip and a magnetic substrate in STM.

## 2. Experimental

### 2.1. EXPERIMENTAL SETUP

The setup used to measure the degree of circular polarization of the fluorescence is illustrated in Fig. 1. A tunneling junction is made between a tip – not necessarily of magnetic material – and the sample under study, e.g. a metallic magnetic thin film. By choosing a convenient bias, resonant light emission can be excited. The helicity of the light is determined with a Pockel's cell and a linear polarizer: the intensities for $\pm\lambda/4$ retardation, $I_+$ and $I_-$, measured by a photomultiplier (PMT in Fig. 1), give the degree of circular polarization by means of the formula



*D.W. Pohl and D. Courjon (eds.), Near Field Optics*, 361–366.

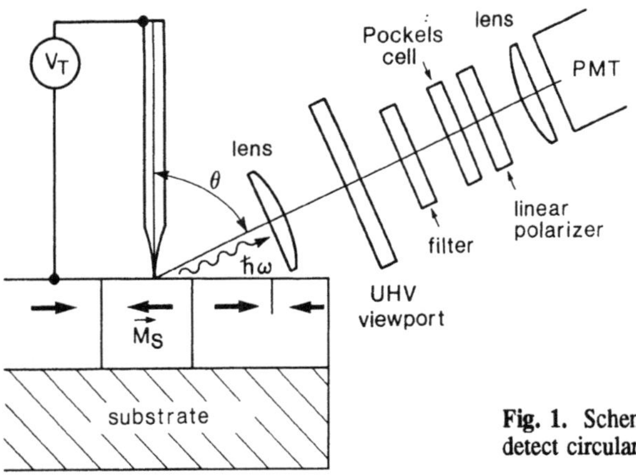


**Fig. 1.** Schematic of the experimental setup used to detect circular polarized photon excitation.

$$\rho = \frac{I_+ - I_-}{I_+ + I_-} \, . \tag{1}$$

$\rho(x,y)$ can be related to the local magnetization $\mathbf{M}(x,y)$ along the observation direction $\mathbf{n}$ by a relation of the type

$$\mathbf{M}(x,y) \cdot \mathbf{n} = k \times \rho \tag{2}$$

where $k$ is a constant. Thus one can see that, for $k \neq 0$, the polarization is expected to change sign when the component of magnetization vector along the optical axis changes from parallel to antiparallel.

2.2. PRELIMINARY RESULTS

An indication that magnetic domains might have already been seen by this effect is given by preliminary experiments involving tunneling with Au or Ni tips on polycrystalline magnetic samples. Figure 2 shows measurements performed using a Ni tip forming a tunneling junction with the surface of a Ni sample. It is found that the polarization of the emitted radiation changes as a function of the tip position and even changes sign in a manner reminiscent of the magnetization profile across a sample with randomly oriented magnetic domains, as expected for the Ni sample which was not purposely magnetized. Note that the polarization and the intensity of the emitted light do not show any apparent correlation, see Fig. 2b. This indicates that the polarization of the light is not critically dependent on the excitation efficiency of the resonance nor seems to be influenced by the topographic features of the sample. This point, crucial for the realization of a magnetic domain imaging instrument, is discussed below in more detail. The degree of the circular polarization is found to be a function of the tip bias and is sensitive to the quality of the surfaces in the tunneling junction. In the following we attempt to clarify possible mechanisms that could give rise to the polarization effects observed. Further indication of magnetic effects are given by experiments which show that the degree of circular polarization of the emitted light changes upon reversal of the magnetization of the tip [3].

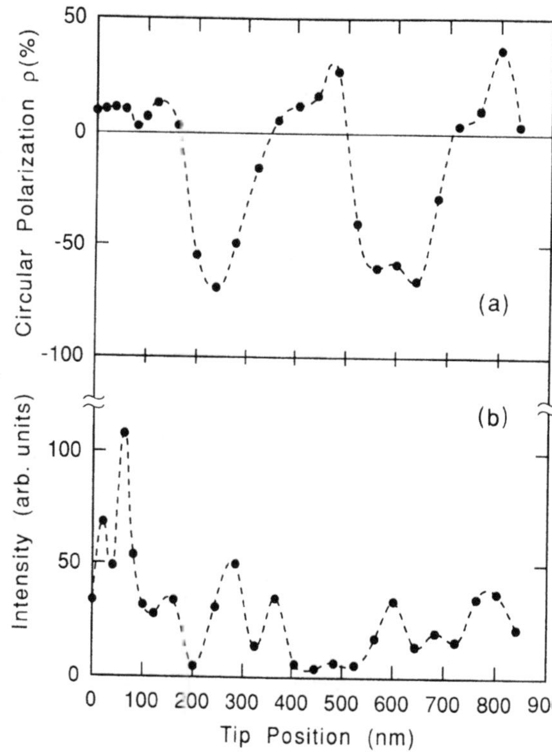


**Fig. 2** (a) Degree of circular polarization of the radiation emitted vs. tip position along a line on the sample's surface and (b) corresponding intensity. The STM experiment was performed with a Ni tip on a Ni polycrystalline sample. Tunneling voltage $V_T = 3$ V, current $i_T = 0.7$ nA.

**Fig. 3** *(below)* Simplified model where the tip is represented by a sphere of radius $R_T$ and the substrate by a surface of local roughness characterized by a local curvature of radius $R_S$.

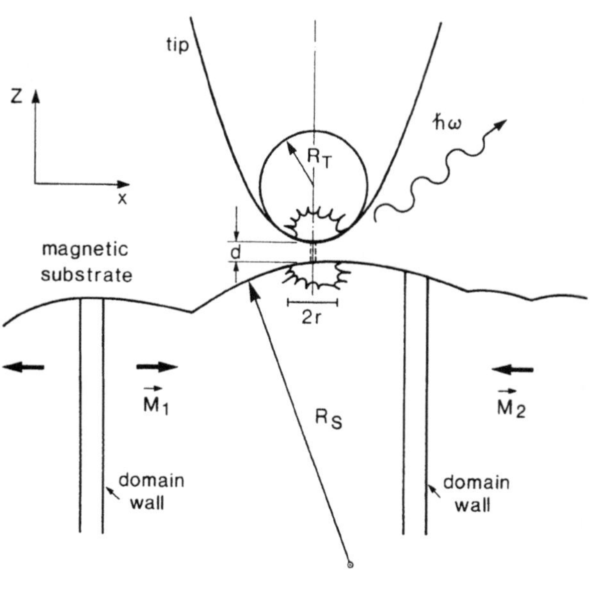


## 3. Discussion

### 3.1. MODEL

The tunneling junction can be modeled representing the tip as a sphere of radius $R_T$ at a distance $d$ from a conducting surface with radius of curvature $R_s(x,y)$, see Fig. 3. The lowest excitation mode of the system corresponds to an oscillating dipole placed perpendicularly to the surface [4]. The measurements actually show that the light also exhibits a linear polarization component oriented along the tip axis, i.e. the radiation is elliptically polarized with the major axis perpendicular to the surface. The emission of linearly polarized light is indeed in accordance with modeling the tip-surface cavity in a simplified manner as described above.

### 3.2. POLARIZATION EFFECTS

3.2.1. *Geometry.* Note that for non-magnetic materials if the electric vector is oscillating perpendicular to the surface the polarization cannot be circularly polarized upon reflection on the tip or substrate surface. Assuming mirror symmetry about the plane containing the tip's axis and the optical detection axis, one does not expect the emission of circularly polarized light due to purely geometric effects [5].

3.2.2. *Spin-Orbit Effects and Symmetry.* There is the possibility of spin-orbit effects giving rise to the emission of circularly polarized light even in the absence of magnetic materials. This has probably been observed in GaAs(110) when tunneling with Ni tips. In that particular case the emission of circular polarization is dominated by the recombination of polarized electrons originating from the ferromagnetic tip [6]. However for increasing tunneling electron energies a non-magnetic polarization background appears which could be related to spin-orbit effects associated with the lack of inversion symmetry in GaAs [7]. Such a mechanism might also be at work in tunneling experiments involving non-magnetic materials where we have also observed polarization effects [2].

3.2.3. *Injection of Spin-Polarized Electrons.* Another effect allowing the detection of magnetic effects by emission of circularly polarized light is the recombination of spin-polarized electrons arising from the magnetic sample. This effect has been demonstrated for the case of a Ni-tip/GaAs(110) tunneling junction [6] where GaAs acts as the optical spin detector. This scheme might also work very well with other materials acting as the spin detector.

3.2.4. *Magneto-Optical Effect.* For magnetic materials one can invoke the magneto-optical effects as the possible origin of the circular polarization of the light emitted. This is actually a plausible explanation since the electrons involved in the excitation mode will certainly experience Lorentz forces due to the sample magnetization. This is the origin of magneto-optical effects. Using the STM, however, we can gain the important advantage of dealing with magneto-optic excitations confined to a region of nanometer size. Hence the polarization will be defined by the magnetic volume within which the excitation is confined. The size of this volume is estimated in the following.

### 3.3. RESOLUTION

In the model depicted in Fig. 3 the resonant excitation giving rise to light emission is confined to a region whose lateral range $r$ is estimated to be [4]

$$r = \sqrt{dR_{\mathrm{T}}}\ . \tag{3}$$

If the substrate is not ideally flat, which is normally the case, $r$ is given by Eq. (3) taking

$$\frac{1}{R} = \frac{1}{R_{\mathrm{T}}} + \frac{1}{R_{\mathrm{s}}}\ , \tag{4}$$

where $R_{\mathrm{s}}(x,y)$ is the radius of curvature of the substrate at the coordinates $(x,y)$ in a plane on the substrate surface. Thus, taking typical values, $R = 49$ nm and $d = 1$ nm, we find from Eq. (1) that $r = 7$ nm. The main point to be made here is that $r$ is a reasonable estimate of the best lateral resolution achievable. The actual resolution expected with the present approach might be somewhat reduced by the wavelength-dependent absorption of light traveling through the medium, which for metals is of the order of 10 nm for light in the visible range. Even then the resolution should be quite usable for characterization of magnetic media. We note that the radiated intensity, in the dipolar mode, is expected to increase as $I \propto R_S^2$ for $R_s \geq 25$ nm, while for $R_S \leq 25$ nm it decreases with increasing $R_s$ due to damping [4,8]. From Eqs. (3) and (4), however, we see that this should not have a strong influence on $r$, i.e. on the resolution especially when $R_s >> R_T$, which is normally the case. As can be seen from Eq. (4) the resolution should actually improve somewhat for rough substrates.

## 4. Final Remarks

In the scheme proposed here the imaging of magnetic domains is realized through the measurement of the polarization of the light emitted by the excitation caused when electrons tunnel between a tip and magnetic sample. Optimal detection of the magneto-optical effect can be achieved set by choosing the proper photon energy by means of a monochromator and by tuning with the tunneling bias for maximum emission intensity [1]. The resolution achievable is expected to be about two orders of magnitude better than the $\lambda/2$ diffraction limit of Kerr microscopy, one order of magnitude better than the resolution of 65 nm we reported for magnetic force microscopy [9], and somewhat better or comparable to the best resolution of 35 nm demonstrated with spin-resolved scanning electron microscopy [10] or Lorentz force microscopy [11], 3-4 nm, defined as in [10]. A comparable resolution to the present technique might be also reached by detecting the Kerr effect in scanning near field optical microscopy [12]. In an earlier report [13] it is proposed that the radiative recombination of circular polarized electrons injected through a ferromagnetic layer epitaxially grown on III-V compounds be used to image magnetic domains. With that technique the film thickness is limited by the penetration depth of the injected electrons, which is typically several nanometers. The use of III-V compounds as an optical detector of spin-polarized electron tunneling has been demonstrated recently [6]. The technique proposed here is complementary and it has the advantage of potentially allowing magnetic domains to be imaged also in much thicker films and surfaces of bulk material.

## Acknowledgments

Many thanks to Ph. Renaud for fruitful discussions with the aim of proving or disproving the possibility of observing magneto-optical effects with the STM-excited light emission from metals. Thanks also to J. Gimzewski for sharing valuable information. Many thanks also to B. Weiss for preparing the Ni polycrystalline sample.

## 5. References

[1] R. Berndt, A. Baratoff, and J.K. Gimzewski, in Scanning Tunneling Microscopy and Related Methods, R.J. Behm et al. (eds.) (Kluwer, Dordrecht, 1990) 269.

[2] S.F. Alvarado and Ph. Renaud, 10th Int'l Conf. STM, Interlaken, Switzerland, August 12-16, 1991.

[3] S.F. Alvarado, to be published.

[4] R.W. Rendell and D.J. Scalapino, Phys. Rev. B 24, 3276 (1981).

[5] There is the possibility of producing circularly polarized light if the mirror symmetry in the plane defined by the observation axis and the tip axis is broken. In such cases the degree of circular polarization is expected to depend on the topographic features of the sample's surface. Evidence for this effect has not been found so far.
[6] S.F. Alvarado, and Ph. Renaud, Phys. Rev. Lett. 68, 1387 (1992).
[7] S.F. Alvarado, H. Riechert, and N.E. Christensen, Phys. Rev. Lett. 55, 2716 (1985); see also references 3-5 therein.
[8] B.N.J. Persson and A. Baratoff, Phys. Rev. Lett. 68, 3224 (1992).
[9] C. Schönenberger and S.F. Alvarado, Z. Phys. B - Condensed Matter 80, 373 (1990).
[10] R. Allenspach and A. Bischof, Appl. Phys. Lett. 54, 587 (1989).
[11] G.R. Morrison, H. Gong, J.N. Chapmann and V. Hrnciar, J. Appl. Phys. 64, 1338 (1988).
[12] E. Betzig and J. Trautman, Science, Vol. 257, 189 (1992).
[13] S.F. Alvarado, IBM Disclosure Bulletin, Vol. 33, No. 11, p. 469, April 1991, TDB SZ889-0043.

# A NEW KIND OF SURFACE WAVE : THE LOCALITON

D. MAYSTRE
Laboratoire d'Optique Electromagnétique
Unité de Recherche associée au CNRS N. 843
Faculté des Sciences et Techniques de St Jérome (Case 262)
13397 MARSEILLE Cédex 13
FRANCE

ABSTRACT. From the new concept of surface localiton, it is shown that a randomly rough surface can localize the light on a small part of its profile. Rigorous numerical calculations show the structure of a localiton. Finally, the possibility of observing such a wave from experimental measurements is suggested.

## 1. Introduction

The study of light scattering by randomly rough surfaces has been widely investigated in the recent years, in particular for investigating the phenomenon of enhanced backscattering of light [1,2]. The possibility of localization of electromagnetic field by particles or by randomly rough surfaces has also been suggested recently [3]. The aim of this paper is to show numerically localization phenomena produced by randomly rough surfaces and to explain intuitively the origin of this phenomenon. In this study, the new notion of surface localiton plays a vital role.

## 2. Definition of a Localized Field: the Localiton

### 2. 1. THE LOCALIZATION.

The first difficulty encountered in the study of localization of light by randomly rough surfaces lies in its definition. If a light source is placed above a metallic plane, the field at the vicinity of the metal is concentrated in a small region close to the source and decreases rapidly outside this region. Of course, this phenomenon cannot be classified as a localization phenomenon, since it is obviously the trivial consequence of the decrease of the field generated by a light source during propagation. Do actual phenomena of localization of light by surfaces exist and how to define such phenomena ?.

### 2. 2 . SURFACE POLARITON AND SURFACE LOCALITON.

We recall in figure 1A what is a surface polariton (or surface plasmon) on a metallic surface. Such a wave is generated by a source of energy placed at infinity on one side of the surface and decreases exponentially during propagation. Obviously, such a surface wave is not localized, due to the exponential increase of the field in the direction of the source. In order to define the localization phenomenon, we conjecture that a surface wave called surface localiton is able to exist without any light source in space, the field of this localiton tending to zero on both sides of the surface (figure 1B). In other words, a surface localiton is not generated by any source placed in any direction, including the direction of the mean plane of the rough surface.



*D.W. Pohl and D. Courjon (eds.), Near Field Optics, 367–376.*

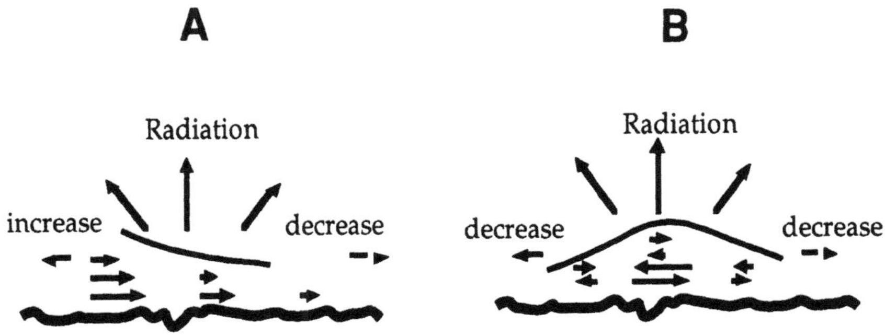


Figure 1: Definition of the surface localiton (figure 1B) and comparison with the surface polariton (figure 1A)

This definition seems to be in complete contradiction with physical intuition: since there is no source which can compensate the losses by radiation in the air or absorption in the metal, the amplitude of the field associated with a localiton must decrease and thus cannot be harmonic. The answer to this contradiction is quite obvious: if it exists, the surface localiton must correspond to a complex frequency $\omega$ with a negative imaginary part, in such a way that the time dependence in $\exp(-i\omega t)$ is exponentially decreasing. Let us notice that a complex value of $\omega$ implies complex values of the wave number in the air $k = \omega/c$ and of the wavelength $\lambda = 2\pi/k$ (but the imaginary part of $\lambda$ will be positive).

2. 3. THE TEMPERED POLARITON.

In the following, we try to deduce the existence of a surface localiton continuously from the existence of the surface polariton of figure 1a. In a first step, we consider the simplest case of a plane metallic surface. It is well known that a plane interface between air and metal is able to support the propagation of p-polarized surface polaritons. It has been noticed that a surface polariton is not localized since its amplitude tends to infinity in the direction of the source. Let us show that this exponential damping along the plane interface can be eliminated. With this aim, it is worth noticing that a surface polariton has not a Fourier transform in the direction of the propagation axis. Mathematically speaking, the surface polariton is not a tempered distribution (let us recall that, according to the classical notation of the theory of distributions, a tempered distribution has a Fourier transform in the space of distributions). Let us recall that a straightforward calculation based on Maxwell equations shows that the surface polariton propagating along the x axis of a plane metallic surface (x z plane) of complex index $\nu$ is a p-polarized field with the magnetic field H $\hat{z}$ given by

$$H = B\, e^{(i\alpha x + i\beta y)} \quad \text{in the air} \tag{1}$$

$$H = B\, e^{(i\alpha x - i\gamma y)} \quad \text{in the metal} \tag{2}$$

with

$$\alpha = \frac{\pm k\nu}{\sqrt{1+\nu^2}} \tag{3}$$

$$\beta = \frac{-k}{\sqrt{1+\nu^2}} \tag{4}$$

$$\gamma = \frac{k\nu^2}{\sqrt{1+\nu^2}} \tag{5}$$

B being an arbitrary complex number.

Since k is real and ν complex, the surface polariton has a complex constant of propagation. For example, silver has for $\lambda = 0.436\ \mu m$ an index $\nu = 0.47 + 4.84\,i$, in such a way that

$$\frac{\nu}{\sqrt{1+\nu^2}} = 1.02139 + 4.3351\ 10^{-3}\ i \tag{6}$$

thus the surface polariton has a propagation constant α with a small imaginary part, which implies its exponential decrease in the propagation. Now, allowing k to be complex, it can be shown from (3) that α becomes a real number, provided that

$$\mathrm{Arg}\left\{ k \right\} = -\mathrm{Arg}\left\{ \frac{\nu}{\sqrt{1+\nu^2}} \right\} \tag{7}$$

In these conditions, for silver at 0.436 μm,

$$\mathrm{Arg}(\dot{\lambda}) = -\mathrm{Arg}(k) = -\mathrm{Arg}(\omega) = 4.2443\ 10^{-3}\ \text{radians} \tag{8}$$

Since the damping of the field has been removed, its Fourier transform exists in the field of distributions. From (1) and (2), it can be deduced that its Fourier transform is a Dirac function of x. However, such a field is not localized: the modulus of its amplitude remains constant in the propagation. Using the notation of the theory of distributions, this surface wave will be called "tempered polariton". From an intuitive point of view, the tempered polariton may be understood as a surface wave generated by a source with a time-dependent intensity, located at $x = \pm\infty$, in such a way that the damping of the field in the propagation (due to the losses) is exactly compensated by the damping of the excitation. In other words, the time damping of the source and the space damping of the propagating polariton are exactly equivalent: compared with the classical space-damped polariton, the exponential decrease along the x axis with constant amplitude in time is replaced by an exponential decrease in time with constant amplitude along the x axis. We will see that introducing a random roughness transforms it continuously into a surface localiton that does not need any source and that decreases in amplitude at infinity.

## 3. Influence of a Random Roughness on the Behaviour of the Tempered Polariton

### 3. 1. MODEL OF RANDOMLY ROUGH SURFACE

Due to theoretical and numerical reasons, we thought it very convenient to use the model of 'random grating' where the randomly rough surface is represented by a grating of large period $d = 2\pi/K$, this period containing a large number of random asperities. This model is a classical one and has been used by many authors (see for instance the articles of A.A.Maradudin or M. Nieto-Vesperinas in references 1 and 2). One period of such a computer constructed surface is shown in figure 2, the period being equal to 7.6836 μm, the rms σ to 0.045 μm and the correlation length T to 0.06366 μm.

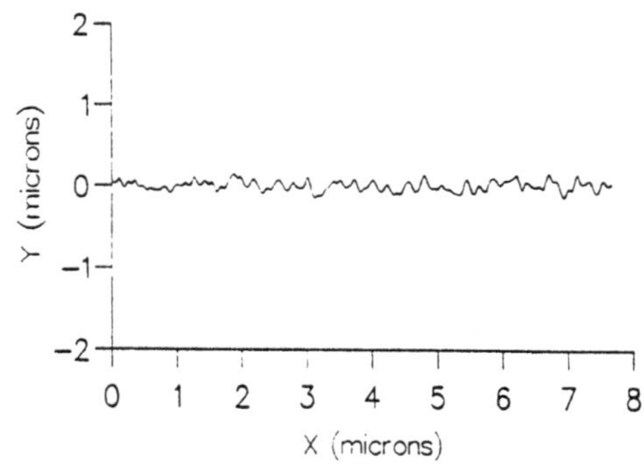


Figure 2: Graph of a period of a random grating constructed on the computer, with a period d = 7.6836 μm, a correlation length of 0.06366 μm and a rms of 0.045 μm.

3. 2 . TEMPERED POLARITON OF A RANDOM GRATING

Starting from the definition of the polariton of the plane metallic surface given by (1) and (2), the polariton of a grating must be written in the form [4]

$$\text{in the air, } H = \sum_{n=-\infty}^{+\infty} B_n e^{i\alpha_n x + i\beta_n y} \tag{9}$$

$$\text{in the metal, } H = \sum_{n=-\infty}^{+\infty} C_n e^{i\alpha_n x - i\gamma_n y} \tag{10}$$

with

$$\alpha_n = \alpha_0 + nK \tag{11}$$

$$\beta_n = \sqrt{k^2 - \alpha_n^2} \tag{12}$$

$$\gamma_n = \sqrt{k^2 v^2 - \alpha_n^2} \tag{13}$$

$$\mathrm{Re}\{\beta_n\} + \mathrm{Im}\{\beta_n\} > 0 \tag{12'}$$

$$\mathrm{Re}\{\gamma_n\} + \mathrm{Im}\{\gamma_n\} > 0 \tag{13'}$$

In this much more complicated case, the calculation of the constants of propagation $\alpha_n$ of the classical polariton cannot be performed analytically like in section 2.3 and requires the use of sophisticated computer codes dealing with scattering by diffraction gratings. The code used in the following was achieved 15 years ago [5]. It solves the boundary value problem of grating using a rigorous integral method and wide comparisons with experimental data and other rigorous theories have shown that its precision is very good, typically better than 1%. It can be shown that the constant of propagation $\alpha_0$ is a pole of the scattering matrix of the grating. The calculation is achieved by searching for a pole of a complex function (the determinant of the S matrix) in the complex plane of $\alpha_0$. By continuity, the value of $\alpha_0$ tends to the analytical value of $\alpha$ for the plane when the grating tends to the plane. Like for the plane, does it exist a tempered polariton of the "random grating"?. If it exists, the exponential behaviour of the field in x must be eliminated, which implies that $\alpha_0$ (thus $\alpha_n$ ) must be real. Moreover, the representation of a randomly rough surface by a grating does not make sense if the period of the grating does not contain a large number of asperities, thus a large number of wavelengths, at least if we are in the resonance domain where the wavelength has the same order of magnitude as the mean width of the asperities. In these conditions, K is very small with respect to k and one can consider that the $\alpha_n$ constitute a continuum (it is exact at the limit when d tends to infinity). Thus, at the limit, the constant $\alpha_0$ contained in (11), which is only a translation constant for the $\alpha_n$, can be removed and we can write

$$\alpha_n = nK \tag{11'}$$

Obviously, with this condition, the field described by the Fourier series (9) and (10) is a tempered polariton. In the following section, intuitive considerations are used to predict whether this tempered polariton will be a localiton or not.

### 3. 3. CONSIDERATIONS OF CONTINUITY.

We represent in figure 3 the modulus of the amplitudes $B_n$ of the tempered polariton of the random grating. Figure 3A shows the limit of these amplitudes as the rms $\sigma$ (mean height of the asperities) tend to zero. In that case, the random grating tends to the plane and we tend to the tempered polariton of the plane. In that limit case, only two kinds of propagating waves can exist in the series

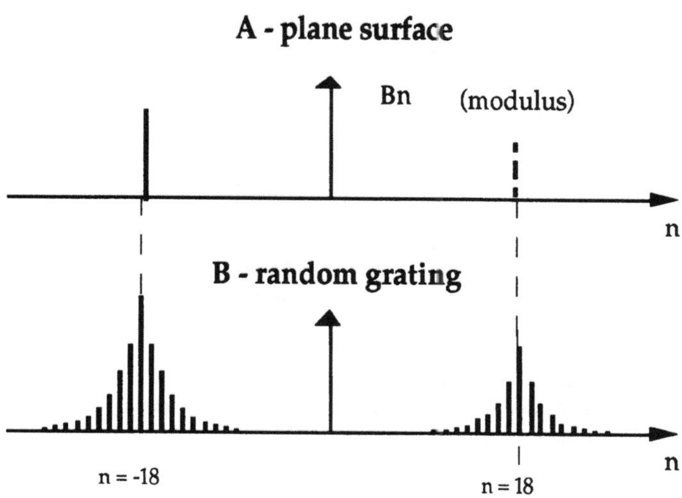


Figure 3: Comparison between the tempered polariton of the plane and that of a random grating

For reason of continuity, the expression of the tempered polariton of a random rough surface given by (9) and (10) must tend towards the expression of the tempered polariton of the plane given by

(1) and (2) when the rms of the random rough surface tends to zero. Thus, it exists an order n of the grating such that

$$nK = k\frac{\nu}{\sqrt{1+\nu^2}} \tag{14}$$

and this order reduces to the tempered polariton of the plane propagating towards $x = +\infty$ when the rms tends to zero. The opposite order (-n) tends to the second tempered polariton. From (14), it can be deduced that, when the rms vanishes, the location of the complex value of $\lambda$ in the complex plane is given by

$$\lim_{\sigma\to 0}(\lambda) = \frac{d}{n}\frac{\nu}{\sqrt{1+\nu^2}} \tag{15}$$

where n is a positive integer. Since d and $\nu$ are fixed, we derive that it exists an infinity of values of $\lambda$ corresponding to the values of n between 1 and infinity. Each of these poles correspond to the two tempered polaritons of the plane. It can be noticed that with d = 7.6836 μm, introducing n = 18 in (15) provides a value of the wavelength equal to 0.436 + 0.00185 i. Assuming that the value of the index remains the same as for $\lambda$ = 0.436 μm, (15) provides us the limit of one of the wavelengths at which a tempered polariton can be supported by the grating.

### 3. 4. INTUITIVE CONSIDERATIONS ABOUT LOCALIZATION OF LIGHT BY A RANDOM GRATING.

Now, let us suppose that the rms of the random grating is increased. It can be guessed that two kinds of coupling phenomena may occur:

- -coupling between the +18th and the -18th orders,
- -coupling between the±18th orders and the closest orders .

This means that for small values of the rms, $B_{\pm 18}$ will remain the most important in modulus, but their amplitudes will be linked, and not independent like in figure 3A. In addition, other $B_n$ will appear, as schematized in figure 3B. It is interesting to notice that this very simple consideration enables us to conjecture that the tempered polariton of the random grating may be localized in order to become a surface localiton. If we bear in mind that the $B_n$ are nothing but the Fourier coefficients of the field at a given ordinate (to within a multiplication by an exponential factor), as shown by (9) and (10), we must deduce that the tempered localiton of the plane cannot be localized since its Fourier coefficients reduce to 1 or 2 components (in other words, it is the Fourier transform of 1 or 2 Dirac functions). On the other hand, the Fourier coefficients of the tempered polariton of the random grating have a support that not reduces to zero, thus the width of the surface localiton may be finite. If we assume for simplicity that the distribution of $B_n$ approaches two gaussian functions, then the width of the localiton may be derived. Of course, the model of random grating does not allow one to see any localization if the width of the localiton exceeds the grating period. The phenomenon of localization may be understood in an other way. In figure 3A, the constant of propagation $\alpha$ of the tempered polariton is fixed while we cannot have any information on its localization since the modulus of its amplitude is constant. On the other hand, the constant of propagation of the tempered polariton of figure 3B is not fixed: the constants $\alpha_n$ belong to two subsets around $\alpha$ and $-\alpha$. In other words, for an observator, we go from a case where the velocity of the polariton is perfectly known (figure 3A) to a case where the velocity can be known with a given and finite precision, depending on the width of the two distributions of $B_n$ in figure 3B. Refering to quantum physics, it may be conjectured that the localization of the tempered polariton, impossible in figure 3A, becomes possible with the random grating of figure 3B.

Obviously, this cannot be considered as a proof of the existence of localitons and this is why we have used the computer code for finding the location of the poles of the S-matrix and for investigating the structure of the tempered polaritons.

## 4. Numerical data

### 4. 1. TRAJECTORIES OF THE WAVELENGTH IN THE COMPLEX PLANE WHEN THE RMS IS INCREASED

In the following, we consider the random grating of figure 2. In order to go continuously from the plane to the random grating, a continuous set of random gratings have been constructed , deduced from that of figure 2 by a similarity with respect to the y axis. Each of those gratings will be characterized by its rms.

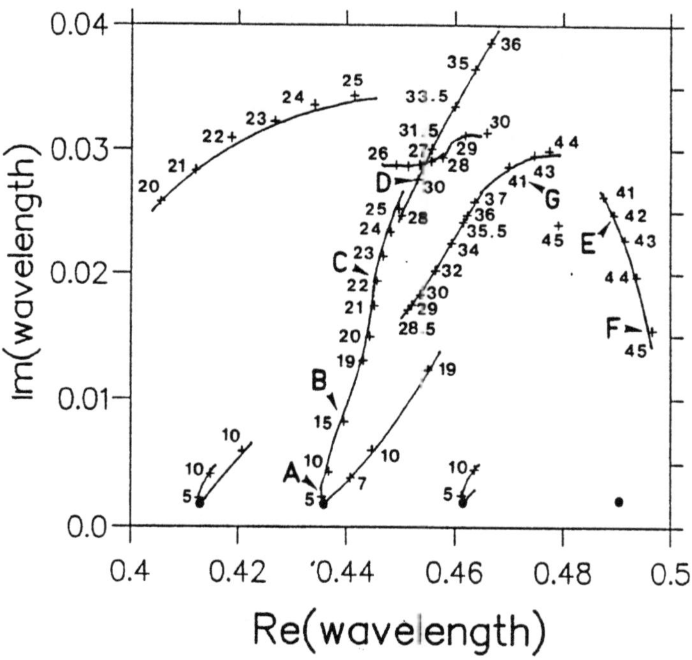


Figure 4: Trajectories of the poles of the S-matrix in the complex plane of the wavelengths. The numbers indicate the rms of the surface in nm. The circles give the analytical values provided by (15), with, from the left to the right, n = 19, 18, 17 and 16.

Figure 4 shows the trajectories of the poles in the complex plane of $\lambda$. These trajectories have been constructed from the analytical values given by (15) by increasing progressively the rms. One may have an idea of the computation time required by these calculations by considering that one of these points needed a computation time of about 90' on an IBM 530 Risc 6000 working with vectorized subroutines (around 30 Mflops). The first remark is that these trajectories look like a set of continuous segments of curve, with discontinuitie between these segments. This feature and the time computation makes very difficult this search and this is a reason why we cannot be sure that all the segments are represented in figure 4 (contrary to what happens for the limit of the pole when the rms tends to 0, we have not any analytic formula at hand for predicting these discontinuities). The trajectories for lowest values of the rms tend to the analytic values deduced from (15) using n = 17, 18 and 19. The analytic value corresponding to n = 16 is also given. It must be remarked that one of the analytic values corresponds to the origine of two trajectories. This is not surprising since we noticed in section 3.3 that one of these values correspond to 2 independent tempered polaritons propagating in opposite directions. For the plane, these polaritons are characterized by the same complex wavelength. This feature does not hold any more when the rms is different from zero.

## 4. 2. SCATTERING AMPLITUDES.

In order to know if one of these tempered polaritons remains linked to a given direction of propagation, we have drawn in figure 5 the calculated values of the modulus of $B_n$ versus n for increasing values of the rms. Obviously, the amplitudes of the positive and negative orders have the same order of magnitude, even for very small rms. These figures agree with our predictions of section 3.4: when the rms is increased, the distribution of $B_n$ is widened in such a way that for rms greater than 0.015 μm, it exists only one distribution going from one side to the other without any interruption. This allows us to predict large losses of the tempered polariton by radiation, the amplitudes of the propagating diffracted orders being very large. It may be conjectured that the large number of continuous segments corresponding to large values of the rms is due to the creation

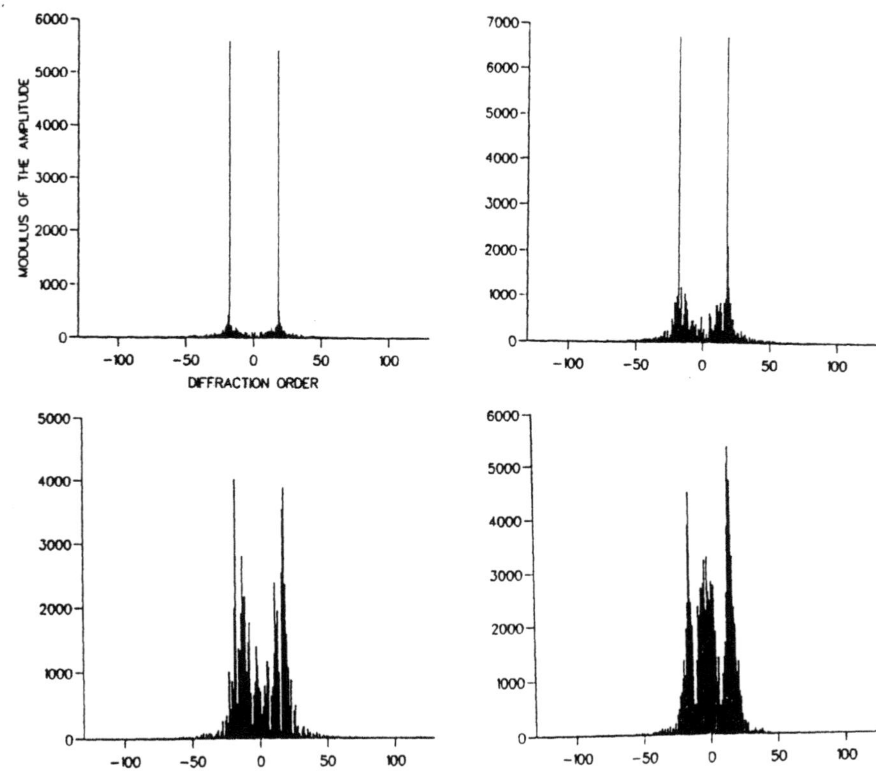


Figure 5: Distribution of the scattering amplitudes of the tempered polaritons corresponding to the points A (top left), B (top right), D (bottom left) and F (bottom right) of figure 4.

of new poles. On the other hand, it does not appear clearly why the trajectories are discontinuous, even though this phenomenon seems very similar to those linked with "stop bands" in various fields of Physics.

4. 3. SURFACE LOCALITON

Finally, we have drawn in figure 6 the square of the modulus of the field versus the abscissa for some of these poles, for an ordinate y corresponding to the top of the highest asperity. Obviously, very strong phenomena of localization are obtained for values of the rms greater than 30 nm, even though no general rule can be stated (computations made for points C , E and G of figure 4 show that for instance, localization is better for point D than for G ). It is important to notice that the supports of the surface localitons are not identical, even though most of these polaritons are close to the center of the figure. The correlation length does not exceed 5 wavelengths at point F.
It can be conjectured that the notion of threshold of rms for localization, which seems to be obvious in figure 6, is in fact a consequence of our grating model: greater values of the period should allow

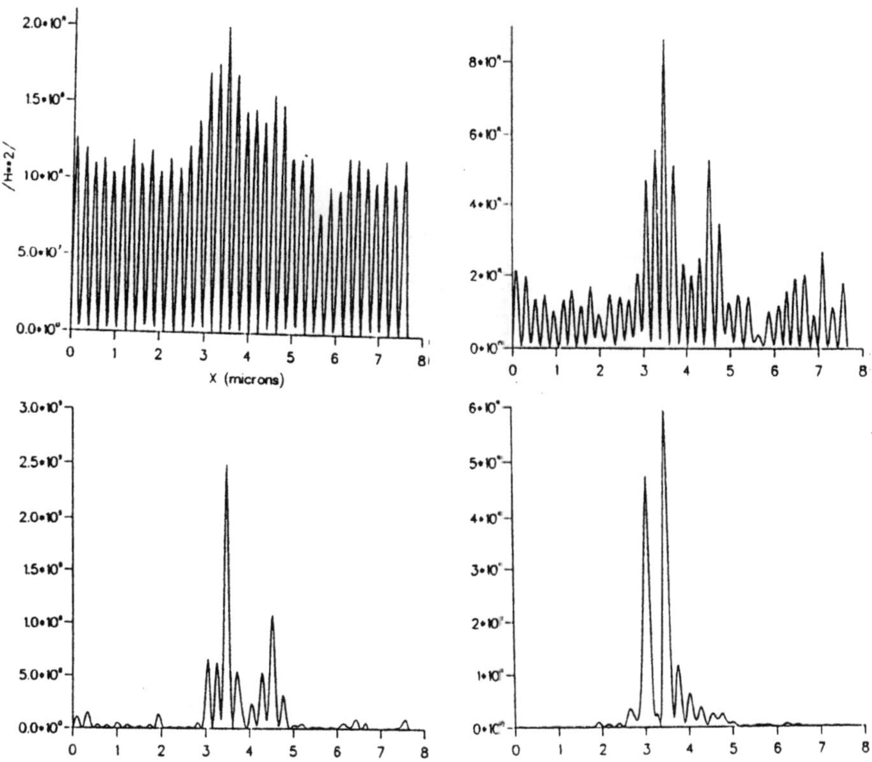


Figure 6: Square of the modulus of the field at the top of the surface versus the abscissa, for points A (top left), B (top right), D (bottom left) and F (bottom right) of figure 4.

us to exhibit localization phenomena for lower rms. In fact, since the repartition of Bn is enlarged as soon as the rms differs from 0, it may be conjectured that localitons exist for any non-null value of the rms.

4. 4. HOW COULD A LOCALITON BE OBSERVED.

Let us suppose that a random rough surface is illuminated by a transient field which vanishes after $t = t_0$. It can be shown using the saddle point method that the asymptotic time-dependence of the field when t tends to infinity is imposed by the location of the poles of the S-matrix in harmonic regime. This means that for an incident transient field (for instance with a gaussian time-dependence) having a Fourier transform with respect to t centered around a given frequency $\omega$, the time dependence of the field after $t = t_0$ will tend, after a transitory regime for small values of $t-t_0$, towards a sum of exponential terms in $\mathbf{f_p}(x,y) \exp(-i\omega_p t)$. Here, $\mathbf{f_p}$ is the field corresponding to the p-th localiton of the surface profile, pole of the S-matrix for $\omega = \omega_p$, the sum being restricted to the poles $\omega_p$ close to $\omega$. This entails that for large values of t, the asymptotic behavior of the field is given by the localiton having the smallest imaginary part.

So, the transient field after illumination tends to be localized around particular points of the surface (centers of the localitons) which radiate energy at infinity and decrease exponentially with different time-constant. If the rms is large enough, the random rough surface observed from infinity should look like an ensemble of localized light sources. This could provide a means for an experimental evidence of localitons in Optics or (much easier) in the microwaves region.

## 5. Conclusion.

Starting from the notion of tempered polariton and searching poles of the S-matrix in the complex plane of the wavelength, we have been able to exhibit strong localization phenomena by random rough surfaces. The calculations have been made for metallic surfaces illuminated by p-polarized light in the visible region but we conjecture that, for p-polarization, the generalization to perfectly conducting metals should not make problems since in that case, the surface polaritons tend to surface waves[4]. On the other hand, it is well known that for s-polarized light, surface waves generated by cavity resonances are restricted to deep surfaces.

The notion of localiton, deduced from the search for poles in the complex plane of wavelengths could be generalized easily to an ensemble of particles. This could lead to the definition of volume localitons.

The study of the consequences of the existence of surface localitons on the phenomena of absorption and enhanced backscattering are in progress.

## References

1. Maystre, D.and Dainty, J.C. (1991) Modern analysis of scattering phenomena, Adam Hilger, New York .

2. Nieto-Vesperinas, M. and Dainty, J. C. (1990) Scattering in volumes and surfaces, North-Holland, Amsterdam.

3. Thouless, D. J. (1980) Phys. Rep., 67, p. 5.

4. Maystre, D. (1982) 'General study of grating anomalies', in A.D. Boardman (ed.), "Electromagnetic surface modes", John Wiley & sons, Chichester, p. 661-724 .

5. Maystre, D. (1978) 'A new general integral theory for dielectric coated gratings', J. Opt. Soc. Amer., 68, p. 490-495.

# DIFFRACTION GRATINGS AS COMPONENTS FOR PHOTON SCANNING TUNNELING MICROSCOPE IMAGE INTERPRETATION

**M. NEVIERE and P. VINCENT**
*Laboratoire d'Optique Electromagnétique, URA CNRS n° 843,*
*Faculté des Sciences de St-Jérôme, Boîte 262,*
*13397 MARSEILLE CEDEX 20*

## EXTENDED ABSTRACT

Since the invention of the Scanning Tunneling Microscope (STM) [1-2], several types of local probe microscopes have been developed, among which the Atomic Force Microcope [3] and the Photon Scanning Tunneling Microcope (PSTM)[4-6]. The basic idea which sustains both STM and PSTM is that, if a servo-mecanism is able to move the probe near the observed surface in such a way to maintain the signal constant, the trajectory of the probe tip should reproduce the shape of the surface. This idea comes from the fact that near a plane interface the current intensity produced by Tunnel Effect in Quantum mechanics or the electromagnetic field collected by the tip of the fiber in the Optical Tunnel Effect decreases exponentially with the distance of the tip. The question is whether this fundamental behavior is maintained when the plane surface is replaced by a corrugated one.

In order to try to answer this question, we replace the observed corrugated surface by a dielectric grating, which is a simple model leading to a two dimensional problem. Then the Differential Theory of Gratings [7] allows computing field intensity maps everywhere and comparing their shape with the grating profile, whatever the lighting conditions and polarization of incident wave may be. It is found that the equal intensity lines have a shape generally very different from the one of the observed surface and that, for low value of the intensity, they may even present a S like shape [8,9] or have a quasi elliptical shape, while the observed



*D.W. Pohl and D. Courjon (eds.), Near Field Optics*, 377–378.

surface is a simple sinusoidal or lamellar grating. However these complicated near field structures disappear when the groove spacing is small enough in order to ensure that no propagating order exists near the probe.

The formalism allows introducing a fiber end in our model by a collection of periodic tips and studying the influence of these tips on the equal intensity lines. The perturbation turns out to be generally small, but a resonance may appear for a particular position of the tips, which seems to double the periodicity of the observed sample. This phenomenon is not linked with a coupling between neighbouring tips, since it remains when we suppress 4 tips over 5, i.e. when the tip period becomes 5 times greater than the grating period, a situation for which the coupling between tips is established to be negligible. However, the resonance can be reduced by increasing the distance between tip and sample surface.

The excitation of a surface plasmon near a thin lamellar silver grating leads to very complicated field maps. But since all intensity lines present a great change in intensity when the angle of incidence is slightly varied near the critical resonance angle which excites the surface plasmon, PSTM turns out to be an efficient tool for detecting such a surface wave.

**REFERENCES**

[1] G. Binnig and H. Rohrer, Helv. Phys. Acta, 55, 726 (1982)
[2] G. Binnig and H. Rohrer, Surf. Sci. 126, 236 (1983)
[3] G. Binning, C.F. Quate and Ch. Gerber, Phys. Rev. Lett, 56, 930 (1986)
[4] R.C. Reddick, R.J. Warmack, T. Ferrell, Phys. Rev. B 39, 767 (1989)
[5] D. Courjon, K. Sarayeddine, M. Sajer, Opt. Commun., 71, 23 (1989)
[6] F. de Fornel, J.P. Goudonnet, L. Salmon, E. Lesniewska, Proc. SPIE Paris, April, vol.1139, 77 (1989
[7] P. Vincent, "Differential Methods" dans "Electromagnetic Theory of Gratings", Springer-Verlag (Heidelberg, 1980)
[8] J.P. Goudonnet, M. Nevière, "Analysis of images of periodic structures obtained by photon scanning tunneling microscopy", dans "Application and Theory of Periodic structures", J.M. Lerner et W.R. Mc Kinney, eds., chap.15, p.130-139 (SPIE, 1991)
[9] J.P. Goudonnet, L. Salmon, F. de Fornel, E. Bourillot, P. Adam, M. Nevière, P. Guérin, "Recent progress in photon scanning tunneling microscopy", dans "Scanning Probe Microscopics, 1639, p.12-23, (SPIE, 1992)

# OPTICAL NEAR-FIELD INTERACTION: ON THE LOCAL FIELD INSIDE A QUANTUM TIP

O. KELLER
*Institute of Physics, University of Aalborg*
*DK-9220 Aalborg Øst, Denmark*

ABSTRACT. Using a selfconsistent scattering-theory approach, the local field inside a mesoscopic tip interacting with a number of point dipoles placed on top of a flat surface is studied. For tips exhibiting quantum size effects it is demonstrated how the basic integral equation for the local field can be transformed into a physically appealing set of linear algebraic equations among the relative strengths of the fields generated by the quantum transitions in the mesoscopic tip. The eigenmode resonances and the multipole dynamics of the system are investigated. Quantum tips consisting of subparticles without wavefunction overlap are studied, and it is shown that the present theory in the appropriate limit contains the results of previously established phenomenological point-particle models.

## 1 Introduction

It is of central importance in optical near-field microscopy to understand on the microscopic level the electrodynamic interaction between the probe tip and the surface particles, ultimatively atoms or molecules. Since the size of the probe tip normally is large compared to that of a single atom it is necessary to take into account in a selfconsistent manner the internal electrodynamics of the tip. For tips of macroscopic size this is usually done by using a well known long wavelength electrostatic expression for the polarizability of a homogeneous sphere. Such an approach inherently has a number of fundamental shortcomings, and for mesoscopic tips it tends to be meaningless [1], [2]. The use of mesoscopic tips certainly is of interest for optical near-field microscopy due to the fact that the resolution of a near-field microscope is limited by the sharpness of the local-field resonances in the system, e.g. in the tip. In a mesoscopic tip pronounced resonances should be attainable. In previous works, the internal dynamics of the probe tip has been studied on the basis of the assumption that the tip can be modelled by an assembly of electromagnetically interacting point dipoles emitting electric-dipole and possibly magnetic-dipole plus electric-quadropole and higher order multipole fields [3] - [7]. While such an approach might be adequate for mesoscopic tips consisting of electronically weakly coupled units (a dielectric medium or molecular microcrystallites) it certainly would be too simple for systems, e.g. metal probes, with strong electronic coupling across the domain of the tip. Furthermore, the point-dipole model generally is not able to take into account optical quantum size phenomena, known to be important in most mesoscopic systems. Based on this fact, and because I believe



D.W. Pohl and D. Courjon (eds.), Near Field Optics, 379–390.

that mesoscopic probes as such, e.g. quantum wells, small metallic droplets, and single molecules, will be of substantial interest for the future development of near-field optics, my group has undertaken theoretical studies of the local-field problem in mesoscopic quantum tips.

In the present paper the basic framework for a local-field description of the electrodynamics of a mesoscopic quantum tip is established. Taking as a starting point a selfconsistent scattering-theory formalism [2], [8] the local field inside a mesoscopic tip interacting with a set of electric point-dipoles placed on a plane surface having a sharp boundary is investigated. For simplicity, and conceptual clarity, it is assumed that the microscopic conductivity response of the quantum tip is dominated by its nonlocal paramagnetic part. In the presence of only the paramagnetic contribution to the response the local field inside the mesoscopic tip can be expressed in terms of a set of unknown coefficients. Using an adequate supertensor (vector) notation the prevailing but unknown fields on the surface dipoles are eliminated in favour of these unknown coefficients. In turn, the basic integral equation for the local field is replaced by a set of linear, algebraic equations among the unknown coefficients adequate for a numerical analysis. The eigenmode condition for the quantum tip-surface dipole-surface system is established, and the multipole dynamics of the mesoscopic quantum particle is studied. Since the eigenmode condition of the present work constitutes a generalization of that previously examined for a point-like tip [7] the so-called configurational resonances [7] are put in perspective. The reduction obtained in the dimensionality of the algebraic matrix problem when the quantum tip consists of subparticles without overlap of the electronic wavefunctions is mentioned. Finally, it is shown how the present theory in the appropriate limit is reduced to the phenomenlogical point-particle theory.

## 2 Local Field Inside a Mesoscopic Tip

Let us consider the situation depicted in Fig. 1. A monochromatic electromagnetic field of angular frequency $\omega$ is incident upon a system composed of a (flat) surface, a number ($N$) of point dipoles placed on top of the surface, and a (quantum) tip of finite size. To determine in a selfconsistent manner the local field inside the quantum tip, one takes as a starting point the following formal expression for the amplitude, $\vec{E}(\vec{r};\omega)$, of the prevailing electric field at the space-point $\vec{r}$ [2]:

$$\vec{E}(\vec{r};\omega) = \vec{E}^B(\vec{r};\omega) - i\mu_0\omega \int \overleftrightarrow{G}^B(\vec{r},\vec{r}\,';\omega)\cdot\vec{J}(\vec{r}\,';\omega)d^3r' \ . \tag{1}$$

In this equation $\vec{E}^B(\vec{r};\omega)$ denotes the background ($B$) field, i.e. the sum of the incident field and the field reflected from the surface in the absence of the surface dipoles and the quantum tip, and $\overleftrightarrow{G}^B(\vec{r},\vec{r}\,';\omega)$ is the associated background propagator. This propagator (a dyadic electromagnetic Green function) describes the electromagnetic field propagation from a source point located at $\vec{r}\,'$ to a point of observation at $\vec{r}$ in a so-called pseudovacuum consisting of the vacuum-surface system, alone. For our fundamental considerations $\vec{E}^B(\vec{r};\omega)$ and $\overleftrightarrow{G}^B(\vec{r},\vec{r}\,';\omega)$ can be considered as prescribed quantities to be obtained from

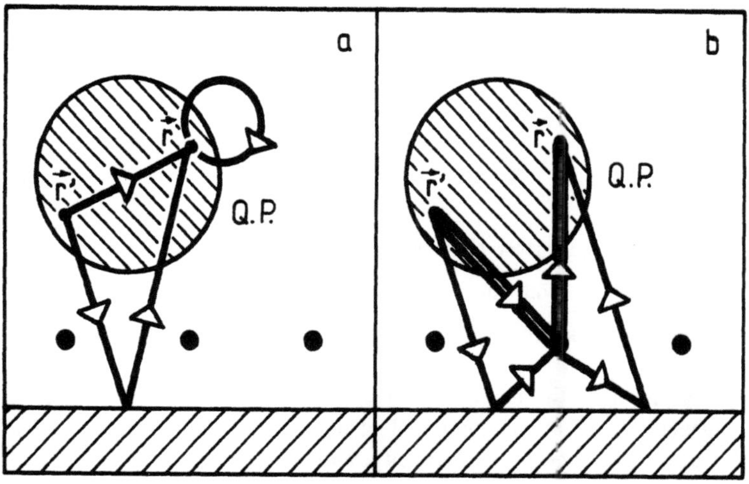


*Fig. 1. Schematic diagrams illustrating the various channels of electromagnetic wave propagation from a source point at $\vec{r}'$ to an observation point at $\vec{r}$, both points being located inside the quantum particle (tip) [Q.P]. In Fig.1a, the channels shown are those which do not involve scattering on the point dipoles (PD) of the surface (S) [PD are indicated by blackened circles]. Fig.1b, shows the four channels ($\vec{r}' \rightarrow PD \rightarrow \vec{r}, \vec{r}' \rightarrow PD \rightarrow S \rightarrow \vec{r}, \vec{r}' \rightarrow S \rightarrow PD \rightarrow \vec{r}, \vec{r}' \rightarrow S \rightarrow PD \rightarrow S \rightarrow \vec{r}$) involving scattering on the point dipoles. The scattering strengths of the point dipoles and the source strengths at the various points inside the quantum particle are calculated in a selfconsistent manner, which takes into account multiple scattering effects to infinite order.*

the electrodynamics of the bare vacuum-surface system, and as such the theory developed below is not limited to flat sharp-boundary surfaces, say. From a practical point of view, explicit expressions for $\vec{E}^B$ and $\overleftrightarrow{G}^B$ of course can be established for simple geometrics and for simple models of the electrodynamics of the surface (medium), only. The quantity $\vec{J}(\vec{r}';\omega)$ is the light-induced current density distribution inside the quantum tip (in the following often named the quantum particle ($Q.P.$)) and on the point dipoles at the surface. Thus, within the framework of linear response theory

$$\vec{J}(\vec{r}';\omega) = \sum_{j=1}^{N} \overleftrightarrow{S}_j(\omega) \cdot \vec{E}(\vec{r}_j;\omega)\delta(\vec{r}' - \vec{r}_j) + \int_{Q.P} \overleftrightarrow{\sigma}(\vec{r}', \vec{r}'';\omega) \cdot \vec{E}(\vec{r}'';\omega)d^3r'' , \quad (2)$$

where $\overleftrightarrow{S}_j(\omega)$ is the microscopic conductivity tensor of the $j$th point dipole located at position $\vec{r}_j$, and $\overleftrightarrow{\sigma}(\vec{r}, \vec{r}';\omega)$ is the generally nonlocal conductivity response tensor of the quantum particle describing the response to the selfconsistent field prevailing across the particle. By combining Eqs. (1) and (2), and leaving out $\omega$ from the notation, one obtains the following equation for the prevailing field:

$$\vec{E}(\vec{r}) = \vec{E}^B(\vec{r}) + \sum_{j=1}^{N} \overleftrightarrow{T}(\vec{r}, \vec{r}_j) \cdot \vec{E}(\vec{r}_j) + \int_{Q.P.} \overleftrightarrow{K}(\vec{r}, \vec{r}') \cdot \vec{E}(\vec{r}')d^3r' , \quad (3)$$

with

$$\overleftrightarrow{T}(\vec{r},\vec{r}_j) = -i\mu_0\omega \overleftrightarrow{P}^B(\vec{r},\vec{r}_j)\cdot \overleftrightarrow{S}_j , \tag{4}$$

and

$$\overleftrightarrow{K}(\vec{r},\vec{r}\,') = -i\mu_0\omega \int_{Q.P.} \overleftrightarrow{G}^B(\vec{r},\vec{r}\,'')\cdot \overleftrightarrow{\sigma}(\vec{r}\,'',\vec{r}\,')d^3r'' . \tag{5}$$

Since it has been assumed from the outset that the surface particles were point dipoles, it is necessary *a priori* to leave out from the propagator $\overleftrightarrow{G}^B$ the self-field part, and thus retain only the propagating part, denoted by $\overleftrightarrow{P}^B$, in the relevant terms, cf. Eq. (4) and the discussion in Ref. [2].

To proceed from here one need the explicit expression for the conductivity response tensor, $\overleftrightarrow{\sigma}(\vec{r},\vec{r}\,')$. In general, $\overleftrightarrow{\sigma}(\vec{r},\vec{r}\,')$ is composed of a local diamagnetic part and a nonlocal paramagnetic part. For conceptual simplicity, and because of the possibility of having interlevel resonance, we shall assume that the paramagnetic term is the dominating one. Thus, within the framework of the random-phase-approximation one has

$$\overleftrightarrow{\sigma}(\vec{r},\vec{r}\,') \simeq \overleftrightarrow{\sigma}^{para}(\vec{r},\vec{r}\,') = \frac{2i}{\omega}\sum_{m,n} \frac{\rho^F_{mm} - \rho^F_{nn}}{\hbar\omega + E_m - E_n} \vec{\jmath}_{nm}(\vec{r}\,)\vec{\jmath}_{mn}(\vec{r}\,') , \tag{6}$$

where $\rho^F_{mm}$ and $\rho^F_{nn}$ are the on-diagonal matrix elements of the thermal equilibrium [free $(F)$] density matrix operator belonging to the states $m$ and $n$, $E_m$ and $E_n$ are the associated level energies, and $\vec{\jmath}_{\alpha\beta}$ is the transition current density from state $\alpha$ to state $\beta$. By making use of the explicit expression for $\overleftrightarrow{\sigma}(\vec{r},\vec{r}\,')$, Eq. (3) can be rewritten in the form

$$\vec{E}(\vec{r}\,) = \vec{E}^B(\vec{r}\,) + \sum_{m,n} \beta_{mn}\vec{F}_{nm}(\vec{r}\,) + \sum_{j=1}^{N} \overleftrightarrow{T}(\vec{r},\vec{r}_j)\cdot \vec{E}(\vec{r}_j) , \tag{7}$$

where the $\beta_{mn}$'s, given by

$$\beta_{mn} = \int_{Q.P.} \vec{\jmath}_{mn}(\vec{r}\,')\cdot \vec{E}(\vec{r}\,')d^3r' , \tag{8}$$

at this stage are unknown constants, and

$$\vec{F}_{nm}(\vec{r}\,) = 2\mu_0 \frac{\rho^F_{mm} - \rho^F_{nn}}{\hbar\omega + E_m - E_n} \int_{Q.P.} \overleftrightarrow{G}^B(\vec{r},\vec{r}\,'')\cdot \vec{\jmath}_{nm}(\vec{r}\,'')d^3r'' . \tag{9}$$

The $\vec{F}_{nm}(\vec{r}\,)$'s in principle are known vectorial functions of $\vec{r}$. Note also that $\vec{F}_{nm} = \vec{0}$ for $n = m$. Damping effects can be accounted for phenomenologically by replacing $\omega$ by $\omega + i\nu$ [$\nu$: damping constant] in the denominator of Eq. (9).

To obtain the local field $\vec{E}(\vec{r}\,)$ from Eq. (7) we need to calculate the prevailing fields, $\vec{E}(\vec{r}_j)$ $(j = 1, 2, \cdots N)$, on the sites of the dipoles and the $\beta_{mn}$ constants. We achieve our goal in two steps. Firstly, we calculate the $\vec{E}(\vec{r}_j)$'s in terms of the unknown $\beta_{mn}$'s. Secondly, we determine the $\beta_{mn}$'s from a set of linear and inhomogeneous algebraic equations.

Introducing the supervectors

$$\vec{\mathcal{E}} = \begin{pmatrix} \vec{E}(\vec{r}_1) \\ \vec{E}(\vec{r}_2) \\ \vdots \\ \vec{E}(\vec{r}_N) \end{pmatrix} \; ; \; \vec{\mathcal{E}}^B = \begin{pmatrix} \vec{E}^B(\vec{r}_1) \\ \vec{E}^B(\vec{r}_2) \\ \vdots \\ \vec{E}^B(\vec{r}_N) \end{pmatrix} \; ; \; \vec{\mathcal{F}}_{nm} = \begin{pmatrix} \vec{F}_{nm}(\vec{r}_1) \\ \vec{F}_{nm}(\vec{r}_2) \\ \vdots \\ \vec{F}_{nm}(\vec{r}_N) \end{pmatrix} \tag{10}$$

and the supertensor

$$\overleftrightarrow{\mathcal{T}} = \begin{pmatrix} \overleftrightarrow{T}(\vec{r}_1,\vec{r}_1) & \cdots & \overleftrightarrow{T}(\vec{r}_1,\vec{r}_N) \\ \vdots & & \vdots \\ \overleftrightarrow{T}(\vec{r}_N,\vec{r}_1) & \cdots & \overleftrightarrow{T}(\vec{r}_N,\vec{r}_N) \end{pmatrix} \tag{11}$$

[with the definitions $\overleftrightarrow{T}(\vec{r}_1,\vec{r}_1) = \overleftrightarrow{T}(\vec{r}_2,\vec{r}_2) = \cdots = \overleftrightarrow{T}(\vec{r}_N,\vec{r}_N) = \overleftrightarrow{0}$, cf. Eq. (4) and the remarks below Eq. (5)], one finds in supervector (tensor) notation by letting $\vec{r}$ in Eq. (7) in turn be equal to $\vec{r}_1, \vec{r}_2, \cdots \vec{r}_N$ the relation

$$\vec{\mathcal{E}} = \vec{\mathcal{E}}^B + \sum_{m,n} \beta_{mn} \vec{\mathcal{F}}_{nm} + \overleftrightarrow{\mathcal{T}} \cdot \vec{\mathcal{E}} \,. \tag{12}$$

In terms of the unknown $\beta_{mn}$'s the solution for $\vec{\mathcal{E}}$ is

$$\vec{\mathcal{E}} = [\overleftrightarrow{\mathcal{U}} - \overleftrightarrow{\mathcal{T}}]^{-1} \cdot [\vec{\mathcal{E}}^B + \sum_{m,n} \beta_{mn} \vec{\mathcal{F}}_{nm}] \,, \tag{13}$$

where $\overleftrightarrow{\mathcal{U}}$ is a super unit tensor. By introducing a supertensor

$$\overleftrightarrow{\mathcal{T}}_0(\vec{r}) = \begin{pmatrix} \overleftrightarrow{T}(\vec{r},\vec{r}_1) & \cdots & \overleftrightarrow{T}(\vec{r},\vec{r}_N) \\ 0 & \cdots & 0 \\ \vdots & & \vdots \\ 0 & \cdots & 0 \end{pmatrix} , \tag{14}$$

and by making use of the expression for $\vec{\mathcal{E}}$ in Eq. (13), Eq. (7) takes the form

$$\vec{E}(\vec{r}) = \vec{E}^B(\vec{r}) + \sum_{m,n} \beta_{mn} \vec{F}_{nm}(\vec{r}) + \overleftrightarrow{\mathcal{T}}_0(\vec{r}) \cdot [\overleftrightarrow{\mathcal{U}} - \overleftrightarrow{\mathcal{T}}]^{-1} \cdot [\vec{\mathcal{E}}^B + \sum_{m,n} \beta_{mn} \vec{\mathcal{F}}_{nm}] \,. \tag{15}$$

Finally, by inserting this formula for $\vec{E}(\vec{r})$ into Eq. (8) one obtains the basic set of linear, algebraic equations among the unknown $\beta_{mn}$'s, i.e.

$$\beta_{mn} - \sum_{o,p} N_{po}^{mn} \beta_{op} = H_{mn} \tag{16}$$

with the abbreviations

$$N_{po}^{mn} = \int_{Q.P.} \vec{\jmath}_{mn}(\vec{r}) \cdot \vec{F}_{po}(\vec{r}) d^3r + [\int_{Q.P.} \vec{\jmath}_{mn}(\vec{r}) \cdot \overleftrightarrow{\mathcal{T}}_0(\vec{r}) d^3r] \cdot [\overleftrightarrow{\mathcal{U}} - \overleftrightarrow{\mathcal{T}}]^{-1} \cdot \vec{\mathcal{F}}_{po} \,, \tag{17}$$

and

$$H_{mn} = \int_{Q.P.} \vec{\jmath}_{mn}(\vec{r}) \cdot \vec{E}^B(\vec{r})d^3r + [\int_{Q.P.} \vec{\jmath}_{mn}(\vec{r}) \cdot \overleftrightarrow{T}_0(\vec{r})d^3r] \cdot [\overleftrightarrow{\mathcal{U}} - \overleftrightarrow{\mathcal{T}}]^{-1} \cdot \vec{\mathcal{E}}^B . \quad (18)$$

Once the $\beta_{mn}$'s have been obtained from Eq. (16), the prevailing local field can be determined everywhere in space by means of Eq. (15).

The relations in Eqs. (15) - (18) form an adequate starting point for a quantitative numerical calculation of the optical near-field (and intermediate and far-field) interaction between a quantum tip and an assembly of surface particles. Instead of embarking on such a quantitative analysis, which would lead us far beyond the scope of this paper, let me emphasize in the subsequent sections a few qualitative aspects, and let me show how the present theory in the appropriate limits contains the result of previously published microscopic theories.

## 3 Eigenmode Condition

The eigenmodes of our system can be determined by looking for selfsustaining solutions for the local field. For a selfsustaining solution to exist, $\vec{E}(\vec{r})$ must be different from zero in the absence of the background field. By denoting the selfsustaining (or resonant) field by $\vec{E}_{res}(\vec{r})$ it follows from Eq. (3) that this field must satisfy the equation

$$\vec{E}_{res}(\vec{r}) = \sum_{j=1}^{N} \overleftrightarrow{T}(\vec{r},\vec{r}_j) \cdot \vec{E}_{res}(\vec{r}_j) + \int_{Q.P.} \overleftrightarrow{K}(\vec{r},\vec{r}\,') \cdot \vec{E}_{res}(\vec{r}\,')d^3r' . \quad (19)$$

In turn, this implies that the resonant values of the $\beta_{mn}$'s named $\beta_{mn}^{res}$, must fulfil a set of homogeneous and linear algebraic equations, viz.

$$\beta_{mn}^{res} - \sum_{o,p} N_{po}^{mn} \beta_{op}^{res} = 0 \ , \quad (20)$$

cf. Eq. (16) [$\vec{E}^B(\vec{r}) = 0$ gives $H_{mn} = 0$ for all $m,n$]. In order for the set of equations in (20) to have a nontrivial solution, the determinant of the system must be zero, i.e.

$$Det\left\{\delta_{mn,op} - N_{po}^{mn}\right\} = 0, \quad (21)$$

where $\delta_{mn,op}$ is the Kronecker delta. The condition in Eq. (21) is the eigenmode (or resonance) condition for our system. If the resonance condition is fulfilled the ratios between the various $\beta_{mn}^{res}$'s and a given one, $\beta_{12}^{res}$ say, can be obtained. In turn, the associated resonant field can be determined to within an arbitrary constant. Let me stress here that the resonance condition in Eq. (21) is the *real* resonance condition for the entire system. Inside the relation in Eq. (21) is hidden resonances of the various subsystems, e.g. the configurational resonances of the system of surface dipoles, the bulk and surface plasmon resonances of the vacuum-surface system, and the single-particle (level transitions) and collective resonances of the quantum particle. Whether or not these subsystem resonances play a significant role for the resonances of the entire system among other things depends on how strongly coupled the subsystems are.

From a physical point of view the eigenmode condition in Eq. (21) is basically different from that usually employed in near-field optics. Thus, in the conventional approach the eigenmode condition is obtained from a spatial discretization of the integral equation in (1), the discretization being based on the assumption of completely localized electronic states. In the present work, a new and more general eigenmode condition (Eq. (21)) based on the discrete nature of the bound energy eigenstates of the quantum tip has been established.

# 4 Multipole Dynamics

At optical frequencies it is usually correct to assume that the background field varies slowly in space across the orbitals of the quantum particle. In turn, this means that it is justified to replace $\vec{E}^B(\vec{r})$ in Eq. (18) by a Taylor series expansion around the center of the quantum particle. Letting the center be located at $\vec{r} = \vec{r}_0$, one obtains

$$\int_{Q.P.} \vec{j}_{mn}(\vec{r}) \cdot \vec{E}^B(\vec{r}) d^3r = [\int_{Q.P.} \vec{j}_{mn}(\vec{r}) d^3r] \cdot \vec{E}^B(\vec{r}_0)$$

$$+[\int_{Q.P.} \vec{j}_{mn}(\vec{r})(\vec{r} - \vec{r}_0) d^3r] : \vec{\nabla}\vec{E}^B(\vec{r}_0) + \cdots \quad (22)$$

It is important to emphasize here that it usually is safe to make the Taylor expansion in the background field whereas in general it would be incorrect for a mesoscopic particle to make such an expansion in the local field. Thus, one should avoid from the very beginning of the analysis, i.e. in Eq. (3), to make a series expansion of $\vec{E}(\vec{r}')$ around $\vec{r}_0$ in the term $\int_{Q.P.} \overleftrightarrow{K}(\vec{r}, \vec{r}') \cdot \vec{E}(\vec{r}') d^3r'$. The expansion in Eq. (22) essentially establishes the relation between the present formalism and those based at the outset solely on electric-dipole transitions in the quantum particle. Hence [9],

$$\int_{Q.P.} \vec{j}_{mn}(\vec{r}) d^3r = \frac{i}{\hbar}(E_n - E_m)\vec{P}_{mn} , \quad (23)$$

and

$$\int_{Q.P.} \vec{j}_{mn}(\vec{r})\vec{r} d^3r = \frac{i}{\hbar}(E_n - E_m) \overleftrightarrow{Q}_{mn} - \frac{e}{2m} \overleftrightarrow{U} \times \vec{L}_{mn} , \quad (24)$$

where

$$\vec{P}_{mn} = -e \int_{Q.P.} \psi_n^*(\vec{r})\vec{r}\psi_m(\vec{r}) d^3r , \quad (25)$$

$$\overleftrightarrow{Q}_{mn} = -\frac{e}{2} \int_{Q.P.} \psi_n^*(\vec{r})\vec{r}\vec{r}\psi_m(\vec{r}) d^3r , \quad (26)$$

and

$$\vec{L}_{mn} = \frac{\hbar}{i} \int_{Q.P.} \psi_n^*(\vec{r})(\vec{r} \times \vec{\nabla})\psi_m(\vec{r}) d^3r , \quad (27)$$

are the electric-dipole ($ED$), electric-quadropole ($EQ$), and magnetic-dipole ($MD$) (or angular-momentum) transition matrix elements associated with an electronic $m \rightarrow n$

transition, respectively. Above, the wavefunction of the $i$th state has been denoted by $\psi_i$ ($i = m$ or $n$), and the electron charge by $-e$. The quantity $\overleftrightarrow{U}$ is the unit tensor. By comparing Eqs. (22)-(24) it is realized that the term associated with $\vec{E}^B(\vec{r}_0)$ in Eq. (22) stems from $ED$-transitions, the factor to $\vec{\nabla}\vec{E}^B(\vec{r}_0)$, apart from $ED$-contributions, originates in $EQ$ and $MD$-transitions, and so on.

Once the local fields inside the quantum particle and on the surface dipoles have been obtained, the prevailing field can be calculated everywhere in space from Eq. (7). Since, for observation points outside the quantum particle $\overleftrightarrow{G}^B(\vec{r},\vec{r}'')(=\overleftrightarrow{P}^B(\vec{r},\vec{r}''))$ is a slowly varying function of $\vec{r}''$ across the domain of the quantum particle, the functional form of $\vec{F}_{nm}(\vec{r})$ in Eq. (9) can be simplified utilizing a Taylor expansion of $\overleftrightarrow{G}^B(\vec{r},\vec{r}'')$ in $\vec{r}''$ around $\vec{r}_0$. Thus, for a symmetric propagator

$$\int_{Q.P.} \overleftrightarrow{G}^B(\vec{r},\vec{r}'')\cdot\vec{\jmath}_{nm}(\vec{r}'')d^3r'' = [\int \vec{\jmath}_{nm}(\vec{r}'')d^3r'']\cdot\overleftrightarrow{G}^B(\vec{r},\vec{r}_0)$$

$$+[\int_{Q.P.} \vec{\jmath}_{nm}(\vec{r}'')(\vec{r}''-\vec{r}_0)d^3r''] : \vec{\nabla}_0\overleftrightarrow{G}^B(\vec{r},\vec{r}_0) + \cdots \; . \tag{28}$$

By inserting the above expansion into Eq. (9) and this in turn into Eq. (7) it is realized (cf. Eqs. (23) and (24)) that the scattered field from the selfconsistently driven quantum particle is described within the framework of the standard multipole radiation theory.

In cases where the quantum tip and the relevant surface dipoles are located within a volume of linear extension much less than the characteristic wavelength(s) of the background field, it is often sufficient to assume that $\vec{E}^B$ is constant across the system. In consequence, $H_{mn}$ of Eq. (18) is simplified to

$$H_{mn} = \frac{i}{\hbar}(E_n - E_m)\vec{P}_{mn}\cdot\vec{E}^B(\vec{r}_0) + [\int_{Q.P.} \vec{\jmath}_{mn}(\vec{r})\cdot\overleftrightarrow{T}_0(\vec{r})d^3r]\cdot[\overleftrightarrow{\mathcal{U}} - \overleftrightarrow{\mathcal{T}}]^{-1}\cdot\vec{\mathcal{E}}_0^B \; , \tag{29}$$

where $\vec{\mathcal{E}}_0^B = (\vec{E}^B(\vec{r}_0), \vec{E}^B(\vec{r}_0), \cdots, \vec{E}^B(\vec{r}_0))$.

Although from a basic standpoint there is no need for approximating neither the background field nor the electromagnetic propagator (cf. the discussion in section 2), the multipole procedure described in this section serves three important purposes. First of all, it stresses the fact that a multipole expansion is expected to be valid even in the case of strongly delocalized electronic states, because the expansion is made in the background field and not in the prevailing local field. This principle recently has turned out to be extremely important for a correct interpretation of the optical reflection spectra of ultrathin metallic films, i.e. two-dimensional metallic quantum well systems [10], [11]. Secondly, the local-field theory described in section 2 allows one to derive the background field multipole formalism from basic principles, a fact apparently not appreciated previously. Thirdly, the labour connected to numerical calculations of local-field effects are considerably reduced using the accurate and correct multipole scheme.

## 5 Quantum Tip of Subparticles without Wavefunction Overlap; Point-Particle Approach

In some cases, e.g. for a dielectric quantum tip or a tip consisting of a few weakly bound molecules, the effective dimension of the matrix $N_{po}^{mn}$ in Eq. (16) can be drastically reduced so that the quantitative numerical calculation of the local field *a priori* becomes much easier. To realize this let us compare the energy-level diagram for the entire quantum tip with that of a system of subparticles without mutual wavefunction overlap (Fig. 2). If the quantum particle in consideration has $N$ energy levels, one has $N(N-1)$ unknown $\beta_{mn}$'s in Eq. (16) [the diagonal elements $\beta_{nn}$ are zero]. Now, if the tip consists of $M$ electronically decoupled subparticles and if subparticle number $i$ has $N_i$ energy levels, it readily appears that the number of unknown constants in Eq. (16) will be $\sum_{i=1}^{M} N_i(N_i-1)$. With the constraint $N=\sum_{i=1}^{M} N_i$, the dimension of the matrix $N_{po}^{mn}$ is reduced when the quantum particle can be considered to consist of a number of electronically independent subparticles. As an example, if $N=4$ and $N_1=N_2=2$ the number of unknown constants in Eq. (16) is reduced from $4(4-1)=12$ to $2(2-1)+2(2-1)=4$. Thus, in this case the replacement of a four-level quantum particle by two two-level subparticles strongly reduces the dimensionality of the matrix problem. Also symmetry can lead to a reduction of the effective dimensionality of the problem at hand.

The idea of treating the tip as a set of electronically independent but electromagnetically coupled units has previously been suggested by Girard and Bouju [5], [6]. However, the analysis of Girard and Bouju [5], [6] deviates conceptually from the present one, because these authors from the outset assume that the subparticles can be treated as electric

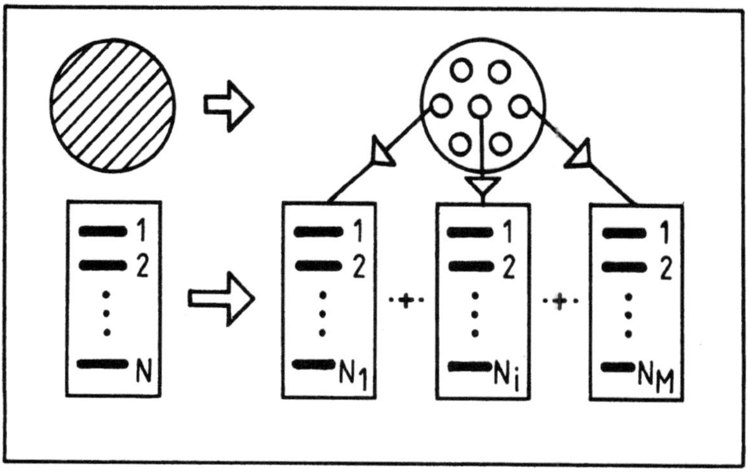


*Fig. 2. Schematic diagram illustrating the transition from a tip consisting of a single quantum particle having $N$ energy levels to a tip composed of $M$ electronically decoupled subparticles of which the ith one has $N_i$ energy levels. With the constraint $N=\sum_{i=1}^{M} N_i$, the number of unknown constants in Eq. (16) is reduced by an amount $\sum_{i(\neq j)=1}^{M}\sum_{j=1}^{M} N_iN_j$.*

point dipoles. Whether or not a finite size particle (or subparticle) in a given case can be considered as a point-like entity is a basic and difficult question of interest in its own right. To indicate the challenge one is confronted with let us rewrite Eq. (3) as follows:

$$\vec{E}(\vec{r}) = \vec{E}^{ext}(\vec{r}) + \int_{Q.P.} \overleftrightarrow{K}(\vec{r},\vec{r}\,') \cdot \vec{E}(\vec{r}\,') d^3r' , \tag{30}$$

where

$$\vec{E}^{ext}(\vec{r}) = \vec{E}^{B}(\vec{r}) + \sum_{j=1}^{N} \overleftrightarrow{T}(\vec{r},\vec{r}_j) \cdot \vec{E}(\vec{r}_j) \tag{31}$$

is the external (*ext*) field driving the quantum particle. The formal solution of Eq. (30) is given by

$$\vec{E}(\vec{r}) + \int_{Q.P.} \overleftrightarrow{\Gamma}(\vec{r},\vec{r}\,') \cdot \vec{E}^{ext}(\vec{r}\,') d^3r' \tag{32}$$

where the field-field response tensor $\overleftrightarrow{\Gamma}(\vec{r},\vec{r}\,')$ has to be obtained from the integral equation

$$\overleftrightarrow{\Gamma}(\vec{r},\vec{r}\,') = \overleftrightarrow{U} \delta(\vec{r}-\vec{r}\,') + \int_{Q.P.} \overleftrightarrow{K}(\vec{r},\vec{r}\,'') \cdot \overleftrightarrow{\Gamma}(\vec{r}\,'',\vec{r}\,') d^3r'' . \tag{33}$$

In the electric point-dipole approach, the basic quantity is the integral (over the quantum-particle domain) of the induced current density, i.e.

$$\int_{Q.P.} \vec{J}(\vec{r}) d^3r = \int\int_{Q.P.} \overleftrightarrow{\sigma}(\vec{r},\vec{r}\,') \cdot \vec{E}(\vec{r}\,') d^3r' d^3r . \tag{34}$$

By inserting in this equation the relation in Eq. (32), and by assuming that the external field is constant across the quantum particle one finds

$$\int_{Q.P.} \vec{J}(\vec{r}) d^3r = [\int\int\int_{Q.P.} \overleftrightarrow{\sigma}(\vec{r},\vec{r}\,') \cdot \overleftrightarrow{\Gamma}(\vec{r}\,',\vec{r}\,'') d^3r'' d^3r' d^3r] \cdot \vec{E}^{ext}(\vec{r}_0) . \tag{35}$$

Now, it is obvious that one can make the identification

$$\overleftrightarrow{S}_0(\omega) = \int\int\int_{Q.P.} \overleftrightarrow{\sigma}(\vec{r},\vec{r}\,') \cdot \overleftrightarrow{\Gamma}(\vec{r}\,',\vec{r}\,'') d^3r'' d^3r' d^3r , \tag{36}$$

and

$$\vec{E}^{ext}(\vec{r}_0) = \vec{E}(\vec{r}_0) . \tag{37}$$

In turn, this means that the light-induced current density distribution in Eq. (2) can be written

$$\vec{J}(\vec{r}\,';\omega) = \sum_{j=0}^{N} \overleftrightarrow{S}_j(\omega) \cdot \vec{E}(\vec{r}_j;\omega) \delta(\vec{r}\,' - \vec{r}_j) . \tag{38}$$

Provided that the last term in Eq. (33) is negligible (the rigorous answer to under which circumstances this assumption is reasonable in fact is a part of the basic challenge for understanding the optical polarizability of small paramagnetic particles, [1], [12] - [15]) so that $\overleftrightarrow{\Gamma}(\vec{r}',\vec{r}'') \simeq \overleftrightarrow{U}\,\delta(\vec{r}'-\vec{r}'')$, Eq. (36) takes the form

$$\overleftrightarrow{S}_0(\omega) = \int\int_{Q.P.} \overleftrightarrow{\sigma}(\vec{r},\vec{r}')d^3r''d^3r' . \tag{39}$$

By inserting the expression for $\overleftrightarrow{\sigma}(\vec{r},\vec{r}')$, taken from Eq. (6), into this equation and by utilizing Eq. (23) one obtains

$$\overleftrightarrow{S}_0(\omega) = \frac{2i}{\hbar^2\omega}\sum_{m,n}(E_n-E_m)^2\frac{\rho^F_{mm}-\rho^F_{nn}}{\hbar\omega+E_m-E_n}\vec{P}_{nm}\vec{P}_{mn} , \tag{40}$$

i.e. precisely the phenomenological result of the electric point-dipole approach.

# References

[1] Zaremba, E. and Persson, B.N.J. (1987) "Dynamic polarizability of small metal particles", Phys.Rev. **B35,** pp. 596-606.

[2] Keller, O. (1992) "Nonlocal electrodynamics of an atom in front of a normal or phase-conjugating mirror", J.Quantum Nonlinear Phenomena (in press).

[3] Labani, B., Girard, C., Courjon, D., and Van Labeke, D. (1990) "Optical interaction between a dielectric tip and a nanometric lattice: implications for near-field microscopy", J.Opt.Soc.Amer. **B7,** pp. 936-943.

[4] Girard, C. and Courjon, D. (1990) "Model for scanning tunnelling optical microscopy: A microscopic self-consistent approach", Phys.Rev. **B42,** pp. 9340-9349.

[5] Girard, C. and Bouju, X. (1991) "Coupled electromagnetic modes between a corrugated surface and a thin probe tip", J.Chem.Phys. **95**, pp. 2056-2064.

[6] Girard, C. and Bouju, X. (1992) "Self-consistent study of dynamical and polarization effects in near-field optical microscopy", J.Opt.Soc.Amer. **B9,** pp. 298-305.

[7] Keller, O., Xiao, M., and Bozhevolnyi, S. (1992) "Configurational resonances in optical near-field microscopy: A rigorous point-dipole approach", Surf.Sci. (in press).

[8] Keller, O. (1988) "Propagator study of the selvedge field in linear and nonlinear nonlocal jellium optics", Phys.Rev. **B38,** pp. 8041-8060.

[9] Keller, O. (1990) "Moment expansion of the optical second-harmonic response tensor of condensed media", phys.stat.sol.(b) **157,** pp. 459-471.

[10] Alieva, E.V., Firsov, E.I., Kuzik, L.A., Yakovlev, V.A., and Pudonin, F.A. (1991) "Optical properties of superthin niobium films deposited on crystalline quartz", Phys.Lett. **A152,** pp. 89-95.

[11] Keller, O. and Liu, A.. (1992) "Optical diamagnetic response of a niobium quantum well", Phys.Lett. **A167,** pp. 301-305.

[12] Wood, D.M. and Ashcroft, N.W. (1982) "Quantum size effects in the optical properties of small metallic particles", Phys.Rev. **B25**, pp. 6255-6274.

[13] Kreibig, U. and Genzel, L. (1985) "Optical absorption of small metallic particles", Surf.Science **156**, pp. 678-700.

[14] Puska, M.J., Nieminen, R.M., and Manninen, M. (1985) "Electronic polarizability of small metal spheres", Phys.Rev. **B31**, pp. 3486-3495.

[15] Ruppin, R. (1992) "Optical absorption by a small sphere above a substrate with inclusion of nonlocal effects", Phys.Rev. **B45**, pp. 11209-11215.

# A SCANNING NEAR-FIELD OPTICAL MICROSCOPE WITH OPTICAL DISTANCE CONTROL

M. HRYNEVYCH, D.J. BUTLER, K.A. NUGENT, AND A. ROBERTS
*School of Physics*
*University of Melbourne*
*Parkville, Victoria, 3052*
*Australia*

ABSTRACT. An all optical scanning near-field microscope including vertical distance regulation is described. The probe-sample distance is maintained to within an accuracy of approximately 1nm using a profilometer based on the principles of confocal microscopy. The profilometer is incorporated into a system where subwavelength resolution is obtained by scanning with an aperture in a metal-coated tapered optical fibre. The final system will be capable of producing both confocal and the higher resolution near-field images along with an accurate profile of the scanned surface using only optical techniques. Intensity maps of the near-zone fields produced from the cleaved face of an optical fibre are also presented. No other optical technique is capable of directly sensing, with such high resolution, evanescent optical fields. Hence, this method is an invaluable aid in measurements of mode field diameters and the development of optical fibre sensors.

## 1. Introduction

The area of Near Field Microscopy has attracted a great deal of attention over the last decade. It is well-known [1] that evanscent fields produced by diffracting structures decay rapidly with distance away from the structure. Thus, in order to avoid the incorrect interpretation of near-field data, one of the major challenges in the development of a near-field optical microscope has been the accurate control of the distance from the detector or source to a sample which is not necessarily smooth. If this distance control is not achieved to within a few percent (i.e. a few nanometres), strong fluctuations in the detected light could be erroneously interpreted as variations in the refractive index of the sample. A number of different distance control mechanisms have been developed, including the recent use of surface force feedback. [2] In the School of Physics at the University of Melbourne we have been engaged in the development of a Scanning Near-field Optical Microscope (SNOM) which has an optical confocal distance control system. This microscope combines the two most important recent developments in optical microscopy. As well as controlling the vertical position of the probe to within 1nm, this technique will also provide confocal images of a specimen simultaneously with the higher resolution near-field images. It will also supply high resolution profiles of the surfaces of the samples being studied.

The aim of our research effort is ultimately to apply SNOM to high resolution imaging of biological samples and to the study of evanescent fields around diffracting and lightguiding structures. The increasing interest in fibre optic sensors means that it is crucial to have a good knowledge of the way the near-zone fields around optical fibres behave. Only then can the effect of the interaction between these fields and external influences on the light propagating within the fibre be accurately determined. An understanding of these influences is central to the use of fibres as sensors.

In section 2 we discuss the development of the confocal distance regulation system and then, in section 3, we discuss the incorporation of this system into a near-field microscope. Our work



*D.W. Pohl and D. Courjon (eds.), Near Field Optics*, 391–398.

involved with optical fibres is presented in section 4. Finally, in section 5, we present our conclusions and the directions we see for further work.

## 2. The Confocal Profilometer

### 2.1 THE CONFOCAL MICROSCOPE

Confocal microscopy has been the subject of intense work over the past decade [3] as it has the interesting properties of rejecting, rather than merely defocusing, the light from depth planes other than the one of interest and providing superior lateral resolution to that found using a conventional widefield microscope. The confocal technique has now found rather widespread acceptance by the biological community and confocal microscopes are now commercially available from a number of manufacturers.

The depth discrimination property allows the surface profile of reflective, scattering or fluorescent objects to be obtained simultaneously with a high resolution confocal image. In this section we describe a confocal surface profiling approach that has sufficient resolution to allow its application to near field microscopy.

### 2.2 THE CONFOCAL PROFILOMETER

The technique described here consists of the modified confocal microscope shown in Figure 1, in which a surface is profiled using the properties of the intensity distribution about the focal point of a lens. [4] A collimated laser beam is focused onto the object, where initially the

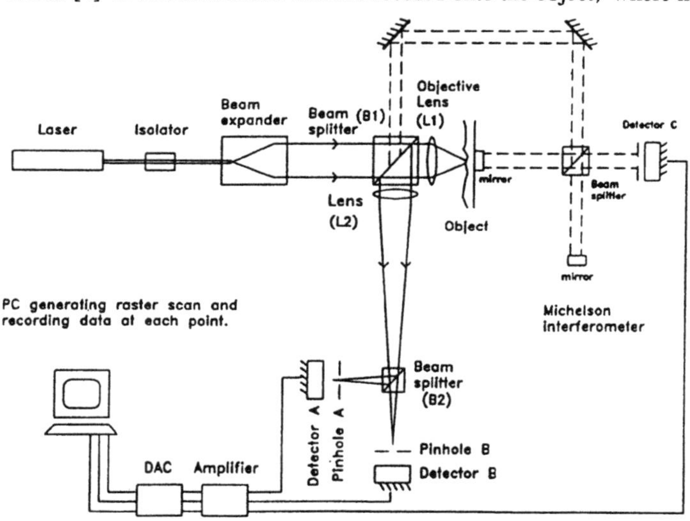


**Figure 1:** The confocal profilometer. The pinholes are located axially in front and behind the focal spot in their respective copies of the focal region that are provided by the beamsplitter B2. The difference in intensity detected at these two pin-holes provides an accurate measure of surface profile of the object. A Michelson interferometer shown with dotted lines in the upper right-hand part of the diagram was used to calibrate the profilometer difference signal .

system is set so that the focal spot falls onto the surface to be profiled. Light reflected or scattered from the surface returns through the objective and is directed via a beam-splitter (B1) into a long focal length lens. A second beam-splitter (B2) produces two copies of the focal distribution. Pinholes are placed on either side of the focal point at locations where the gradient of the transmitted intensity is a maximum. As the object is scanned, the lens-object distance changes with surface height variation, and causes the imaged focal distribution to be displaced along the optic axis so that the pinholes measure different intensities. This results in a differential signal between the two pinholes which, for uniform reflectivity, is a sensitive indication of the object's displacement. (Figure 2) For a small range of object displacements, this differential signal is linear. Over the same region, the sum of the pinhole intensities is approximately constant. The difference signal is also sensitive to reflectivity changes. By normalizing the difference signal, therefore, with the sum signal the effect of varying reflectivity is, to a good approximation, eliminated. This is only one of a number of techniques which can be used to achieve effectively the same result. Others include using a pinhole which vibrates axially to sample the intensity distribution. The method we have chosen is the most straightforward to implement and its accuracy is not significantly less than other, more complex, techniques.

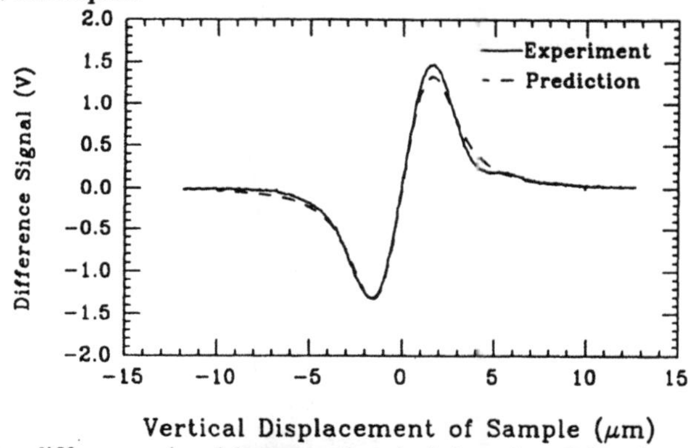


**Figure 2:** The difference signal between the pin-holes, A and B, plotted as a function of vertical displacement. The theoretical curve is that for a Gaussian beam. The linear region is clearly visible over the central two microns of the vertical displacement. The zero displacement coincides with the centre of the linear region where the difference signal is null.

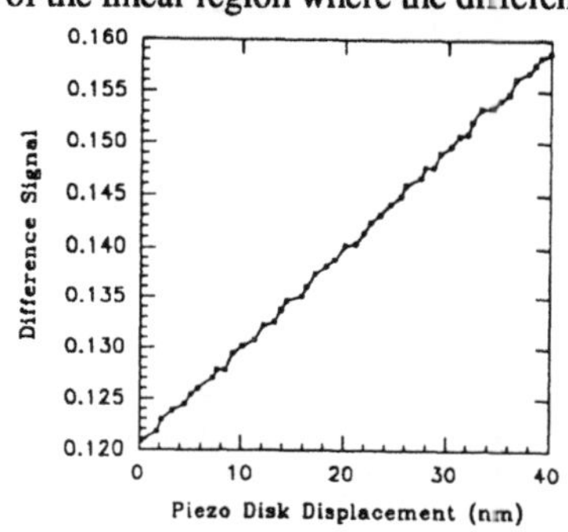


**Figure 3:** The calibration curve of the profilometer near the central part of the linear region. The estimate of the height resolution was taken as the range of scatter of points at a constant difference signal and was found to be approximately 1nm.

The system was calibrated by observing the normalized differential signal as a function of the voltage applied to a piezoelectric disk. A mirror mounted on the back of the disk enabled the displacement to be determined by using a Michelson interferometer simultaneously with the profilometer. The data from the interferometer was analysed to provide precise displacement data. Figure 3 shows the result of this calibration, and it is clear that the data is highly linear over this short range. Note that the instument needs to be calibrated every time it is readjusted to take into account, for example, small shifts in the absolute positions of the pinholes. The data used in Figure 3 was not taken at the same time as that of Figure 2 and, hence, the calibration constant is slightly different in each case. The fundamental factor limiting the resolution of this device is the noise on the output signal; the noise in Figure 3 corresponds to an equivalent displacement of about 1nm. Thus we are led to conclude that our device has a limiting resolution of 1nm.

## 3. A scanning near-field optical microscope with optical distance control

### 3.1 THE ALL OPTICAL SNOM

The confocal profilometer has been shown to have a vertical resolution of approximately 1nm. As the height of the surface can be found, the next requirement is the location of the probe with respect to the focal spot. A useful place to position the probe is just over the focal spot; this has the advantage providing a bright source of illumination for the probe as well as that for profile measurement.

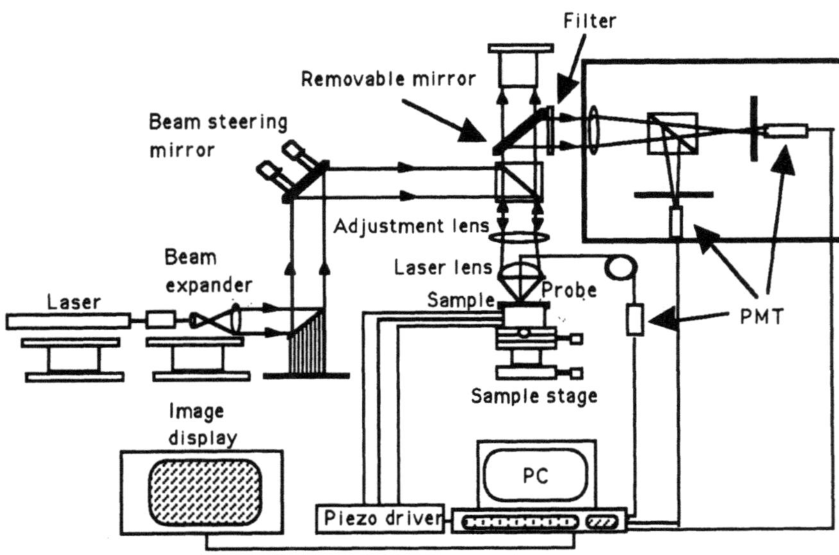


**Figure 4:** The combined SNOM and confocal profilometer. Similar in principle to the confocal profilometer, this instrument has the near-field probe located in the focus of the objective lens of the profilometer, so that the probe tip is at a set distance from the focal spot. This has the advantage of a simple design that makes use of the focal brightness to illuminated the probe and maintains correct probe height by tracking the focal spot over the sample's surface during scanning.

In our combination of a SNOM with the confocal profilometer we have elected to position the probe mechanically through the lens; along its optical axis. The final result is shown diagrammatically in Figure 5. This is basically the confocal profilometer with a probe inserted through the objective lens. Initially we are using a laser achromat with an f-number of 2. Sample scanning is achieved by mounting the sample on a piezo-electric tube with four external electrodes, on which voltages are applied to produce the necessary deviations. The probe is a drawn aluminised optical fibre that is located within the lens with a section of fine stainless steel tubing; this provides a means of stabilising the probe as well as securing the fibre into position. These fibres are made using a technique of drawing the fibre while it is heated with the output of a carbon dioxide laser. The fibre tip is positioned at the required surface height by locating it at the desired distance from the focal spot; this alignment procedure is explained below.

In operation the system will use the profilometer output to provide a feedback signal to the piezo-electric tube sample holder so that it can lengthened or shortened as required to keep the focal spot at the sample surface. In this way this instrument will allow for the possibility of obtaining: a near-field image using the probe; a confocal image from the confocal microscope; and a surface profile from feedback on the pinhole difference signal. It will have the capability of forming a confocal microscope image and then we can select an area in which to produce a higher resolution image using the SNOM.

### 3.2 THE ALIGNMENT PROCEDURE

For such an instrument to operate the alignment of the probe to the focal spot is critical. The relatively crude adjustment of placing the focal spot to within a fraction of a micron of the probe tip (with the aid of a microscope) is achieved by the micrometer adjustments on the beam steering mirror, in the lateral plane to the probe, and adjustments to the xyz sample stage to set the height.

To obtain a further increase in the precision of the alignment this combination of SNOM and confocal profilometer allows adjustments to be made in the probe-focal spot distance with the use of a long focal length meniscus lens situated before the objective lens. Adjustments in the spacing between these lens produces very fine changes in the effective focal length of this lens system, The adjustment is guided by firstly calibrating the instrument so that the correspondence between the difference signal and displacement is known on a mirror surface. Then, by shining light down the fibre, with the profilometer laser turned off, the fibre tip becomes effectively a point source. The detection of the difference signal at the pin-holes of this source provides a measurement of the height of the probe above the mirror surface. Then , turning on the profilometer laser it will be possible to adjust the position of the meniscus lens until a null difference signal is obtained, indicating that the focus is on the mirror surface as required.

## 4. Measurements on optical fibres

### 4.1 FIBRE PARAMETERS

As mentioned in the introduction, the use of optical fibres as sensors of the surrounding environment hinges on a good understanding of the evanescent fields around the fibre and the way external influences affect these fields. This, along with the fact that the characterization of the output of optical fibres is vital if efficient coupling of light into other devices is to be performed, means that the study of evanescent fields around optical fibres and other lightguiding structures is extremely important. In this section, we describe how this can be performed using scanning near-field microscopy. In particular, measurements of the electric field at the output face of a cleaved fibre are extremely useful. Once an accurate description of the output of the fibre has been obtained, expressions relating beam parameters to macro- and micro-bending losses [5] can be used to calculate loss coefficients. In addition, it is possible to characterize the refractive index profile of the fibre using near-field measurements [6], and, from this, it may be possible to estimate any variations in the uniformity of the profile which may affect the beam losses within the fibre. Techniques have been developed over the past decade to determine the various parameters which can be used to describe the output beam. However, no measurements of the near-field of the fibre output have been performed directly, although model experiments have been performed using microwaves.[7] So called 'near-field' measurements have, in the past, generally involved the scanning of a magnified image of the output at the end of the fibre.

As outlined above, many measurements of fibre output have been made in the past. In order to standardize these measurements, a number of 'spot-size parameters' have been defined. (For a review of these see Reference 5) The spot size is defined as the $e^{-1}$ radius of the power in the fibre. In order for this definition to be meaningful, the fields must be radially symmetric and alternative spot size definitions are required for non-circularly symmetric beams.

The radial dependence of the power radiated by the fibre depends on where in the output field the power measurements are taken. Thus, there are a number of different Mode Field Diameters (MFD) which can be defined, depending on how far from the radiating fibre end the measurements were taken. Initially, two MFDs were defined: the Near Field (NF) MFD, $d_n$, and the Far Field (FF) MFD, $d_f$. These are based on near and far zone electric field measurements respectively. Other MFDs have been defined. The so-called Gaussian MFD relates the overlap between the actual field in the fibre and a Gaussian distribution and another MFD, known as $d_\infty$, is defined as $d_\infty = \frac{4}{\sqrt{b^2 - k_1^2}}$ where b is the fundamental mode propagation constant and $k_1$ ($=k_0n_1$) is the cladding wavenumber. This last MFD has proved useful in the estimation of macro- and microbending losses since expressions exist for the loss coefficients in terms of $d_\infty$ and $d_n$.

The precise measurement of near field electric field distributions can also be used to determine the refractive index profile of the fibre itself. Measurements of the near-zone electric field are numerically differentiated and by substituting the values obtained back into the scalar wave equation $Q(r) = k_0^2 n^2(r) - b^2$ can be obtained. Since Q(r) is proportional to $n^2(r)$, knowing the absolute refractive index value at one point in the fibre, for example, in the cladding, enables the determination of n(r).

A number of different techniques have been employed for the measurement of the various diffraction patterns necessary for the calculation of the mode field diameters. These are outlined in Reference [9]. In the following section we will describe using scanning near-field microscopy for determining the near field optical output of the fibre.

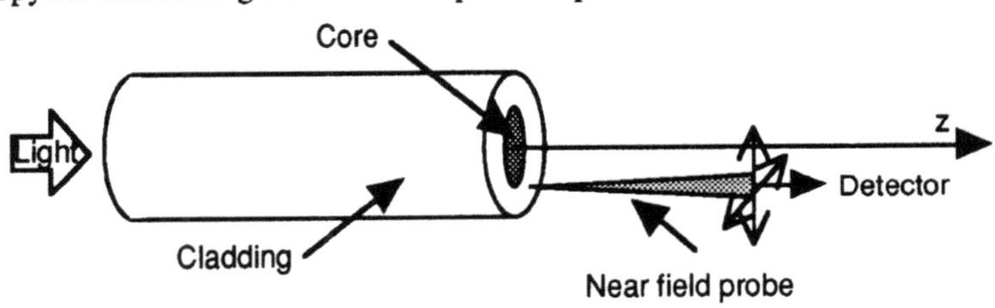


**Figure 5:** Experimental arrangement for near-field measurements of radiation from optical fibres.

4.2 RESULTS

The experimental arrangement for the measurements of the fibre output is as shown in Figure 5. Three-dimensional volume scans of the light emitted from the end of a step-index fibre were taken. The significant structure in the fibre output which we have investigated so far has been on the order of 1μm. A single mode (at 633nm) fibre (NEWPORT F-SV) was used. This fibre has a core and cladding diameters of 4μm and 125μm, respectively and a numerical aperture of 0.1. The light source was a He-Ne laser with a wavelength of 632.8nm. A mode stripper was used and the length of the fibre was approximately 40m. Measurements were taken at separations of 50nm in all planes. The probe consisted of a tapered optical fibre coated with aluminum but leaving a small aperture (50-100nm) at the tip. The lateral scan was accomplished by means of a piezoelectric tube and the vertical distance from the fibre face (z-direction) was controlled using a stepper motor.

Figure 6 shows an isometric plot of the intensity distribution parallel to the end of the fibre at a distance of less than 500nm. The data clearly shows the anticipated Gaussian structure. The intensity distribution in a plane containing the axis of the fibre is shown in Figure 7. This measurement has not previously been performed with the exception of an earlier microwave analogue experiment.

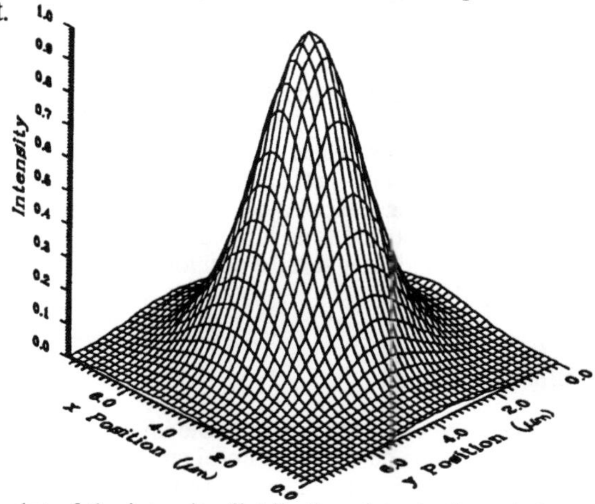


**Figure 6:** Isometric plot of the intensity distribution close to the exit face of the single mode fibre.

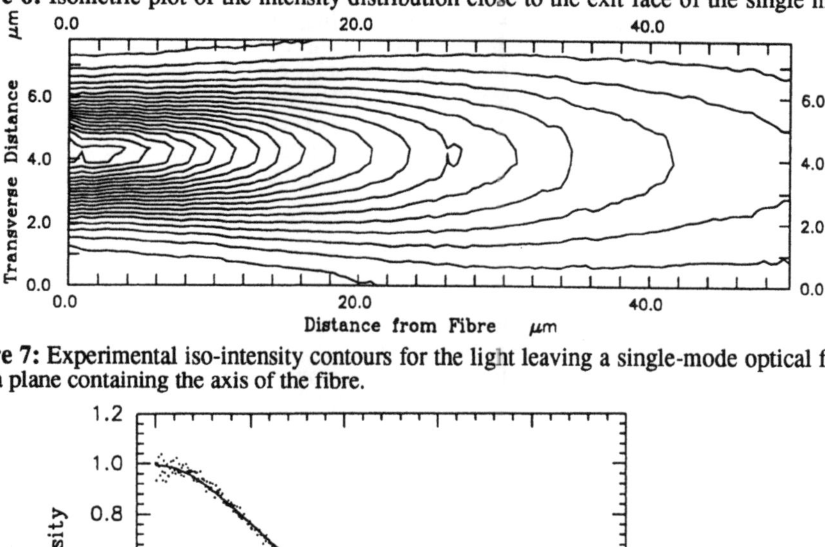


**Figure 7:** Experimental iso-intensity contours for the light leaving a single-mode optical fibre over a plane containing the axis of the fibre.

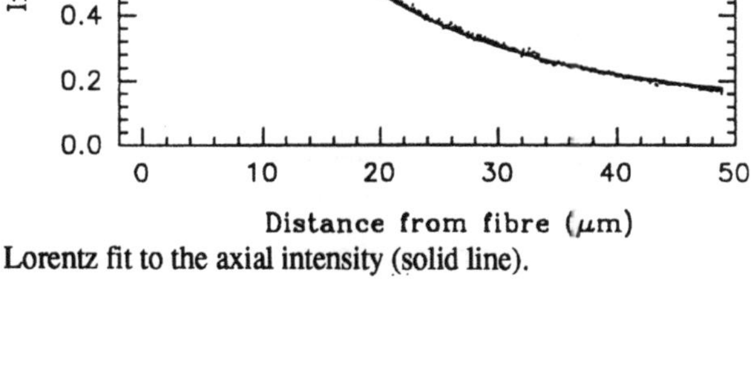


**Figure 8:** Lorentz fit to the axial intensity (solid line).

As foreshadowed, an important application of measurements of this form is the determination of the various modal field diameters. A near field mode field diameter may be found by fitting a straight line to a logarithmic plot of the near-zone electric field.[7] Least squares fitting of this data gives a mode field diameter of 3.482±0.016 microns. This compares favourably with a value of 3.62μm assuming that the core diameter was exactly 4μm. The precision of this first attempt is comparable to the best measurements made using other techniques [9] and is sufficiently good that it should be feasible to perform direct determinations of the refractive index profile .

The sample-probe separation was estimated by assuming that the output of the fibre was approximately Gaussian in shape and propagated approximately as a Gaussian. This assumption means that the z dependence (see Figure 8) on axis possesses a Lorentz profile. A Lorentz profile was fitted to the data, and from it the distance of the probe from the cleaved fibre end deduced. This technique indicated that the position of closest approach of the probe to the fibre end is approximately 100nm.

## 5. Conclusions and future work

This paper has outlined the progress we have made in the union of a SNOM with a confocal profilometer as a distance controlling device. A confocal profilometer has been assembled and shown to have a height resolution of 1nm and currently is used in surface profiling in other areas of the group's research interests. Excellent progress has been made to date in the assembly of the combined instrument and shows signs of fulfilling expectations. Current work on the combined instrument is concentrated on improving the quality of the probes and of the objective lens. In particular, probes consisting of metal coated tapered optical fibres are currently being developed and have improved the accuracy of measurements made.

Along with the development of the microscope, work is continuing on the measurement of the near-zone fields radiated from a cleaved face of an optical fibre. These measurements permit the determination of quantities such as the mode field diameter with unprecedented resolution.

## 6. References


1. Roberts A. (1989) 'Near-zone fields behind circular apertures in thick, perfectly conducting screens', Journal of Applied Physics 65, 2896-2899.
2. Betzig, E. and Trautmann, J.K. (1992) 'Near-field optics: microscopy, spectroscopy, and surface modification beyond the diffraction limit', Science 257, 189-195.
3. Wilson, T. and Sheppard, C. (1984) 'Theory and practise of scanning optical microscopy', Academic Press, London.
4. Butler, D.J., Horsfall, A., Hrynevych, M., Kearney, P.D., and Nugent, K.A. 'Confocal profilometer with nanometric vertical resolution', (in preparation)
5. Artiglia, M., Coppa, G., di Vita, P., Potenza, M. and Sharma, A. (1989) 'Mode field diameter measurements in single mode optical fibres', Journal of Lightwave Technology LT-7, 1139-1152.
6. Sladen, F.M.E., Payne, D.N., and Adams, M.J. (1976) 'Determination of optical fibre refractive index profiles by a near-field scanning technique', Applied Physics Letters 28, 255-258.
7. Albrecht, J. and Neumann, E.-G. (1979) 'Simulation of the near-field of single-mode fibres by means of a microwave model', Microwaves, Optics and Acoustics 3, 109-114.
8. Butler, D.J., Nugent, K.A., Roberts, A., and Kearney, P.D. (1991) 'Imaging of near-field diffraction patterns with sub-wavelength resolution in the optical region'. Proceedings of SPIE 1556, 19-26.
9. Coppa, G., di Vita, P. Rossi, U. (1976) 'Characterisation of single-mode fibres by near-field measurement', Electronics Letters 293-294.

# PROGRESS IN VIBRATING STYLUS NEAR FIELD MICROSCOPY

Adrian Korpel, Scott Samson and Kurt Feldbush
*Dept. of El. and Comp. Eng., University of Iowa,*
*Iowa City, IA 52242, USA.*

## Abstract

In our proposed method of near field optical microscopy, the near field is probed by a vibrating solid stylus and the radiation downstream from the stylus is measured. The underlying idea is that at the sharp stylus tip or edge, evanescent fields are converted into traveling waves which are subsequently detected conventionally. The area of perturbation, i.e. the resolution, is determined by the vibration of the stylus and labeled by the frequency or frequencies of vibration.

## Introduction

In most methods of near field optical microscopy, the electric field close to the object is detected by a tapered waveguide-like structure with a tip aperture much smaller than the wavelength of light. {See the references contained in [1]). Although aware of the progress that has been made in fabricating such probes, we became interested in simpler, solid, non-transparent probes after noticing the ease with which the profile of a laser beam can be measured by a vibrating knife edge instead of a slit [1]. We also were attracted by the convenience of achieving variable resolution by simply changing the amplitude of vibration. Finally, we were intrigued by the possibility that an obscuration type of probe, such as a knife edge, might be applied to radiation other than light.



*D.W. Pohl and D. Courjon (eds.), Near Field Optics, 399–406.*

The first problem to solve was to extend the knife edge operation to two dimensions. This we did by using two orthogonal knife-edges vibrating at different frequencies, and detecting the difference frequency. As we will show, the detected radiation in such a system appears (in the simplest possible theory) to originate from a point source located at the intersection of the two knife edges. This leads naturally to the concept of a vibrating knife edge corner formed out of one piece of material. An obvious generalization of the technique is the use of a vibrating stylus, where the word 'stylus' is used in a very general sense so as to include shapes other than needle-like.

At this conference it has become clear that various groups [3-5] are performing or have proposed solid stylus techniques, although the stylus may not always be vibrating [4]. On the other hand, waveguide-type probing has also been combined with vibration [6].

Our preliminary experiments are not aimed at detecting submicron detail per se, but rather at demonstrating the general feasibility of our approach, while successively increasing the resolution. We will give a brief review of our principle of operation, with special emphasis on various methods of two-dimensional signal processing. This will be followed by a discussion of the maximum attainable signal-to-noise ratio and some thoughts on the mechanism of evanescent-to-plane wave conversion.

## Principle of operation

Fig.1a shows, in much simplified form, the principle of operation of the most usual methods of near field optical microscopy: a tapered waveguide of some sort probes the field - in particular the evanescent field - and converts the optical energy at the small tip into waveguide modes the power in which is then measured at the other end of the waveguide. (Conversely, the tapered waveguide may be used as a means to illuminate the object by a sub-micron spot.) Although the illumination shown here is incident perpendicularly, it is obvious that other angles may be used, in particular angles beyond critical incidence.

In our technique [1,2], the generic version of which is shown in Fig.1b, the field is not probed by a narrow-tip waveguide, but by a a vibrating solid stylus The total power in the far field above the stylus is measured, and the part characteristic of the frequency or frequencies of the stylus is taken to represent that part of the total

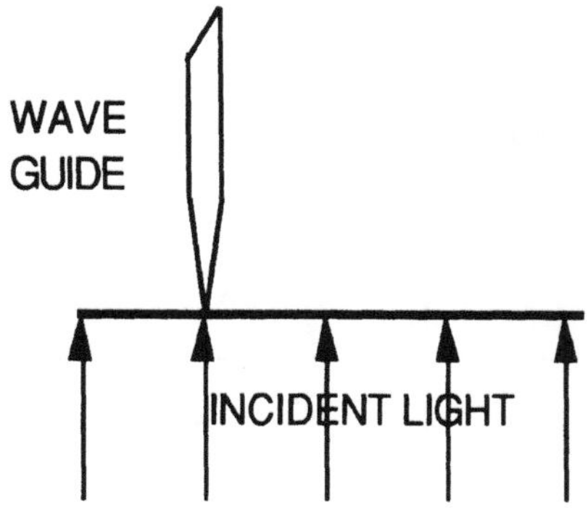


Fig. 1a

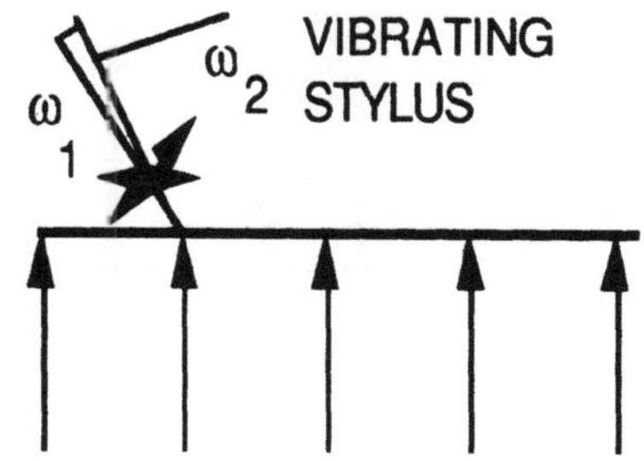


Fig. 1b

Fig. 1. a) Conventional near field probing. b) Probing by vibrating solid stylus.

field covered by the vibrating tip. Thus it is plausible to assume that the resolution should be determined by the tip radius and the amplitude of vibration. Actual physical measurements on gratings in a one-dimensional configuration confirm the second part of this conjecture in a semi-quantitative way [1]. In two dimensions it has been confirmed qualitatively by inspecting the detail in images of small lettering, and quantitatively by a simulation experiment [2].

In our experiments the generic stylus of Fig. 1b takes the shape of a flat knife edge corner, as shown in Fig. 2a. A laser beam (circle) is focused on the knife edge which is placed parallel and in close proximity to the object.The latter is modeled by the transparency $\tau(x,y)$ where $\tau$ denotes the intensity transmission factor. In the actual analysis, we use the complimentary configuration of Fig.1b. In the experiments, too, this configuration is used, by employing two mutually perpendicular vibrating knife edges.

The knife edge corner is vibrating simultaneously in the X and Y directions with amplitudes $w_1$ and $w_2$ and frequencies $\omega_1$ and $\omega_2$ respectively. The instantaneous position of the corner is given by

$$x_t = x_s + \Delta x_t = w_1 \cos(w_1 t) \tag{1}$$

$$y_t = y_s + \Delta y_t = w_2 \cos(w_2 t) \tag{2}$$

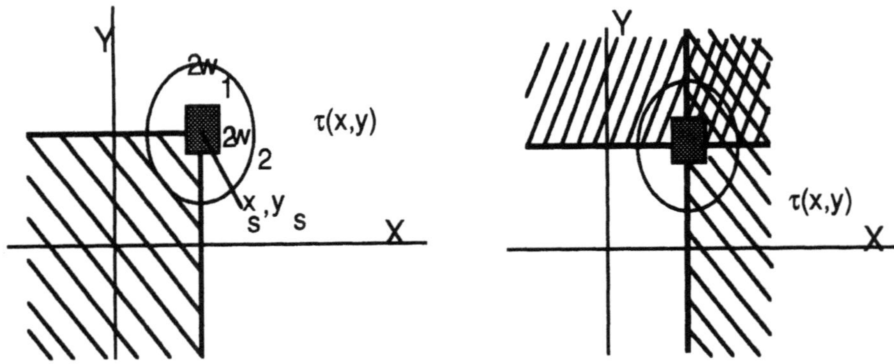


Fig. 2a

Fig. 2b

Fig.2. a) The vibrating knife edge corner. b) Complimentary configuration used for analysis.

In the simplest mode of operation, the total power in the field transmitted past the knife edge is measured while the transparency is being moved underneath to provide scanning. The time varying parts of the current at $\omega_1$, $\omega_2$ and $\omega_1 \pm \omega_2$ are selected by a lock-in amplifier. In the calculations to follow, we use simple model in which the knife edge corner (situated at zero distance from the object) geometrically obscures a part of the incident field and the total power is measured. A simple, scalar perturbation calculation [2] of the various harmonic components, as detected by a photo detector with sensitivity $\alpha$ Ampères/Watt, gives the following results:

$$i_1(t) = \alpha \int_{-\infty}^{y_s} I_b(0, y-y_s)\tau(x_s, y)dy \times w_1 \cos(\omega_1 t) \tag{3}$$

$$i_2(t) = \alpha \int_{-\infty}^{x_s} I_b(x-x_s, 0)\tau(x, y_s)dx \times w_2 \cos(\omega_2 t) \tag{4}$$

$$i_3(t) = \frac{\alpha}{2} I_b(0,0)\ \tau(x_s,y_s) \times\ w_1 w_2 \cos[(\omega_1+\omega_2)t] \tag{5}$$

$$i_4(t) = \frac{\alpha}{2} I_b(0,0)\ \tau(x_s,y_s) \times\ w_1 w_2 \cos[(\omega_1-\omega_2)t] \tag{6}$$

It will be seen that the sum and difference frequency components carry direct information about the intensity at the knife edge corner. The components at $\omega_1$ and $\omega_2$ provide information integrated over the direction normal to the corresponding vibration. If the illumination is uniform, i.e. if $I_b$ is constant, then post-detection differentiation of the recorded amplitudes of these components will also give the desired information. If $I_b$ is not constant (as is likely to be the case in practice), then some artifacts appear in the differentiated image. This is discused in detail in ref. [2].

It is to be expected that a more sophisticated model will be needed to analyze the effects of partial coherence, finite receiving aperture, polarization, separation between the knife edge and the object, finite rounding of the knife edge corner, conductivity and thickness of the knife edge, etc. To the extent that the knife edge corner acts as an ideal absorber, the technique need not be limited to optical fields only. There is no fundamental reason why it should not be applicable to, say, acoustics (ultrasonics), X-rays or electrons.

## Signal to noise ratio

In the ideal case the operation is shot noise limited and the (power) signal to noise ratio is given by [2]:

$$S/N = \frac{\alpha}{4\pi^2}\frac{P_t}{eB}\left(\frac{w_1 w_2}{w_0^2}\right)^2 \tag{7}$$

where $P_t$ is the total power of the incident light, e the electron charge, B the bandwidth and $w_0$ the waist of the incident gaussian beam. It has been assumed that $\tau(x,y) = 1$. The waist $w_0$ is fundamentally limited to about the wavelength of the incident radiation, of course, and this sets a lower bound to the size of the observable detail. For a $\approx$ 0.4 A/W, $P_t$ = 3 mW, $w_1/w_0 = w_2/w_0 = 0.01$, B = 100 Hz, we find that S/N = 43 dB. Recent experiments have shown that our present operation is not shot noise limited, but that intrinsic, multiplicative laser noise is the limiting factor. Both with a diode laser and a He-Ne laser, the noise has a 1/f character that begins to flatten out at about 1000-4000 Hz. The temporal character of the He-Ne noise spectrum is characterized by spikes that drift through the low frequencies below about 200 kHz. These may be due to beating between the laser modes. Measurements show that in the example given above, the intrinsic noise from a He-Ne laser would reduce the signal to noise ratio by about 38db. The situation for a laser diode appears to be somewhat better, but more measurements have to be performed. At any rate, one of our immediate goals is to suppress laser fluctuations by feedback or cancel them by some form of signal processing using the laser beam itself as a reference.

## Evanescent wave conversion

As mentioned before, the underlying mechanism of our technique is the conversion of evanescent wave to propagating waves at the knife edge corner. Our calculations so far have been based on the obscuration of the light in the plane of the knife edge, and are equally valid (apart from signal to noise considerations) for coherent or incoherent light. The plane wave spectrum of the light received, however , could be very different in these two cases, and this would presumably show up if spatial filtering is used (either deliberately to improve the S/N ratio or inadvertently through finite aperture effects). It is therefore of interest to calculate the homogeneous plane wave spectrum $A(k_x,k_y)$ for coherent illumination. For simplicity we have assumed that the light spot focused on the knife edge corner has a square cross section of $d \times d$ and a uniform amplitude $E_0$. Using the electrical engineering convention [7] in which $A(k_x,k_y) = \mathcal{F}^{-1}\{E(x,y)\}$, with $\mathcal{F}^{-1}$ the inverse Fourier transform operator, we find, for $\tau(x,y) = 1$:

$$A(k_x,k_y) \approx$$

$$E_0\left(\frac{d}{2}\right)^2 \exp[jk_x(x_s-\frac{d}{4})]\exp[jk_y(y_s-\frac{d}{4})]\operatorname{sinc}\left(\frac{k_x d}{4\pi}\right)\operatorname{sinc}\left(\frac{k_y d}{4\pi}\right)$$

$$+ E_0\left(\frac{d}{2}\right)\operatorname{sinc}\left(\frac{k_y d}{4\pi}\right)\exp[jk_y(y_s-\frac{d}{4})]\exp(jk_x x_s)\Delta \mathbf{x_t} \qquad \rightarrow \omega 1$$

$$+ E_0\left(\frac{d}{2}\right)\mathrm{sinc}\left(\frac{k_x d}{4\pi}\right)\exp\left[jk_x\left(x_s-\frac{d}{4}\right)\right]\exp(jk_y y_s)\Delta \mathbf{y_t} \qquad \rightarrow \quad \omega_2$$

$$+ E_0\exp(jk_x x_s + jk_y y_s)\Delta \mathbf{x_t}\Delta \mathbf{y_t} \qquad \rightarrow \quad \omega_1 \pm \omega_2 \qquad (8)$$

where higher order terms in $\Delta x_t$ and $\Delta y_t$ have been left out. It is instructive to consider the right hand terms in eq. (8) separately.

The first term represents the DC background. It has the usual sinc dependence on aperture size, and the phase term indicates that the center of the source is located at $x_s$-d/4, $y_s$-d/4, i.e. in the center of the unobscured light spot. The second term is at frequency $\omega_1$ and representative of the vibration in the X direction. It shows the sinc dependence in $k_y$ only, and appears to originate from the point $x_s$, $y_s$-d/4. The third term, representing movement in the Y direction, has a similar interpretation with x and y interchanged. The last term is the most interesting one as it contains the sum and difference frequencies which, as we have seen, lead to the desired information directly. Note that this term appears to originate from the knife edge corner at $x_s$, $y_s$, *as if the latter were a point source.*

To extract all the information contained in the latter signal, we would ideally have to heterodyne it with a reference point source that is also located at the origin. What we actually do in our experiment is to simply measure the total power contained in (8). This gives two contributions at the sum and difference frequencies. The first contribution arises from the heterodyning of the first (DC) term with the fourth ($\omega_1 \pm \omega_2$) term, the second contribution arises from the heterodyning of the third ($\omega_1$) term and the third ($\omega_2$) term. It is readily seen that both contributions are equal and that they have a similar sinc character as the DC reference signal (but with an effective aperture of 2d because of the phase off-set), indicating that d must be made as small as possible in order to extract the maximum information from the sum and difference frequency signal (It is assumed in this reasoning that the total incident power $E_0^2d^2$ is to remain constant). This is also implicit in (5) and (6), because there it is the intensity $I_b$ that determines the amplitude of the detected signal. However, the present analysis throws light on the angular distribution of the desired terms, and is the one that should be used in determining the effects of spatial filtering. We are presently investigating this in some detail.

## Acknowledgment

This research is supported by the Army research office under grants DAAL-03-91-G-0014 and DAAL-03-92-G-0207.

## References

1. Korpel, A., Mehrl, D. J. and Samson, S. (1990) 'Beam profiling by vibrating knife edge: implications for near field optical microscopy', Int. Jnl. of Imaging Systems and Technology 2, 203-208.

2. Korpel, A., Samson, S. and Feldbush, K. (1992) 'Two-dimensional operation of a scanning optical microscope using a vibrating knife edge corner', Int. Jnl. of Imaging Systems and Technology, in press.

3. Wickramasinghe, K., IBM Research, T. J. Watson Research Center, P. O. Box 218, Yorktown Heights, NY 10598, USA, private communication.

4. Specht, M., Pedarnig, J. D., Heckl, W. M., and Hänsch, T. W., 'Scanning plasmon near field microscope', Workshop on Near Field Optics, Besançon/Arc & Senans, France, October 26-28, 1992.

5. Boccara, A. C., ESPCI Lab. Optique, 10 Rue Vauquelin, 75005 Paris, France, private communication.

6. M. Vaez-Iravani, 'Comparative imaging of samples using simultaneous force and optical microscopy', Workshop on Near Field Optics, Besançon/Arc & Senans, France, October 26-28, 1992.

7. Korpel, A., Lin, H. H. and Mehrl, D. J. 'Convenient operator formalism for Fourier optics and inhomogeneous and nonlinear wave propagation', (1989), J. Opt. Soc. Am. A 6, 630-634.

# DEVELOPMENT OF A NEW OPTICAL TUNNELING MICROSCOPE

A. DE MARCO, R. MICHELETTO[1], A. TRABUCCO, P. VIOLINO
*Department of Experimental Physics*
*Turin University*
*Via Pietro Giuria 1*
*10125 Torino*
*Italy*

ABSTRACT. An optical tunneling microscope with subnanometric resolution is described. This photon tunneling instrument is extremely sensitive to surface profile along the longitudinal axis, while its lateral resolution is comparable to that of a classical optical microscope. It operates by monitoring the totally reflected light and then it is rather insensitive to the physical properties of the sample, except profile and complex refractive index. It can scan over a surface of 75x75 μm and the sample roughness can be up to about 100 nm. The instrument was characterized by using Langmuir-Blodgett films with bilayers of known height; repeated scans have shown a good reproducibility, of about 0.5 nm.

## 1. Introduction

We have developed a new type of optical scanning tunneling microscope with subnanometer longitudinal resolution. It does not use the near field optics approach; its basic idea was first proposed by Bonazzola *et al* (1985). It is a three-dimensional scanning instrument but its lateral resolution is diffraction-limited and then comparable to that of a classical optical microscope. It operates by monitoring the totally reflected light and consequently it can scan over thin or thick, transparent or opaque, conducting or non-conducting samples, also fully immersed in water. In fact, as the light is collected before it reaches the sample, many physical properties of the sample are almost irrelevant.

## 2. Apparatus

The core of the experimental apparatus is shown in fig. 1. A light beam (633 nm) impinging perpendicularly upon one of the equal lateral surfaces of a 90° glass prism is focussed at the center of its largest lateral surface, ground into a spherical shape, where

[1] Department of Physical Sciences, Ancona University, Italy



D.W. Pohl and D. Courjon (eds.), Near Field Optics, 407–410.

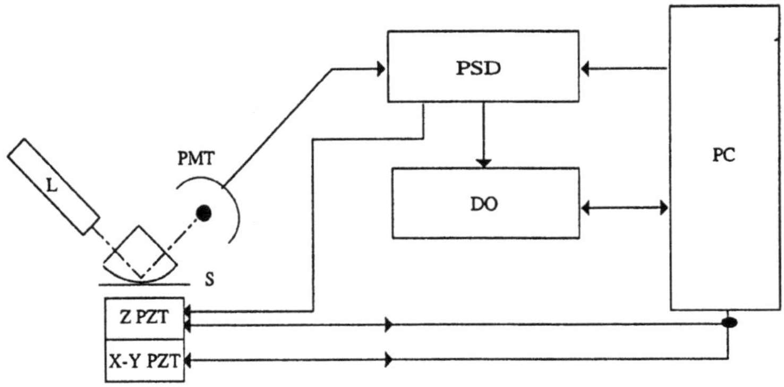


Figure 1. Experimental apparatus. L: Laser source, S: Sample, PZT: Piezoelectric translators, PMT: Photomultiplier tube, PSD: Lock-in amplifier, DO: Analog to digital converter and buffer memory, PC: Personal computer for instrument managing.

it undergoes total reflection. The reflected beam leaves the prism and then it is collected by a photomultiplier tube (PMT). The sample, lying in front of the spherical surface, is moved along the longitudinal axis by a piezoelectric translator (PZT) towards the light spot on the spherical surface. When tunnel effect occurs the fraction $R$ of the reflected surface reaching the PMT is a well known function of the distance between the glass prism and the sample and of the refractive indexes of the three media involved, as shown in fig. 2.

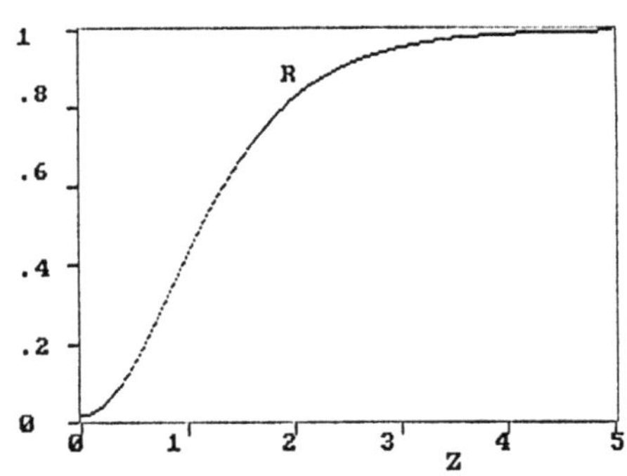


Figure 2. Reflectivity in the optical tunnel effect, as a function of the thickness of the intermediate medium (expressed in units of $\lambda/2\pi$). Incidence angle: 45°, Refractive indexes: 1.74, 1 and 1.58, Light polarized in perpendicular mode.

In order to increase the sensitivity along the longitudinal axis the sample is attached to a further small PZT driven by the reference signal of a lock-in amplifier, thus obtaining a small periodic displacement of the sample along that direction. The PMT signal is then

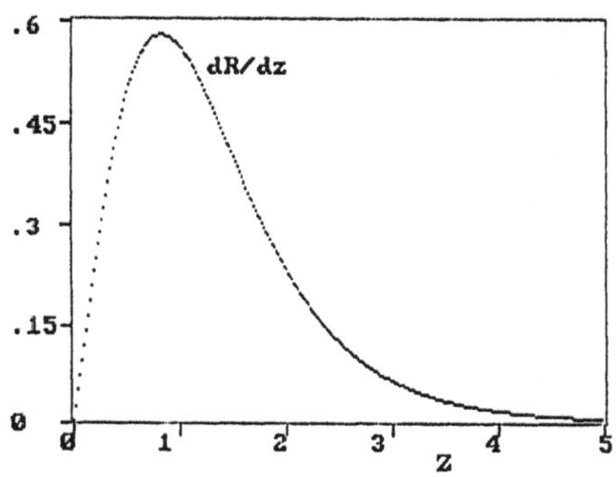


Figure 3. Derivative of the reflectivity with respect to $z$, as in fig. 2.

analyzed by the lock-in amplifier whose output is proportional to $\frac{dR}{dz}$, $z$ being the longitudinal displacement (fig. 3). We usually work on the descending part of this curve, i.e. at a distance of about 150 nm. When the sample is moved along the $x$ and $y$ axes a three dimensional profile of the sample can be reconstructed by processing the output of the lock-in amplifier. The PZTs moving the sample along the three axes are hysteresis-free and their range is 15 μm along the longitudinal axis and 75 μm along the other two axes.

## 3. Results

We investigated Langmuir-Blodgett films where sharp steps of known height were present by doing one-dimensional scans across such steps. A series of such scans is shown in fig.

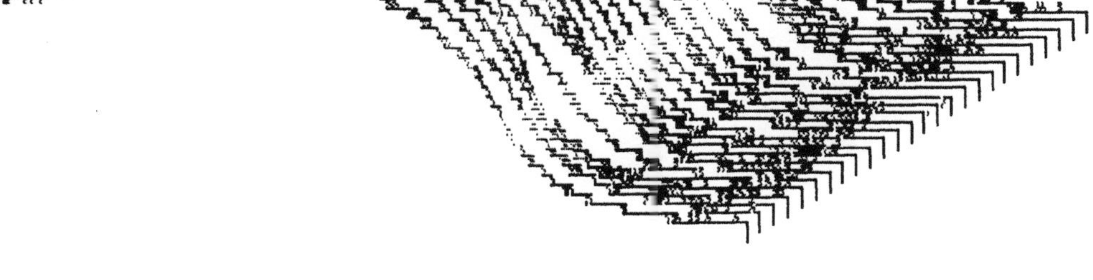

Figure 4. Several scans across a step in a Langmuir-Blodgett film (barium stearate bilayer, thickness 5.03±0.5 nm) showing a good reproducibility.

4. In this case the refractive index is constant throughout, and the obtained profile is therefore only linked to the sample topography. By comparing the signal we get across a step and its known height we can estimate the resolution and the reproducibility of the instrument. The noise-equivalent displacement in the longitudinal direction turns out to be less than 0.1 nm. After repositioning and observing the same step in another region, however, the reproducibility is a bit worse: the average differences between measurements is equivalent to about 0.5 nm.

The longitudinal resolution is therefore slightly better than that obtained by Guerra (1990) with a non-scanning instrument. The lateral resolution is worse, but we did so far no attempt to optimize it, and it can certainly be improved. Our instrument is relatively simple, since it uses only standard scanning elements and no special optics, except the convex prism.

This instrument was first presented by De Marco *et al* (1990) and is described in further detail in a forthcoming paper (De Marco *et al* 1992), where possible developments are also discussed.

## 4. References

Bonazzola, G.C., Ciocchetti, G., Conte, E., De Marco, A., Maringelli, M. and Trabucco, A. (1985) 'Possible application of the optical tunnel effect to membrane biophysics', Eur. Biophys. J. **12**, 51-55

De Marco, A., Maringelli, M., Trabucco, A. and Violino, P. (1990) 'The optical scanning tunneling microscope: Status and perspectives', Basic Appl. Histochem. **34**, 315

De Marco, A., Micheletto, R., Trabucco, A. and Violino, P. (1992) 'Development of an optical scanning tunneling microscope', Optics Communications, in press

Guerra, J.M. (1990) 'Photon tunneling microscopy', Appl. Optics **29**, 3741-3752

## Subject Index

**Author Index**

GPSR Compliance
The European Union's (EU) General Product Safety Regulation (GPSR) is a set of rules that requires consumer products to be safe and our obligations to ensure this.

If you have any concerns about our products, you can contact us on

ProductSafety@springernature.com

In case Publisher is established outside the EU, the EU authorized representative is:

Springer Nature Customer Service Center GmbH
Europaplatz 3
69115 Heidelberg, Germany

www.ingramcontent.com/pod-product-compliance
Ingram Content Group UK Ltd.
Pitfield, Milton Keynes, MK11 3LW, UK
UKHW020131290926
15471UKWH00003B/411
* 9 7 8 9 4 0 1 1 1 9 7 9 5 *